Studies in Systems, Decision and Control

Volume 87

Series editor

Janusz Kacprzyk, Polish Academy of Sciences, Warsaw, Poland
e-mail: kacprzyk@ibspan.waw.pl

About this Series

The series "Studies in Systems, Decision and Control" (SSDC) covers both new developments and advances, as well as the state of the art, in the various areas of broadly perceived systems, decision making and control- quickly, up to date and with a high quality. The intent is to cover the theory, applications, and perspectives on the state of the art and future developments relevant to systems, decision making, control, complex processes and related areas, as embedded in the fields of engineering, computer science, physics, economics, social and life sciences, as well as the paradigms and methodologies behind them. The series contains monographs, textbooks, lecture notes and edited volumes in systems, decision making and control spanning the areas of Cyber-Physical Systems, Autonomous Systems, Sensor Networks, Control Systems, Energy Systems, Automotive Systems, Biological Systems, Vehicular Networking and Connected Vehicles, Aerospace Systems, Automation, Manufacturing, Smart Grids, Nonlinear Systems, Power Systems, Robotics, Social Systems, Economic Systems and other. Of particular value to both the contributors and the readership are the short publication timeframe and the world-wide distribution and exposure which enable both a wide and rapid dissemination of research output.

More information about this series at http://www.springer.com/series/13304

Aleksander Sładkowski
Editor

Rail Transport—Systems Approach

Springer

Editor
Aleksander Sładkowski
Department of Logistics and Industrial Transportation
Silesian University of Technology
Katowice
Poland

ISSN 2198-4182 ISSN 2198-4190 (electronic)
Studies in Systems, Decision and Control
ISBN 978-3-319-51501-4 ISBN 978-3-319-51502-1 (eBook)
DOI 10.1007/978-3-319-51502-1

Library of Congress Control Number: 2016960770

Printed on acid-free paper

This Springer imprint is published by Springer Nature
The registered company is Springer International Publishing AG
The registered company address is: Gewerbestrasse 11, 6330 Cham, Switzerland

Preface

Rail transport is one of the most important sectors for the economies of many developed countries. Its importance can and should increase in the future. This is explained quite simply. In comparison to aviation, rail takes longer, but is significantly less expensive. Marine and river transport are great alternatives, but they are significantly slower than rail transport in moving goods to their destination. In addition, most importantly, transport via by ship often leaves goods short of their final destination, requiring further road or rail transport.

Rail transport's most serious competition comes from truck transport. However, truck transport loses its advantage when delivering cargo over long distances, particularly those in excess of 400 km. In addition, there are important environmental and safety aspects. Again, in this regard, rail transport has an edge.

Thus, developed countries are interested in developing rail transport as an alternative to road transport. This entails having an effective system for delivering goods and passengers on various continents, particularly within Eurasia. It is obvious that the development of this mode of transport cannot be considered in isolation from other transport modes. For example, for the transport of passengers over long, transcontinental distances, air transport has significant advantages. However, the best way to get people and to and from airports is by rail.

Similarly, the transport of large quantities of goods, such as containers, is best carried out by sea or inland waterway, particularly if there are no significant time constraints. But transport to and from the port is also often most efficient via rail. Thus, rail transport may be considered a subsystem of the intercontinental transport system.

Rail transport is a highly complex system consisting of rolling stock, transport infrastructure (track facilities, energy supply systems, numerous buildings, etc.), administration, management and control services. It should be added that the manufacturer of all these elements and fulfilling their maintenance (plants, repair shops, depots, etc.), the relevant educational institutions, which are required to prepare personnel for some extent related to the rail transport.

Each element in this scheme is also a specific subsystem. For example, if we consider an electric locomotive, it consists of a number of highly responsible subsystems (control systems, power system, brake system, heating system, pressurized air supply system, suspension systems, motor systems, etc.). It should be noted that if we want to consider traffic safety, it is necessary to evaluate not only the interaction of the individual subsystems of the locomotive, but also its interaction with external systems, namely, the track, contact wire, signaling and communication systems, etc. Thus, it becomes apparent that rail transport can be considered only with a systems approach. Moreover, regardless of the level at which one or the other system is considered, an acceptable solution can be found only with the use of a systems approach.

In the present monograph, the authors show how scientists of various countries solve specific rail transport problems using the systems approach. In particular, the book describes the experience of scientists from Romania, Germany, Czech Republic, UK, Russia, Ukraine, Lithuania and Poland. It should be noted that these countries' rail systems have historical differences. In particular, these countries have railways with different gauges, signaling and communication systems, energy supply systems, and, finally, political systems, which affect their approaches to rail management.

Despite the fact that most of the authors work at universities, the monograph is directly aimed at solving essential problems facing the rail industry in different countries. In some cases problems are solved, transforming ideas into concrete technical, economic or organizational solutions. In other cases, potential solutions for problems are identified.

Structurally, the monograph is divided into two parts. Part I provides a systematic analysis of rail transport and its maintenance. Part II is devoted to rail transport infrastructure and management development. Particular attention is paid to security issues.

The book is written primarily for professionals involved in various problems of rail transport. Nevertheless, the authors hope that this book may be useful for rail manufacturers, for technical staff and managers of rail transport operators, and for students of transport specialties, as well as for a wide range of readers who are interested in the current state of transport in different countries.

Katowice, Poland Aleksander Sładkowski

Contents

Part I
Means of Transport and Maintenance

Effects of Braking Characteristics on the Longitudinal Dynamics of Short Passenger Trains

Cătălin Cruceanu, Camil Ion Crăciun and Ioan Cristian Cruceanu

Abstract This original study investigates comprehensively the effects on longitudinal dynamics of short trains determined by usual combinations of different brake types that currently equip passenger railcars in operation: disc brake in fast-action and cast iron brake block system in fast high power action mode. The interest on such research is determined by the differences between the braking characteristics determined by two major aspects: the specific dependency of friction coefficient between wheel and cast iron blocks, respectively, brake discs and pads on velocity and normal forces applied on frictional couples, as well as by the brake cylinders air pressure evolution during the process. Preliminary studies and results of numerical simulations performed on single railcars submitted to braking constitute reasonable qualitative argumentation of the present research. Taking into account certain assumptions in order to eliminate, as much as possible, any aspect potentially disturbing the direct influence of braking characteristics, an original longitudinal dynamics simulation program was developed. The filling characteristics of air brake distributors were experimentally determined on static computerized testing system in the Braking Laboratory of the Faculty of Transports in University POLITEHNICA of Bucharest. The data files were adequately implemented into the simulation program. The results of numerical simulations indicate that braking characteristics have major influence on longitudinal dynamics of the train. We identified three patterns of in-train force evolutions, depending on the configuration of the braking systems featuring each vehicle of the train. Regarding operational aspects, the arrangement of railcars with different brake systems in the train composition affects the longitudinal dynamic reactions. We recommend measures to increase traffic safety and passengers comfort.

C. Cruceanu (✉) · C.I. Crăciun
University Politehnica of Bucharest, 313 Splaiul Independenţei, Bucharest, Romania
e-mail: c_cruceanu@yahoo.com

C.I. Crăciun
e-mail: craciun_camil@yahoo.com

I.C. Cruceanu
S.C. Atelierele CFR Grivita S.A., 359 Calea Griviţei, Bucharest, Romania
e-mail: ioan.cruceanu@gmail.com

A. Sładkowski (ed.), *Rail Transport—Systems Approach*, Studies in Systems, Decision and Control 87, DOI 10.1007/978-3-319-51502-1_1

Keywords Railway vehicle · Longitudinal dynamics · Train · Braking · Simulation program · Disc brake · Cast iron block brake · Fast-acting mode · High power brake

1 Introduction

Train braking is a critical and complex process, with relevant impact on the safety of the traffic. Traditionally, railway vehicles are equipped with compressed air systems, known as a UIC pneumatic brake [1]. The operational principle implies that the system is ready-for-use and brakes are released when working pressure is applied to the entire brake pipe of the train. Braking is commanded and controlled by pressure reduction in the brake pipe, while the signal for release increases pressure in the same brake pipe.

An important operational problem is the limited transmission speed of braking initiation. The brakes actuate successively in the length of the train so that, while the first vehicles are slowing down, others are trying to push, still unbraked, from the rear. To mitigate such phenomena, engineers developed specific diminishing in the rate of air pressure variation in the brake cylinders, accepting, however, an inevitable slight decrease in braking performance.

This operational mode, which does not allow simultaneous application of the vehicles' brakes, creates conditions, during transitional stages immediately following the pneumatic braking command, to generate important longitudinal in-train reactions causing stress to the couplers, affecting passenger comfort and, in certain cases, even the traffic safety. Effects such as the breaking of couplings during braking raised justified concerns, particularly given increased tonnages and running speeds of trains.

Theoretical and experimental studies complemented with computer applications and simulations for the analysis of specific phenomena, e.g., [2–13], indicate a general pattern of longitudinal dynamics of trains submitted to braking actions: the successive actuation of air distributors in the length of the train determines an important initial compression and generates oscillations in the train body. While the maximum pressure in the brake cylinders tend to stabilize, the oscillatory movement is quite rapidly damped and the magnitude of the longitudinal reactions between the vehicles of the train diminish, becoming insignificant and keeping so until the train stops. Trains experience the highest levels of longitudinal dynamic forces usually in the middle, and the compressive forces in couplers are higher than the tensile ones.

It is notable that most studies are focused on long, heavy freight trains with UIC brake block systems, operating in slow-action mode. The mass and active brake distribution in the train composition, the initial velocity, etc. may contribute to particular aspects that do not affect essentially the presented pattern.

Passenger trains, in general, are shorter, lighter and more uniform in composition, so specific issues arise [11–13]. The main braking system operating in fast-acting mode [1], complementary systems that might be associated [14–17], the random action of wheel slide protection devices in case of impaired wheel-rail adhesion, etc. modify the longitudinal dynamic reactions in the train body as magnitude, evolution in time during the braking process and disposition of compressive and tensile forces between adjoined vehicles.

The main braking systems of passenger vehicles are usually equipped with tread brakes made of cast iron, composite brake blocks, or disc brakes with organic or sintered pads [18] for the vehicles destined for running speeds of 160 km/h or more [19]. Each system determines particular braking characteristics, mainly according to the friction coefficient dependency on velocity. There are noticeable differences [8, 16–18], generating different braking forces evolutions and having important influence on the longitudinal dynamics of the train.

The differences are even more significant in the case of railcars featuring high power brake systems with two stages of brake block pressure on the wheel tread. Such systems operate at high pressure at the beginning of braking, maintained while the speed is greater than a predetermined threshold and followed thereafter by a reduced pressure level until the railcar stops [19].

The main goal of the study is to investigate the influences of different brake systems, in various combinations in the passenger train composition, on the longitudinal dynamics and to evaluate the correspondent effects regarding the magnitude and evolutions of compressive and tensile forces exerted between neighboring vehicles. We considered systems in current use on passenger vehicles: disc brake with organic pads in P(fast-acting) action and cast iron tread brake blocks in P and R (high power) action.

The interest in such original research is justified because, for maximum velocities less than 160 km/h, passenger trains commonly operate vehicles equipped with different braking systems. In order to increase the accuracy of the results, we acquired experimentally data referring to the air pressure evolution in the brake cylinders.

The chapter is organized in six sections. The first one presents a brief state of the art in longitudinal dynamics of trains submitted to braking actions, giving general reasons and operational arguments to justify the main targets of the study. In the second section, the main particularities of the UIC brake system in use on passenger vehicles in operation and specificities of the friction elements equipment according to the maximum designed running speed there are summarized. In the third section, the main theoretical bases of the longitudinal dynamics simulation programs developed for the studies are outlined: the mechanical model of the train, of couplers, braking and resistance forces, as well as certain pneumatic aspects having essential influences on the in-train forces during braking actions. The fourth section consists of a qualitative argumentation of the studied problem, highlighting the particular evolutions of braking forces, velocities, decelerations, generated by the use of different braking equipment. The analysis relies on the results of emergency braking simulations for individual vehicles, performed in similar initial conditions. The fifth section is dedicated to numerical simulations of longitudinal dynamics during braking actions for several representative usual operational configurations of short passenger trains. Consistent to the aim of highlighting exclusively the influence of braking characteristics on the longitudinal dynamics of such trains, the main assumptions and initial data are presented and explained. The conclusions are outlined in the last section.

2 Particularities of UIC Brake System

A UIC brake system is an essential feature for safe and reliable rolling stock braking, available to the operator under all conditions and providing fail-safe operation. As the main braking system, the indirect-action compressed air brake system is compulsory for railway vehicles [1]. It is a continuous, automatic pneumatic brake, which uses air pressure variations both to command braking/releasing actions and to generate the necessary forces to apply on the brake blocks or pads. Controlled retardation forces are available throughout the vehicle's speed range in order to enable speed reductions, stopping at fixed location and remaining stopped on gradients.

The pneumatic command is generally recognized as extremely safe and reliable. The indirect operational principle implies that the system is ready for use and released when a working pressure (generalized 5 bars) is applied in the brake pipe. Braking/releasing actions are commanded and controlled by reduction/increase of the air pressure in the brake pipe.

Vehicles are equipped with brake cylinders that use the compressed air, supplied by the auxiliary reservoir, which constitutes the only source during braking actions. The air brake distributor, which is a complex mechanical-pneumatic equipment, is the essential local device of the system able to provide pneumatic connections between the brake pipe and the brake cylinder, between the auxiliary reservoir and the brake cylinder and between the brake cylinder and atmosphere, according to the pressure changes within the brake pipe.

The specific operating mode of the indirect-action brake gives automatic action property, essential for the traffic safety and makes sufficient use of compressed air for the brake system operations using a single brake pipe of the train. Moreover, it makes available for the passengers a simple and safe braking control for emergencies through the alarm signal. This can be operated by any passenger in a situation in which the emergency brake valve immediately opens the brake pipe, determining its rapid venting and, by consequence, generating emergency braking in order to bring the train to a stand [1].

The UIC brake system is schematically presented in Fig. 1. For details regarding principles, role of main components and operation, see, for instance, [14–16].

In the case of passenger vehicles with a UIC braking system, in operational use is compulsory the fast-acting (or P, or passenger) action mode, characterized by a filling time of 3–5 s and a releasing time of 15–20 s [1]. As a friction-based system, the usual equipment consists either of cast iron or composite brake blocks acting on the wheel tread, or with composite pads acting on both lateral surfaces of brake discs.

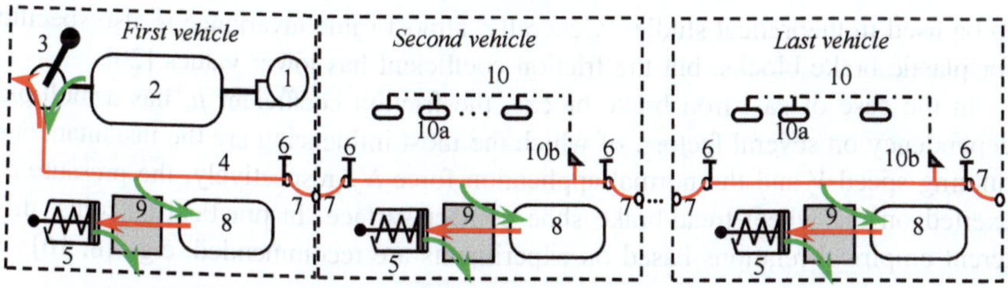

Fig. 1 Schematic of UIC (indirect-action compressed-air) brake system: *1* compressor; *2* main reservoir; *3* driver's brake valve; *4* brake pipe; *5* brake cylinder; *6* air isolating cock; *7* hose coupling; *8* auxiliary reservoir; *9* air brake distributor; *10* emergency signal; *10a* emergency signal handle; *10b* emergency signal valve

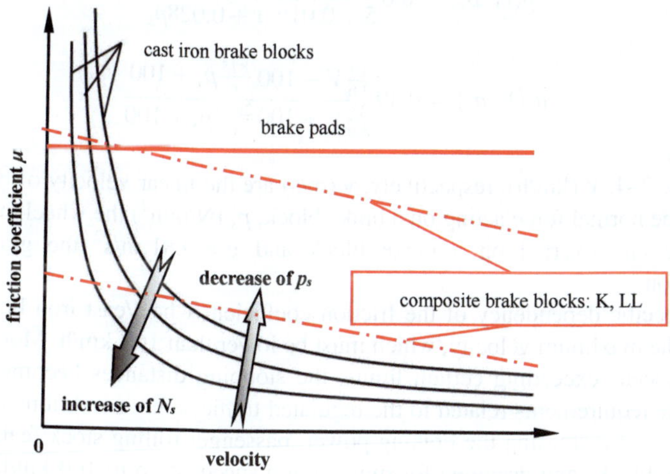

Fig. 2 Dependency of friction coefficient on different parameters: velocity, application force N_s, contact pressure p_s

It is worth noting the important differences in values and in evolution during braking action, of the friction coefficient between the mentioned materials. The friction coefficient between cast iron brake blocks and wheels strongly depends on several parameters: the most significant are the instantaneous velocity of the vehicle, the normal acting force on each shoe and the contact pressure, while the use of plastic (composite) materials for brake shoes or pads enables an approximatively invariance of the friction coefficient in normal conditions (see Fig. 2).

In the case of a classical brake disc with composite pads according to [18], the friction coefficient μ_d can be considered to be constant during the braking, having a recommended mean value of

$$\mu_d = 0.35 \tag{1}$$

to be used in theoretical studies. Generally, almost same invariance is also specific for plastic brake blocks, but the friction coefficient has lower values [20].

In the case of cast iron brake blocks, the friction coefficient μ_s has a multiple dependency on several factors, of which the most influencing are the instantaneous running speed V and the normal application force N_s, respectively, the pressure p_s exerted on the wheel tread/brake shoe contact surface. In practical calculus, different empirical relations based on experiments are recommended, e.g., [8, 16]:

$$\mu_s(V, N_s) = 0.6 \frac{V + 100}{5V + 100} \frac{\frac{16}{g} N_s + 100}{\frac{80}{g} N_s + 100} \tag{2}$$

$$\mu_s(v, p_s) = 0.6 \frac{1 + 0.01v}{5 + 0.01v} \frac{1 + 0.0056 p_s}{1 + 0.028 p_s} \tag{3}$$

$$\mu_s(V, p_s) = 0.49 \frac{\frac{10}{3.6} V + 100}{\frac{35}{3.6} V + 100} \frac{\frac{875}{g} p_s + 100}{\frac{2860}{g} p_s + 100} \tag{4}$$

In Eqs. 2–4, V (km/h), respectively, v (m/s) are the linear velocity of the wheel, N_s (kN) the normal force acting on a brake block, p_s (N/mm^2) the wheel tread/brake block pressure exerted on a brake block and g = 9.81 m/s^2 the gravitational acceleration.

The specific dependency of the friction coefficient wheel/cast iron brake block restricts the maximum velocity, which must be lower than 160 km/h. Moreover, for running speeds exceeding certain limits, the stopping distances become too long against the requirements related to the regulated traffic safety conditions [21]. With the purpose of increasing the braking power, passenger rolling stock featuring cast iron brake blocks and designed for running at a speed of up to 160 km/h, must be equipped with high power action brakes.

The high power action brakes have the particularity of working with two stages of applied forces on the brake blocks, depending on the actual velocity of the vehicle. At the beginning of braking, a high pressure in the cylinders maintains constant as long as the speed is greater than a threshold velocity. As soon as it falls below the predetermined value, a changeover of the air pressure within the brake cylinders becomes effective and the low pressure is kept steady until the vehicle has stopped. Consequently, the vehicle benefits from a higher average braking force during the process (see Fig. 3), in the limits of the same wheel/rail adhesion.

3 Train Model

Longitudinal train dynamic research commonly targets the evaluation of the magnitude of in-train forces, in order to improve the safety of the traffic and, when referring to passenger trains, to diminish the comfort alteration during braking.

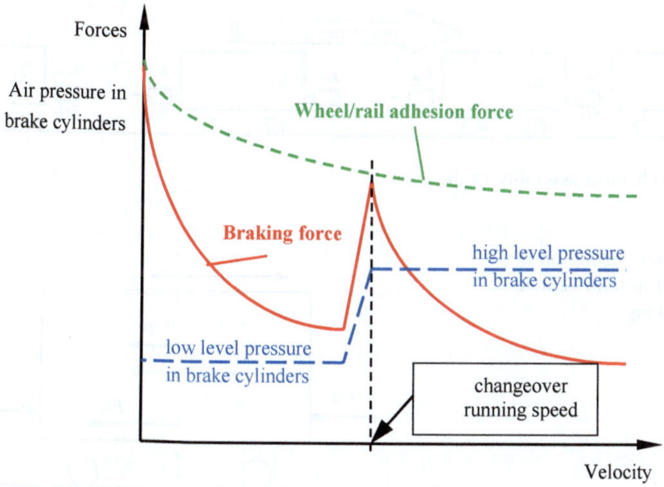

Fig. 3 Principle of operation for high power brake (cast iron brake blocks)

In such studies, researchers consider both the motions along the track of the train as a whole and of the individual component vehicles, usually neglecting the vertical and lateral movements.

The longitudinal dynamic of the train depends on numerous parameters, e.g., operational (train composition, length, load distribution, velocity), specificities of the track (grade, curvature), brake systems characteristics and response (type, brake delay time, actuation particularities, brake forces time evolution), vehicle design, technical and constructive characteristics.

The braking process itself consists of a succession and overlapping of mechanical and pneumatic processes evolving with different intensities in the length of the train.

Consequently, it is required a comprehensive approach, based on adequate embedded complex models, taking into account the evolution of the different aspects that are contributing to the achievement of safe reductions of running speed and stopping of trains in operation.

3.1 Mechanical Model of the Train

In mechanical perception, neglecting the vertical and horizontal dynamics—usual assumption in such studies—the train can be considered an elastic-damped lumped system consisting in n individual rigid masses m_i representing each vehicle, connected through elements having well defined elastic k_i and damping c_i equivalent characteristics in respect to the individual couplers' device particularities (see Fig. 4).

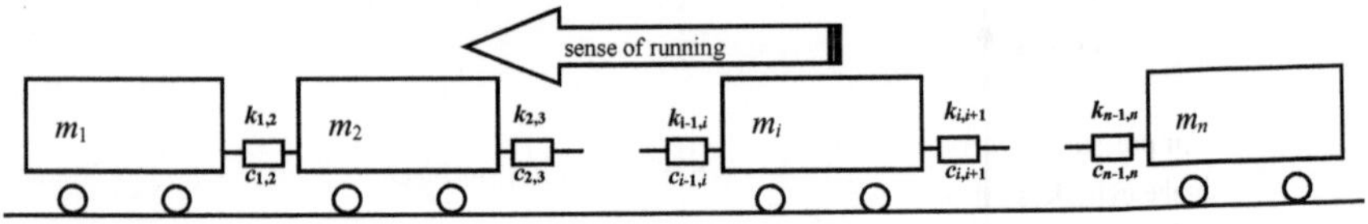

Fig. 4 Mechanical assembly of the train

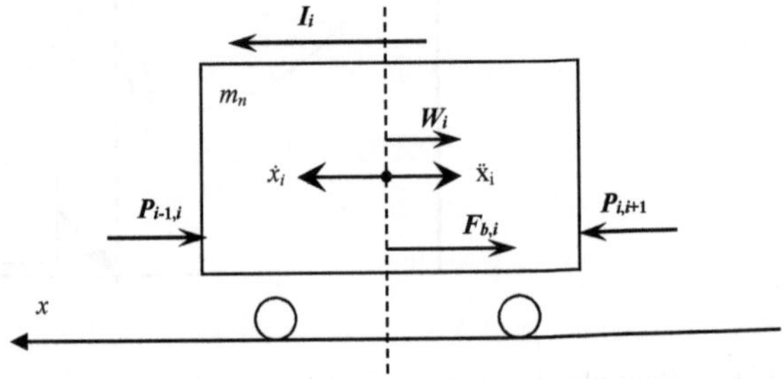

Fig. 5 Forces acting on a generalized in-train vehicle during braking

Considering the multibody formulation, for the generalized "i"-th vehicle of the train, the forces governing the displacement x_i are the braking forces $F_{b,i}$ and the resistances W_i as retardation forces, the forces $P_{i-1,i}$ and $P_{i,i+1}$ acting in the connections with neighbored vehicles and the inertial forces I_i (see Fig. 5).

The correspondent differential equation describing the longitudinal movement of the vehicle is

$$m_i\ddot{x}_i + P_{i,i+1} - F_{b,i} - W_i - P_{i-1,i} = 0 \tag{5}$$

The present study follows the classical approach of theoretical studies on in-train longitudinal forces, consisting of a mechanical cascade-mass-point unidirectional model (see, for instance, [5, 9, 22]). Each vehicle is considered a rigid body having at front ends couplers based on combined use of draw-gears and a pair of buffers that are linking to neighbored vehicles. The shock and traction devices have specific elastic and damping characteristics, generally dependent on the relative displacement Δx_i and on the relative speed $\Delta\dot{x}_i$ between adjoining vehicles.

Applied to the vehicles of the train, Eq. 5 conducts to a nonlinear differential equation system of second order. Considering $i \in [1, n]$ and attaching the boundary conditions $P_{0,1} = P_{n,n+1} = 0$, it was obtained the equation system describing the movements of the mechanical system presented in Fig. 6.

An important aspect to deal with in conceiving the mechanical model of the train is represented by the couplers, which ensure the link between the railway vehicles. The couplers have specific elastic and damping characteristics, according to the remarkable effects regarding the protection of the vehicle's structure, the loading's integrity and the passengers comfort.

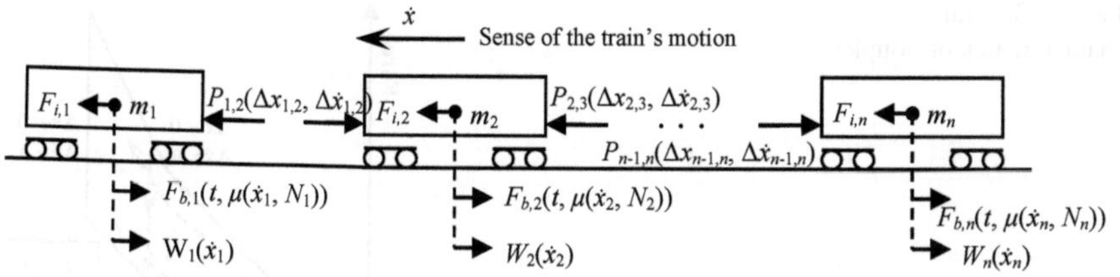

Fig. 6 Mechanical model of the train

Generally, the traditional couplers widely used in Europe are composed of a pair of lateral buffers, a traction gear and a coupling apparatus at each extremity of the vehicle. Their characteristics have significant influences for the longitudinal dynamics of the train, with running stability implications. There are different constructive solutions, the specificities being consistent with the requirements determined by mass, potential collision shocks and passengers comfort, etc.

According to the particular constructive and operational characteristics, the behavior of buffer and draw-gear devices is quite complex, due to several nonlinear phenomena, e.g., variable stiffness-damping, hysteretic properties, preloads of elastic elements, draw-gear compliance, clearance between the buffers discs, etc.

The problem of modelling the rail vehicles couplers has been approached in several ways. A comprehensive review regarding the modelling of friction draft gear is presented in [22].

Typical constructive solutions for buffers and traction devices rely on metallic elastic rings (RINGFEDER type) based on friction elements to fulfil the required damping effects. The general characteristics force-displacement depends mainly on the stroke Δx representing in fact the relative displacement between adjoined vehicles and on the relative velocity $\Delta \dot{x}$ associated to its sign.

Because of dry friction presence, in operation, the device is blocked as long as the external forces do not overcome the internal static friction forces. In addition, the forces in compressive and tensile stroke are significant different.

In the present study, in order to avoid intricacies determined by the discontinuities of friction forces between elastic rings in RINGFEDER elements, a smoothening approach was preferred [11, 12]. It is notable that such models have the advantage of a much simpler simulation, as well as limitations, for example the fact that the bodies in contact tend to drift until the resultant force vanishes, which does not exactly correspond to the operational effects [23].

The model for the shock and traction apparatus considering the nonlinear characteristics is shown in Fig. 7.

The forces in couplers were evaluated assuming an algebraic sum of elastic and friction forces.

Fig. 7 General characteristics of couplers

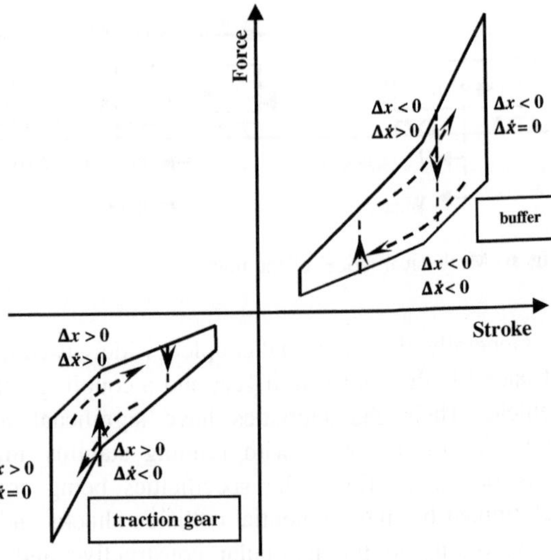

According to functional specific action of the elastic elements comprising friction elements, there were determined the following equations [24]:

$$P_{ij}(\Delta x_{i,j}, \Delta \dot{x}_{i,j}) = \begin{cases} k_b \Delta x_{i,j} + c_b |\Delta x_{i,j}| tanh(u\Delta \dot{x}_{i,j}), & \text{for } \Delta x_{i,j} < 0 \\ 0, & \text{if } \Delta x_{i,j} = 0 \\ k_t \Delta x_{i,j} + c_t |\Delta x_{i,j}| tanh(u\Delta \dot{x}_{i,j}), & \text{for } \Delta x_{i,j} > 0 \end{cases} \quad (6)$$

In Eq. 6 k and c were denoted as equivalent constants depending on the elasticity of the elements and on the friction characteristics between the metallic rings inside. The supplementary indexes "b" and "t" are referring to buffer and traction devices, respectively. In addition, $\Delta x_{i,j}$ represents the stroke, $\Delta \dot{x}_{i,j}$ the relative speed for the considered element of the couplers between i and j adjacent vehicles of the train. The scalar factor $u \gg 1$ determines how fast the tan$h$ function changes from near −1 to near +1.

Generally, in order to describe the operation of the couplers as accurate as possible, the constants can be established on experimental bases. In the present study, the constants k and c were determined by taking into account the characteristic diagrams of RINGFEDER-type buffers built in Romania by ICPVA-SA that equip trailed passenger vehicles [25, 26]. Keeping the international prescriptions [27, 28] and according to [17], the correspondent values are presented in Table 1.

More details about the actual model, preliminary work and tests to achieve and define this solution can be found in [12, 13, 24].

Table 1 Main data in simulation of traction and collision devices

Specific constants for buffers	$k_b = 2.8 \times 10^6$ N/m
	$c_b = 1.4 \times 10^6$ N/m
Specific constants for traction devices	$k_t = 5.46 \times 10^6$ N/m
	$c_t = 2.43 \times 10^6$ N/m
Scaling factor	$u = 10^4$

3.2 *Braking Forces*

Braking forces are most important in longitudinal dynamics and their evolution is essential for the whole process, both in terms of braking distance and of magnitude of in-train forces developed between adjoined vehicles.

The main braking systems of railway vehicles are friction-dependent, so the braking force can be classically described considering the friction coefficient μ and the normal forces N_j exerted by the brake rigging on each of the j brake blocks on the wheel tread, respectively, on each of the j pads acting on the lateral surface of the brake disc:

$$F_{\mathrm{b}} = \sum_j \mu N_j \tag{7}$$

According to the typical mechanical construction of main brake systems, the normal forces N_j are generated from the brake cylinder. The resultant force F_p on the brake cylinder piston rod is amplified and transmitted to the brake blocks or pads through the brake rigging, characterized by the amplification ratio i_{a} and a certain mechanical efficiency η_{br}.

In the case of disc brakes with individual automatic slack adjuster device incorporated in the brake cylinder piston rod, as currently in use, the braking force can be determined:

$$F_{\mathrm{b}} = \mu\, n_{\mathrm{bc}} F_p i_{\mathrm{a}} \frac{2r_{\mathrm{m}}}{D_0} \eta_{\mathrm{br}}, \tag{8}$$

where n_{bc} represents the number of brake cylinders of the considered vehicle, D_0 is the wheel diameter and r_{m} the medium friction radius of the brake discs.

In the case of fitting with brake blocks actioned by symmetrical brake rigging with automatic slack adjuster mechanism on the main brake bar, typical solution in the case of passenger vehicles destined for running speeds of less than 160 km/h, the braking force is

$$F_{\mathrm{b}} = \mu\left(F_p i_{\mathrm{c}} - R_{\mathrm{sa}}\right) i_{\mathrm{v}} n_{\Delta} n_{\mathrm{bc}} \eta_{\mathrm{br}} \tag{9}$$

In Eq. 9, R_{sa} represent the resistance forces due to the automatic slack adjuster device, i_c the central brake rigging and i_v the brake rigging's vertical levers amplification ratio, n_Δ the number of triangular axels.

The force developed at the brake cylinder piston rod F_p depends on the instantaneous air pressure p_{bc} as main variable, the piston diameter d_{bc} and the resistance forces F_R due to the brake cylinders back spring, summed up, in the case of individual automatic slack adjuster device incorporated in the brake cylinder piston rod, with the correspondent resistance forces:

$$F_p(t) = \frac{\pi d_{bc}^2}{4} p_{bc}(t) - F_R \tag{10}$$

It was assumed that certain terms and factors defining constructive and functional characteristics are constant for the same vehicle during braking. Defining a correspondent generic equivalent term ξ and keeping in mind that that resultant force F_p on the brake cylinder piston rod is practically in direct dependency on the actual air pressure within the brake cylinders p_{bc}, as shown in Eqs. 8–10, can be summarized a generalized expression:

$$F_b(t) = \xi p_{bc}(t) \mu \tag{11}$$

Presented in that form, Eq. 11 highlights the variable factors that define the main sources of nonlinearities, which characterize the brake force.

Explanations regarding the mathematical representation of friction coefficients for usual cases were presented in Sect. 2.

It is important to mention one more important operational aspect: an effective braking response is expected only after the force exerted by the increasing air pressure acting on the brake cylinder piston overcomes the resistance forces due to the brake cylinders back spring, the friction forces between the piston seal and cylinder and performs all displacements determined by the released positions of brake riggings, wear, elasticity etc.

Usually, this is taken into account by considering a minimum necessary pressure level of approximately 0.4 bar within the brake cylinder to rely on the initiation of real braking force of the vehicle. The effects of this particular aspect on longitudinal dynamics of trains cannot be neglected, as comprehensively shown in [13].

Once the pressure gets the commanded maximum level—for trailed vehicles submitted to emergency braking action is regulated to $p_{bc,\ max} = 3.8 \pm 0.1$ bar [1] —it remains steady during the process, unless the action of wheel slip prevention devices in the case of poor adhesion. An exception is for high power brake system. In that case, immediately after the threshold velocity is reached, the air pressure in brake cylinders drops to the low level $p_{bc,\ low} = 1.9 \pm 0.1$ bar [1], which afterwards maintains constant (see Fig. 3).

Consequently, knowing the air pressure evolution $p_{bc}(t)$ in the cylinders during the braking process and the maximum designed brake force $F_{b,max}$, respectively, the maximum normal force on a cast iron brake block $N_{max/block}$ corresponding to emergency braking, the instantaneous braking force can be described accordingly:

$$F_{b-disc}(t) = \begin{cases} 0 & \text{if } p_{bc}(t) < 0.4 \text{ bar, else} \\ \frac{p_{bc}(t)}{p_{bc,max}} F_{b,max} \end{cases} \tag{12}$$

or

$$F_{b-block}(t) = \begin{cases} 0 & \text{if } p_{bc}(t) < 0.4 \text{ bar, else} \\ n_{bb} \frac{p_{bc}(t)}{p_{bc,max}} N_{max\,/\,block}\, \mu_s(V, N_s(p_{bc}(t))) \end{cases} \tag{13}$$

In Eq. 13, n_{bb} counts for the number of the brake blocks of the considered vehicle.

It is important to mention that, due to the wheel—rail adhesion dependency, the maximum braking force has to be limited, in order to prevent over-braking. So, considering m_v the mass of the vehicle and

$$\mu_a = \frac{0.33}{1 + 0.011V} \tag{14}$$

the wheel-rail adhesion coefficient in normal conditions, the maximum braking force of the vehicle that can be provided by the UIC braking system is

$$F_{b,max} = \mu_a m_v g \tag{15}$$

According to the type, in Eq. 15, V (km/h) represents either the constructive running speed of the vehicle equipped with brake discs or composite brake blocks, or 0 km/h if cast iron brake blocks are used.

Consequently, in the case of vehicles designed for a specific running speed V_{max} and fitted with brake discs or composite brake equipment, the maximum brake force can be described as follows:

$$F_{b-disc}(t) = \begin{cases} 0 & \text{if } p_{bc}(t) < 0.4 \text{ bar, else} \\ \frac{p_{bc}(t)}{p_{bc,max}} m_v g \mu_a(V_{max}) \end{cases} \tag{16}$$

and for a cast iron block brake system equipment:

$$F_{b-block}(t) = \begin{cases} 0 & \text{if } p_{bc}(t) < 0.4 \text{ bar, else} \\ 0.33 \frac{p_{bc}(t)}{p_{bc,max}} \frac{m_v g}{n_{bb}} \mu_s(V, N_s(p_{bc}(t))) \end{cases} \tag{17}$$

In Eqs. 13 and 17, the function μ_s defining the block/wheel thread friction coefficient can be defined according to Eqs. 2, 3 or 4.

3.3 Resistance Forces

Usually in theoretical approaches, propulsion resistances W_p, acting permanently whenever a vehicle/train is moving in alignment on flat ground and the supplementary resistances W_s are separately considered.

The latter are intermittent opposing forces, acting only in certain conditions (grades, curves, wind, etc.) and overlapping the propulsion resistances, which are added up algebraically when appropriate [2, 29, 30].

Typically, propulsion resistance is described based on Davis' equation:

$$W_p = a + b\dot{x} + c\dot{x}^2 \tag{18}$$

where a is the journal resistance coefficient, fixed in quantity, not a function of velocity but depending on the vehicle's weight and number of axles; the second term is mainly dependent on flanging friction and, therefore, speed dependent, but coefficient b is usually small and sometimes is missing in some empirical formulae; the third term c takes into account the air resistance and is a function of the square of the speed.

The calculation of rolling stock resistance is still dependent on empirical formulae. The constants in Eq. 18, depending on type, constructive and operational characteristics of each vehicle, are experimentally determined.

It is important to mention that often it is more practical, for computational reasons, to refer to the specific propulsion resistance w_p (N/kN), defined in relation to the weight of the vehicle:

$$w_p = \frac{W_p}{m_v g} \tag{19}$$

In the present study, the specific propulsion resistances are determined based on Eq. 18, using the coefficients $a = 1.649$, $b = 0$ and $c = 2.4995 \times 10^{-4}$, which are considered appropriate for the majority of Romanian passenger vehicles in operation [29, 30]:

$$w_p = 1.694 + 2.4995 \times 10^{-4} V^2 \tag{20}$$

V (km/h) represents the running speed.

3.4 Pneumatic Aspects

In the case of UIC brakes, commands are transmitted along the train as pressure reference and the braking system of each vehicle specifically interacts with the complete pneumatic plant of the train.

Two important aspects are involved in the longitudinal behavior of the train: the moments when each air distributor begins to command the filling of the brake cylinders and the subsequent evolution of the air pressure in the brake cylinders. Accordingly, there are two different problems that are usually emphasized: the propagation of braking signal along the brake pipe and the response of the distributor of each vehicle [3–6, 31, 32].

Such processes, specific for UIC air brake system, are difficult in modelling due to the complexity of cumulated phenomena that occur. The fluid system approach involves taking into account airflows in brake pipe, reservoirs, calibrated orifices and other pneumatic resistances.

Moreover, pneumatic processes, by themselves complex, are conditioned by, or determine, mechanical displacements of pistons, opening or closing of certain valves pressure, controlled by pre-tensioned springs.

All these specific pneumatic and mechanical problems have to be integrated into the model to approach the trains' brake system operation in order to simulate adequately the distribution of brake cylinder pressure evolution along the train according both to the filling characteristics and to the transmission speed of brake action. An important parameter in this case is, e.g., the sensitivity of the brakes related to the decrease of the pressure in the brake pipe, which, according to [1], must be such that the brake is activated within 1.2 s if the normal working pressure drops by 0.6 bar in 6 s.

Regarding the air pressure increase in brake cylinders, this is controlled by the air brake distributor of each vehicle, subsequently reaching the limit of sensitivity.

For the present study, consistent with the main target referring to short passenger trains, the brake actuating moments were determined according to the length of the vehicles and their position in the train, considering a constant braking propagation rate in length of the train.

Regarding the filling characteristics of air brake distributors, we considered appropriate to base on our experimentally determined data, adequately implemented into the simulation program.

4 Preliminary Study and Qualitative Argumentation

In operation, classical passenger vehicles are usually equipped with UIC pneumatic brake systems [1] acting tread brake blocks on wheels (for maximum constructive velocities of less than 160 km/h) or pads on brake discs. In the case of cast iron shoes, vehicles designed for 140 km/h have high power brake with double pressure stage according to velocity (R type [19]). Coaches characterized by large differences between laden and tare weights, for example double-deck coaches, luggage and/or postal vans, are equipped with variable load braking system operating as fast-acting brakes (P type [1]), the same as for disc brakes.

As pointed out in the previous section, longitudinal dynamics of trains is highly determined by the difference in instantaneous forces acting on vehicles, among which the braking ones are most influential.

Actually, passenger trains operate in various brake systems compositions and therefore is expected that differences in braking characteristics have important influence with respect to the magnitude and time evolution of longitudinal dynamic reactions between vehicles.

The aim of this section is to make qualitative evaluations on the effects of these differences, relative to single operated passenger vehicles, equipped with different braking system and submitted to emergency braking in identical conditions on a straight, flat, clean, dry track. The interest was mainly focused on the variance of instantaneous velocity and deceleration, since it is obvious that, accordingly, the longitudinal dynamics is going to be affected when coupled together in the same train.

The studies were performed by the means of an original vehicle braking performances simulation program based mainly on the Newton-Euler formalism.

Denoting x the space traveled in braking action, $m_e = \rho\, m_{\mathrm{v}}$ the equivalent mass of the vehicle (the higher than one factor ρ is accounting for the mechanical effects of the inertia of rotating masses [21]) and considering the aforementioned characteristics of the track, the equation of movement is

$$F_{\mathrm{b}}(t, x, p_c) + \frac{m_{\mathrm{V}} g}{1000} w_{\mathrm{p}}(t, \dot{x}) = m_e \ddot{x} \tag{21}$$

In Eq. 21, $F_{\mathrm{b}}(t, \dot{x}, p_{\mathrm{bc}})$ correspond to Eqs. 12 and 13, according to the brake type equipment considered and w_{p} is given by Eq. 20. In the case of cast iron block brakes, in Eq. 13 the friction coefficient is computed in this study based on Eq. 2.

Equation 2 was integrated in MATLAB using the ode 45 solver.

In order to increase the precision of the simulations results and to emphasize the effects of the brake cylinders air pressure evolution during the braking as accurate as possible, experimental acquisitioned data were implemented.

Experiments were performed on the computerized brake systems acquisition data in the Laboratories of Faculty of Transport in University POLITEHNICA of Bucharest. Details about the stands and the computerized systems can be found in [33]. There are static rigs, equipped with mechanical and pneumatic assembly of classical braking system, in use for passenger vehicles designed for 140 km/h. Braking commands were set using a KD-2 type brake valve.

There were determined the filling characteristics for KE 1b and KEs air brake distributors in current use on Romanian rail vehicles. Pressure transducers used for data acquisition have a sample rate of 0.02 s. Recorded evolutions of the air pressure in brake cylinders are presented in Fig. 8.

In order to emphasize mainly the effects of different braking systems in operation, certain conditions were imposed in conducting the simulations. The most important considered in this study were:

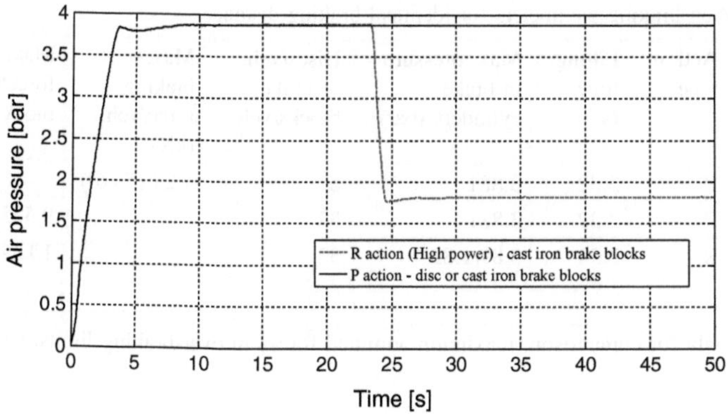

Fig. 8 Experimental record of the air pressure evolution in brake cylinders during emergency brake action

- identical coaches in terms of design, dimensions, weight;
- each vehicle is submitted to an emergency braking action from the same initial running speed;
- identical filling time and filling characteristic of the braking cylinders, in respect to the international prescriptions [1];
- the braking system of each vehicle performs the same braking power in terms of identical stopping distance, in agreement with [21].

According to the usual equipment of coaches, there were studied three cases corresponding to the following UIC pneumatic braking systems configurations:

- disc brake using composite pads in respect to [18], controlled by KE-1 type air distributor operating in fast-acting mode [1]. This equipment is specific for railway passenger vehicles destined for 160 km/h [19];
- phosphoric cast iron brake blocks (grade P10 [34]) according to [20], controlled by KEs type air distributor operating in high power mode [19]. This equipment is specific for railway passenger vehicles destined for 140 km/h [19];
- phosphoric cast iron brake blocks (grade P10 [34]) according to [20], controlled by KE-2 type air distributor operating in fast-acting mode [1]. This equipment is specific for railway passenger vehicles destined for less than 160 km/h [19] and characterized by important differences between laden and tare weights.

In simulations was considered each vehicle having a mass of 50 t and the factor accounting for rotary inertia $\rho = 0.04$ [21]. The initial velocity was set to 140 km/h. A stopping distance of 948 m was imposed for all coaches.

Using the facilities of the simulation program, there were firs established the maximum necessary braking force (brake disc), respectively, the maximum normal force on a brake block, in respect to the imposed condition to fit the previously defined braking distance.

Table 2 Main braking parameters for identical braking distance

Brake system type	Action type	Filling time (s)	Max. pressure in brake cylinders (bar)	Discs/veh. or brake blocks/veh	Max. braking force/veh. (kN)	Max. normal force/brake block (kN)
Disc	P	3.32	3.881	8	42.3 (*58.6*[a])	
Cast iron brake block	P	3.32	3.881	32		9.525
	R	3.34	3.89/1.82	32		10.34 (*12.65*[a])

[a]In italics, only for comparison, maximum admitted forces in over-braking limits, using Eqs. 16 and 17

The main numerical results are presented in Table 2, along with basic initial characteristic braking parameters.

Upon a brief analysis of the results, it is worth noting that in identical conditions the use of cast iron brake blocks in fast-acting operation of UIC systems determines the lowest braking power quantified by the longest stopping distance of 948 m. Actually, this was the reason of adopting the imposed braking distance in the present study. In both other cases, a higher braking power is proven by the fact that the maximum braking or normal force/brake block are still far from wheel/rail adhesion limitations (see Table 2).

Implementing the above presented data in the simulation program, we determined the braking characteristics and the main kinematic parameters evolution (velocity, deceleration) corresponding to the three cases, in order to assess the possible effects on longitudinal dynamics.

Comparing the resultant braking characteristics presented in Fig. 9a, the differences in braking forces during stopping are obvious. More than that, even though the braking distance is the same (948 m) in all of cases, the instantaneous differences in velocity (see Fig. 9b) and decelerations (see Fig. 9c) are obvious. The details presented in figures highlight the evolution of various parameters in different periods of the braking process.

Close attention should be given to the first phases of braking, during the filling time of cylinders, known as essential in longitudinal dynamics of trains. It is also expected to have important influence the transition process from high to low pressure in cylinders, specific for the operation of high power brake system around 50 km/h.

Nevertheless, one can observe that the duration of the braking process is different, depending on the operational particularities of each braking system.

All these results indicate clearly that, if these vehicles are to be combined in a train, the magnitude of in-train forces is definitely expected to be higher than in a uniform train composition.

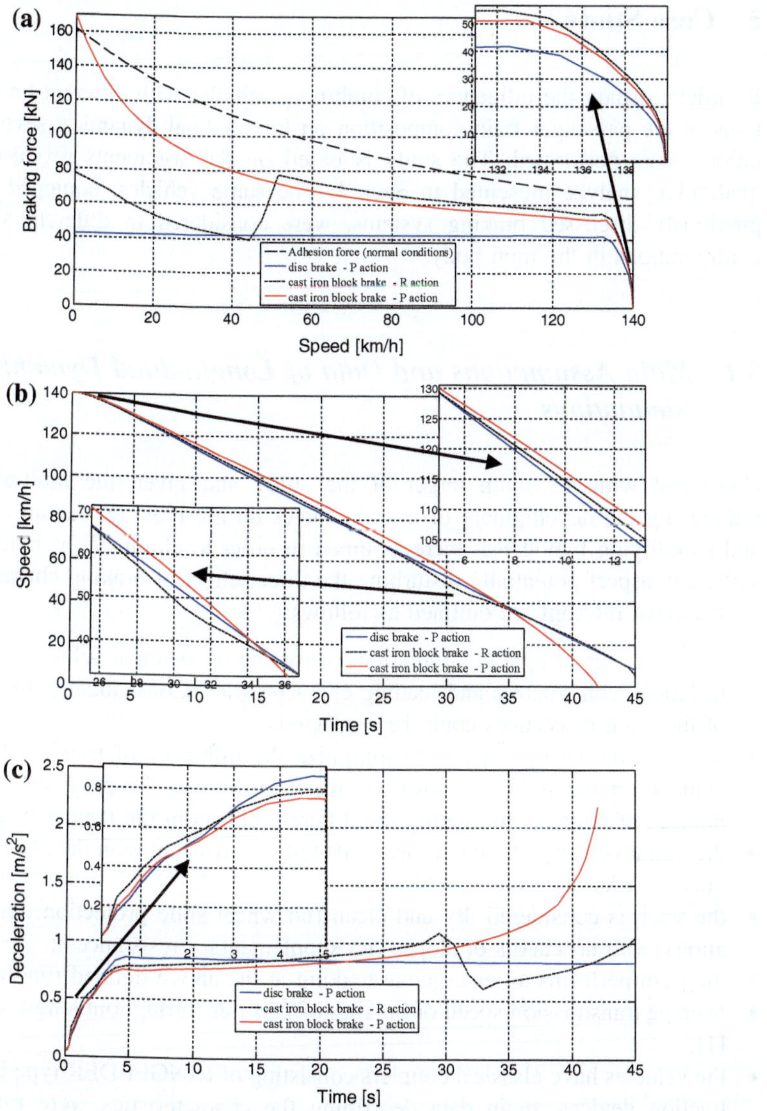

Fig. 9 Comparison between the simulation results for 50 t vehicle submitted to emergency braking from 140 km/h, with different brake systems and fitting the same stopping distance: **a** braking characteristics, **b** velocity time history, **c** deceleration time history

5 Case Study

In order to study the influences of combining vehicles with different braking systems in the passenger train composition on longitudinal dynamics, several simulations were performed. The study is based on the arguments revealed by the qualitative analyze presented in Sect. 4. The same vehicles, equipped with the previously discussed braking systems, were considered in different significant configurations in the train body.

5.1 Main Assumptions and Data of Longitudinal Dynamics Simulations

Consistent with the main target of the study and given the multiple factors influencing the development of in-train forces during braking, certain constraints and simplifying hypotheses were assumed in order to eliminate, as much as possible, any aspect potentially disturbing the direct effect of braking characteristics.

The most relevant are outlined as follows:

- train in a very simple configuration, consisting of identical vehicles in terms of technical, constructive and loading characteristics. Consequently, the influence of the main resistances could be neglected;
- a small number of railcars, to minimize the influence of train's length. Four vehicles in the train compositions were considered, keeping also a relevant number of couplers to identify the different effects in the length of the train;
- the initial velocity of the train was 140 km/h, consistent with the use of cast iron brake blocks in certain combinations;
- the track is considered dry and clean (no wheel slide protection devices actuation), with no curves or slopes (no supplementary resistances);
- the train performs an emergency braking at the above defined running speed;
- average transmission speed of braking of 250 m/s throughout length of the train [1];
- the vehicles have classical couplers consisting of RINGFEDER type buffers and traction devices, main data describing the characteristics were presented in Table 1;
- four-axle passenger vehicles of 50 t mass and 25 m length are used;
- according to the brake system equipment of each vehicle, the correspondent main input data are presented in Table 2;
- screw couplings tightened up, with no clearance between the buffer discs.

The simulations were carried out for representative brake system combinations encountered in passenger trains operation, as summarized in Table 3.

Table 3 Simulations schedule

Case/acronym	Configuration of brake systems in the train's composition				
	Type of	Vehicle 1	Vehicle 2	Vehicle 3	Vehicle 4
Case 1 DDDD	Brake	Disk	Disk	Disk	Disk
	Action	P	P	P	P
Case 2 PPPP	Brake	Brake block	Brake block	Brake block	Brake block
	Action	P	P	P	P
Case 3 RRRR	Brake	Brake block	Brake block	Brake block	Brake block
	Action	R (50[a])	R (50[a])	R (50[a])	R (50[a])
Case 4 RRRR1234	Brake	Brake block	Brake block	Brake block	Brake block
	Action	R (51[a])	R (50.33[a])	R (49.66[a])	R (49[a])
Case 5 DRRR	Brake	Disk	Brake block	Brake block	Brake block
	Action	P	R (50[a])	R (50[a])	R (50[a])
Case 6 DRRR234	Brake	Disk	Brake block	Brake block	Brake block
	Action	P	R (51[a])	R (50[a])	R (49[a])
Case 7 RDDD	Brake	Brake block	Disk	Disk	Disk
	Action	R (50[a])	P	P	P
Case 8 DDDR	Brake	Disk	Disk	Disk	Brake block
	Action	P	P	P	R (50[a])
Case 9 DPPP	Brake	Disk	Brake block	Brake block	Brake block
	Action	P	P	P	P
Case 10 PPPD	Brake	Brake block	Brake block	Brake block	Disk
	Action	P	P	P	P
Case 11 RPPP	Brake	Brake block	Brake block	Brake block	Brake block
	Action	R (50[a])	P	P	P
Case 12 PPPR	Brake	Brake block	Brake block	Brake block	Brake block
	Action	P	P	P	R (50[a])

[a]Running speed (km/h) when high/low pressure regime changes during braking action

For greater clarity, certain complementary explanations are required, mainly regarding the cases involving the high power brake system.

The predetermined speed threshold for changeover between the two stages of stabilized pressure in the cylinders during braking is 50 km/h [19].

Relative to this nominal value, deviations are permitted up to ±1 km/h. Therefore, consistent with the settings of the specific device—usually mechanical centrifugal governors—on each vehicle, the air pressure level change in the brake cylinders can happen randomly within the running speed limits of 51–49 km/h.

Hence, while the running speed of the train decreases in the specified range, important instantaneous differences in braking forces would occur between the vehicles.

In order to include into the study these particular operational aspects connected to the high power brake system, the following situations were considered:

- the centrifugal governors of all vehicles of the train are set to the nominal value (50 km/h)—case 3 (*RRRR*);
- the centrifugal governors of the vehicles of the train are set from the head to the end of the train, to velocities equally dispersed, in decreasing order values, within the limits of the accepted range—in cases 4 (*RRRR1234*) and 5 (*DRRR234*).

The original longitudinal dynamics simulation program was conceived in MATLAB on the theoretical basis presented in Sect. 3 and is adaptable to different train compositions (as number, type, constructive and operational characteristics of component vehicles). It is possible to simulate any combination of classical braking system in operation within the train composition.

Comprehensive information regarding the longitudinal dynamics simulation program can be found in [12, 16, 24].

It is worth noting that, given the initial evolution of the process—defined by Eqs. 12 and 13—and in respect to the aforementioned assumptions regarding the resistances, during the time elapsed between the braking command and the moment of reaching the minimum air pressure level of 0.4 bar in the brake cylinders on the first vehicle of the train, there is a lack of retardation forces. Consequently, no longitudinal dynamic reactions occur in the train body and, in the present study, the stated timeframe is neglected in the simulation program.

As important input data of the simulation program, the filling characteristics of brake cylinders can be implemented either as experimentally acquired data files or as mathematical time-dependent functions. In the latter situation, precision depends on the accuracy of interpolation, but the solving process is more time-efficient.

In the present study, experimental acquisitioned data files are used, as indicated in Sect. 4.

The following are the main outputs of the simulation program: the in-train compression and tensile forces evolution, distribution and magnitude, relative displacements, brake forces, distance and speed time evolution. It is worth mentioning that, according to usual convention in studies regarding the longitudinal dynamics of trains, the compression forces are considered positives, while the tensile ones negatives.

5.2 Results and Comments

Particularly given the special concerns on the traffic safety in rail transport, the main numerical results are primarily focused on the magnitude of the longitudinal dynamic reactions, compression and tensile forces, and their correspondent location in the train body.

These are summed up in Table 4.

Table 4 Results of simulations: maximum in-train forces

Uniform composition of the train in terms of braking system				
Case acronym[a]	DDDD	PPPP	RRRR	RRRR1234
Compression (kN)	10.766	12.569	13.505	13.505
Coupler	3–4	3–4	3–4	3–4
Tensile (kN)	*2.892*	*1.883*	*2.004*	*11.5*
Coupler	*3–4*	*3–4*	*3–4*	*2–3*
Disc (P action) + cast iron brake blocks (R action)				
Case acronym[a]	DRRR	DRRR234	RDDD	DDDR
Compression (kN)	12.065	12.065	21.959	10.994
Coupler	3–4	3–4	1–2	3–4
Tensile (kN)	*22.071*	*22.081*	*13.395*	*22.143*
Coupler	*3–4*	*3–4*	*3–4*	*2–3*
Cast iron brake blocks (P action) + other brake systems				
Case acronym[a]	DPPP	PPPD	RPPP	PPPR
Compression (kN)	11.623	57.105	13.061	35.142
Coupler	3–4	3–4	3–4	3–4
Tensile (kN)	*57.253*	*4.789*	*35.169*	*3.905*
Coupler	*1–2*	*3–4*	*1–2*	*1–2*

[a]See Table 3

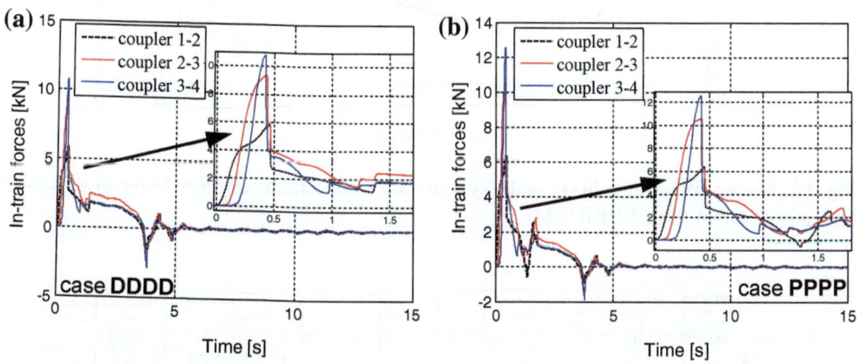

Fig. 10 In-train forces in uniform brake system train composition: **a** brake disc (case DDDD), **b** cast iron brake block system (case PPPP)

The studied cases are grouped into three categories, consistent with the operating practices regarding the usual brake systems combinations in passenger trains. They constitute also representative situations of in-train forces evolutions.

For studies and analyses, more adequate and useful are the diagrams of in-train forces evolution during the braking. Some of the most relevant are presented in Figs. 10, 11, 12, 13, 14 and 15.

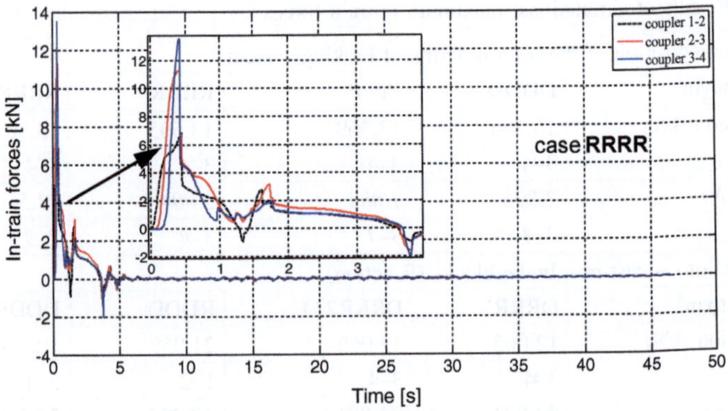

Fig. 11 In-train forces in uniform brake system train composition: cast iron brake block system, high power action (case RRRR)

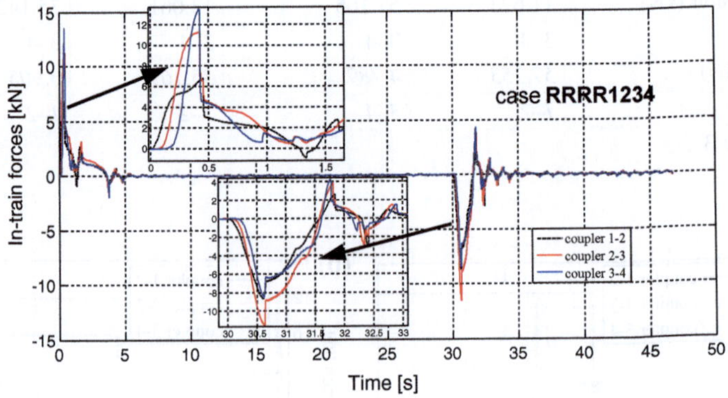

Fig. 12 In-train forces in uniform brake system train composition: cast iron brake block system, high power action (case RRRR1234)

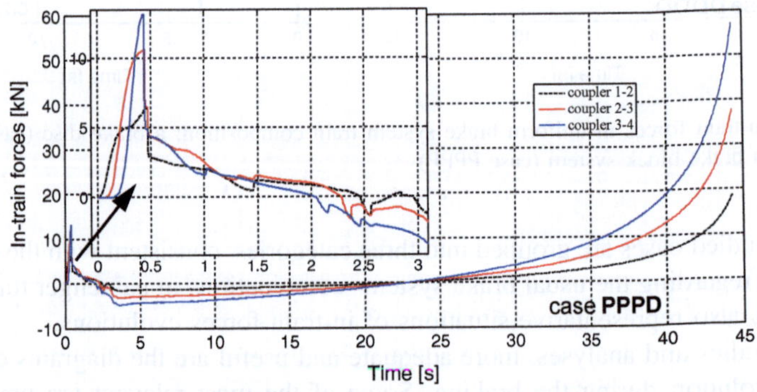

Fig. 13 In-train forces in mixed brake system train composition: cast iron brake block system, fast-acting and disc brake (case PPPD)

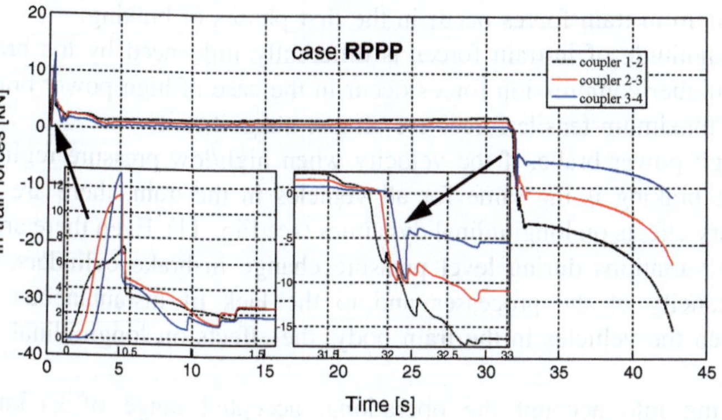

Fig. 14 In-train forces in mixed brake system train composition: cast iron brake block system, high power and fast-acting (case RPPP)

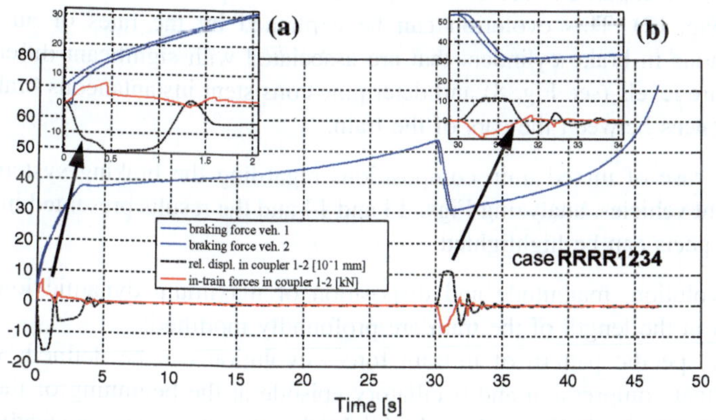

Fig. 15 Braking forces, relative displacement and in-train forces between first two vehicles of a train in RRRR1234 case

In the case of uniform train compositions regarding the braking systems of the vehicles, analyzing Figs. 10, 11 and 12, it may be remarked that:

- time evolution of dynamic longitudinal forces in long of the train are similar and in accordance with classical patterns: important compression and oscillatory movements of the vehicles within the train body during the first phases of the braking process, until steady air pressure is established in all brake cylinders, rapidly diminishing due to the strong damping capacity of the couplers. Further, while all forces acting on the vehicles remain constant in the given assumptions, no in-train forces occur until the train stops;

- maximum in-train forces occur in the first phases of braking;
- the magnitude of in-train forces is differently influenced by the brake system type: higher compression forces occur in the case of high power brake system, while maximum tensile forces are almost insignificant;
- for high power brake, if the velocity when high/low pressure regime changes during braking is the same for all vehicles in the train, there are no supplementary effects on longitudinal dynamics (see Fig. 11). Even there are important forces variations during level pressure change in brake cylinders, due to the simultaneity of the processes and to the lack of instantaneous differences between the vehicles in the train body, the effects on longitudinal dynamic is null;
- if taking into account the operational accepted range of ±1 km/h around 50 km/h for the moments of high/low pressure regime changes during braking, simulations show, as expected, that a second oscillatory process is initiated in the train body. The magnitude of the longitudinal reactions between the vehicles is higher compared to those developed in the first phases of the braking process (see Fig. 12). This evolution can be explained by the rates of air pressure variations in brake cylinders that are associated with significant differences in pressure levels (see Fig. 8) and determine consistent instantaneous brake forces differences between the cars of the train.

In the case of mixed train compositions regarding the braking systems of the component vehicles, analyzing Figs. 13 and 14 and the results presented in Table 4, certain aspects can be highlighted:

- the evolution, magnitude and disposition of maximum dynamic longitudinal forces in the length of the train are profoundly modified;
- a new specific pattern of in-train forces evolution can be defined: while the important compression and oscillatory episode at the beginning of the braking process remain similar to those highlighted in previous cases, a steady increase of the in-train forces characterizes the behavior of the system until the train stops. This particular evolution is caused by the continuous difference in braking forces between the vehicles equipped with disc brakes and those featuring cast iron brake blocks (see Fig. 2). In these cases, the longitudinal dynamic forces have a continuous increase tendency and, as a consequence of any brake systems combination comprising cast iron brake blocks, the maximum values of compression or tensile forces arise at the end of the braking process, just before the train stops;
- at the end of the braking process, the in-train forces are several times higher than those experienced by the train in the first phases of the braking process;
- any brake system combination comprising high power cast iron brake block generates a second oscillatory process in the train body during the brake cylinders pressure transition from the high to the low level. Concurrently, it can be ascertained that the magnitude of in-train forces is lower than in the precedent case (see, for instance, Figs. 13 and 14). This effect is explainable as a

consequence of specific evolution of braking forces whilst the running speed is lowering under 50 km/h (see Figs. 2 and 8), inducing a decrease in instantaneous difference of forces acting on the component cars of the train.

An important feature of our longitudinal dynamics simulation program is the possibility of simultaneous outlining of various parameters evolution during braking.

As an example, the simultaneous visualizing of the time history of braking forces between two adjacent vehicles, combined with the representation of relative displacement and the forces acting in the correspondent coupler, is extremely useful in explaining and analyzing the in-train forces evolution.

For instance, Fig. 15 presents the evolution of above-mentioned parameters between the first and the second cars of the train during the emergency braking process in the case of uniform high power cast iron brake block system composition in the case of different velocity thresholds for high/low pressure level change (case RRRR1234, see also Fig. 12).

Detail *a* in Fig. 15 clearly shows the beginning of the braking process: during the first 0.1 s, only the first railcar begins braking, so a rapid compression of the buffers occurs and, consequently, compression forces have a rapid increase. After the braking system of the second railcar actuates, the relative differences in braking forces begin to decrease. The assembly is still compressed, but the reduction of relative displacements diminishes the in-train forces between the two vehicles.

The diagrams show the concordance of the sense of relative displacements with the compression or tensile dynamic forces in coupler 1–2 of the train. Obviously, the whole process is also influenced by the behavior of the other railcars.

Detail *b* in Fig. 15 represents the period of transition (between 30 and 31.3 s of the braking process) from high to low pressure in the brake cylinders of the first railcars and the correspondent multiple effects on relative displacements and forces in coupler.

An overview on the effects of combining different braking systems in four-railcar passenger trains regarding the magnitude of in-train forces during emergency braking is presented in Fig. 16.

It appears that tensile forces are more sensitive to braking systems combinations than the compressive ones. The latter keep an almost constant level, except the

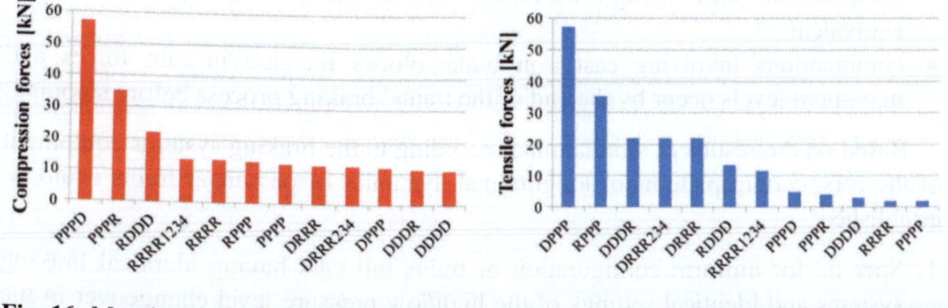

Fig. 16 Magnitude of in-train forces during emergency braking process for different combinations of braking systems

PPPD, PPPR and RDDD combinations, cases that determine a high increase of maximum compression forces in the train body. For the presented situation of a four-railcar train, the tensile longitudinal dynamic forces are almost insignificant in the presence of a single type of brake system.

6 Conclusions

The present study investigates comprehensively the influence of different braking characteristics on longitudinal dynamics of short passenger trains. There were analyzed the action of disc and of cast iron brake block systems, as high power and fast-acting operational mode. The individual braking power is set so that each railcar has the same braking distance in similar conditions; preliminary results of simulations reveal differences in instantaneous braking forces, running speeds and decelerations, according to specific characteristics of brake equipment.

Longitudinal dynamics simulations were carried out for various usual operational combinations in a four-railcar passenger train subjected to emergency braking from an initial running speed of 140 km/h.

The main conclusions, in the limits of the assumptions, hypothesis and initial conditions presented in Sect. 5.1, can be summarized as follows:

- the operational characteristics of the braking systems have important influences on the longitudinal dynamics of trains;
- generally, compression forces are significantly higher than the tensile ones, except certain brake system combinations;
- usually, higher levels of in-train forces are attained during the first braking phases;
- the time evolution of dynamic longitudinal forces between neighboring railcars indicates that the highest magnitudes are experienced in the second half of the train, except certain brake system combinations;
- brake systems combinations in the train generates significant higher levels of longitudinal dynamic forces in the train body;
- tensile forces are more influenced by the diversity of brake systems than the compression ones, though maximum values are, for certain combinations, equivalent;
- combinations involving cast iron brake blocks increase in-train forces and maximum levels occur by the end of the trains' braking process before stopping.

Based on the results of simulations, according to the braking systems equipment of the cars, certain patterns of longitudinal dynamics of passenger trains could be established:

1. Specific for uniform configuration of trains (all cars having identical braking systems and identical settings of the high/low pressure level changeover in the case of high power system), defined by important compression and oscillatory

episode at the beginning of the braking process, rapidly diminished due to the strong damping capacity of the couplers. Maximum compression and tensile in-train forces occur in the initial phases of the braking process;
2. Specific for high power braking equipment with random settings of the high/low pressure level changeover within the accepted limits (as usual in operation), defined by a second oscillatory process initiated in the train body during pressure modification in brake cylinders when the running speed drops below 50 km/h. The magnitude of the longitudinal reactions between the cars is higher compared to those developed in the first phases of braking process;
3. Specific for any combination including cast iron brake blocks, defined by important compression and oscillatory episode at the beginning of the braking process, followed by a steady increase of the in-train forces until the train stops. Depending on the disposition of different braking systems in the train composition, in the last part of braking, the in-train forces might be of the compression or tensile nature. The maximum in-train forces occur at the end of the braking process.

It is also worth noting that we consider that the originally developed longitudinal dynamics simulation program proves to be an efficient and useful instrument for studying and research, helpful in the investigation and better understanding of the complex process of train braking.

Based on the analysis, keeping a uniform composition of the train in terms of braking systems is recommended. If not always possible, then one should avoid at least the most disadvantageous combination in the train: disc brakes and cast iron block brake systems, in fast-acting operational mode.

However, we must admit that the magnitude of in-train forces obtained in the studied cases is not high enough to affect traffic safety, but deceleration variations, mainly during oscillatory episodes, may affect passenger comfort. In any case, one must keep in mind that certain operational conditions are still to be investigated: longer trains, different railcars, masses, maximum braking designed power for each railcar of the train, etc.

We consider the present study as an initial investigatory one, the results indicating that more investigations are necessary to develop and better clarify the presented domain.

References

1. UIC leaflet 540 (2006) Brakes—air brakes for freight trains and passenger trains, 5th edn. Nov 2006
2. Cole C (2006) Longitudinal train dynamics. In: Iwnicki S (ed) Handbook of railway vehicle dynamics. Taylor & Francis, London, pp 239–278
3. Wu Q, Luo S, Cole S (2014) Longitudinal dynamics and energy analysis for heavy haul trains. J Mod Transp 22(3):127–136

4. Belforte P, Cheli F, Diana D, Melzi S (2008) Numerical and experimental approach for the evaluation of severe longitudinal dynamics of heavy freight trains. Veh Syst Dyn 46(S1):937–955
5. Pugi L, Fioravanti D, Rindi A (2007) Modelling the longitudinal dynamics of long freight trains during the braking phase. In: 12th IFToMM world congress, Besançon, France, 18–21 June
6. Cantone L (2011) TrainDy The new Union Internationale des Chemins de Fer software for freight train interoperability. Proc Inst Mech Eng F J Rail Rapid Transit 225:57–70
7. Ansari M, Esmailzadeh E, Younesian D (2009) Longitudinal dynamics of freight trains. Int J Heavy Veh Syst 16(1/2):102–131
8. Nasr A, Mohammadi S (2010) The effects of train brake delay time on in-train forces. Proc Inst Mech F J Rail Rapid Transit, 224(6): 523–534
9. Zobory I, Békefi E (1995) On real-time simulation of the longitudinal dynamics of trains on a specified railway line. Period Polytech Transp Eng 23:3–18
10. Lifen D, Jilong X (2010) Research on the effect of traction tonnage on train longitudinal impact. Key Eng Mater 450:466–469
11. Cruceanu C, Crăciun C (2013) About longitudinal dynamics of classical passenger trains during braking actions. AMM 378:74–81
12. Cruceanu C, Oprea R, Spiroiu M, Crăciun C, Arsene S (2009) A model for the dynamic longitudinal reactions in the body of a braked passenger train. In: Proceedings of the 2009 international conference on computer science, WORLDCOMP'09, CSREA Press, Las Vegas, NV, pp 58–64
13. Cruceanu C, Crăciun C (2014) Influence of application time regulated limits on longitudinal dynamic forces in passenger short trains during braking process. In: Mathematical applications in modern science—proceeding of the 19th international conference on applied mathematics, Istanbul, pp 136–145
14. Hasegawa H, Uchida S (1999) Braking systems. JRTR 20(June), 52–59
15. Sharma RC, Dhingra M, Pathak RK (2015) Braking systems in railway vehicles. IJERT 4 (1):206–211
16. Cruceanu C (2012) Train braking. In: Perpinya X (ed) Reliability and safety in railway InTech, pp 29–74
17. Cruceanu C (2009) Brakes for railway vehicles. MatrixRom, Bucharest (in Romanian)
18. UIC leaflet 541-3 (2010) Brakes—disc brakes and their application. General conditions for the approval of brake pads, 7th edn, July 2010
19. UIC leaflet 546 (1967) Brakes—high power brakes for passenger trains, 5th edn, Jan 1967
20. UIC leaflet 542 (2013) Brake parts—interchangeability, 6th edn, Feb 2013
21. UIC leaflet 544-1 (2013) Brakes—braking performance, 5th edn, June 2013
22. Wu Q, Cole C, Luo S, Spiryagina M (2014) A review of dynamics modelling of friction draft gear. Veh Syst Dyn 52(6):733–758
23. Oprea RA, Cruceanu C, Spiroiu M (2013) Alternative friction models for braking train dynamics. Veh Syst Dyn 51(3):460–480
24. Crăciun C, Mazilu T (2014) Simulation of the longitudinal dynamic forces developed in the body of passenger trains. Ann Fac Eng Hunedoara Int J Eng XII(3):19–26
25. Sebeşan I, Copaci I (2008) Theory of elastic systems in railway vehicles. MatrixRom, Bucharest (in Romanian)
26. Copaci I (1998) Experimental tests for railway vehicles. Univ. "Aurel Vlaicu", Arad (in Romanian)
27. UIC leaflet 528 (2007) Buffer gear for coaches, 8th edn, Sept 2007
28. UIC leaflet 520 (2003) Wagons, coaches and vans—draw gear—standardisation, 7th edn, Dec 2003
29. Chiriac G (2002) Contributions to the rational use of movement energy of vehicles. Ph.D. thesis, Technical University "GH. ASACHI", Iaşi (in Romanian)

30. Ghițescu D, Călugăru D, Donciu T et al (1971) V. Handbook for railway traction calculations. Ministry of Transport and Telecommunications—Institute for Studies and Research in Transport, Bucharest (in Romanian)
31. Cantone L, Karbstein R, Müller L, Negretti D, Tione R, Geißler H-J (2008) Train dynamic simulation—a new approach. 8th world congress on railway research, Seoul, Korea, 2008
32. Wei W, Lin Y (2009) Simulation of a freight train brake system with 120 valves. Proc Inst Mech F J Rail Rapid Transit 223(1):85–92
33. Crăciun C, Cruceanu C (2011) Brakes for railway vehicles—experimental applications. MatrixRom, Bucharest (in Romanian)
34. UIC leaflet 832 (2004) Technical specification for the supply of brake-shoes made from phosphoric iron for tractive and trailing stock, 3rd edn, Jan 2004

The Behavior of the Traction Power Supply System of AC 25 kV 50 Hz During Operation

Radovan Doleček, Ondřej Černý, Zdeněk Němec and Jan Pidanič

Abstract This paper studies the AC 25 kV 50 Hz traction power supply system, which is used, in particular, in the Czech Republic. Nowadays, railway operation is very complicated and sophisticated, both from the viewpoint of railway infrastructure and means of transport. New technologies, devices and standards bring new problems for rail operation, including coupling to surrounding elements in the traction power supply system, transient effects during the recuperation mode of traction vehicles, the influence of neighboring track contact lines when disconnecting contact lines, etc. The main findings of these problems are detailed in this paper.

Keywords Traction power supply systems · Transient effects · Recuperation · Short-circuit

1 Introduction

Rail transport uses new technologies and devices in its infrastructure and transport systems for improving the safety, comfort, and speed of services. Therefore the AC 25 kV 50 Hz traction power supply system, which, in addition to other necessary

R. Doleček (✉) · O. Černý
JPTF, Department of Electrical and Electronic Engineering and Signaling in Transport, University of Pardubice, Studentska 95, 530 02 Pardubice, Czech Republic
e-mail: radovan.dolecek@upce.cz

O. Černý
e-mail: ondrej.cerny@upce.cz

Z. Němec · J. Pidanič
FEEI, Department of Electrical Engineering, University of Pardubice, Studentska 95, 530 02 Pardubice, Czech Republic
e-mail: zdenek.nemec@upce.cz

J. Pidanič
e-mail: jan.pidanic@upce.cz

A. Sładkowski (ed.), *Rail Transport—Systems Approach*, Studies in Systems, Decision and Control 87, DOI 10.1007/978-3-319-51502-1_2

systems for signaling, safety and communication systems, is used in the Czech Republic, has some specific characteristics during operation. One of these is transient effects, which can lead to non-standard dangerous conditions in the traction systems. It is necessary to maintain the standard requirements for all devices and equipment during their operation. For this reason, related engineering departments at the University of Pardubice have focused their research activities on studying the behavior of the traction system during faults or selected operation states. Therefore, it was necessary to analyze the particular elements and parts of the AC 25 kV 50 Hz traction power supply system and their couplings. The above-mentioned Czech traction power supply system is a little different from other traction systems. This system uses a specific filter-compensation device (FCD) in traction substations.

2 Configuration of the Traction Power Supply System

The general configuration of the Czech AC 25 kV 50 Hz traction power supply system with rail vehicles for transport in Fig. 1 contains the following:

- contractor feeding line of 110 kV,
- traction substation with FCD,
- catenary (the whole structure—contact line) and
- electric vehicles.

The electromagnetic compatibility is discussed more and more. Therefore, usage of the FCD in traction substations is necessary. This FCD is utilized for power factor corrections and to reduce current harmonics caused by older electric

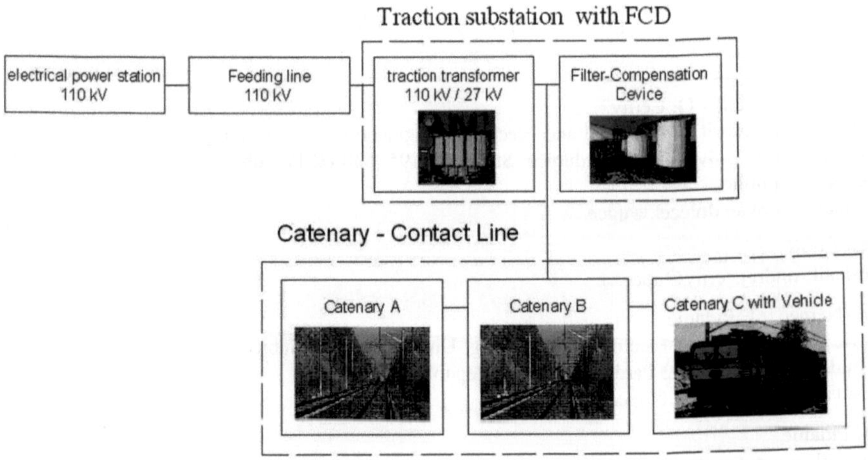

Fig. 1 The AC 25 kV 50 Hz traction power supply system in the Czech Republic

single-phase locomotives with diode converters, which are still very common on Czech railways, see Figs. 2 and 3. This FCT during transient effects also influences the behavior of the traction system, see Fig. 4.

Fig. 2 Locomotive type 242 with a diode rectifier, (DPF ≅ 0.84) and locomotive type 263 with a thyristor-diode bridge, (DPF ≅ 0.77)

Fig. 3 Locomotive type 210 with a thyristor controller and uncontrolled rectifier, (DPF ≅ 0.76) and locomotive type 362 with an input converter filter, (DPF ≅ 0.87)

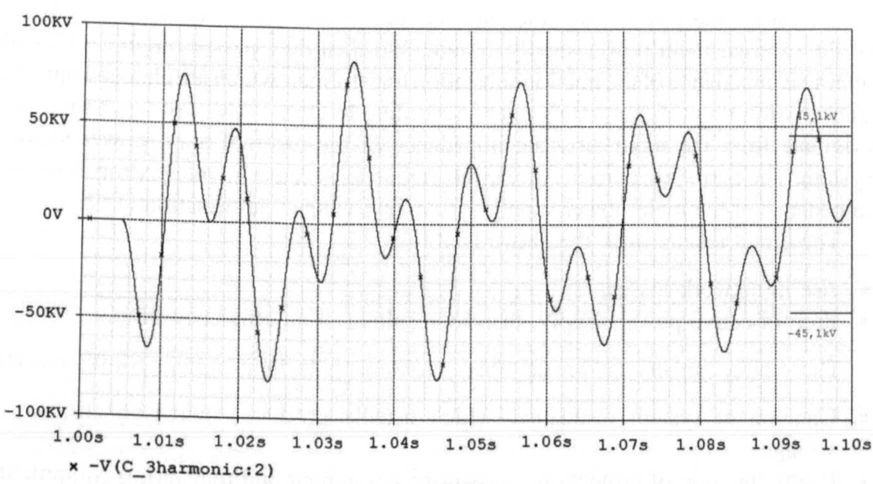

Fig. 4 Locomotive pantograph showing voltage at the 3rd harmonic

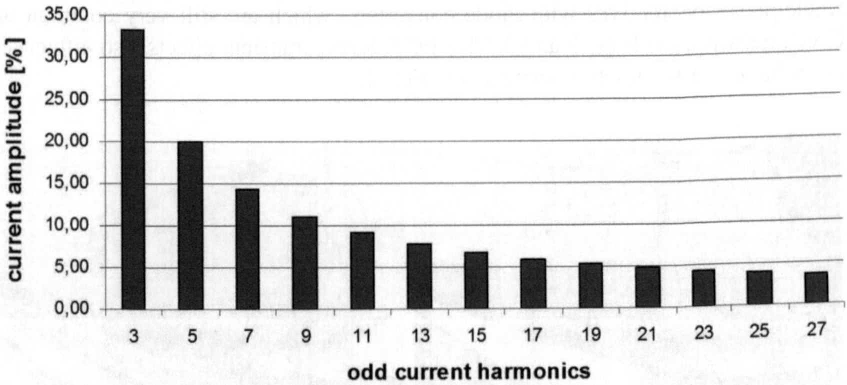

Fig. 5 The current harmonic spectrum of the diode rectifier

Electric locomotives with diode rectifiers generate all odd current harmonics (i.e. 3rd, 5th, 7th, and so on), see Fig. 5. The current harmonics pass through a catenary, traction substation transformer, a 110 kV contractor feeding line and then to the main power supply system.

The harmonics spectrum (i.e. harmonic numbers), which depends on types of rectifier connections, is formulated by equation

$$N = (p \cdot n) \pm 1, \quad (1)$$

where

p is pulse number of rectifier (–)
n integer number 1, 2, … n

The harmonics pass through the catenary section, independently of the impedance of the external main power supply system, and then they pass through a series of alternate impedance of the traction substation transformer, which are changed only by transmission ratio of this transformer. According to Ohm's law, voltage harmonics originate from the input alternate impedance of the external main power supply system. The harmonic currents of various frequencies cause a voltage drop from the impedance of the main power supply system and voltage deformation [1, 2].

The direct results from the above lead to a

- rise in network losses,
- drop of active power supply, thereby resulting in an efficiency drop.

These direct results can create others problems with their system configurations

- Creation of system resonance, which usually produces increased current or voltage.
- Faulty function of protections, measuring equipment and registering equipment.
- Interference of telecommunication equipment and control circuits.

The requirements for the FCD are given according to [1–3]:

- Adjust to inductive power factor of fundamental harmonic of traction consumption of single-phase locomotives to the required value of contractor (i.e. with inductive character) in connecting the point of traction substation to guarantee sufficient compensating power.

Minimize transfer of current harmonics at the 3rd, 5th and, perhaps even, the 7th harmonic numbers to corresponding components in voltage of the connecting point of the traction substation were under the minimum values required by the contractor.

- Guarantee input impedance of the traction substation as complex (i.e. including catenary capacity and traction consumption of single-phase locomotives) for the operating frequency of the centralized ripple control of the contractor was higher than the required value (i.e. the level needed to prevent a drop of this operating frequency f_{CRC} = 216.67 Hz at the connecting point of the traction substation).
- These conditions have to be realized in all traction load ranges of the traction substations, according to the principle of a single-way power supply of the catenary section.

The FCD is designed under these conditions, see Fig. 6. The FCD contains two parallel series LC branches of the 3rd and the 5th harmonics with parallel connecting decompensation branches. The LC branches' tuning does not consist of order number of harmonic exactly, but it consists of low-order values as n_3 = 2.90–2.95 and n_5 = 4.98–5.00. The required sufficient total input impedance Z = 500–900 Ω for f_{CRC} are realized by the suitable option of C_3 and C_5 values in branches; this ensures that they are dependent on each other. A disconnecting switch connects the 5th harmonic LC branch, thereby the filtration requirement, which has to be started at the lowest number of harmonic, is carried out. The FCD structure provides to the 7th harmonic LC branch.

The decompensation branch comes with or without a reducing transformer, thyristor phase controller and decompensation chokes. Decompensation is handled by a decompensation choke, which is controlled. Thus control is realized with inductive power factor DPF = 0.98 of input power, which is measured at the

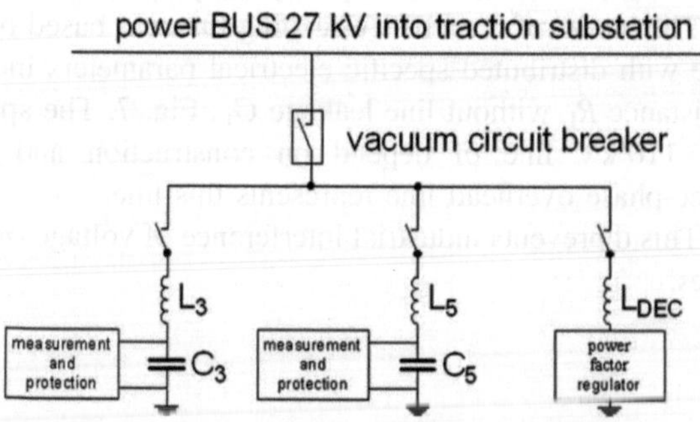

Fig. 6 The structure of the FCD

connection point of traction power supply substation. The creation of additional harmonics (i.e. primarily 3rd harmonics) into voltage of 27 kV bus is increased by partial controlling of the decompensation branch controller. The sum of both the 3rd harmonics controller and system are obtained amid congestion of the 3rd harmonic LC branch. Thus, LC branch tuning for FCD is under frequency 150 Hz.

3 Mathematical Model of Traction Power Supply System

Transient effects are analyzed at the linear systems that are defined equations. It was necessary to avoid building a physical model of a traction power supply system, because it would be costly or lead to the loss of process monitoring ability and the transient effects of circuit behavior. Therefore, we chose the program PSpice for mathematical simulation. PSpice utilizes input data to create a mathematical model of traction circuit elements, which represent the AC 25 kV 50 Hz traction power supply system. It was necessary to create quality models that represent real devices, because the simulation results could be distorted by unsuitable substitutions or simplifications of the traction circuit. The mathematical simulation results can be as exact as model elements and describe only effects that present using models. A creation of quality models that represent real devices well is the most important and most complicated problem of simulations of electronic circuits. Therefore mathematical models of the traction circuit were made for all parts of the AC 25 kV 50 Hz traction power supply system in the Czech Republic, as described below.

3.1 Model of Contractor Feeding Line of 110 kV

The 110 kV contractor feeding line and the catenary have the same character as that of a homogenous line with distributed electrical parameters. They can be considered to be a long electric line, which can be substituted in a two-port network as π-element or T-element with distributed electrical parameters or electrical long line with the following parameters: series specific resistance R_s, series specific inductance L_s, parallel specific capacity C_s, parallel specific leakage G_s [4].

The model of the 110 kV feeding line was based on a long homogenous electric line with distributed specific electrical parameters inductance L_L, capacity C_L and resistance R_L without line leakage G_L, Fig. 7. The specific electrical parameters of the 110 kV line of depend on construction and materials used. A standard three-phase overhead line represents this line.

This d prevents industrial interference of voltage or current and non-symmetrical lines.

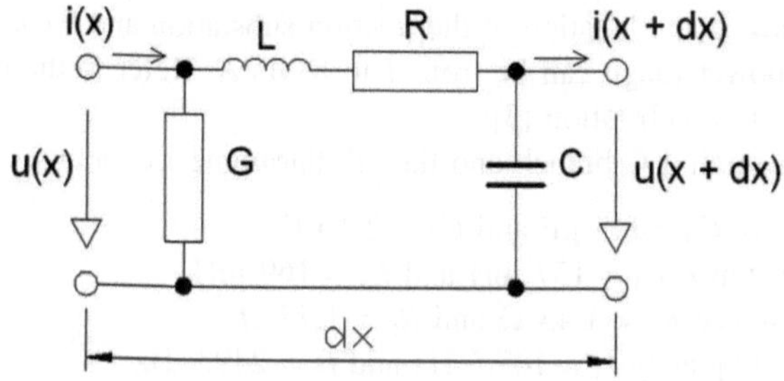

Fig. 7 The model of the 110 kV contractor feeding line for 50 Hz

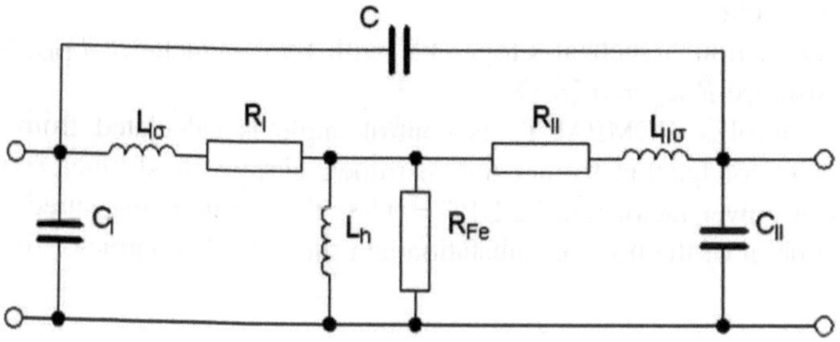

Fig. 8 The model of the traction transformer of 110/27 kV with 10 MVA for 50 Hz

3.2 *Model of Traction Substation*

The traction substation contains the 110/27 kV traction transformer with 10 MVA and the FCD. The 110/27 kV traction transformer can be presented only by one series for 50 Hz. The inductance L_{TT} is given a short-circuit voltage for the traction transformer and series resistance R_{TT}, which represents active losses. The values of alternate series inductance depend on the used tap of the transformer, because the transformer ratio can be a little bit different for each transformer, see Fig. 8. These transformers have a wide regulation range of output voltage (i.e. 2 × 8 taps), which can be changed under power. Current harmonics pass through the traction transformer, and they are changed only by the used winding ratio.

The characteristic parameters of model transformer are

- short-circuit active losses 53 kW,
- series inductance $L_{TT} = 24$ mH,
- substitute resistance $R_{TT} = 0.39\ \Omega$.

Two series LC branches of the 3rd and the 5th harmonic and the decompensation branch represent the model of the FCD. The tuning of the LC branches is not adjusted to the number of the harmonic exactly, but it has to set at a lower value. This adjustment of the LC branches is necessary, because harmonics from the 110 kV feeding line could overload these LC branches. The FCD final parameters are set according to the location of the traction substation and local parameters, so that the FCD power range can be from 1 to 8 MVA. Refer to the example of the FCD in the traction substation [5]:

The 3rd harmonic LC branch and the 5th harmonic LC branch

- Total capacity C_3 = 8.5 μF and C_5 = 2.4 μF
- Choke inductance L_3 = 137 mH and L_3 = 169 mH
- Choke resistance R_3 = 1.43 Ω and R_5 = 1.77 Ω
- Resonance frequency f_3 = 147.5 Hz and f_5 = 249.9 Hz

Decompensation branch

- Reducing transformer 27/6 kV
- Air-core choke
- Decompensation branch at site 27 kV with total inductance L_{DEC} = 0.596 H and resistance R_{DEC} = 6.24 Ω
- Phase controller COMPACT, its control angle is calculated from values of instrument voltage transformer and instrument current transformer, so in order to values of power factor will be DPF = 0.98. This value is measured at the connecting point of the traction substation and the 110 kV contractor feeding line.

3.3 Model of Catenary

The catenary is also an electrical homogenous line with distributed electrical parameters, and it can be presented as a long electrical line [6, 7]. This presumption can be taken, because sections of the catenary are longer in comparison with sections of the station catenary. As mentioned previously, the model of the homogenous line has four parameters: series specific resistance R_{CL}, series specific inductance L_{CL}, parallel specific capacity C_{CL}, and parallel specific leakage G_{CL}. The G_{CL} leakage of the catenary and G_{CL} leakages of other lines, which are connected with the catenary, are left out of the calculations, because they have very high values. Line insulators provide excellent isolation of the contact line [8, 9].

Specific resistance R_{CL} and specific inductance L_{CL}, which are dependent on frequency, enter into the calculation. The current, which passes through the conductor, is pushed out on conductor surface (i.e. skin-effect) by the increasing frequency. Then the useful section of conduction (i.e. effective section of conduction) is decreased and specific resistance R_{CL} is increased. The current is decreased by the skin-effect, so the loop area decreases, too. The specific inductance L_{CL} decreases until the definite frequency, where remains constant, see Fig. 9. The specific capacity C_{CL}, which consists of the capacity of all conductors that have traction voltage, is measured by the returned line, which represents the ground. Its numerical values will depend on the number of conductors, their height, their external diameter and the configuration of the neighboring of electrified railway track (tunnel, railway cutting, railway embankment, station and so on).

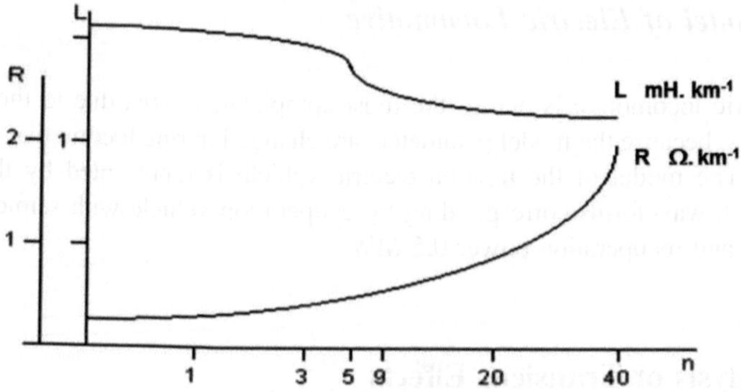

Fig. 9 The dependence of *L* and *R* parameters on frequency

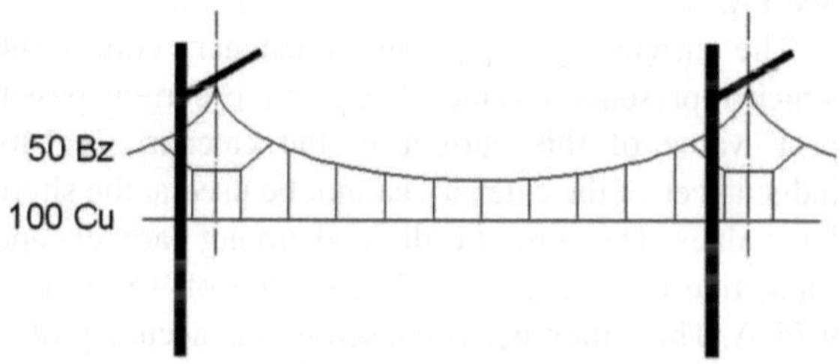

Fig. 10 The main structure of the catenary

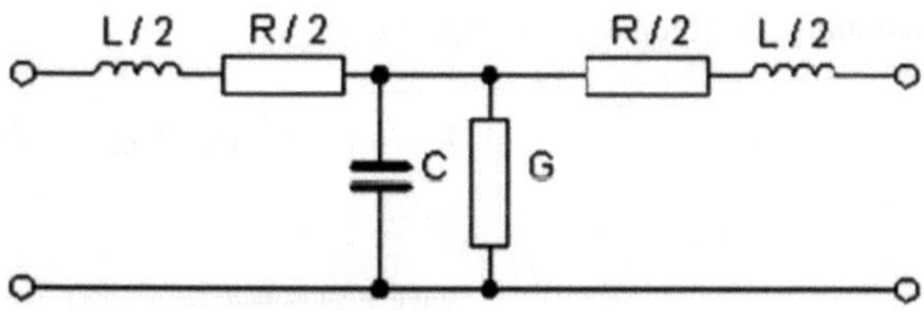

Fig. 11 The model of the catenary

The model of the catenary with a standard structure of 100 Cu + 50 Bz, see Fig. 10, with an intensive line has the same characteristics as a homogenous long electric line with distributed electrical parameters, but there are main parameters (i.e. resistance, inductance, capacity and leakage), see Fig. 11. This presumption can be taken, because sections of the catenary are longer in comparison with the sections of the station catenary. The parameter values used for this model are (for the whole length of catenary)

- series specific resistance $R_{CL} = 0.4\ \Omega\ km^{-1}$,
- series specific inductance $L_{CL} = 1.0\ mH\ km^{-1}$,
- parallel specific capacity $C_{CL} = 20\ nF\ km^{-1}$ (with intensive line),
- parallel specific leakage $G_{CL} = 0\ S\ km^{-1}$.

3.4 Model of Electric Locomotive

The electric locomotive is one of the most complicated parts, due to the variable parameters, because the model parameters are changed during locomotive operation [10–13]. The model of the traction electric vehicle is represented by the power source with waveforms corresponding to recuperation vehicle with semiconductor converter and recuperation power 0.5 MW.

4 Analysis of Transient Effects

4.1 Short-Circuit at Traction Substation

See Fig. 12.

The current $I_{CL\_beginning}$ in the catenary comes out from the initial value 9.8 A, which represents a value of capacitive current passing through the catenary. The peak value of this current in the catenary is 146 A, see Fig. 13. The series inductances of the catenary cannot be used at the shorted current, because they have low values. The current is divided among parallel capacities. The initial value of the capacitive current has simulated value of 9.8 A. It has a higher calculated value, 9.75 A. This difference is caused by the accuracy of simulation models. The arrival of the wave at the end of the catenary, which is not mismatched, is reflected. The time of the passing wave (i.e. delay time) to one direction for selected section of catenary is given by the equation

$$\mathrm{TD} = l_{\mathrm{CL}} \cdot \sqrt{L_{\mathrm{CL}} \cdot C_{\mathrm{CL}}} = 206\ \mu\mathrm{s}, \tag{2}$$

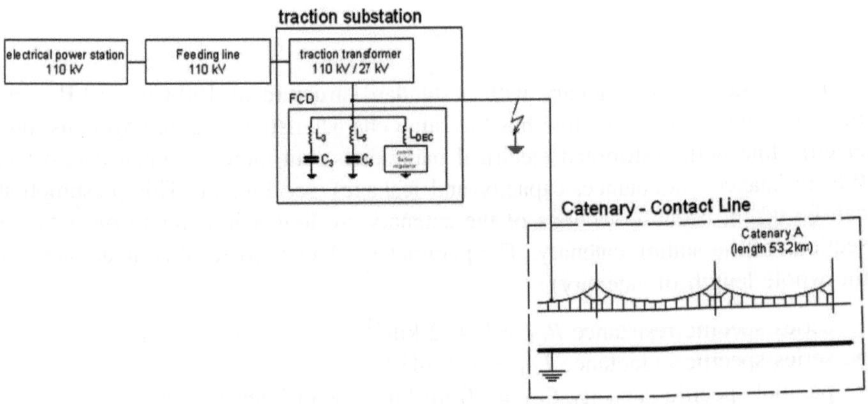

Fig. 12 The circuit diagram at short-circuit at the traction substation

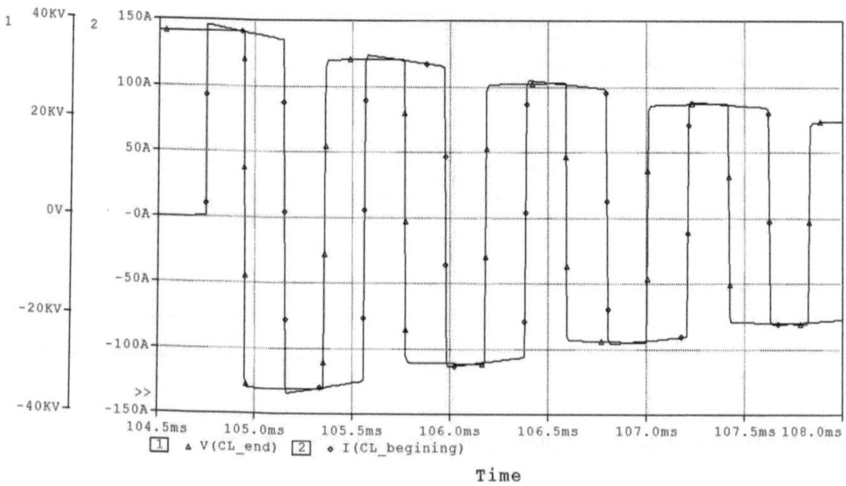

Fig. 13 The voltage at the end of the catenary and the current in catenary after short-circuit at the traction substation

where

l_{CL} is selected section length of catenary (km), $l_{CL} = 53.2$ km
L_{CL} series specific inductance of catenary (mH km^{-1}), $L_{CL} = 1.0$ mH km^{-1}
C_{CL} specific capacity of catenary (nF km^{-1}), $C_{CL} = 15$ nF km^{-1}.

The time of the passing wave in the catenary takes 412 μs both directions (i.e. from the beginning of the catenary to the end of the catenary and from the end of the catenary to the beginning of the catenary). Voltage V_{CL_end} at the end of catenary comes out from initial value 38.9 kV and this value does not get again, Fig. 13. The whole effect subsides after 1.6 s. The character of short-circuit and passage time of the wave are the same for various types of FCD connections.

4.2 Short-Circuit at the End of Catenary

See Fig. 14.

The current $I_{CL\_beginning}$ in catenary comes out from the initial value 9.8 A and its peak value gets 1.19 kA. The voltage $V_{CL\_beginning}$ at the beginning of the catenary comes out from the initial value 38.9 kV. Its value gets 90.3 kV in time c. 620 ms which depends on section of catenary (in this case) due to reflection of wave at the end of the open catenary then voltage falls consecutively, Fig. 15. Theoretically, peak voltage values at output of the traction substation can be achieved threefold the peak value for traction voltage (i.e. 116.7 kV). This value is obtained in triple the time of the wave passage. The catenary as a long line is ended inductance, which is represented substitutional inductance of traction transformer

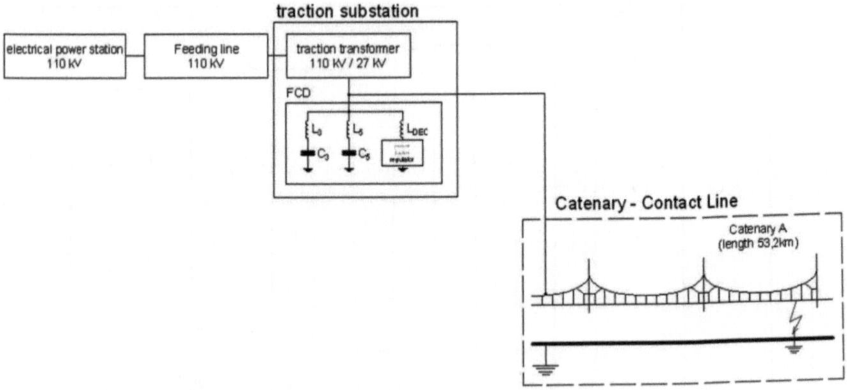

Fig. 14 The circuit diagram at short-circuit of the end of the catenary

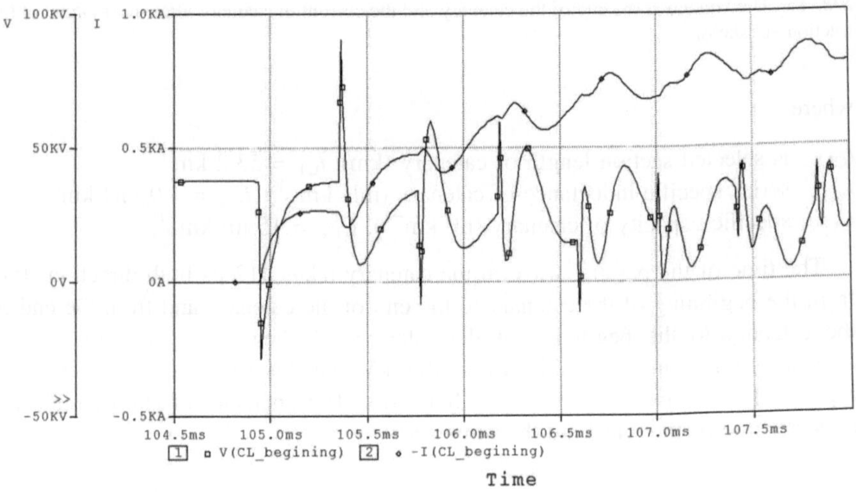

Fig. 15 The voltage at the end of the catenary and the current in the catenary after short-circuits at the end of the catenary

L_{TT} = 24 mH. Internal impedance of source (i.e. 38.9 kV) can be considered as zero impedance. This inductance seems to be an infinite impedance during a few milliseconds [14].

Wave reflection impedance of the traction substation does not depend on the number of LC branches of the FCD, because the traction substation in terms of the catenary, consists of the following parallel inductances: inductance of the traction transformer L_{TT} = 24 mH, inductance of the 3rd harmonic LC branch L_3 = 137 mH and inductance of the 5th harmonic LC branch L_5 = 169 mH.

The wave comes to the open end of the homogenous line it reflects with the same polarity as the original wave. Unlike the original wave, the wave coming to the shorted end of the homogenous line is reflected with a reversed polarity. The reflection wave has the same polarity and the same amplitude. After short-circuiting, it is possible to suppose a constant value of voltage during a few milliseconds at the traction substation. Wave passage time is the sum of the following:

- first passage time of wave from the shorted end of the catenary to the traction substation,
- second passage time of wave from the traction substation to the shorted end of the catenary,
- third passage time of wave back to the traction substation.

4.3 Short-Circuit at the End of Catenary as One Section

See Figs. 16, 17 and 18.

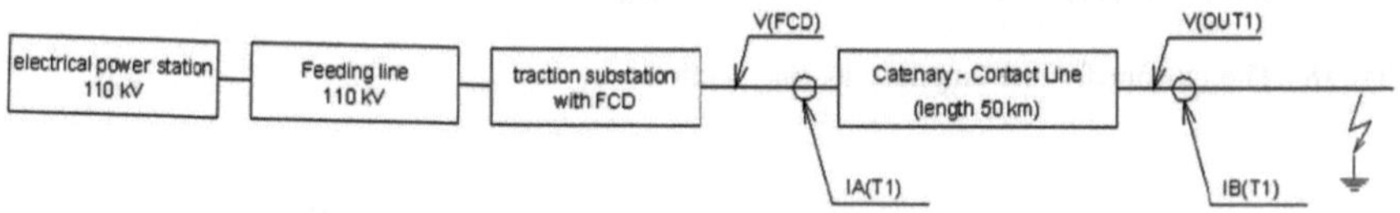

Fig. 16 The circuit diagram at short-circuit of the end of the catenary as one section

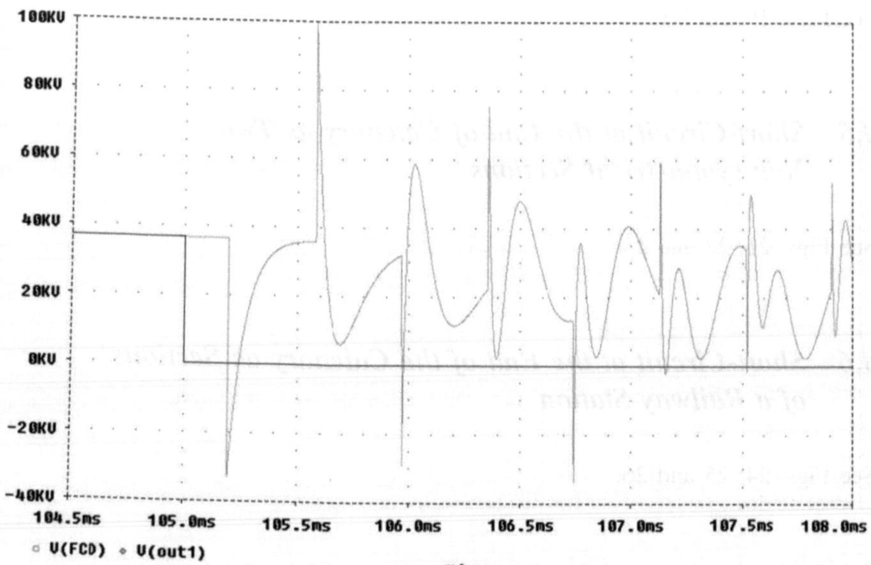

Fig. 17 The voltage for the catenary as one section

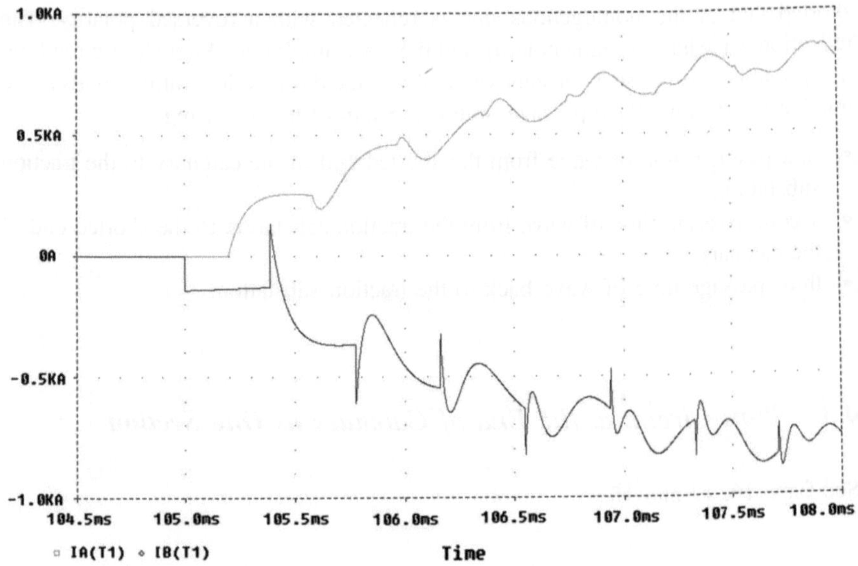

Fig. 18 The current for the catenary as one section

4.4 *Short-Circuit at the End of the Catenary as Two Symmetrical Sections*

See Figs. 19 and 20.

4.5 *Short-Circuit at the End of Catenary as Two Non-symmetrical Sections*

See Figs. 21, 22 and 23.

4.6 *Short-Circuit at the End of the Catenary as Sections of a Railway Station*

See Figs. 24, 25 and 26.

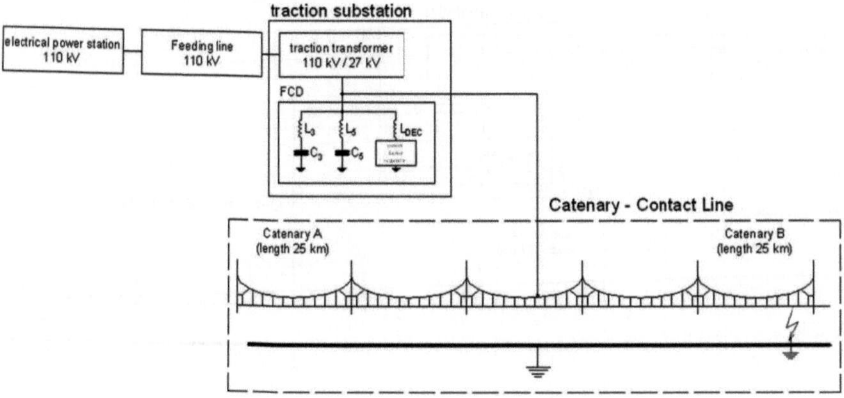

Fig. 19 The circuit diagram at short-circuit of the end of the catenary as two symmetrical sections

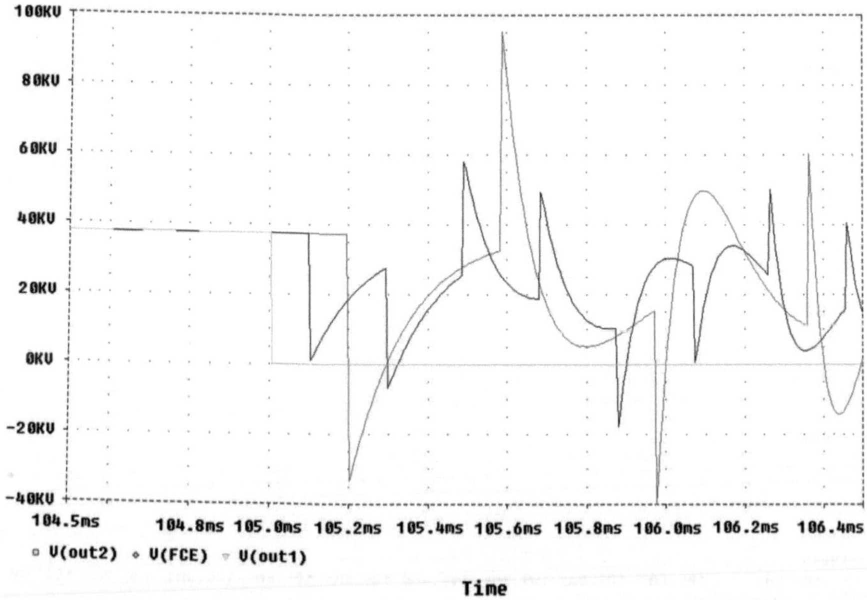

Fig. 20 The voltage at the traction substation and the voltage at the end of catenary A after a short-circuit at the end of catenary B

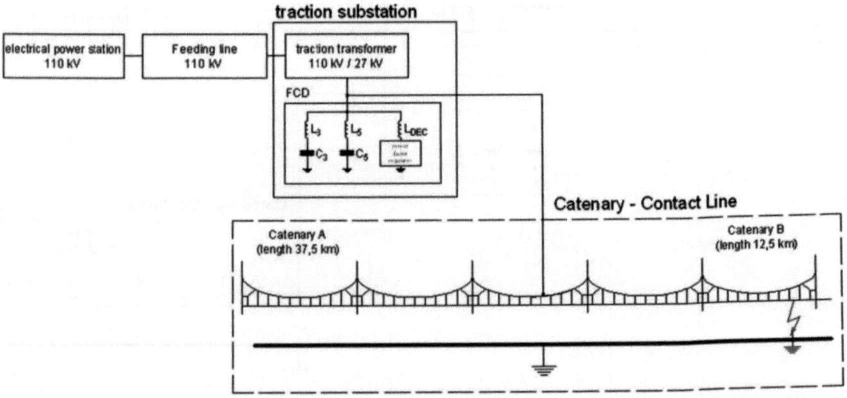

Fig. 21 The traction circuit diagram by short-circuit at the end of the catenary as two non-symmetrical sections

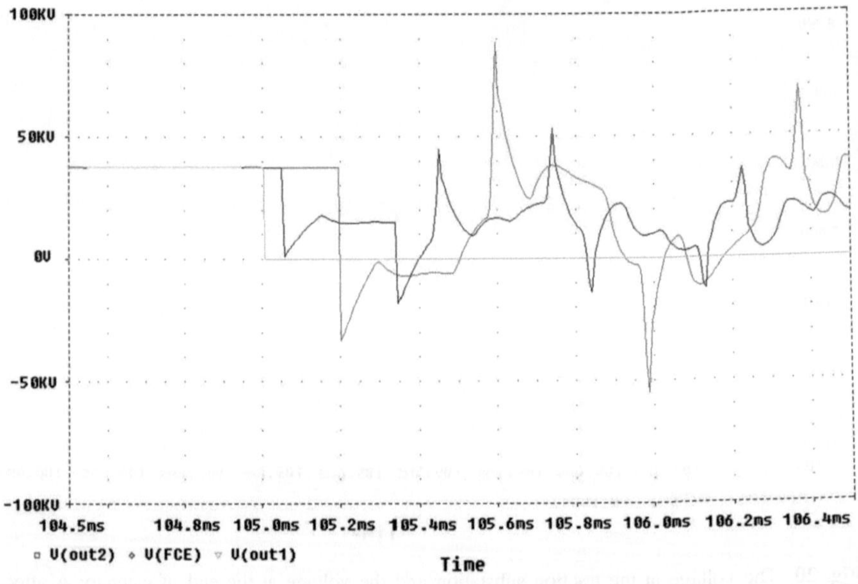

Fig. 22 The voltage at traction substation and the voltage at the end of the catenary A after short-circuit at the end of the catenary B

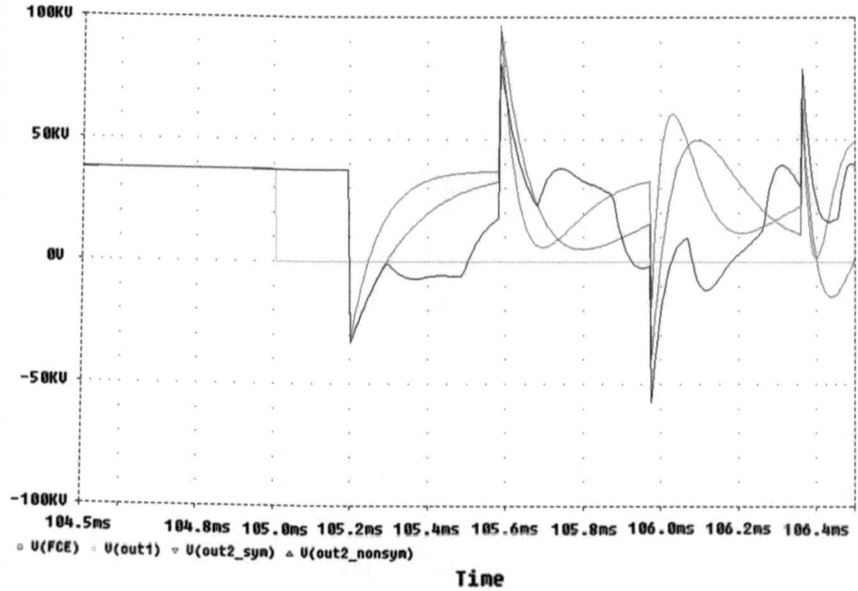

Fig. 23 The comparison of voltage for symmetrical and non-symmetrical sections

4.7 Findings of Transient Effects During Short-Circuits

- The electrical values of output voltage of the traction substation depend on the original conditions of short-circuit at the catenary. The short-circuit origin is always maximized by traction voltage. The presented cases of short-circuits are made without any traction consumption.
- Regarding overvoltage, the worst case is when the catenary is represented by one section. In this case, the voltage surge is reflected at the traction substation. Theoretically, traction substation voltage can rise up to threefold higher than traction voltage because of wave interference. The catenary as a long lossy line is ended by inductance represented by inductance of the traction transformer $L_{TT} = 24$ mH. Internal impedance of the 38.9 kV source can be considered as zero impedance. This inductance seems to be infinite impedance during a few milliseconds. The reflection of the voltage surge on impedance of the traction substation does not depend on the number of LC branches of the FCD, because the traction substation from the viewpoint of catenary consists of parallel inductances which are: traction transformer inductance with $L_{TT} = 24$ mH, inductance of the 3rd harmonic LC branch with $L_3 = 137$ mH and inductance of the 5th harmonic LC branch with $L_5 = 169$ mH. The real peak of voltage at output of traction substation is about 80 kV due to line losses that are higher than in the simulation model.

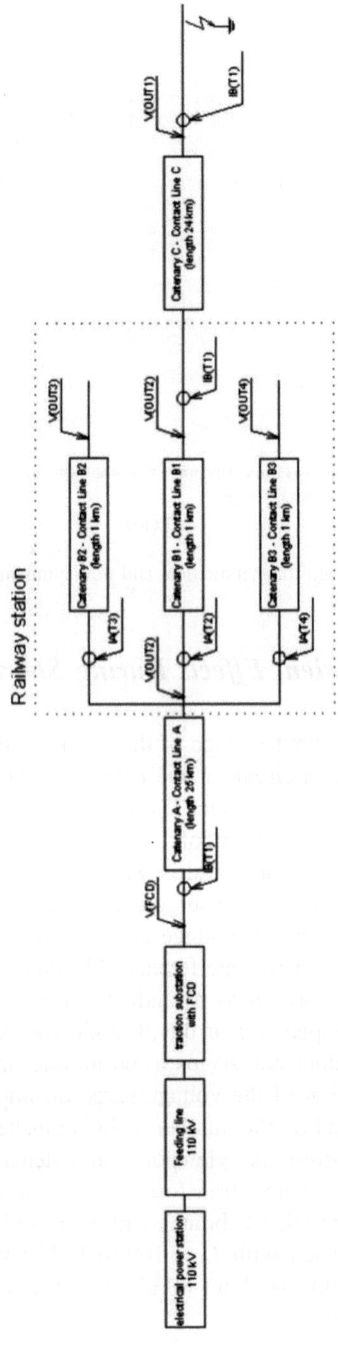

Fig. 24 The circuit diagram at short-circuit of the end of the catenary with railway station

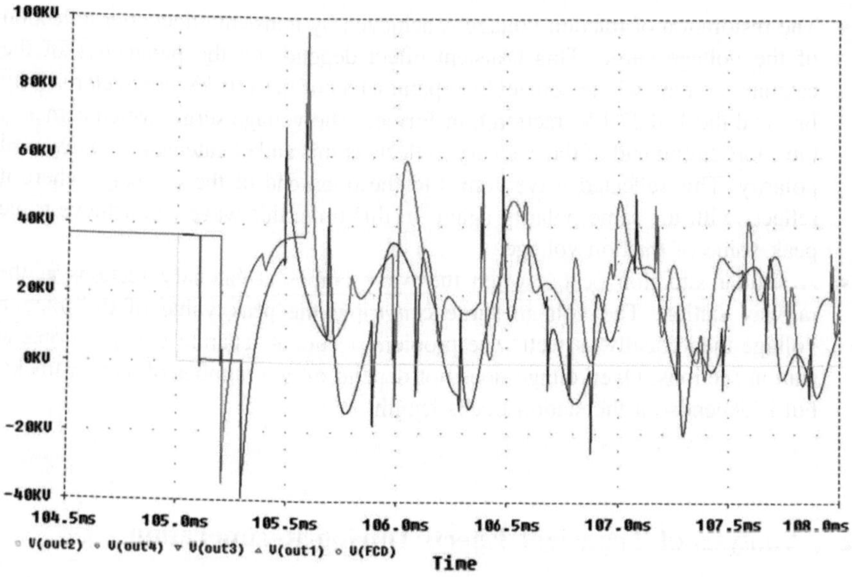

Fig. 25 The voltage for catenary sections with a railway station

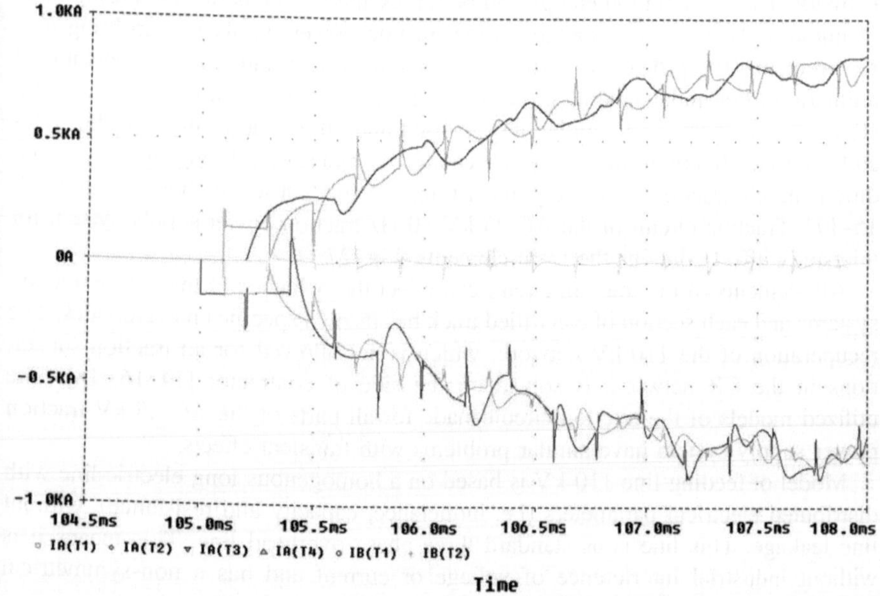

Fig. 26 The current for catenary sections with a railway station

- The restoration of traction voltage is achieved by transient effect after reflection of the voltage surge. This transient effect depends on the parameters of the catenary and the whole contact line, parameters of the 110 kV contractor supply line and the 110/27 kV traction transformer. The voltage surge, which comes at the short-circuit end of the catenary, reflects at this end of catenary with inversed polarity. This reflected wave comes to the open end of the catenary, where it reflects with the same polarity again. In this time, this wave can achieve triple peak value of traction voltage.
- A similar situation occurs when the wave comes at catenary sections at the railway station. The voltage surge can triple the peak value of the traction voltage theoretically as well. The problem occurs at separate voltage surges at station sections. Overvoltage does not depend on the number of station tracks, but it depends on the station tracks length.

5 Analysis of Transient Effects During Recuperation

The efficiency of traction energy usage is increasingly discussed in rail transport. This is related to the usage of energy from recuperation braking, which can be consumed by an electric rail vehicle or transformed to the 110 kV contractor mean network. The recuperation energy can be represented by units of MW during units of minutes. These energy sources in the traction power supply system bring new requirements for protection settings for maintaining standards and operational regulations. For understanding of behavior of traction circuits during these transient effects the mathematical simulations are used and allow one to monitor the states and points of traction circuits, which cannot be monitored during operation conditions in real traction circuit (e.g. identifying the root causes of short-circuits) [13, 15–17]. Traction circuit of the AC 25 kV 50 Hz traction power supply system for this study effect contains the main elements, Fig. 27.

All elements of the traction circuit can affect the behavior of the whole traction systems and each section of electrified track has its own specific characteristics. The recuperation of the 110 kV network, which is not allowed for all traction substations in the CR network, is solved on the side of contractor [14–16, 18]. The utilized models of the traction circuit made for all parts of the AC 25 kV traction power supply system have similar problems with transient effects.

Model of feeding line 110 kV is based on a homogenous long electric line with distributed electrical parameters (i.e. inductance, capacity and resistance), without line leakage. This line is as standard three-phase overhead line. This means it is without industrial interference of voltage or current and has a non-symmetrical character of line.

The model of the traction substation contains the 110/27 kV traction transformer with 10 MVA and filter-compensation device (FCD). The model of the 110/27 kV traction transformer can be presented by a series inductance L_{TT} in energetic

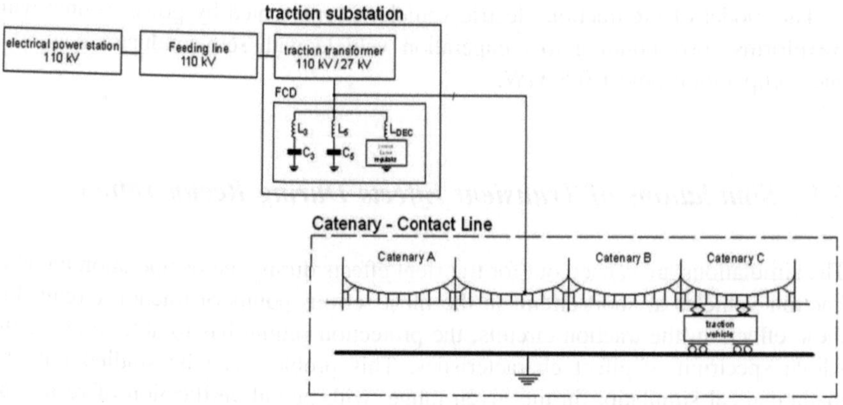

Fig. 27 The traction circuit of the AC 25 kV traction power supply system with recuperation locomotive

harmonic area, which is given short-circuit voltage of traction transformer and series resistance R_{TT} represents active losses. In the case of higher harmonics, it is necessary to add a passive filter with inductance and series resistance. The current harmonics pass through traction transformer and they are changed by using only a winding ratio. The characteristic parameters of the model transformer are short-circuit active losses of 53 kW, series inductance L_{TT} = 28 mH, resistance R_{TT} = 0.42 Ω.

The two series LC branches of the 3rd and the 5th harmonics and a decompensation branch represent the FCD model placed into the traction substation. The tuning of the LC branches is not adjusted to the number of the harmonic exactly, but it has to be adjusted for lower values. This adjustment of LC branches is necessary because harmonics from 110 kV feeding line could overload these LC branches. The following are the parameter values used for this model: capacity C_3 = 8.8 μF, inductance L_3 = 132 mH, R_{L3} = 1.3 Ω resonance frequency f_3 = 147.7 Hz and C_5 = 2.5 μF, inductance L_5 = 163 mH, R_{L5} = 1.7 Ω resonance frequency f_5 = 249.5 Hz. The decompensation branch has a reducing transformer 27 kV/6 kV, air-core choke and semiconductor controller which control power factor angle near to DPF = 0.98.

The model of catenary with standard structure of 100 Cu + 50 Bz with intensive line has the same character as homogenous long electric line with distributed electrical parameters, but there are considered all main parameters (i.e. resistance, inductance, capacity and leakage). This presumption can be taken because sections of catenary are longer in comparison with sections of station catenary. The following parameter values are used for this model: series specific resistance R_C = 0.3 Ω km^{-1} series specific inductance L_C = 0.8 mH km^{-1} and parallel specific capacity C_C = 20 nF km^{-1}.

The model of the traction electric vehicle is represented by power source with waveforms corresponding to recuperation vehicle with semiconductor converter and recuperation power 0.5 MW.

5.1 Simulations of Transient Effects During Recuperation

The simulations are carried out for transient effects during the recuperation mode of traction vehicles at short-circuit in the most serious points of traction circuit. For these effects in the traction circuits, the protection setting has to able to cover the whole spectrum of effect characteristics. This problem can be studied only by mathematical simulation in the given range, with partial verification of results by real measurements. The output data of the simulation program represent voltage and current waveforms. The critical states are deduced from knowledge of individual waveforms. The input parameters for protection settings of traction circuit can be gained by the analysis of these states [19]. When there is a short-circuit between a traction vehicle in recuperation mode and a traction substation, rapid exchanges of energy occur between the traction vehicle in recuperation mode and the short-circuit starting point as well as between the traction substation with FCD and the short-circuit starting point. At the beginning, the short-circuit is fed from two points and the energy from the recuperation vehicle is absorbed in units of ms, see Fig. 28.

In the case of the energy from the traction substation, the current wave is absorbed in tenths of ms (without protection off). During this time, a large amount of energy is transferred through the catenary; for example, the FCD supplies almost 20 kJ during the first units of ms to this short-circuit.

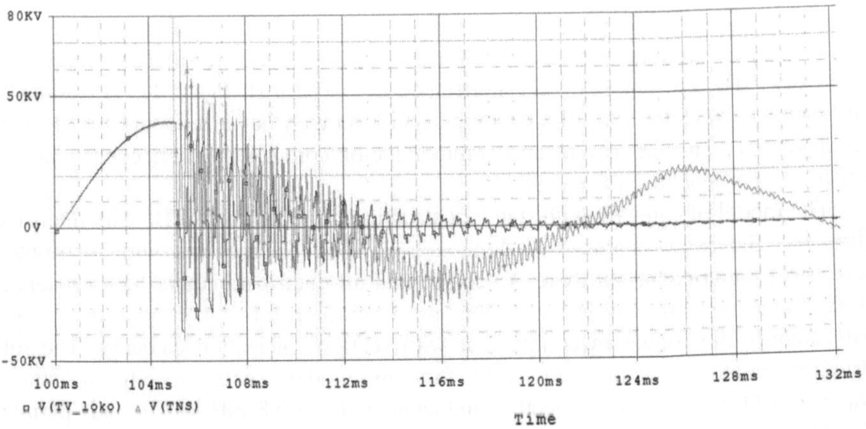

Fig. 28 The voltage at traction substation and voltage at pantograph of locomotive with recuperation

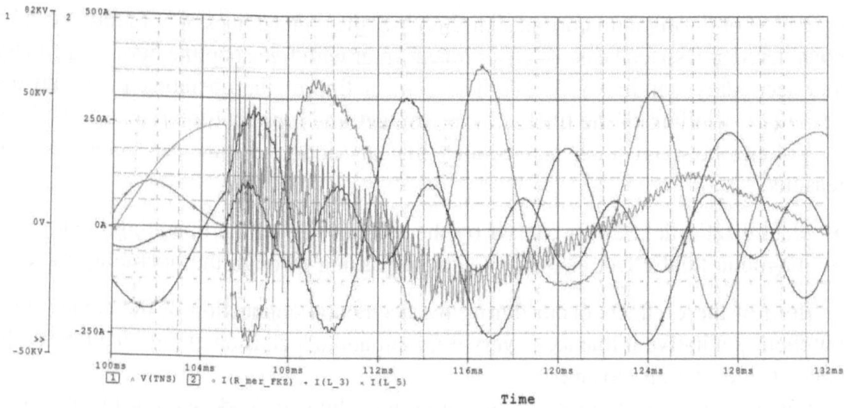

Fig. 29 The voltage and current at FCD

The energy transferred from the network to the traction power supply system is very high for this case. At the short-circuit, the characteristics of the 110/27 kV traction transformer and the impedance of lines are the only limiting circuit elements. The resistive impedances are decreased when the traction system achieves high efficiency, but these impedances are suitable for the short-circuit, when one considers the lowering of the short-circuit current and harmonics by modern electric vehicles propagating in the network (Fig. 29).

5.2 *Findings for Transient Effects During Recuperation*

The critical states are deduced from knowledge of individual waveforms. The input parameters for adjustment protections of traction circuit were obtained by the analysis of these states. The simulation diagrams, which are represented by voltage and current waveforms, can be also used as a main tool for particular project of traction substation of protection settings process. The design of the protections utilizes the traction substation design with FCD or without FCD.

6 Analysis of Voltage and Current at Contact Line at Track Closure

For disconnecting the contact line (i.e. track closure), of the AC 25 kV 50 Hz traction power supply system, it is necessary to connect the line with rail at both the front and at the end of the protected work area by two grounded rods. For this problem, it is necessary to analyze the influence (i.e. coupling ways) of the neighboring track contact line under the conditions that occur during the common operation or during fault current. For this case, the contact lines of both tracks can be considered as a special case of an air transformer with free coupling between primary and secondary winding [20, 21]. The contact line of the operating track and the secondary winding of the track closure contact line represent the primary winding.

Voltage between track and contact line at section closure can occur for two reasons:

- electrostatic field effect caused by the contact line of the operating track,
- magnetic field effect caused by current passing through the operating track.

Both of these effects occur during the accidental connection of the contact line, but their qualitative influence is very different and depends on circuit diagram of the contact line section closure.

For an electrical connection of the contact line of work section closure, it is necessary to take into consideration three circuit diagrams that arise at section closure.

- The contact line closure is disconnected from the feeder, but it is not connected to the rails; it hangs as an insulated line above the rails, see Fig. 30.
- This contact line is connected at one end to the rail, see Fig. 31.
- This contact line is connected at both ends to the rail, see Fig. 32.

It is possible to consider the other circuit diagram represented by a partial parallel connection of line at the contact line closure at the double track, but this is the special case of full paralleling [22]. The circuit diagram of the contact line closure is changing progressively during the protection of the work section in the above-mentioned three circuit diagrams, according to ground rods location by workers. The Eqs. (3) and (4) come from a substitute circuit diagram of the above-mentioned situation, which is valid for

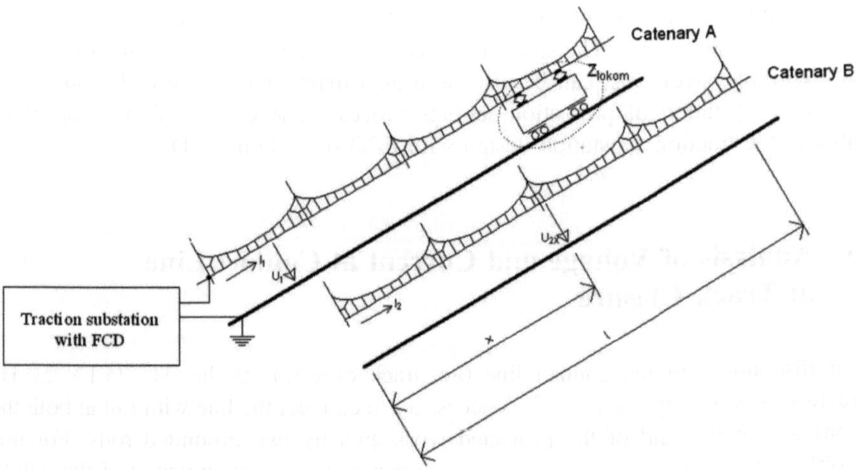

Fig. 30 The isolated contact line

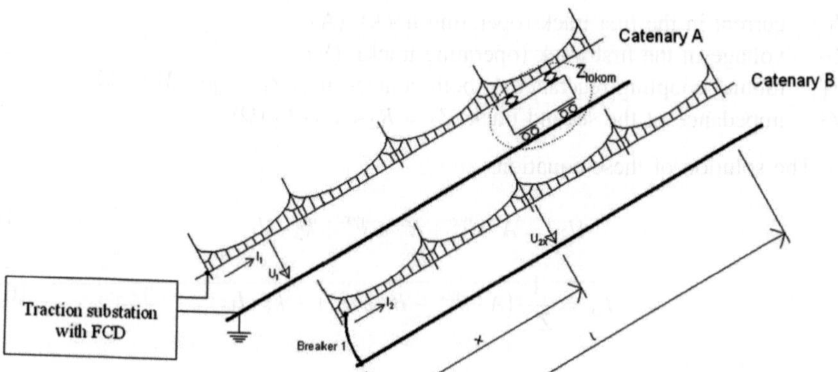

Fig. 31 The contact line one-sidedly connected to the rail

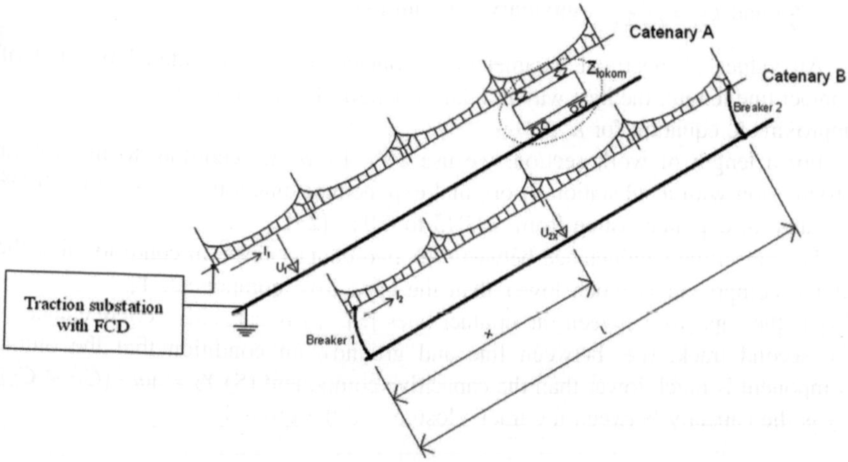

Fig. 32 The contact line with double-sided connection connected to the rail

$$-\frac{\mathrm{d}U_{2x}}{\mathrm{d}_x} = Z_{12} \cdot I_1 + Z_2 \cdot I_{2x} \tag{3}$$

$$-\frac{\mathrm{d}I_{2x}}{\mathrm{d}_x} = -Y_{12} \cdot U_1 + Y_2 \cdot U_{2x}, \tag{4}$$

where

U_{2x} voltage in point from the second track (track closure), (V)
I_{2x} current in the same point of the second track (track closure), (V)

I_1 current in the first track (operating track), (A)
U_1 voltage in the first track (operating track), (V)
Z_{12} mutual coupling reactance of both contact lines $Z_{12} = j\omega \cdot M_{12}$, (Ω)
Z_2 impedance of the second track, $Z_2 = R_2 + j\omega \cdot L_2$ (Ω).

The solution of these equations

$$U_{2x} = A \cdot \varepsilon^{j\gamma x} + B \cdot \varepsilon^{-j\gamma x} + k_2 \cdot U_1 \tag{5}$$

$$I_{2x} = \frac{1}{Z_{2v}} \left(A \cdot \varepsilon^{j\gamma x} - B \cdot \varepsilon^{-j\gamma x} \right) - k_1 \cdot I_1, \tag{6}$$

where

A, B integrating constants, values of which are given by boundary conditions of electrical circuits, (–)

Z_{2v} wave reactance of the second track $Z_{2v} = \sqrt{\frac{Z_2}{Y_2}}$, (Ω)

γ constant of propagate wave $\gamma = \sqrt{Z_2 \cdot Y_2}$, (–)

$k_1 = \frac{Z_{12}}{Z_2}$ and $k_2 = \frac{C_{12}}{C_{12} + C_2}$ auxiliary constant (–).

All values of electrical parameters of contact line are considered per unit of contact line length, the best way is 1 km of length. Furthermore, it is necessary the approximate equation for $\varepsilon^{\gamma x}$.

For a length of work section, we use $\varepsilon^{ax} \doteq 1 + a \cdot x$, according to the Taylor progression with a substation of original exponential function with accuracy 0.1% in range of exponent value from −0.045 to 0.045 [23].

Y_{12} capacitive conductance between the two contact lines on condition that the ohmic component is much lower than the capacitive component, $Y_{12} = j\omega \cdot C_{12}$, C_{12} is the capacity between the contact lines [S], Y_2 is capacitive conductance of the second track, (i.e. between line and ground), on condition that the ohmic component is much lower than the capacitive component (S) $Y_2 = j\omega \cdot (C_{12} + C_2)$, C_2 is the capacity between the track closure and the ground.

6.1 Isolated Contact Line

Boundary conditions for this circuit diagram come from the fact that the current cannot pass through at the ends of isolated section and then $I_{20} = I_{2\ell} = 0$. Under this presumption and equation usage (5) with using approximate equation for $\varepsilon^{\gamma x}$, we obtain

$$U_{2x} \doteq k_2 U_1 = \frac{C_{12}}{C_{12} + C_3} U_1 \tag{7}$$

$$I_{2x} \doteq 0 \tag{8}$$

The findings for isolated contact line

- Voltage between isolated contact line and rail is achieved by only capacitive distribution independently on the position at contact line.
- Resistance of this voltage source is very high (i.e. low values of capacity), so this voltage is "soft" and under load is decreasing.
- No current passes through isolated contact line.

In operation, this mentioned voltage occurs between point of ground rod, which is connected to the rail, and contact wire during their approach.

6.2 Contact Line One-Sidedly Connected to the Rail

The presumption for deduction of boundary conditions

- Voltage in the point of contact line connected to the rail is zero.
- Current is not passing through on the opposite end of contact line, which is isolated.

Under presumption that connection between contact line and rail is localized to origin of length coordinate x and length of contact line is l and then $U_{20} = I_{2\ell} = 0$. For voltage and current in any location of section closure, we obtain:

$$U_{2x}{}^{\cdot} = -Z_{2v} \cdot k_1 \cdot \gamma \cdot I_1 \cdot x = -\mathrm{j}\omega M_{12} \cdot I_1 \cdot x \tag{9}$$

$$I_{2x}{}^{\cdot} = -k_2 \frac{U_1}{Z_{2v}} \cdot \gamma(\ell - x) = -\mathrm{j}\omega C_{12} \cdot U_1(\ell - x) \tag{10}$$

The findings for line one-sidedly connected to the rail

- Voltage, which is induced in the contact line of the second track, is directly proportional to the length of the one-sided connected section and then directly proportional to the current that passes through the operating contact line of the second track.
- Current, which passes through the point of connection of the second track of the contact line to the rail, is obtained only from the mutual capacity of both lines and the voltage of the operating line.

In operation, this voltage occurs in the place where the worker places the second ground rod (i.e. on the opposite end of the work section closure from where the ground rod is already placed). In the case of higher currents in the neighboring operating section, this voltage can reach dangerous values. Current, values of which are deduced from Eq. (10), passes through the ground wire of the first rod and it has no practical meaning from the viewpoint of its negligible magnitude.

6.3 Contact Line Two-Sidedly Connected with Rail

The boundary conditions for this case arise from the circuit diagram, so that voltage at both ends of examined line is zero (i.e. in the place of both earth rods) and then $U_{20} = U_{2\ell} = 0$. By using Eqs. (5), (6) and the usual simplification of exponential function, we obtain the following:

$$U_{2x} \doteq k_2 \cdot U_1 \left(1 - \frac{2 - \gamma \cdot \ell}{2 - \gamma \cdot \ell}\right) = 0 \tag{11}$$

$$I_{2x} \doteq -\mathrm{j}\omega C_{12} \cdot U_1 \cdot \frac{\ell - 2x}{2 - \gamma \cdot \ell} - k_1 \cdot I_1 \tag{12}$$

The findings for line double-connection to the rail:

- The voltage is zero in the entire length of section between the ground rods at the points of track closure.
- Current, which passes through this part of the contact line, consists of two components: inductive and capacitive.
- Inductive components depend only on k_1 and are directly proportional to the current of the operating track. It depends neither on the length of the section nor location of the point at track closure.
- Capacitive component depends on voltage of contact line of operating track. It is zero in the center of the examined section (i.e. in the middle of the ground rods) and increases linearly to both ground rods but with different polarity.
- Inductive components are applied significantly in the operation, whereas capacitive components have no impact (i.e. its values are insignificant). Inductive components can reach the high values that require very good condition of the ground wire cable at the ground rods.

6.4 Findings for Contact Line Power Closure

For numerical values of the above-mentioned equations, it is necessary to use numerical values of a particular constant of the contact line. For this reason, results from measuring are utilized [23]. The value for impedance of one track of contact line was measured $Z_2 = 0.222 \div 0.485j = 0.534\angle 65.5^\circ\ \Omega\,\mathrm{km}^{-1}$. The reactance of mutual coupling of both contact lines is $Z_{12} = 0.133\,\Omega\,\text{km}^{-1}$. This value is given by Eq. (9) inducted voltage at single-sided connection to the line.

The measured voltage is $U_2 = 0.133\,\text{V}\,\text{A}^{-1}\,\text{km}^{-1}$. The capacity of the contact line to the ground was measured $C_2 = 0.0155\,\mu\text{F}\,\text{km}^{-1}$. The capacity between the contact lines, which was calculated by roles of potentials between conductors, is $C_{12} = 0.0136\,\mu\text{F}\,\text{km}^{-1}$. It is possible to calculate other coefficients from numerical values by their definition equations $\gamma = 2.21 \times 10^{-3}$, $Z_{2v} = 242\,\Omega\,\text{km}^{-1}$, $k_1 = 0.250$ and $k_2 = 0.467$.

Voltages and currents for three circuit diagrams were calculated by equations:

- Voltage for isolated contact line by Eq. (7), when the voltage in the operating track is 25 kV is $U_2 = 11.7\,\text{kV}$. This value does not depend on length of section closure and in its any possible place. Internal resistance of this source is given by capacity C_{12} and its value is hundreds, kΩ which decreases with length of section.
- Inducted voltage for one-sided contact line with rail at the opened end (i.e. the high value at this point), which is calculated by Eq. (9), is $U_2 = 0.133\,\text{V}\,\text{A}^{-1}\,\text{km}^{-1}$. Current passes through the point of connection (i.e. contact line and rail), which is calculated by Eq. (10) at 25 kV in the operating track, is $I_2 = 0.107\,\text{A}\,\text{km}^{-1}$. This value is negligible.
- Voltage for the two-sided contact line with rail is calculated by Eq. (11). Voltage in any possible place of operating contact line is zero. Induced component of current passing through to the point of connection of the contact line closure with rail by (12) is $I_2 = 0.250\,\text{A}\,\text{A}^{-1}$. The capacity component of this current is

7 Conclusion

New technologies in rail, including devices for infrastructure and transport, have meant improved safety, comfort, and speed of services. It has also brought new problems for the AC 25 kV 50 Hz traction power supply system with a specific filter-compensation device in traction substations in the Czech Republic. These problems have resulted in some specific behavior issues during operation modes. For these reasons, research activities during last several years in engineering departments at the University of Pardubice have been focused on studying the behavior of the traction system during faults or selected operation states, for example, transient effects and coupling to surrounding elements. These states can be studied in particular by mathematical simulation without building physical models in the given range, with partial verification of results by real measurements. The main findings and the results are explored in each part of this article.

Acknowledgements The research is supported by the Technology Agency of the Czech Republic under grant No. TE01020038.

References

1. Hlava K (1996) The limitation of FCD effect to centralized ripple control signal of electric energy contractor, part no. 1, 2. Report no. D 237 4026, TÚDC department EMC Prague (in Czech)
2. Hlava K (1994) Design of addition of filter-compensation device to traction substations at Czech Railways, Partial report Z 0024 003, „Control branch of filter-compensation equipment“ (BK 22 459), Praha (in Czech)
3. Standard PNE 38 2530 (1994) Central ripple control, transmitter and receiver (in Czech)
4. Burrtscher H Laboratory model to examine extension and superposition of high frequency at railway network. Co-operator at institution for AIE, ETH Zurich, ORE A 122 Part no. 3.2, Work program (in German)
5. Hlava K (2005) Analysis of conditions of FCD for traction substation of Czech Railways Modřice, Prague, Report No. 11 (in Czech)
6. Hlava K (2004) Electromagnetic compatibility of railway devices. University of Pardubice, ISBN 80-7194-637-0 (in Czech)
7. Bazelyan ME, Raizer PYu (1998) Spark discharge. New York CRC Press LLC, USA. ISBN 0-8493-2868-3
8. Standard EN 50 122-2/A1 Railway devices—stationary tractive devices—part no. 2. Protecting measures from effects of dispersion currents witch are bring out by DC traction systems
9. Standard ČSN 34 93 25 Ceramic insulators. Insulators for traction line of railways (in Czech)
10. Gao L, Xu Y, Xiao X, Liu Y, Jiang P (2008) Analysis of adverse effects on the public power grid brought by traction power system. In Proceedings of electric power conference (EPEC). Vancouver BC, pp 1–7
11. Wang J, Xue L, Liu Y, Li M, Pan L, Li S, Wang S, Liu Y (2008) The interaction research between public grid and traction power supply system. In: Proceedingsof electricity distribution (CICED'08). Guangzhou, China, pp 1–8
12. Na R, Song H, Tian-Lin L, Xiao-Jun P (2014) Study on modeling and simulation of rail transit traction power supply system in urban power distribution system by using PSCAD. In: Proceedings of electricity distribution (CICED'14). Shenzhen, China, pp 695–699
13. Sopov VI, Biryukov VV, Prokushev YA, Rylov YA (2008) Electric transport vehicle power-supply system analysis with various traction networks topologies. In: Proceedings of strategic technologies (IFOST'08). Novosibirsk, Tomsk, pp 462–464
14. Nahvi M, Edminisfer J (2003) Electric circuits. McGraw-Hill print, USA. ISBN 0-07-139309-2
15. Zhengqing H, Yuge Z, Shuping L, Shibin G (2011) Modeling and simulation for traction power system of high-speed railway. In: Proceedings of power and energy engineering conference (APPEEC). Wuhan, China, pp 1–4
16. Pee-chin T, Poh ChL, Holmes DG (2005) Optimal impedance termination of 25 kV electrified railway systems for improved power quality. IEEE Trans Power Syst 20(2):1703–1710
17. Lixiang S, Xing Z, Min L, Hongying P (2014) Modeling and influence research of traction power system based on ADPSS. In: Proceedings of power system technology (POWERCON). Chengdu, China, pp 193–198
18. Minwu CH, Qun-Zhan L, Guang W (2009) Optimized design and performance evaluation of new cophase power system. In: Proceedings of power and energy engineering conference (APPEEC'09). Wuhan, China, pp 1–6

19. Zhao T, Wu M (2011) Electric power characteristics of all-parallel AT traction power supply system. In: Proceedings of Transportion, Mechanical and Electrical Engineering (TMEE'11). Changchun, China, pp 895–898
20. Zhao W, Zou J, Wang J (2010) Study on harmonic detection methods in traction power supply system. In: Proceedings of power and energy engineering conference (APPEEC'10). Chengdu, pp 1–4
21. Wildi T (2006) Electrical machines, drives and power system, 6th edn. Pearson, Prentice Hall, USA. ISBN 0-13-196918-8
22. Paynter TR (2006) Introductory electronic device and circuits, 7th edn. Pearson Prentice Hall, New Persey, USA. ISBN 0-13-171641-7
23. Hlava K (1968) Voltage and current ratio at closure track of AC 25 kV traction system. Železniční doprava a technika, sv. 16, č. 4, pp 73–75, (in Czech)

Systems Approach to the Analysis of Electromechanical Processes in the Asynchronous Traction Drive of an Electric Locomotive

Pavel Kolpakhchyan, Alexander Zarifian and Alexander Andruschenko

Abstract Improving the efficiency of modern electric locomotives calls for new approaches to design. At the design stage, the dynamic properties of the mechanical parts and the behavior of electrical equipment and control systems in various modes of operation (e.g., start and acceleration, traction regime, coasting movement, wheel-slide protection) must be evaluated. These objectives can be achieved using a systems approach to the analysis of electromechanical processes in asynchronous traction electric locomotives. To solve these problems, a complex computer model based on the representation of an traction drive with Alternating Current (AC) induction motors as a controlled electromechanical system is developed. A description of methods applied in modeling of traction drive elements (traction motors, power converters, control systems), as well as of mechanical parts and of "wheel–rail" contact, is given. The control system provides individual control of the traction motors, and focuses on the results of dynamic processes modeling in various modes of electric locomotive operation. Mathematical modeling methods were used to investigate the dynamic characteristics of the electric locomotive EP20 under various conditions: movement in a straight line, in curves and in the turnouts.

Keywords Rail transport · Electric locomotive · Asynchronous traction drive · Computer modeling of electromechanical processes

---

P. Kolpakhchyan (✉) · A. Zarifian
Rostov State Transport University, Rostov-on-Don, Russia
e-mail: kolpahchyan@mail.ru
URL: http://www.rgups.ru

A. Zarifian
e-mail: zarifian@mail.ru
URL: http://www.rgups.ru

A. Andruschenko
Engineering Center "Rail technology", Novocherkassk, Russia
e-mail: andrewa62@mail.ru
URL: http://www.tmholding.ru

A. Sładkowski (ed.), *Rail Transport—Systems Approach*, Studies in Systems, Decision and Control 87, DOI 10.1007/978-3-319-51502-1_3

1 Introduction

Current methods for improving the traction properties of an electric locomotive typically involve the use of brushless traction motors and improvements to their control systems.

The use of brushless traction motors (three-phase AC induction motors) complicates the system of electric power conversion and is characterized by a high degree of interaction and mutual influence of processes between the elements of the traction drive. This requires in-depth research at the design stage, including the study of interconnected electromechanical processes arising at various modes of functioning, and establishment of optimal control algorithms.

One of the perspective methods for studying processes in a traction drive is the use of mathematical modeling. Especially effective is the use of computer models in the design stage, as the early stages of a new electric locomotive design begets answers to many questions, resulting in not making complex and costly prototypes [1–3]. We draw attention to the fact that in the cited papers and in this chapter, the full-size model of a locomotive having multi-motor traction drive is considered. In perspective, the evaluation of a locomotive's energy efficiency at the realization of various control algorithms must be obtained. Another objective is to research the dynamic processes in various modes of the electric locomotive operation (start and acceleration, traction regime, coasting movement, wheel-slide protection, etc.).

The use of an asynchronous traction drive (ATD) significantly impacts the main indicators of electric locomotives. The asynchronous traction motor power is about one and a half times higher than the collector traction motor power of the locomotive. The shaft torque of the asynchronous traction motor exceeds the shaft torque of collector traction motor by 20–40% [1–3].

The increased power and shaft torque of the asynchronous traction motors make high demands for durability and ensuring the resource of the traction mechanical transmission (gear, clutch, traction motor bearings and gear, etc.). Asynchronous traction motors have a higher speed than the collector traction motors. For example, future passenger locomotives will be designed with traction motor rotation of up to 3600 rpm. The increased rotation speed of the asynchronous traction motor shaft requires the installation of new bearings, high-precision manufacturing and careful dynamic balancing of rotating parts.

With the application of the ATD, the weight and volume of electrical equipment, which is located in the body, increases. The difference in weight is partially offset by a lighter asynchronous traction engine, but weight redistribution occurs between the body and the bogies. The weight of the body with equipment increases and the weight of bogies decreases, and the height of the gravity center of the body with electric equipment and of the locomotive increases, which negatively affects dynamics and the impact on the track, particularly in curved track. This issue is relevant for high-speed electric locomotives with ATD. Therefore, at the design stage, effective measures must be implemented to ensure standard dynamics and impact on the track. It is necessary to choose optimal specifications of spring suspension and damping systems to reduce the weight of steel structures, which should provide the required weight indices and strength of the locomotives.

Bogies with arched side frames must be applied in the construction of electric locomotives with ATD to lower the height of the center of gravity of the body with equipment, and to increase the volume of the back to accommodate the equipment in it. Arched side frames have a more complex stress–strain state than frames with a horizontal top sheet. The stress–strain state of the frame must be investigated at the design stage in order to provide strength and fatigue endurance.

As the passenger electric locomotive's speed increases, it is necessary to provide an effective braking system that will ensure braking without wheel sliding, and that will provide a running wheel service life of more than 1.0 million kilometers. A disc brake meets these requirements. It eliminates the thermal and mechanical stresses on the wheel-rolling circle and increases its service life. However, the unsprung load borne by mounted wheels significantly increases, which adversely affects the interaction between the vehicle and track.

The torque of the asynchronous traction motor has a complex spectral composition. Therefore, the design of the bogie must prevent matches of vehicle component frequency and frequency of torque pulsation on the shaft of an asynchronous traction motor so as to prevent resonance phenomena in the elements of the traction drive and in its attachment to the frame, which can lead to their destruction.

Traditionally, the design involves the development of a locomotive underframe and a breadboard model, its testing and further refinement. This approach involves significant input, and increases the timescale of electric locomotive development. The use of computer models as the controlled electromechanical system is useful for accelerating the design process. These models must take into account the interaction of the locomotive underframe, traction motors and control system. The new EP20 passenger electric locomotive designed for use on Russian railways was designed following this approach. As such, the conceptual issues of joint modeling and modeling results analysis of traction drive and of a locomotive underframe were used while designing this locomotive.

The use of computer models of electromechanical processes in traction drives takes on particular significance during the development of the control system, allowing its structure and parameters to be developed in parallel with the rest of the equipment. This provides an opportunity to take into consideration the features of the traction drive functioning revealed during the design of its individual elements or entire subsystems.

A modern electric locomotive with asynchronous traction motors is a complex dynamic system composed of interacting subsystems—electrical, mechanical and control. In addition, the capacity of the traction power system is comparable to the capacity of the electric locomotive. For an adequate representation of historical parameters, it is necessary to use a system analysis of processes in the electric locomotive based on the representation of the AC traction drive as a controlled electromechanical system. In accordance with this approach, it is necessary to develop a complex mathematical model comprising an interacting model of subsystems outlined above to solve the problems described above.

2 Problem Formulation

The development of the locomotive underframe and traction drive are realized in parallel, which is a feature of using complex mathematical models in the early stages of design. At this preliminary stage, we know the structure and parameters of the power transformers, the traction motor parameters, the regulatory principles and the control system settings. Therefore, to develop a complex mathematical model of the electric locomotive EP20, we used data known at the time of creation of the model. Further comparison of simulation results with the test data showed that the electromechanical processes in the electric locomotive were adequately described by the suggested mathematical model, despite some differences in the parameters of the traction motors, and the structural and regulatory principles implemented in the control system.

Structural and regulatory principles of traction drives are the same regardless of whether one or the other type of traction motor is used. This is why the questions of developing the mathematical model of traction drive will be further considered using the example of a traction drive with asynchronous traction motors (ATM), as it is the most common [4] at the present time.

Figure 1 is a functional diagram of an electric locomotive traction drive with an asynchronous traction motor, where 1 = traction power-supply system; 2 = energy converter; 3 = traction motors; 4 = the mechanical part of the electric locomotive; 5 = a place of "wheel–rail" contact; 6 = control system.

Mathematical modeling of the processes occurring in ATD in dynamic modes is associated with a number of features:

- ATD is a complex dynamic system, composed of interacting subsystems—electrical, mechanical and control. In addition, the capacity of traction substations is comparable to that of an electric locomotive; thus, for adequate representation of the processes, co-simulations are needed.
- The variation in rotor speed is quite wide—from zero to several thousand rpm.
- The power semiconductor devices in circuits of static converters have nonlinear characteristics.

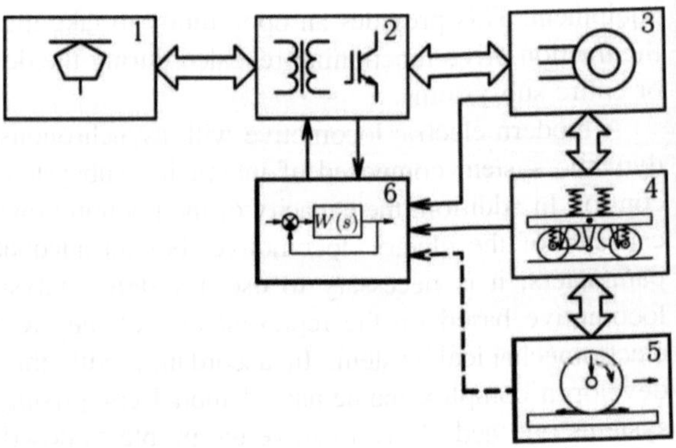

Fig. 1 Functional diagram of the traction drive of an electric locomotive with asynchronous traction motors

- ATM are highly-used machines, during the study of processes in them, the saturation of the magnetic circuit must be taken into account.
- The speed of the processes occurring in the individual subsystems of the traction drive differs by several orders of magnitude.
- The load of a traction drive is determined by the creep force (adhesion) in "wheel–rail" contact, which has essentially nonlinear dependence on the speed of wheel slip, and has hysteresis.
- When designing the traction drive, various structures of electric energy converters with different algorithms and laws of control can be used; moreover, it is possible to use different versions of the mechanical parts.

Complex systems with nonlinear characteristics and load have traditionally been analyzed using simulation methods, enabling investigation of the existing or newly created system options. This is particularly important during the design stage. Therefore, the computer model must be complete in the sense of solving the principal problems, and must be convenient to use, reliable and adaptive.

To ensure an accurate description of traction drive processes, consideration must be given to the transient mechanical processes associated with the development and decay of wheel spin and wheel sliding, together with the electromagnetic processes in traction motor circuits, given their high degree of mutual influence. As mentioned above, the hallmark of the traction electric drive is a strong correlation of processes in the circuits of all electric traction motors. Motors influence each other's work through both the mechanical and the electrical part of the drive, and their interaction takes place at different levels. Therefore, it is important to know what we are designing: wheels-motor single unit, bogie, a single section or several sections operating in parallel.

It is obvious that modeling only the wheels-motor unit cannot provide an answer to many questions related to the mutual influence of various elements of the traction electric motor. Applications of such models are largely limited by factors such as the preliminary calculations of the engine parameters and design, specification schemes of static converters and control system synthesis. A number of factors that have a significant impact on the current processes are not taken into account, although such a model is good enough to derive the specific nature of traction motor load in the form of a friction pair. Such a model is not viable for the study of electromechanical processes in traction electric motors, because it does not allow for the special characteristics of the combined action of its elements.

Analysis of a traction electric drive via simulation of a bogie does not provide the level of accuracy necessary for considering the interaction between the individual components of the system at different levels. This may be, for example, analysis of the electric locomotive movement under transient conditions or at starting, control system operation under conditions of spinning or sliding wheels, or the interaction of the electric locomotive and traction power supply systems. Therefore, modeling of electromechanical processes that occur in one or more sections operating in parallel provides the best results, because only in such a case can we observe the full picture of electromechanical processes in the traction electric drive of the electric locomotive. Thus the use of a systems approach to the study of traction electric drives allows us to consider electromechanical processes for all types of electric locomotives.

Integrated analysis of the traction electric drive as an electromechanical system involves the creation of ATM models, energy converters, control systems, mechanical sections of the electric locomotive, and processes in "wheel–rail" contact. The developed models should allow research in all modes of electric locomotives, including starting and movement at a low speed, taking into account all the main factors affecting the representation accuracy of the process.

3 The Mathematical Model of an Electric Locomotive with Asynchronous Traction Drive

3.1 The Mathematical Model Structure

The complex computer model of ATD as a controlled electromechanical system (CES), presented in this paper, is a further development of previously created models [1–3].

The electric locomotive is considered as a CES which includes the mechanical part as a multi-body system, the electric part (energy conversion devices and traction motors) and the control system (control algorithms and their realization).

There are direct and feedback communications between the processes in the mechanical and electric parts. Values of electromagnetic torque at the traction motor shafts are included in the right-hand side of the motion equations; values of wheel set angular velocities obtained from the mechanical part are included in the electromagnetic equations [5].

Electric locomotive EP20 is an electric locomotive with a dual power supply, meant to operate on an railroads with alternate current (AC) or direct current (DC) electrification systems: 25 kV AC, 50 Hz and 3 kV DC (EN 50163). ATM supply is carried out with the individual converters, consisting of an autonomous voltage source converter and a 4q-S converter (rectifier dedicated to energy recovery). When powered from the AC contact system, converters are connected to the transformer traction windings. When powered from the DC contact system, the power is supplied to the converter DC link (inverter input). For this reason, the voltage in the DC link is 2800 V.

Figure 2 shows the structure of a six-axle locomotive with individual control of traction motors (individual axle drive) as a CES. The structure of the electrical part includes a human interface, main locomotive control system and electric power conversion devices: main transformer, 4q-S converter, DC-link, self-commutated voltage inverter (VI) and ATM. The structure of the mechanical part includes the car body, three bogies and suspension elements.

The processes in separate subsystems and elements of the locomotive are modeled using various methods. The authors would like to emphasize that their

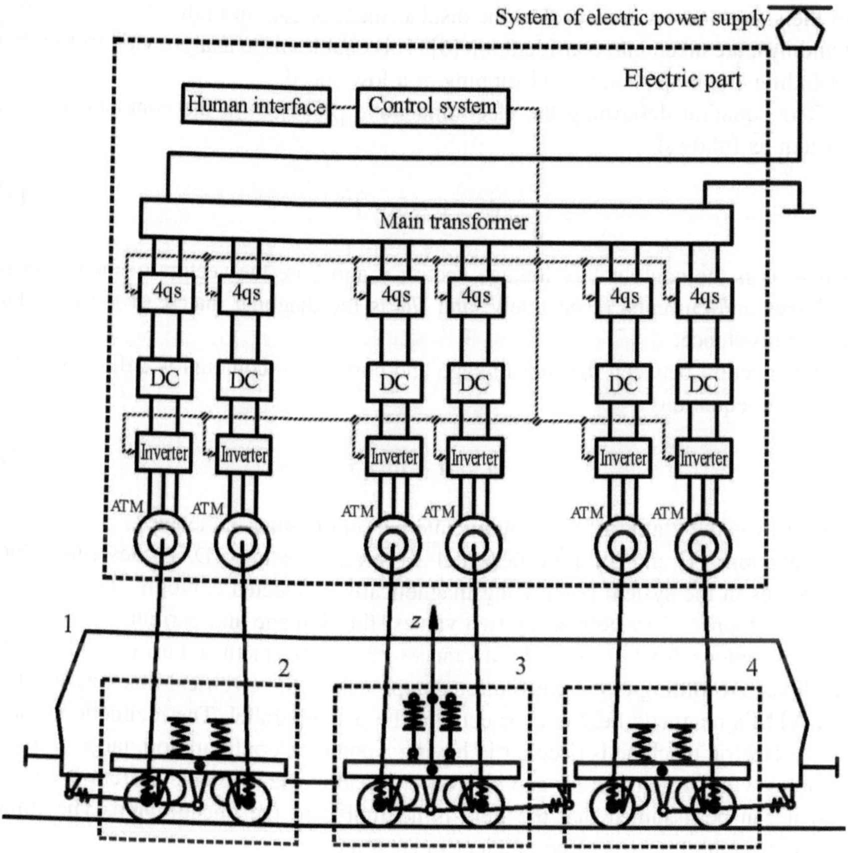

Fig. 2 Electric locomotive as a controlled electromechanical system: *1* car body; *2–4* bogies

main aim was to develop an adequate model of the rather complicated CES based on well-known modern methods and approaches.

3.2 The Modeling of the Asynchronous Traction Motor

The transformation of electromagnetic energy into mechanical energy is carried out in the ATM. The ATM is a key element of the traction drive; the parameters and design features of the motor have a direct impact on the technique and economic characteristics of the electric locomotive.

The model is based on the presentation of the ATM as a system of magnet-connected contours. Determination of the motor's magnetic system parameters is carried out by field theory methods (using the finite element method). In the mathematical model, the saturation and non-homogeneity of the active layer of the stator and rotor, as well as the displacement of current in the stator and rotor windings, are taken into consideration [6]. This enables us to analyze various modes including stopping/starting and running at a low speed.

The equation describing the electromagnetic processes in the contours can be written as follows:

$$\dot{\boldsymbol{\psi}} = \mathbf{u} - \mathbf{r} \times \mathbf{i}, \tag{1}$$

where $\dot{\boldsymbol{\psi}}$ is the contour flux linkage vector; $\mathbf{u}$ and $\mathbf{i}$ are the vectors of the contour voltages and currents, respectively; and $[R]$ is the diagonal matrix of the contour active resistance.

Connection between the flux linkage and currents in contours is defined by the algebraic equations

$$\dot{\boldsymbol{\psi}} = \mathbf{u} - \mathbf{m} \times \mathbf{i}, \tag{2}$$

where $\mathbf{m}$ is the matrix of the contour internal and mutual inductances.

Equations (1) and (2) are differential algebraic equations (DAE) describing the processes in the system comprising magnetically connected contours. The state of each contour is characterized by two values: flux linkage and current.

The contour flux linkage and inductances matrix are calculated using field theory methods. According to the conventional approach of the electrical machine theory, the ATM's magnetic field is considered to be plane-parallel. The magnetic system of the electric machine is peculiar in having a complex configuration, large number of areas with current and air-gap clearance. Besides, the magnetic system is closed and it can be assumed that the field is negligible in the outside area. The finite element method proves to be efficient in such a case.

The developed model can adequately represent the processes in the ATM under steady-state and transient conditions, including low-speed and emergency modes.

To analyze the processes in ATM at angular speeds above 10–15% of the nominal speed, it is expedient to use models based on the Park–Gorev equations [6].

3.3 The Modeling of the Processes in an Energy Conversion System

The energy conversion system provides the operation of the electric locomotive by feeding from a catenary network of 3-kV DC and of 25-kV, 50-Hz AC.

The energy conversion system is modeled as an electric circuit that consists of a main transformer, input converters (4q-S), a DC link and self-commutated voltage inverters (VI) for feeding the traction motors [4]. The principal scheme of the power circuit of an electric locomotive that feeds from the AC network is shown in Fig. 2.

The 4q-S converters connected to the traction transformer windings function in parallel to the general filter of the DC link. The inverter works as the filter load and feeds the ATM. The processes in the electric traction drive of all three bogies under the condition of feeding from the main transformer are considered in the model.

The locomotive is equipped with three power converters; each one feeds the motors of a two-axle bogie. Each power converter includes 4q-S converters, a DC link and VI. The individual control of traction motors (individual axle drive) was realized (see Fig. 2).

The principal scheme of a power circuit of an electric locomotive bogie that feeds from the contact AC network is shown in Fig. 3. Two 4q-S converters connected to the traction transformer windings are functioning in parallel to the general filter of the DC current link. The inverter functions as the filter load and feeds two ATM. The processes in the electric traction drive of all three bogies under the condition of feeding from the main transformer were considered in the model.

The ATM DTA-1200A for electric locomotive EP20 had the following characteristics [3]:

Nominal rating power	$P_n = 1470$ KW;
Line voltage	$U_{ll} = 2183$ W;
Power line frequency	$f_n = 89$ Hz;
Stator winding resistance at a temperature of 20 °C	$R_s = 0.0207\ \Omega$;
Leakage inductance of stator winding	$L_s = 0.00086$ H;
Rotor winding resistance at a temperature of 20 °C	$R_r = 0.01988\ \Omega$;
Leakage inductance of rotor winding	$L_r = 0.000716$ H;
Mutual inductance	$L_m = 0.01753$ H;
Excitation current	$I_m = 129$ A;
Number of pairs of poles	$2p = 6$.

The parameters of traction transformer are following [3]:

Design voltage of power winding	$U_C = 25$ kW;
Design voltage of traction windings	$U_{TR} = 2100$ W;
Design power of traction windings	$S_{TR} = 2100$ VA.

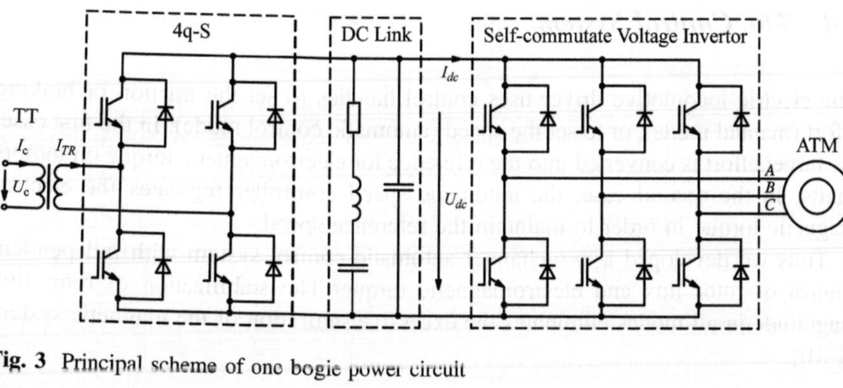

Fig. 3 Principal scheme of one bogie power circuit

The processes in the power circuits are described by the automatically generated DAE derived from Kirchhoff's law base:

$$f(\mathbf{x}_{n+1}, \dot{\mathbf{x}}_{n+1}, t_{n+1}) = 0, \tag{3}$$

where $\mathbf{x}_{n+1}$ is the state variable of the electric circuit (currents through inductances, voltage at condensers) at time point t_{n+1}. The solution to Eq. (3) is carried out with the backward differentiation formula

$$\mathbf{h}(\mathbf{x}_{n+1}) = \dot{\mathbf{x}}_{n+1} = -\frac{1}{\Delta t}\sum_{i=0}^{k} \alpha_i \mathbf{x}_{n+1-i},$$

where α_i is the backward differentiation formula coefficients; Δt is the step size.

Substitution of this formula in Eq. (3) yields [7]

$$\mathbf{x}_{n+1}^{(j+1)} = \mathbf{x}_{n+1}^{(j)} - \left[\mathbf{J}(\mathbf{x}_{n+1}^{(j)})^{-1} f\left(\mathbf{x}_{n+1}^{(j)}, \mathbf{g}_{n+1}^{(j)}, t_{n+1}\right)\right].$$

The Jacobian matrix for the function $f(\mathbf{x}_{n+1}, \mathbf{g}(\mathbf{x}_{n+1}), t_{n+1})$ at $\mathbf{x}_{n+1} = \mathbf{x}_{n+1}^{(j)}$ is derived as follows:

$$\mathbf{J}\left(\mathbf{x}_{n+1}^{(j)}\right) = \left.\frac{\partial f(\mathbf{x}, \dot{\mathbf{x}}, t)}{\partial \mathbf{x}}\right|_{\substack{\mathbf{x}=\mathbf{x}_{n+1}^{(j)} \\ \dot{\mathbf{x}}=\mathbf{g}(\mathbf{x}_{n+1}^{(j)})}} - \left.\frac{\alpha_0}{h}\frac{\partial f(\mathbf{x}, \dot{\mathbf{x}}, t)}{\partial \dot{\mathbf{x}}}\right|_{\substack{\mathbf{x}=\mathbf{x}_{n+1}^{(j)} \\ \dot{\mathbf{x}}=\mathbf{g}(\mathbf{x}_{n+1}^{(j)})}}.$$

The locomotive traction characteristics are mostly defined by the structure, principles and algorithms of the traction drive control system. A number of specific demands are usual for the control system of the ATM. The regulation of the ATM torque should be carried out without pulsations and self-exciting oscillations and its performance should be sufficient to prevent wheel slippage. It is necessary to strive for reduced torque ripple and electrical losses in the elements.

3.4 The Control System

The electric locomotive driver uses control handles to set the traction or braking effort (manual mode), or to set the speed (automatic control mode). In the first case, the target effort is converted into the reference for electromagnetic torque on motors shafts. In the second case, the automatic speed controller regulates the electromagnetic torque in order to maintain the reference speed.

Thus we developed a two-channel automatic control system with independent control of rotor flux and electromagnetic torque. The stabilization of rotor flux magnitude in all modes eliminates the excessive saturation of the magnetic system [8–10].

Processes in a squirrel-cage asynchronous motor are considered in a d–q rotating coordinate system. The d–q rotating coordinate system associated with rotor. The angular speed ω relative to the α–β fixed coordinate system is calculated as

$$\omega = 2\pi \frac{\mathrm{d}}{\mathrm{d}t}\theta(t),$$

where $\theta(t)$ is the angle between α–β and the d–q coordinate system.

If the rotor flux vector $\vec{\Psi}_\mathrm{r}$ is along an axis d, then

$$\Psi_{\mathrm{rd}} = \left|\vec{\Psi}_\mathrm{r}\right|; \quad \Psi_{\mathrm{rq}} = 0.$$

The position of $\vec{\Psi}_\mathrm{r}$ relative to the α–β fixed coordinate system is defined by an angle θ. In this case, the location of vectors is as shown in Fig. 4, where δ is the angle between the vectors of stator current $\vec{I}_\mathrm{s}$ and the rotor flux $\vec{\Psi}_\mathrm{r}$, and φ is the angle between $\vec{I}_\mathrm{s}$ and stator voltage $\vec{U}_\mathrm{s}$.

Processes in the squirrel cage asynchronous motor are described by the equations:

$$\begin{cases} \vec{U}_\mathrm{s} = \vec{I}_\mathrm{s} r_\mathrm{s} + \frac{\mathrm{d}\vec{\Psi}_\mathrm{s}}{\mathrm{d}t} + j\omega\vec{\Psi}_\mathrm{s}, \\ 0 = \vec{I}_\mathrm{r} r'_\mathrm{r} + \frac{\mathrm{d}\vec{\Psi}_\mathrm{r}}{\mathrm{d}t} + j\omega\vec{\Psi}_\mathrm{r}. \end{cases} \tag{4}$$

where r_s and r'_r are the stator and rotor resistances, $\vec{I}_\mathrm{r}$ is the rotor current vector and $\vec{\Psi}_\mathrm{s}$ is the stator flux vector.

The relationship between the currents and fluxes of the stator and rotor are described by the equations:

$$\begin{cases} \vec{\Psi}_\mathrm{s} = L_\mathrm{s}\vec{I}_\mathrm{s} + L_\mathrm{m}\vec{I}_\mathrm{r}, \\ \vec{\Psi}_\mathrm{r} = L_\mathrm{m}\vec{I}_\mathrm{s} + L'_\mathrm{r}\vec{I}_\mathrm{r}. \end{cases} \tag{5}$$

Fig. 4 Location of stator voltage, stator current and rotor flux vectors

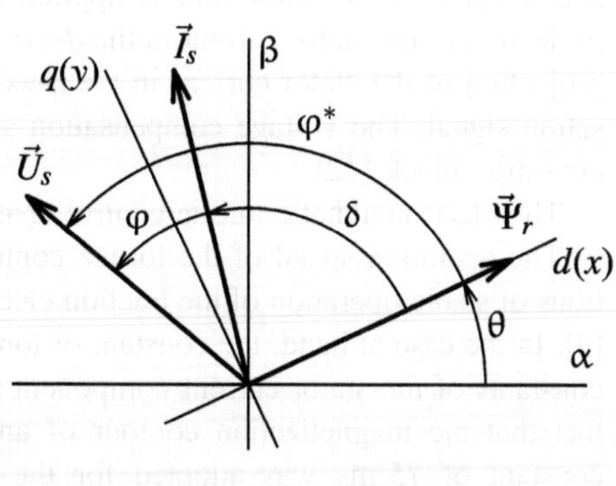

where L_s and L'_r are the stator and rotor inductances, respectively, and L_m is the mutual inductance.

The electromagnetic torque is equal to the imaginary part of the vector product of the rotor flux and the stator current:

$$M_{em} = \frac{3}{2} p \frac{L_m}{L'_r} \mathrm{Im}\left[\vec{\Psi}_r \times \vec{I}_s\right], \tag{6}$$

where M_{em} is the electromagnetic torque and p is the number of pole pairs.

After substitution of the rotor current from (5) to (4), the second equation becomes:

$$\vec{I}_s \frac{L_m r'_r}{L'_r} = \frac{d\vec{\Psi}_r}{dt} \left(\frac{L'_r}{r'_r} + j\omega\right) \vec{\Psi}_r. \tag{7}$$

The obtained equations are the basis of the structure of the vector control system. This is a dual-channel system. The first channel controls the rotor flux by regulating projection of the stator current in the d-axis. The second channel controls the electromagnetic torque by regulating projection of the stator current in the q-axis.

However, the motor is powered by the voltage inverter; therefore, the control system must control the voltage vector [11]. To regulate the voltage vector, channels have a double-circuit structure. The outer circuit sets the value of the current vector, and the inner circuit controls the voltage vector by performing a task of the outer circuit. The voltage vector is calculated by the first equation from (4):

$$\vec{U}_s = r_s \vec{I}_s + \frac{d\vec{\Psi}_s}{dt} + j\omega \vec{\Psi}_s. \tag{8}$$

The structure of control circuits of the stator current is determined based on Eq. (8). A block diagram of the control system with independent control of rotor flux and electromagnetic torque is shown in Fig. 5.

The reference signal of rotor flux and electromagnetic torque on the motor shaft is the input of the control system. The difference between the reference and the actual value of the rotor flux is applied to the input of the flux regulator. The projection of the stator current in the d-axis is the output of the control system. The projection of the stator current in the d-axis is summed with the voltage compensation signal. The voltage compensation signal is generated by a cross-link compensation block [12].

The electromagnetic torque control channel has a similar structure.

The operation speed of the torque control channel is determined by the conditions of stable operation of the traction electric drive in the event of wheel spinning [3]. In the case at hand, the constant of torque control time is 6 ms. Here, the time constants of the stator current component regulation are equal to 3 ms. Given the fact that the magnetization contour of an ATM has large magnetic lag, a time constant of 75 ms was adopted for the actuating path of rotor flux linkages.

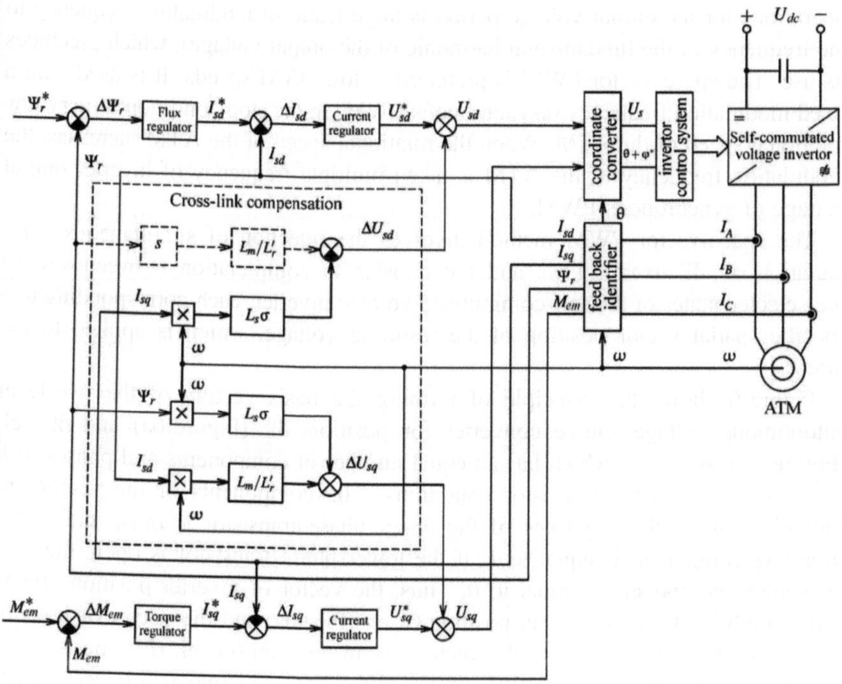

Fig. 5 Block diagram of the control system with independent control of rotor flux and electromagnetic torque

This control system adjustment provides efficient operation of electric traction drive in all studied electric locomotive operation modes.

The method used to forming the inverter output voltage supplying the ATM has a significant impact on the processes in the traction drive. Harmonic composition and electromagnetic torque rate control of the ATM, the losses in the power semiconductor devices and the efficiency of the automatic control system are all dependent on the method of voltage formation. Currently, an autonomous voltage source converter is commonly used; it is collected by the three-phase bridge circuit in insulated-gate bipolar transistors (IGBT). Its scheme is shown in Fig. 3.

There are various methods of forming the output voltage of the self-commutated voltage inverter, and they can be divided into two groups. The first has independent control of each phase; the second has switching between predetermined states of power semiconductor converter modules. To control the autonomous voltage source converter as part of the thermoelectric transducer, modulation is used, with pre-calculated *pulse width modulation* (PWM) relating to the first group of methods and space-vector PWM relating to the second group.

The ability to provide suppression of defined harmonics in the output voltage of the inverter is an advantage of pre-calculated PWM. However, this type of modulation entails significant computational cost when the number of switching operations for the output voltage period is large (ratio of modulation frequency to the frequency of the fundamental harmonic of the output voltage), which precludes its use. The space-vector PWM is preferred at low ATM speeds. It is used with a fixed modulation frequency (asynchronous PWM) in the stop condition of very low rotational speed of the ATM. When the rotational speed of the ATM increases, the modulation frequency of the ATM is taken multiple frequency of inverter output voltage of synchronous PWM.

The space-vector PWM method involves the rejection of simultaneous commutation of all inverter keys and the transfer to commutation between several pre-selected states of the self-commutated voltage inverter, each corresponding to a specific spatial vector position of the resulting voltage, which is applied to the motor.

Figure 6 shows the principle of forming the basis vectors of the two-level autonomous voltage source converter for positions 0° (Figure 6a) and 60° el. (Figure 6b). A vector, which has an equal number of components and phases $(a\ b\ c)^T$, describes inverter transistor conditions. The components of the vector are formed by the following rules: if the upper phase transistor is open, the corresponding component is equal to 1; if the lower phase transistor is open, the corresponding component is equal to 0. Thus, the vector of inverter position $(100)^T$ corresponds to the vector U_0 in position 0° el. (Fig. 6a) and the vector of inverter position $(100)^T$ corresponds to the vector U_{60} in position 60° el. (Fig. 6b).

Vectors for the positions 120°, 180°, 240°, 300° and two 0° el. vectors are similarly formed. They compose the most commonly used basis vectors, which corresponds to the standard six-stroke commutation. Three transistors are simultaneously open: the high one and two low transistors or the two high transistors and the low one. Figure 7 shows the basis vectors and autonomous voltage source converter positions corresponding to them in this case.

Linear and phase load voltage values are determined by ratios [3, 4, 13, 14] on known vectors of converter status:

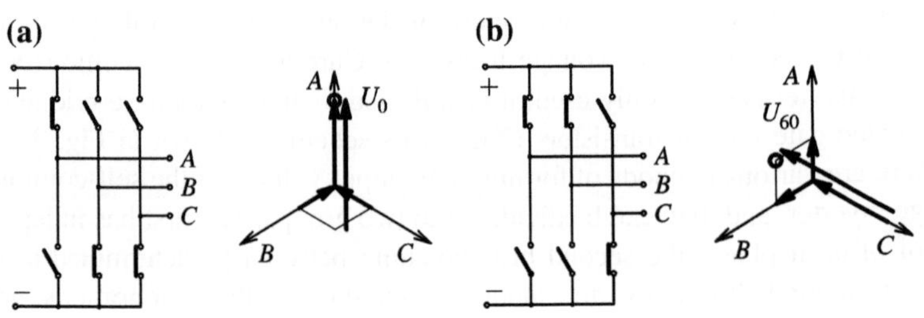

Fig. 6 Forming the basis vectors of the two-level self-commutated voltage inverter **a** for positions 0° el. and **b** 60° el

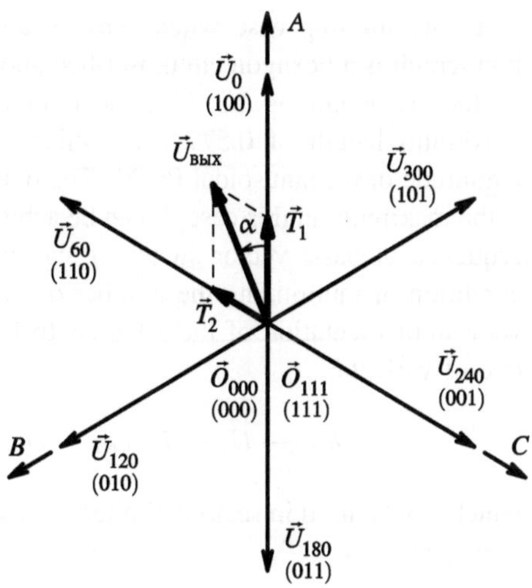

Fig. 7 Basis vectors of a bi-level autonomous voltage source converter

$$\begin{bmatrix} U_{AB} \\ U_{BC} \\ U_{CA} \end{bmatrix} = U_{dc} \begin{bmatrix} 1 & -1 & 0 \\ 0 & 1 & -1 \\ -1 & 0 & 1 \end{bmatrix} \begin{bmatrix} a \\ b \\ c \end{bmatrix},$$

$$\begin{bmatrix} U_A \\ U_B \\ U_C \end{bmatrix} = \frac{1}{3} U_{dc} \begin{bmatrix} 2 & -1 & -1 \\ -1 & 2 & -1 \\ -1 & -1 & 2 \end{bmatrix} \begin{bmatrix} a \\ b \\ c \end{bmatrix}.$$

It is possible to reproduce any required output voltage vector by switching on the PWM time between two basis vectors of current sector U_x and U_{x+60} and zero vectors O_{000} and $O_{111,}$ when there are eight base vectors, two of which are zero vectors, and the other are shifted in space by 60° el. For example, the formation of the vector U_{out} (see Fig. 7) may be carried out using vectors U_0, U_{60} and a zero vector, which is selected from the condition of a minimum number of transistors switching.

The time of vectors U_0 (T_1), U_{60} (T_2) switching is determined according to the formulas:

$$T_1 = \sqrt{2} T_{PWM} \| U_{out} \| \cos(\alpha + 30°);$$

$$T_2 = \sqrt{2} T_{PWM} \| U_{out} \| \sin(\alpha),$$

where T_{PWM} is the PWM period; $\| U_{out} \|$ is the output voltage vector length; α is the angle between the basis vector U_0 and the output voltage vector.

The zero vector switching time is determined as follows

$$T_0 = T_{PWM} - T_1 - T_2,$$

It must satisfy the following condition: $T_1 + T_2 \leq T_{PWM}$.

In the limiting case when zero vectors are not used, the net voltage vector hodograph is a hexagon circumscribed about the basis vectors. Given that the basis vector has length is 2/3 U_{dc}, the output voltage vector can be formed with a maximum length of 0.577 U_{dc}, which is 15% greater than the output voltage amplitude of the sinusoidal PWM. The output voltage vector length is 0.866, and it is the maximum in this case. It can be achieved by the used set of basis vectors. The sequence of basis vector switching can be different, and it is determined by the condition of minimizing the number of transistors switching and features of hardware implementation of the AES control system. The following sequence can be one of variants:

$$O_{000} \to U_x \to U_{x+60} \to O_{111} \to U_{x+60} \to U_x \to O_{000},$$

which can be used in sectors, limited by vectors U_0–U_{60}, U_{120}–U_{180}, U_{240}–U_{300} and sequence

$$O_{000} \to U_{x+60} \to U_x \to O_{111} \to U_x \to U_{x+60} \to O_{000},$$

The duration of each vector O_{000} switch is $T_0/4$, and the duration of the vector O_{111} switch is $T_0/2$. The duration of the vector U_x and U_{x+60} switches are equal to $T_1/2$ and $T_2/2$, respectively, in the first and the second halves of the PWM period.

A feature of this method of basis vector switching is the change in the vector switching order depending on the sector. The sequence of switching does not depend on the direction of output voltage vector rotation. Another feature is the use of both zero vectors O_{000} and O_{111} during every period of the PWM. The operating time of these vectors is equal; the period begins and ends with O_{000}.

Another pattern of the basis vector switching [3, 13, 14] uses the following sequence:

$$U_x \to U_{x+60} \to O_{\left\{ \begin{matrix} 111 \\ 000 \end{matrix} \right\}} \to U_{x+60} \to U_x,$$

where the zero vector O_{111} is used in sectors U_0–U_{60}, U_{120}–U_{180}, U_{240}–U_{300}, and the vector O_{000}–is used in sectors U_{60}–U_{120}, U_{180}–U_{240} and U_{300}–U_0.

A special feature of this pattern is a fixed order of basis vector switching, regardless of the sector number. Zero vectors are selected from the condition of a minimum number of transistors switching (one component is changed in the converter status vector). However, when changing the direction of output voltage vector rotation, it is necessary to change vectors U_x and U_{x+60} and the zero vector.

One of the major advantages of space-vector PWM is the reduction in the number of switches in one PWM period from six to four, and the decrease in dynamic losses in key elements of the converter.

The meaning of the pre-calculated PWM method consists in pre-calculation of switching moments of power semiconductor modules for providing the suppression of specified harmonics in the inverter output voltage [13, 15–19]. The control of

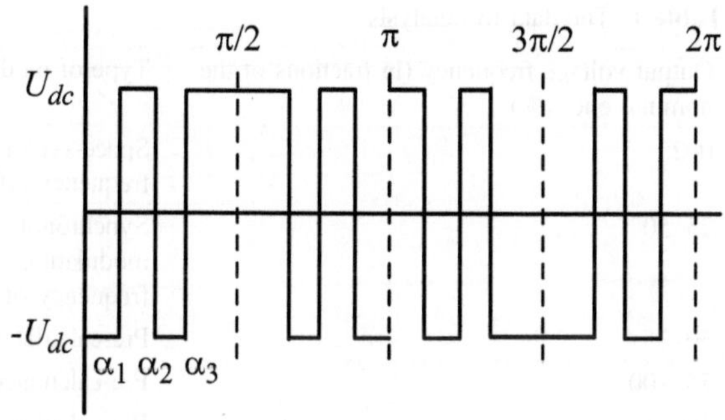

Fig. 8 The output voltage of the voltage source converter (phase potential) using the pre-calculated PWM

each phase is independent. Figure 8 shows the shape of the output voltage (the phase potential toward half a voltage in the DC link).

Expansion of the output voltage in a Fourier series can be represented as follows:

$$U_f(t) = \sum_{n=1} a_n \cos n\,\omega t + b_n \sin n\,\omega t$$

Due to quarter wave symmetry of the output voltage, the even harmonic is absent and only the odd harmonic is present. The amplitude of the nth harmonic is then expressed as

$$b_n = \left(\frac{4}{\pi n}\right)\left[1 + 2\sum_{k=1}(-1)^k \cos n\alpha_k\right] \tag{9}$$

The angles of the shift α_1 to α_m must meet a requirement

$$\alpha_1 < \alpha_2 < \cdots < \alpha_m < \pi/2$$

By setting the required value of the amplitude of the output voltage fundamental harmonic and equating the amplitude of the selected harmonic to zero, we define switching angles using Eq. (9). As an example, for the case shown in Fig. 8, it is necessary to solve a system of equations for determining the switching angles α_1, α_2, α_3:

$$\begin{aligned} &\tfrac{4}{\pi}(1 - 2\cos\alpha_1 + 2\cos\alpha_2 - 2\cos\alpha_3) = U_1; \\ &(1 - 2\cos 5\alpha_1 + 2\cos 5\alpha_2 - 2\cos 5\alpha_3) = 0; \\ &(1 - 2\cos 7\alpha_1 + 2\cos 7\alpha_2 - 2\cos 7\alpha_3) = 0, \end{aligned} \tag{10}$$

where U_1 is the amplitude of the fundamental harmonic.

Nonlinear transcendental equations are thus formed, and after solving these equations, α_1 through α_3 are computed. Triple harmonics are eliminated in the three-phase balanced system, and these are not considered in (10). It is evident that $(m - 1)$ harmonics can be eliminated with "m" numbers of switching angles.

Table 1 The data for analysis

Output voltage frequency (in fractions of the nominal one) (%)	Type of modulation
0–25	Space-vector PWM with fixed modulation frequency (1000 Hz)
25–50	Synchronous space-vector PWM with modulation frequency of up to 1000 Hz
50–75	Pre-calculated PWM, $m = 3$
75–100	Pre-calculated PWM, $m = 2$
100–125	Pre-calculated PWM, $m = 1$
>125	Single-pulse modulation

The simultaneous Eq. (10) are generally not performed in real time of the asynchronous traction drive control. The switching angles for several m values are calculated in advance and are approximated. In the existing control systems, m is odd and it takes values from 1 to 7 [15–19]. A further increase in the switching operations number (modulation frequency multiplicity) does not significantly improve the harmonic content of the inverter output voltage, since the greater the number of the harmonic, the faster its amplitude decreases. In addition, the switching frequency of the power semiconductor devices is limited by the level of thermal losses.

Both modulation types were used for generating an output voltage of inverters supplying the ATM of electric locomotive EP20. In accordance with the preliminary constructional design data, the following algorithms were used in the calculations for generating an output voltage depending on the frequency (Table 1).

Defined parameters of the modulation provide automatic control system operation under selected performance values of electromagnetic torque and flux control channels.

3.5 The Modeling of the Mechanical Part

The mechanical part of the electric locomotive considered as a multi-body structure consists of a car body and three two-axle bogies (see Fig. 1). The electric locomotive's axle formula is Bo-Bo-Bo. The inclined traction rods carry out the transmission of the traction and brake efforts from the bogies to the locomotive car body. The traction motors and reducers are bogie-mounted. In total, the model contains 28 rigid bodies.

The equations of motion are generated based on the Newton–Euler formalism [20]

$$\begin{aligned}\mathbf{M}(\mathbf{q})\ddot{\mathbf{q}} + \mathbf{k}(\mathbf{q}, \dot{\mathbf{q}}) &= \mathbf{Q}(\mathbf{q}, \dot{\mathbf{q}}) + \mathbf{G}(\mathbf{q})^{\mathrm{T}}\lambda, \\ \mathbf{g}(\mathbf{q}) &= 0,\end{aligned} \tag{11}$$

where

$\mathbf{q}, \mathbf{q}, \ddot{\mathbf{q}}$	are the column matrices of Lagrange coordinates, velocities and accelerations;
$\mathbf{M}$	is the mass matrix;
$\mathbf{k}, \mathbf{Q}$	are the column matrices of inertia and applied forces, respectively;
λ	is the vector of Lagrange multipliers corresponding to the cut joints;
$\mathbf{g}$	is the algebraic constraint equations;
m	is the number of constraints; and
$\mathbf{G} = \partial\mathbf{g}/\partial\mathbf{q}^{\mathrm{T}} = \{\partial\mathbf{g}_i/\partial\mathbf{q}_j\}_{i=\overline{1,m}}^{j=\overline{1,n}}$	is the constraint Jacobi matrix.

Equation (9) is DAE. The values to determine are the generalized coordinates $\mathbf{q}(t)$ and the Lagrange multipliers $\lambda(t)$.

Simulation of the mechanical part was carried out in the Universal Mechanism software package [21–23]. The computer animation of the locomotive's movement is shown in Fig. 9.

The model of the mechanical part of electric locomotive EP20 on two-axle bogies (2o-2o-2o wheel arrangement) consists of:

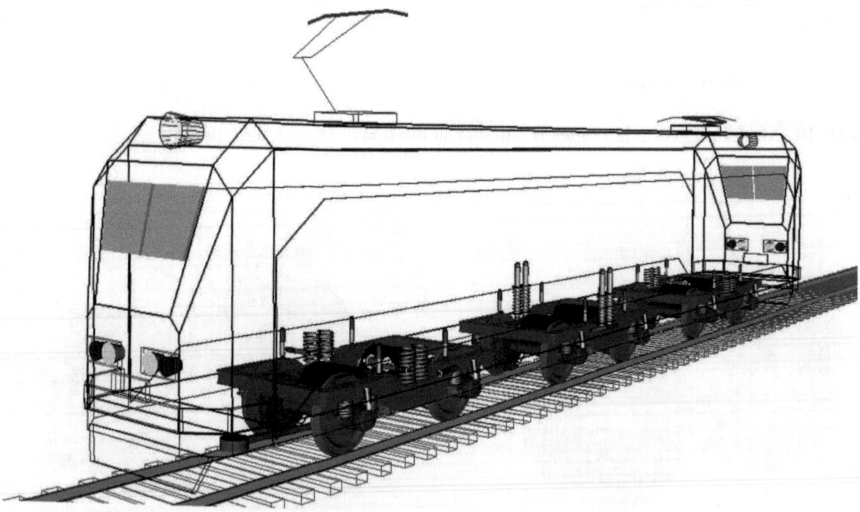

Fig. 9 Model of the mechanical part (computer animation)

- Superstructure
- Three bogie frames
- Three oblique tractions
- Six wheel sets
- Six axle equipment sets (two combined boxes)
- Six gear-motor blocks
- Six rotors
- Six cogged wheels
- Six armature quills

Thus, the model consists of 43 parts. Every part in the system has its own inertial characteristics, i.e., the mass m_i and the elements of the inertia tensor J_i, where i is the number of the part.

A subsystem method was used to create the models. The subsystem of the wheel motor block (WMB) is based on the design of two-axle bogies (Fig. 10). It is shown in Fig. 11.

In modeling, we have used the following mass and inertial characteristics of the vehicle parts of the electric locomotive [23] (Table 2).

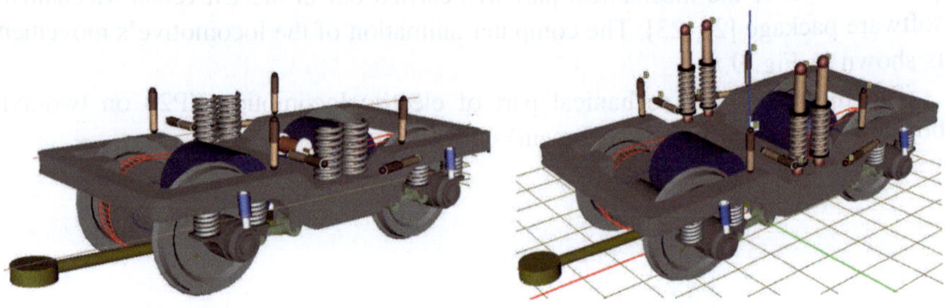

outside bogies central bogie

Fig. 10 Model of two-axle bogies (computer animation)

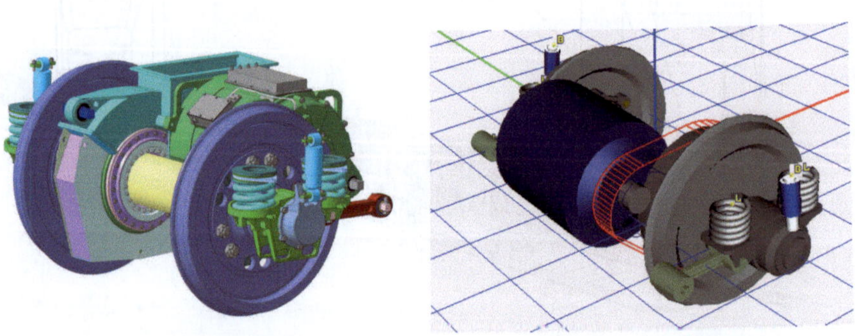

construction computer animation

Fig. 11 Wheel-motor block

Table 2 The parameters of the model

Parameter	Value
Body suspension semi-base (m)	5.78
Body weight (kg)	67000
The height of the car body mass center over rails (m)	2.5
Weight of bogie's frame with equipment without traction electric drive (kg)	4090
Wheel semi-base (m)	1.450
Weight of traction electric drive (kg)	1950
Weight of armature quill (kg)	384

Coasting movement. The modeling of processes during coasting movement was executed by numerical integration of differential equations of motion (11) on the assumption that the torque on the traction motor shafts is zero.

The parameters of the calculation scheme of the locomotive's mechanical parts (dimensions, inertia characteristics, stiffness coefficients of elastic elements, damping coefficients of dampers, etc.) were taken from publications in the technical literature [3, 23–25].

The FASTSIM algorithm [26] and its modification devoted to an unsteady contact model that also provides the correct solution for the case of zero vehicle velocity were used for computation of creep forces. When modeling, the macro geometry of track on which the locomotive is moving (the vertical profile and the plan of the railway) and the presence of the rail's vertical and horizontal micro-irregularities are taken into account. The rail's micro-irregularities were constructed according to the International Union of Railways (UIC) materials.

The mechanical part's full-size model allows us to take into account the redistribution of load between the axles in the traction and braking modes, to investigate the interaction in the side wheel–rail contact during passage of the curves, and so on.

The modeling was done in the velocity range of 10–60 m/s (i.e., from 36 to 216 km/h) with the step of 10 m/s [25].

As can be seen from Fig. 12 (20 m/s), the presence of track micro-irregularities leads to the fact that normal reactions in the wheel–rail contact for all wheels obtain the dynamic components and deviate from their quasi-static values that occur when driving on the track without irregularities.

The spots of wheel–rail contact for all the wheels are shown in Fig. 13 (20 m/s). The shaded area corresponds to the material's adhesion, and the area without coloring to the slip. The resultant force of interaction in the wheel–rail contact is shown as a vector whose value and direction vary during wheel rolling on the rail. The scale in Fig. 13a, b is the same.

In addition, the presence of track micro-irregularities leads to considerable variation in the magnitude and direction of the force within the wheel–rail contact.

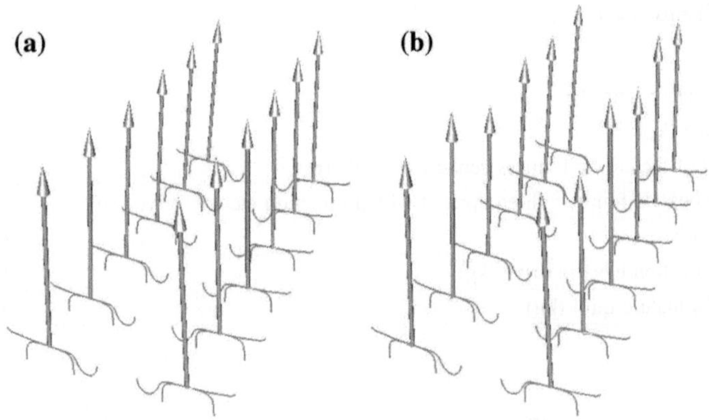

Fig. 12 Normal reactions in the wheel–rail contact during coasting movement for the track without irregularities (**a**) and with irregularities (**b**)

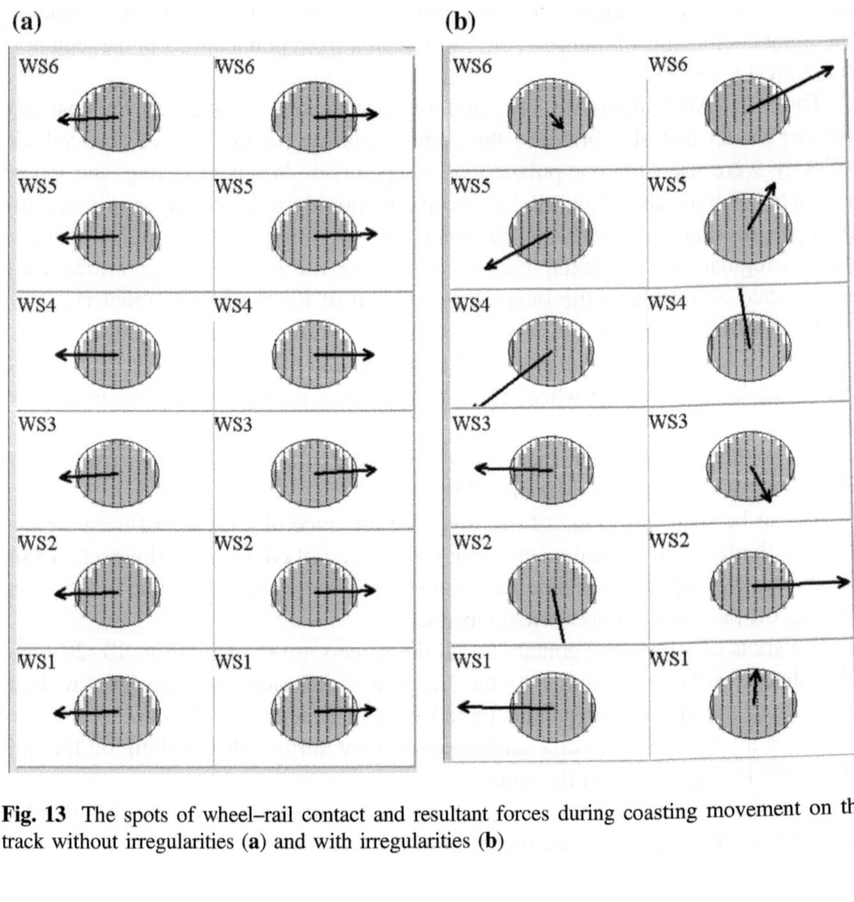

Fig. 13 The spots of wheel–rail contact and resultant forces during coasting movement on the track without irregularities (**a**) and with irregularities (**b**)

Amplitude-frequency characteristics (AFC). The AFC (or frequency response) was built by spectral analysis of the harmonic track profile (perturbation) and the corresponding vibration (response). The horizontal axis shows the frequency perturbation (Hz). The vertical axis shows the dimensionless dynamic coefficient kdyn (amplification ratio). It means in how many times the amplitude of the forced oscillations is greater than static deformation.

Thus, the obtained AFC of the car body bouncing vibrations is shown in Fig. 14a. The AFC for pitching and rolling oscillations of the car body are shown in Fig. 14b, c.

The experimentally determined values of natural frequencies are as follows: for car body bouncing vibrations, 1.89 Hz; for lateral rolling vibrations, 0.64 Hz.

The features of dynamic behavior. We also note some qualitative features of the dynamic behavior, which have been identified in the coasting movement simulation. First of all, the significant increase of the oscillation frequencies with augmenting velocity is evident. However, the vibration amplitudes, starting from a speed of 72 km/h, have only insignificant changes.

The bouncing of the middle bogie at high speeds obtains a greater amplitude than for the front and rear bogies. This is due to the installation of the softer long-stroke springs at the level of the car body suspension of the middle bogie.

In contrast, the pitching of the middle bogie, in comparison with front and rear bogies, is extremely small. The reason for this is the decisive role of the car body pitching vibrations which are transmitted to the extreme bogies more intensively (due to the much greater arms) than to the middle bogie.

As for the bogie's lateral pitching oscillations, at low speed they are determined by the local irregularities of the railway. At high speeds, they are composed of the lateral pitching vibrations of the car body and the high-frequency oscillations of small amplitude.

It should be noted that the results of the coasting movement modeling correspond to the actual tests [23].

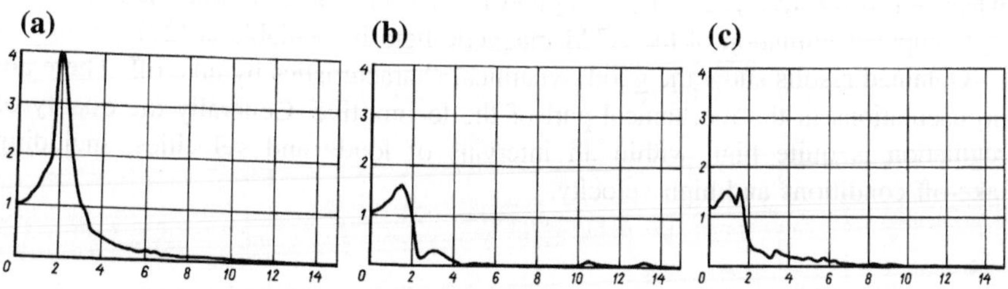

Fig. 14 AFC of the car body vibrations: bouncing (**a**), pitching (**b**) and lateral rolling (**c**)

4 Simulation Results

With the help of the developed computer model of the electric locomotive, the complex estimation of the control system principles and algorithms, the determination of its influence on the functioning of ATM and power conversion devices and the analysis of electromechanical processes at various operation regimes were executed [24, 25].

The analysis of the interaction of the ATD with the mechanical part of the locomotive in the traction mode was performed at various operation modes: the locomotive's start and acceleration, motion in a straight line and in curves with maintaining speed, appearance and elimination of wheel slide, etc.

4.1 Locomotive's Start and Acceleration

Let us consider the simulation results of electromechanical processes in the locomotive during starting and acceleration.

Because of the significant magnetic inertia of the ATM and the existing current and voltage constraints from the inverter, the simultaneous control of moment and flux linkage with a high level of disagreement is practically not possible. Therefore, the control system sets the preset moment value equal to 0 at the point of the electric locomotive commencing, within 0.2 s from the start of operations, which is necessary for termination of the control transience of the rotor flux linkage. The preset moment value is increased to the required value at a predetermined rate, avoiding the undesirable phenomena at the locomotive's mechanical part.

Some simulation results are shown in Fig. 15, where: U_{AB} is the line-to-line voltage of the ATM; I_A is the phase current; M_1^*, $M_{\mathrm{em}1}$ are the preset and realized electromagnetic torques, respectively; ω_1 is the angular velocity of the first ATM's rotor; F_{x1l}, F_{x1r} are the traction forces for the left and right wheels of the leading wheel set, respectively; and $F_{\mathrm{T}1}$, $F_{\mathrm{T}2}$ and $F_{\mathrm{T}3}$ are the forces in the traction rods.

Computer animation of the ATM magnetic field is available at [27].

Obtained results show the good dynamical characteristics by take-off. There are no oscillations in the mechanical part of the locomotive. Generally the quality of regulation is quite high within all intervals of loads and velocities, including take-off conditions and high velocity.

4.2 Movement in Traction Mode with Constant Speed on a Straight Railway Section

The calculations were performed for the case when the locomotive is moving at the constant speed of 80 km/h on a straight horizontal railway section. The railway has

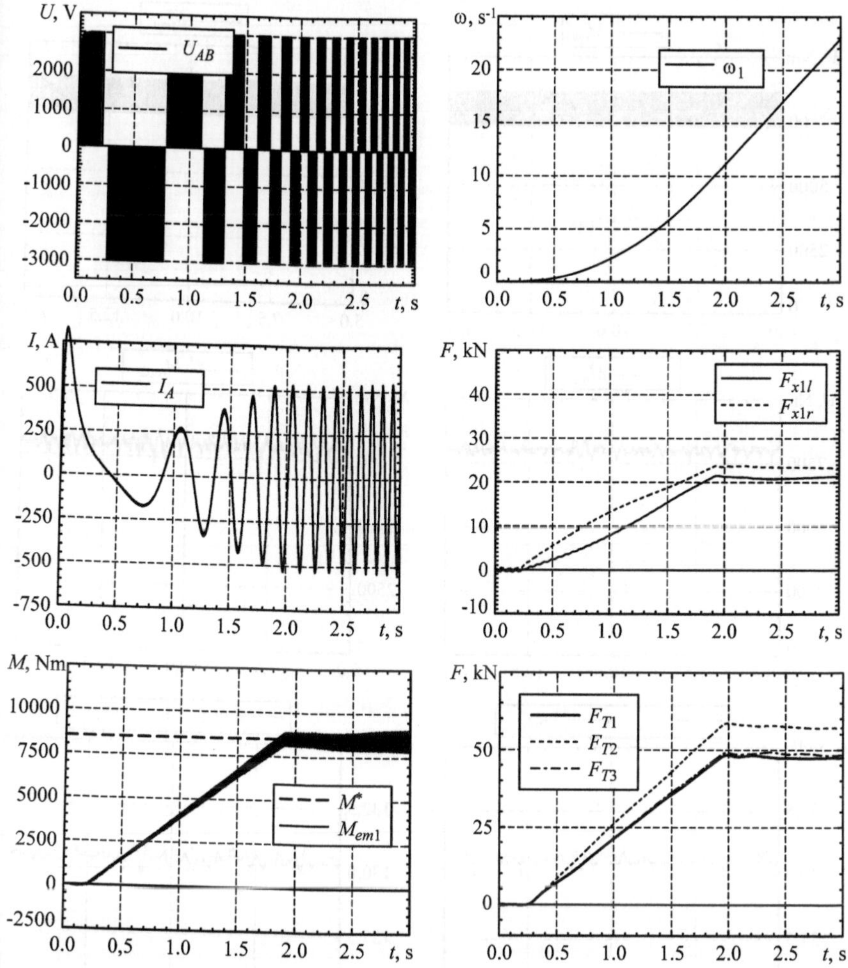

Fig. 15 Electromechanical processes at start and acceleration

micro-irregularities in the vertical and horizontal planes according to the UIC materials. The maximum value of the adhesion coefficient is equal to 0.25.

The power is supplied by AC catenary; the electrical part of the ATD is modeled in accordance with Figs. 16 and 17. The obtained graphics of the electromagnetic torques at the ATM shafts of the front bogie (M_{em1} and M_{em2}), of the torques after the rubber–cord coupling (M_{rcc1} and M_{rcc2}), of rotor angular speeds (ω_{r1} and ω_{r2}) and of the angular speeds of the wheel sets reduced to the rotor (ω_{w1} and ω_{w2}) are shown in Fig. 14. The corresponding results for the middle and rear bogies were also obtained.

The formation of inverter output voltage is performed using space-vector PWM (modulation frequency of 1500 Hz). The principle of automatic regulation for

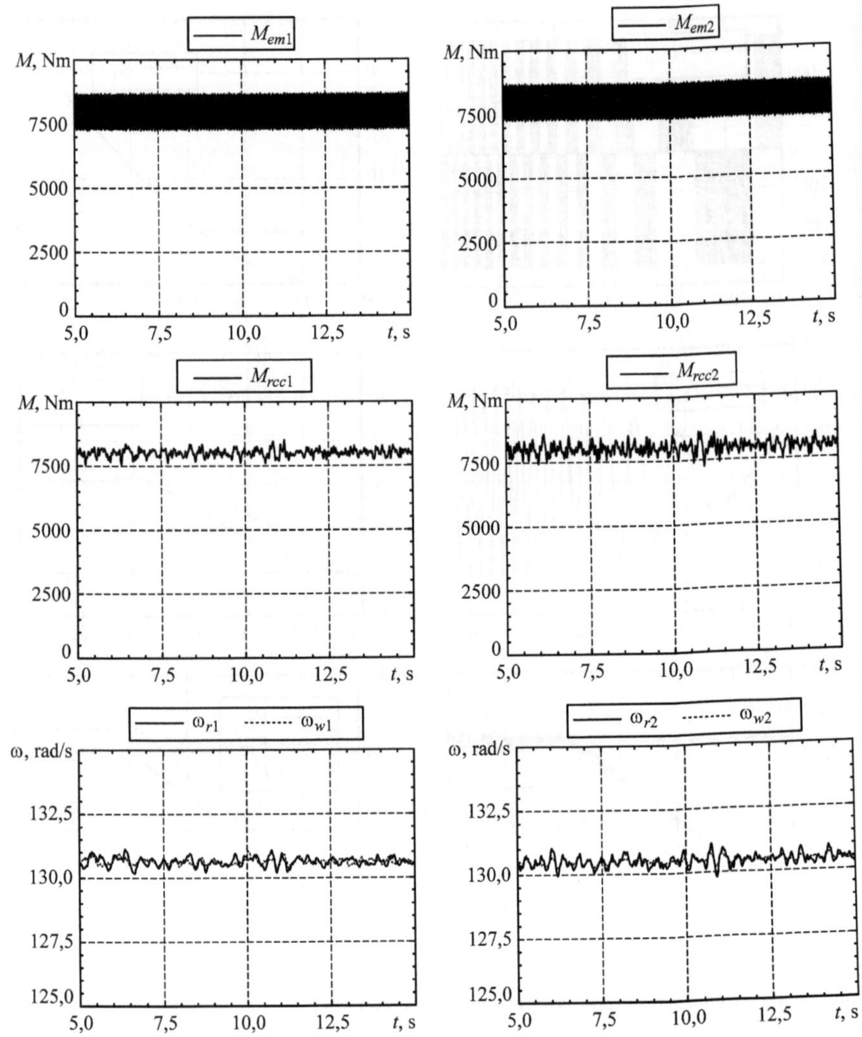

Fig. 16 Dynamic processes during movement in a straight line for the traction mode (*top to bottom*): torques on the ATM shafts and after rubber–cord couplings, angular speed of ATM rotors and of the front bogie wheel pairs

maintaining the constant rotor flux of ADM is realized. The setting of the flux is equal to 3.8 Wb, of the torque—to 8000 Nm, which corresponds to the limit of adhesion.

As can be seen from these results, the high-frequency pulsations of electromagnetic torque on the ATM shaft, arising as a consequence of the power feeding from the inverter, are almost completely absorbed by the rubber–cord coupling. The low-frequency oscillations of the torque are explained by dynamic processes in the locomotive mechanical part, which are a consequence of the impact of an uneven track structure.

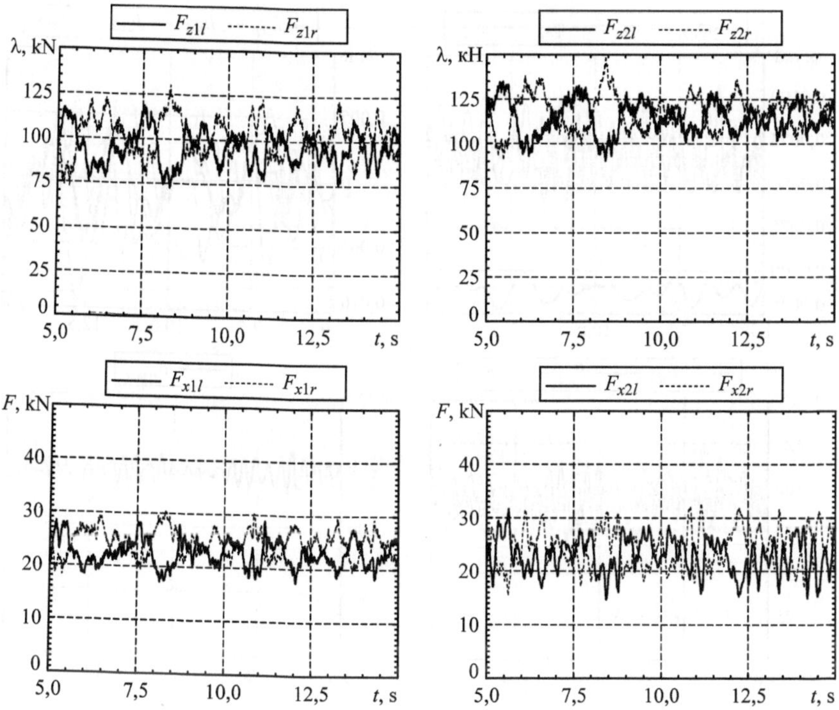

Fig. 17 Dynamic processes during movement in a straight line for the traction mode (*top to bottom*): vertical forces in wheel–rail contact for the wheel sets of the front bogie; traction forces of the left and right wheels

The graphs of vertical forces in the wheel–rail contact for the left and right wheels of the first (Fz1l, Fz1r) and second (Fz2l, Fz2r) wheel sets of the front bogie are shown in Fig. 17. The traction forces for the left and right wheels of the first (Fx1l, Fx1r) and second (Fx2l, Fx2r) wheel sets are also shown.

As a result of dynamic processes occurring when driving on the rail with irregularities, the load of the wheel sets on the rails—and consequently the traction forces—have oscillations of as much as 20–25% from the mean value of these forces (see also Fig. 13). The first wheel set is the most unloaded. It is limiting in terms of the use of adhesion conditions.

Figure 18 shows the graphs of the pitching vibrations for the car body (φ_{cb}) and bogies (φ_{b1}, φ_{b2}, φ_{b2}); the bouncing vibrations for the car body and bogies ($h_{z\ cb}$, $h_{z\ b1}$, $h_{z\ b2}$, $h_{z\ b3}$); the forces in the traction rods (F_{tr1}, F_{tr1}, F_{tr1}) and the total traction force (F_{track}).

The efforts in traction rods deviate about 20% from the mean value. The reason for this is the pulsation of traction forces at the wheel–rail contact points and the relative movement of the mechanical part elements. The inertia of the car body and the superposition of these vibrations for three inclined rods smooth the pulsations of total traction force.

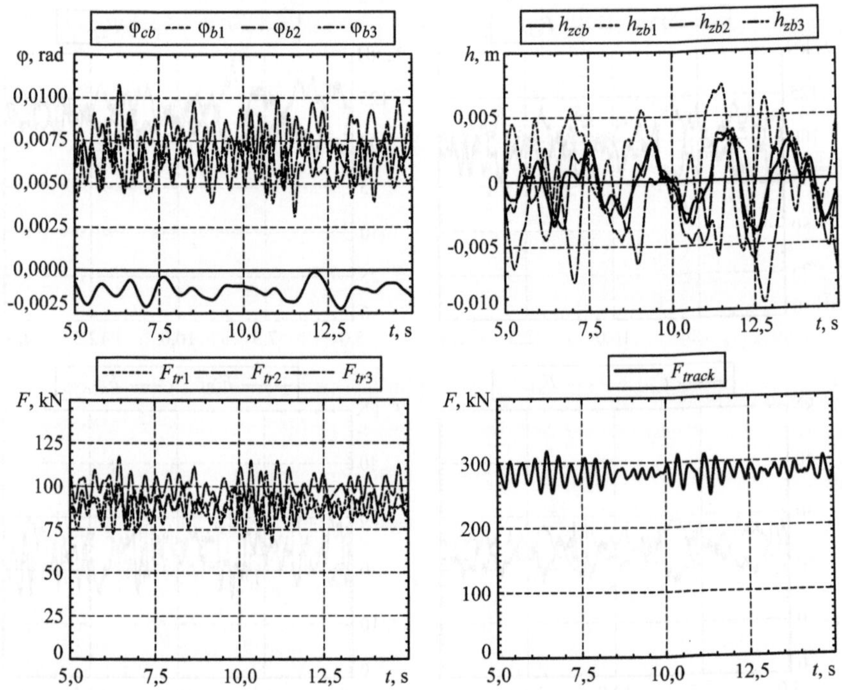

Fig. 18 Dynamic processes during movement in a straight line for the traction mode (*top to bottom*): pitching and bouncing vibrations for the car body and bogies; forces in traction rods; total traction force

In addition to the presented results, the angular and linear displacements of the mechanical part's elements, the forces in springs and dampers, the currents and voltages in the electric power conversion system components and in traction motors, the state variables of control system, etc., were calculated.

4.3 Movement in Traction Mode with Constant Speed on a Curved Railway Section

The calculations were performed for the case when the locomotive is moving at the constant speed 80 km/h on a right curve section. The cant (superelevation) of the outer rail is h = 100 mm.

The length of the track transition curve is 100 m and the length of the constant radius arc is 300 m. The railway has micro-irregularities of rails in the vertical and horizontal planes according to the UIC materials. The maximum value of the adhesion coefficient is taken equal to 0.25.

All parameters of the electrical part and the control system conform to pp. 3.3 and 3.4.

The traction torques after the rubber–cord coupling for the first and second axles of the front bogie (M_{rcc1} and M_{rcc2}) are shown in Fig. 19.

The graphs of normal reactions in wheel–rail contact for the first and second axles of the front bogie (F_{z1l}, F_{z1r}) and (F_{z2l}, F_{z2r}) are presented in Fig. 20; as shown, for the outer rail, the reaction is practically twice that for the inner rail.

The graphs of traction forces for outer and inner wheels of the first (F_{x1l}, F_{x1r}) and second (F_{x2l}, F_{x2r}) axles of the front bogie are shown in Fig. 21; the difference is also almost twice. This is explained not only by the fact that the normal reactions of the outer and inner wheels are different, but also by the excess slip for the inner wheel and by the insufficient slip for the outer one.

The total lateral efforts for the outer and inner wheels of the first (F_{y1l}, F_{y1r}) and second (F_{y2l}, F_{y2r}) axles of the front bogie are shown in Fig. 22. These efforts include both the contact forces in the main wheel–rail contact and the strength in the side contact between the flange of the left wheel and the outer rail head's side surface.

As can be seen from the graphs, the bogie moves in the chord position.

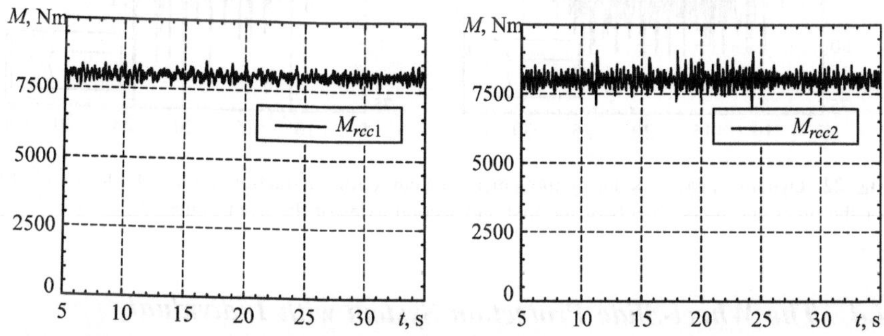

Fig. 19 Dynamic processes during movement in a curved line for the traction mode: torques after the rubber–cord coupling for the first and second axles of the front bogie

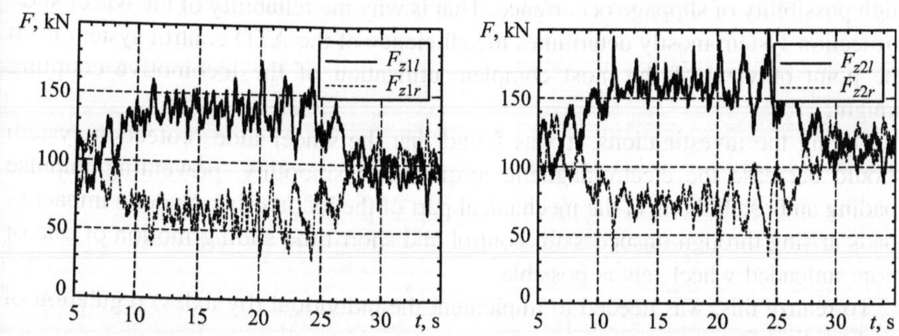

Fig. 20 Dynamic processes during movement in a curve for the traction mode: normal reactions in wheel–rail contact

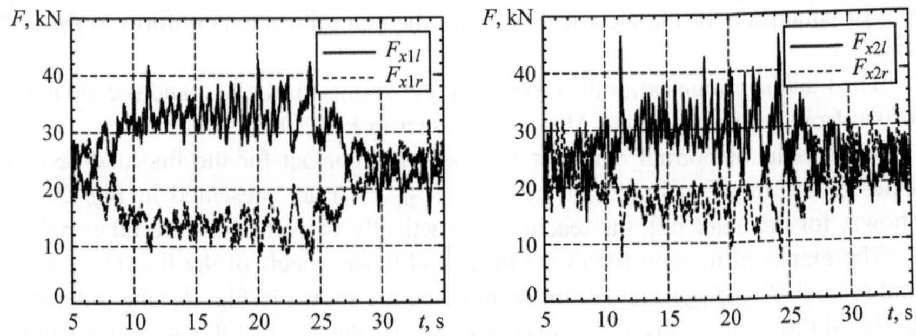

Fig. 21 Dynamic processes during movement in a curve for the traction mode: traction force for outer and inner wheels of the first and second axles of the front bogie

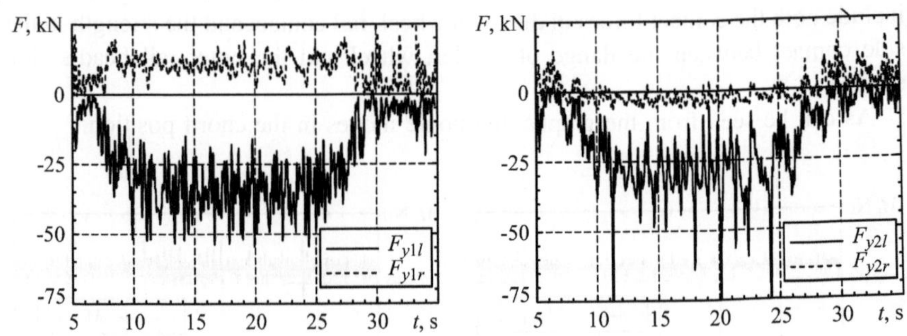

Fig. 22 Dynamic processes during movement in a curve for the traction mode: total lateral efforts for the outer and inner wheels of the first and second axles of the front bogie

4.4 The Wheel-Slide Protection System with Individual Regulation of Traction Forces

The asynchronous traction drive working at a cohesion limit is connected with the high possibility of slippage occurrence. That is why the reliability of the wheel-slide protection system mostly determines the efficiency of the ATD control system from the point of view of the most complete utilization of the locomotive coupling weight.

During the investigations, it was found that the wheel-slide protection system should decrease the electromagnetic torque quite smoothly, preventing impulse loading and oscillations in the mechanical part of the locomotive. Sudden impact of loads arising through unsuccessful control and short-time sliding motion of one or more unloaded wheel sets is possible.

To realize this, was needed to implement the individual (by axles) regulation of total traction force.

Let us consider the results of simulation of a wheel-slide protection system by moving the locomotive on a railway section with poor cohesion conditions. The length of the section exceeds the distance between the extreme wheel sets of the locomotive. Vertical and lateral railway track irregularities were taken into account during the simulation.

Simulation results for the first wheel set are shown in Fig. 23. The following designations are used: ψ_{ws1} is adhesion coefficient in wheel–rail contact of the first wheel set; M_{em1} is electromagnetic torque; ω_{r1} and ω_{w1} are angular velocities of the

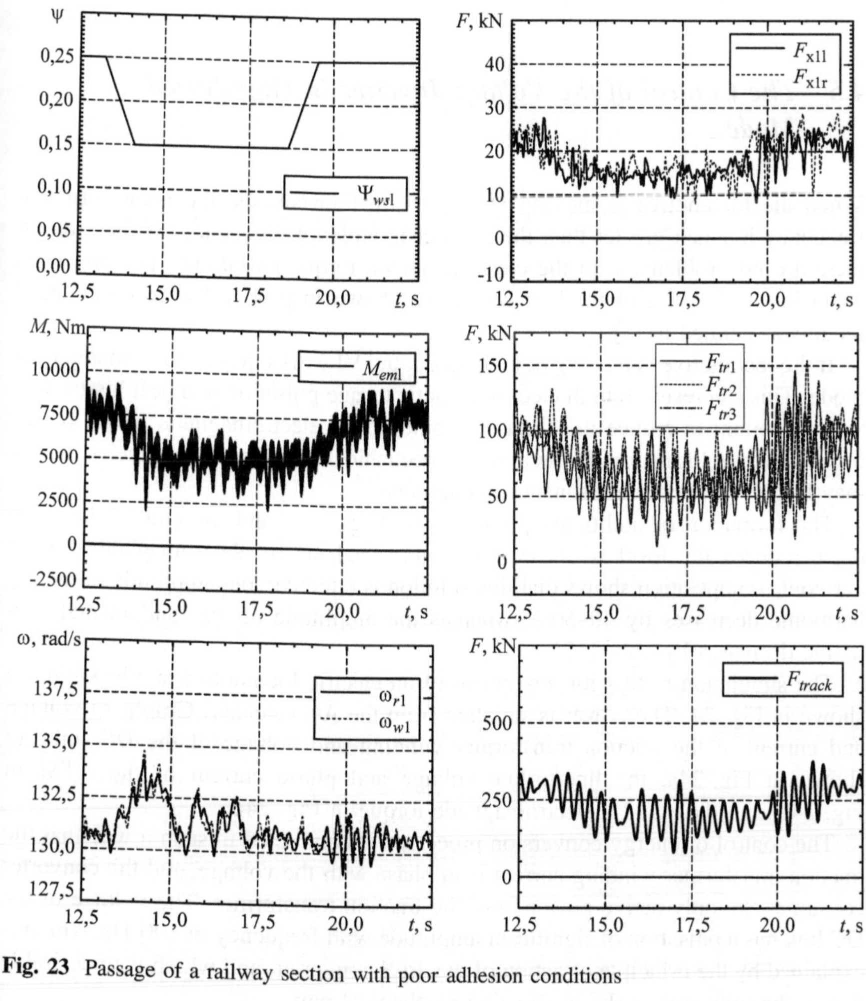

Fig. 23 Passage of a railway section with poor adhesion conditions

first ATM rotor and the first wheel set (reduced to the rotor), respectively; F_{x1l} and F_{x1r} are traction forces of the left and right wheels, respectively; F_{tr1}, F_{tr2} and F_{tr3} are forces in the traction rods; and F_{track} is the total traction force.

The obtained results show that decreasing of the adhesion coefficient leads to increasing of the angular velocity of the first wheel set. Then, the wheel-slide protection system gradually decreases the ATM's torque and the starting slippage is prevented, as is shown in Fig. 23.

When the section with poor adhesion conditions ends, the control system increases the ATM's torque. Thus the principles that are implemented in the control system prevent wheel slippage and provide nearly maximum traction force.

4.5 The Control of the Voltage Inverter in High-Speed Mode

When the locomotive is moving at low or medium speeds, the main harmonic frequency is much greater than the inverter modulation frequency. In these cases, there are no problems with the electromagnetic torque pulsations. The processes occurring at the start of the locomotive are shown in p. 4.1. The medium speed mode was studied in [3].

If the locomotive is moving at high speed, the VI works in one-pulse modulation mode. This can result in high electromagnetic torque pulsation and heat losses. It is very important to reduce the sixth harmonic of the electromagnetic torque pulsations, which is located in the area of the mechanical part Eigen frequencies, which may lead to undesirable resonance phenomena.

The formation of additional pulses at the beginning and the end of the main pulse reduces the level of these pulsations. Analysis of electromagnetic torque harmonic composition shows that this solution is effective: the amplitude of the $6f_1$ harmonic decreases by 40–50%, whereas the amplitude of the fundamental harmonic decreases by only 1–2%.

The simulation results for movement of an electric locomotive at 150 km/h are shown in Fig. 24. The power is supplied from the AC catenary. Graphs of voltage and current of the traction transformer winding and voltage of the DC link are shown in Fig. 24a, the line-to-line voltage and phase current of the ATM in Fig. 24b, and the ATM's electromagnetic torque in Fig. 24c.

The control of energy conversion processes is conducted in such a way that the traction transformer winding current is in phase with the voltage, and the converter consumes the only active power from the traction transformer. The voltage in the DC link has a pulsation of significant amplitude with frequency of 100 Hz, which is explained by the principle of action of the 4q-S converter, and which is undesirable due to the influence to the locomotive mechanical part.

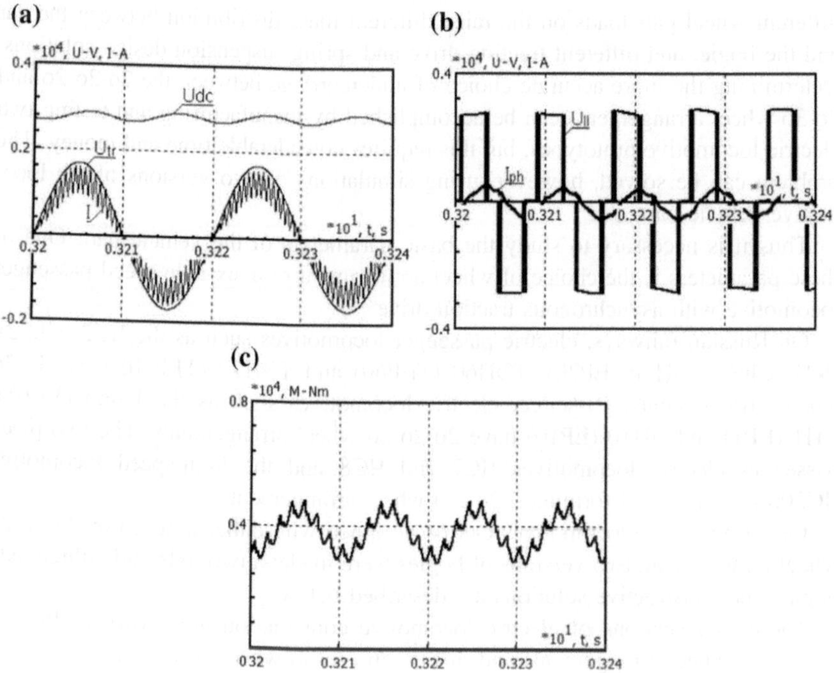

Fig. 24 Voltage and current of the main transformer traction wind, DC link voltage (**a**); ATM line-to-line voltage and phase current (**b**); electromagnetic torque (**c**) in 150-km/h mode

4.6 Analysis of Impact of the Vehicle on the Railway Track, Dynamics and Stability

In accordance with the requirements for Russian railways, a high-speed passenger locomotive at a speed of 200 km/h must have six axles and maximum traction force of 450 kN at the beginning of the movement.

In practice, however, six-axle locomotives with asynchronous traction drives and a speed of 200 km/h have not been developed and operated. Most of the electric locomotives with asynchronous traction drives, which had a speed of 200 km/h, had four axles (2o-2o wheel arrangement). Electric locomotives such as Class 92 and EG 3100, with a design speed of 140 km/h, have a 3-3 wheel arrangement. Electric locomotives such as E90, ESL 9000, ESL 9700 and O'zbekiston with design speeds of 160–175 km/h have a 2o-2o-2o wheel arrangement.

The right choice of wheel arrangement at the design stage is crucial for successful development of high-speed six-axle passenger electric locomotives with asynchronous traction drives and design speeds of 200 km/h. Data obtained by testing locomotives with wheel arrangement wheel truing and 2o-2o-2o wheel arrangements allow a comparative evaluation of the effects on the way. The real benefits of the undercarriage formula are not as obvious, as locomotives have different wheel pair loads on the rails, different mass distribution between the car and the bogie, and different traction drive and spring suspension design solutions. Determining the more accurate choice of undercarriage between the 2o-2o-2o and 3o-3o wheel arrangements can be accomplished by manufacturing and testing two electric locomotive prototypes, but this requires considerable time and money. The problem can be solved, however, using simulations of two versions of the locomotive vehicle part.

Thus it is necessary to study the basic parameters of the vehicle part. One of these parameters is the choice of wheel arrangement of a six-axle speed passenger locomotive with asynchronous traction drive.

On Russian railways, electric passenger locomotives such as the ЧС2 (ChS2), ЧС4 (ChS4), ЭП2к (EP2k), ТЭП60 (TEP60) and ТЭП70 (TEP70) have 3o-3o wheel arrangements. Passenger electric locomotives such as the ВЛ65 (VL65), ЭП1 (EP1) and ЭП10 (EP10) have 2o-2o-2o wheel arrangements. The two-piece passenger electric locomotives ЧС7 and ЧС8 and the high-speed locomotive ЧС200 have an axial formula 2(2o-2o) wheel arrangement.

Because six-axle locomotives can be designed with either a 3o-3o or 2o-2o-2o wheel arrangement, two versions of bogies were made: a two-axle and a three-axle bogie. The constructive solutions are described below.

These two versions of electric locomotive construction are shown in Fig. 25 (2o-2o-2o wheel arrangement) and in Fig. 26 (3o-3o wheel arrangement).

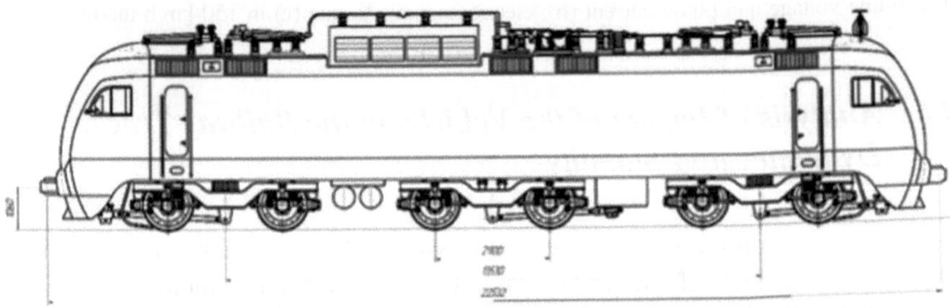

Fig. 25 Electric locomotive on two-axle bogies (2o-2o-2o wheel arrangement)

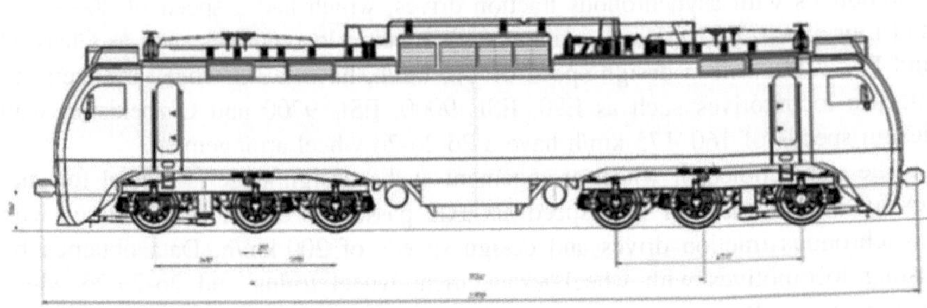

Fig. 26 Electric locomotive on three-axle bogies (3o-3o wheel arrangement)

The main technical characteristics of the vehicle part of an electric locomotive for both versions are shown in Table 3.

Figure 27 shows three-dimensional models of three-axle and two-axle bogies and their components.

Two-axle and three-axle bogies have common design solutions, which feature the application of a wheel-motor unit with integrated traction drive. The traction motor and speed transformer are combined into a single unit gear-motor, which has frame support suspension.

A two-axle bogie consists of two wheel-motor blocks, and a three-axle bogie consists of three wheel-motor blocks that have one axle box link. The bogie's welded frame leans on the wheel sets' axle-boxes through the first-stage suspension springs. Motor-gear blocks consisting of a traction motor and traction gear are mounted on the bogie's frame. The torque is transmitted from the motor-gear block to the wheel set by the transmission mechanism comprising a hollow shaft and two quill drivers. Bogies are equipped with disc brakes with automatic gap control between the brake pad and brake disc, and an automatic parking brake.

Table 3 The main technical characteristics of an electric locomotive

Parameter	Value	
	2o-2o-2o	3o-3o
The length along the automatic coupler axes (mm)	22,532	22,532
Body suspension base (mm)	11,560	13,530
Bogie base (mm)	2900	4720
The diameter of the new wheel (mm)	1250	1250
The load of wheel pairs on rails (kN) (tf)	211 (21.5)	211 (21.5)
Weight of bogie (kg)	17,600	28,500

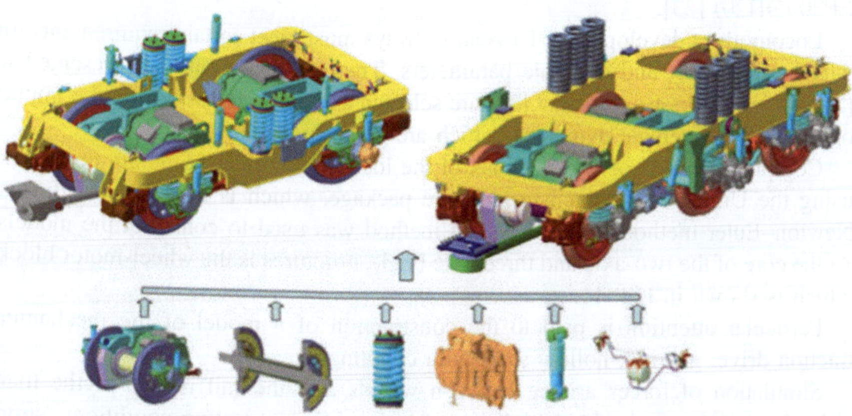

Fig. 27 Two-axle bogie of a 2o-2o-2o vehicle part (*upper left*), three-axle bogie of a 3o-3o vehicle part (*upper right*), and their components

Wheel flange lubrication devices are used to reduce the wear on the wheel set flanges. Cleaning blocks are installed to clean the dirt from the wheels.

The traction and braking are transmitted from the wheel set to the bogie frame by axle box link bars. Oblique traction is attached to the bracket on the bottom sheet of the bogie frame center beam. It transmits traction and braking from the bogie to the body.

A hydraulic damper is installed on each journal box for damping the bogie frame vibrations towards the wheel sets. Four body supports are installed on two-axle bogies. The body supports of the outermost bogie are designed in the form of springs working in compression and shear, while the center bogie supports are designed in the form of compressed elastic swinging rods allowing it to compensate for significant movement of the bogie towards the body when passing curved sections of track.

Six body supports are mounted on the three-axle truck. They are designed as springs working in compression and shear. For damping the vibrations of the body towards the bogie, four vertical hydraulic dampers are installed. Two special hydraulic dampers are mounted for damping the bogie vibration wagging. Two transverse horizontal hydraulic damper are mounted for damping the transverse vibrations of the body toward the outermost bogies with 2o-2o-2o wheel arrangement and three-axle bogies with 3o-3o wheel arrangement; transverse horizontal hydraulic dampers are not mounted on center bogies.

Two versions of electric locomotive motion simulations were carried out in various areas. The study objectives were as follows: assessment of dynamic qualities of the locomotive underframe, the impact on the rail, the need to choose the better of two versions of the locomotive underframe (with 2o-2o-2o and 3o-3o wheel arrangements), and assessing the impact on the rail and the wear of the wheels.

A comparison of the results of the two computer simulation versions of electric locomotive dynamics allows us to choose the wheel arrangement of the locomotive EP20 (ЭП20) [23].

Locomotives developed for Russian railways must meet certain requirements for wheel–rail impact and dynamic parameters. These requirements are presented in [28], so indicators from the list [28] are selected as criteria for assessing the impact on the rail and vehicle dynamics, which are given in Table 4.

Computer models of two versions of the locomotive underframe are developed using the Universal Mechanism software package, which is based on the formal Newton–Euler method. The subsystem method was used to construct the models. At the core of the two-axle and three-axle bogie structures is the wheel-motor block, which is shown in Fig. 11.

Particular attention is paid to the construction of a model of the mechanical traction drive: gearing, hollow shaft with couplings.

Simulation of forces arising between wheels and the rail is one of the main objectives of the study. In accordance with the actual operating conditions, simulation of electric locomotive motion was carried out on R65 rails in straight sections of the path and curves with radii of 350, 650, 1000 and 1500 m in the turnout

Table 4 The criteria for assessing the impact on the rail and vehicle dynamics

Parameters	Standard values
The coefficients of the vertical dynamics, no more than	
– For the first stage of suspension	0.35
– For the second stage of suspension	0.20
The ratio of the frame force to the vertical static axle load when driving the vehicle on straight sections of the path, no more than	0.3
Frame forces in the pointwork, no more than	100 kN
Wheel's side load on the rail by the condition of the strength of separate rail spike fastening, no more than	100 kN
Lateral forces in the pointwork, no more than	120 kN

P65 1/11 during trailing and facing movement. A profile of a new R65 rail has been adopted in accordance with GOST (All-Union State Standard) 8161-75, and the profile of the new wheel according to GOST 11018-2000.

The impact on the rail and vehicle dynamics was assessed based on the following criteria [29-33]:

- The coefficients of the vertical dynamics (for first and second stages of suspension)
- Frame forces
- Wheel lateral load on the rail by the condition of the strength of separate rail spike fastening
- Lateral forces in the pointwork
- Vertical load of the wheels on the rails
- Specific work of friction forces of wheel and rail (taping line and the comb in contact) per meter of the rail

In the design of high-speed locomotives, an important factor is the assessment of stability on the straight sections of the path. In this study, the estimation of motion stability of wheel sets is made in accordance with European Standard EN 14653 *Testing for the acceptance of running characteristics of railway vehicles* on the basis of test results of experienced locomotive. Values of the transverse acceleration measured on the frame of the bogie over box body are at the heart of sustainability assessment.

Simulation was carried out on a straight section of the path up to a speed of 250 km/h, on the curved sections of the path with speed corresponding to the unbalanced acceleration of 0.7 m/s^2, and in turnouts up to a speed of 50 km/h.

Simulation Results on a Straight Track

The following graphs show motion simulation results on the straight track. Figure 28 shows lateral forces of the displacement of rails. Figure 29 shows frame forces. Figure 30 shows vertical forces acting from the wheels on the rails. Figure 31 shows lateral forces acting on the grille shift in line. Figure 32 shows coefficients of vertical dynamics of springs of the first and second suspension stages in a straight line. Figure 33 shows the specific (per meter of rail) work of the friction forces in the contact of the wheels and rails.

The simulation results in the direct part of the path showed that the locomotive with a 2o-2o-2o wheel arrangement had better indicators of dynamics and impact on the rail than the locomotive with a 3o-3o wheel arrangement.

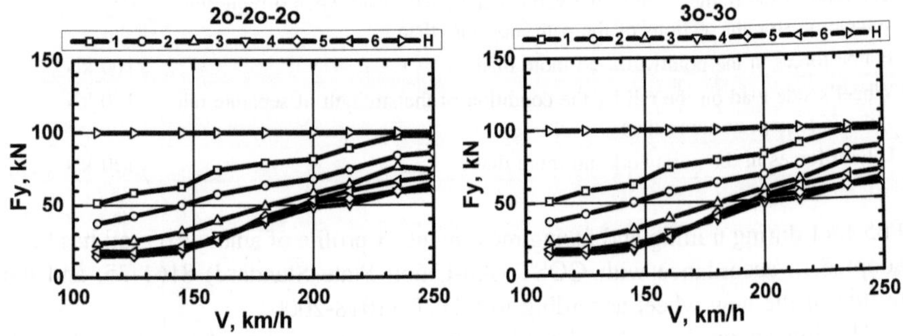

Fig. 28 Lateral forces of the displacement of rails

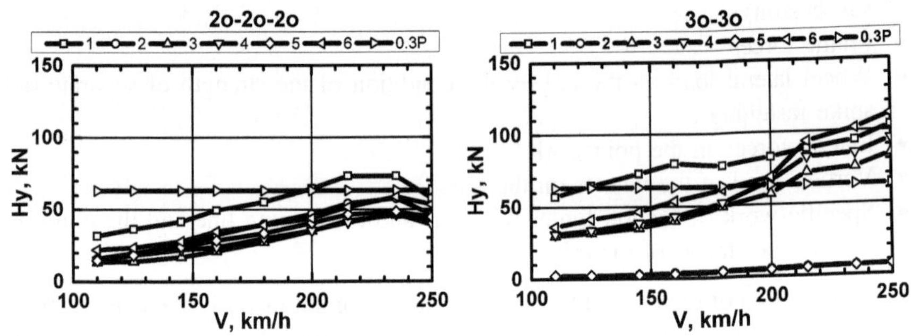

Fig. 29 Frame forces in the straight line

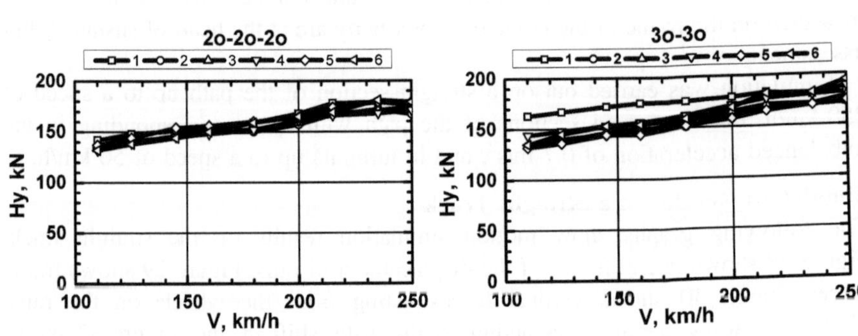

Fig. 30 Vertical forces acting from the wheels on the rails

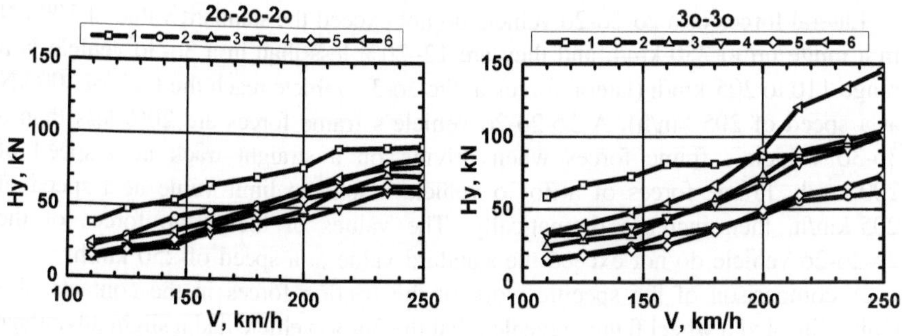

Fig. 31 Lateral forces acting on the grille shift in line

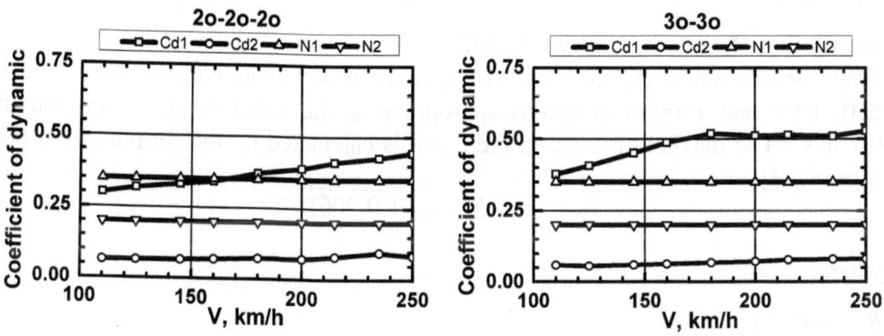

Fig. 32 Coefficients of vertical dynamics of springs of the first and second suspension stages in a straight line

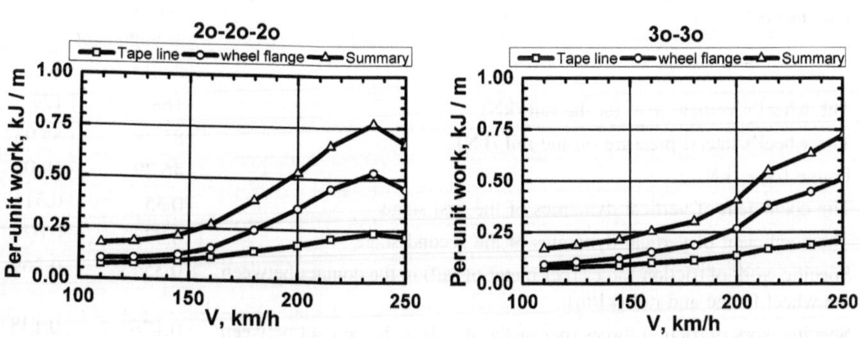

Fig. 33 Specific work of the friction forces (per meter of rail) in the contacts of the wheels and rails

Lateral forces of a 2o-2o-2o vehicle do not exceed the standard value of 100 kN in a range up to 250 km/h and they are 12–20% less than in a 3o-3o vehicle in a range 110 to 205 km/h (lateral forces in the 3o-3o vehicle reach the limit of 100 kN at a speed of 205 km/h). A 2o-2o-2o vehicle's frame forces are 20% less than a 3o-3o vehicle's frame forces when driving on a straight track at a speed of 200 km/h. Frame forces of a 3o-3o vehicle reach the limit value at a speed of 205 km/h, then increase dramatically. The values of the frame forces of the 2o-2o-2o vehicle do not exceed the standard value at a speed of 250 km/h.

A comparison of the specific work of the friction forces in the contacts of a taping line and a wheel flange revealed that the 3o-3o vehicle had a slight advantage over the 2o-2o-2o vehicle when driving on the straight section of rail, as shown in Fig. 33.

Table 5 shows the maximum calculated standard values for speeds up to 200 km/h.

Simulation Results in Curved Track

Traffic simulation in curved track was carried out on curves with a radius of 350, 650, 1000 and 1500 m at speeds appropriate to the unbalanced acceleration of 0.7 m/s^2. The maximum speed in the curve is calculated by the formula from the source [34]:

$$V = 3.6\sqrt{R \cdot (a_{ua} + 0.0061h)};$$

where:

R curve radius, m;
a_{ua} unbalanced acceleration, $a_{ua} = 0.7$ m/s^2;
h maximum canting in a curve, $h = 150$ mm.

Table 5 The maximum calculated standard values for speeds up to 200 km/h

Parameters	Wheel arrangement	
	2o-2o-2o	3o-3o
The wheel's vertical load on the rail (kN)	168	179
The wheel's lateral pressure on the rail (kN)	81.78	85.98
Frame force (kN)	46.79	69.08
The coefficient of vertical dynamics of the first stage	0.35	0.511
The coefficient of vertical dynamics of the second stage	0.073	0.072
Specific work of friction forces (per meter of rail) in the contact between the wheel flange and rail (kJ/m)	0.354	0.269
Specific work of friction forces (per meter of rail) in the contact between the taping line and rail (kJ/m)	0.176	0.138

Table 6 The calculated values of speeds corresponding to the unbalanced acceleration of 0.7 m/s^2 for the different curves

The curve radius (m)	The maximum allowable speed when unbalanced acceleration is 0.7 m/s^2; m/s (km/h)
350	23.8 (85.7)
650	32.44 (117)
1000	40.23 (145)
1500	49.27 (177)

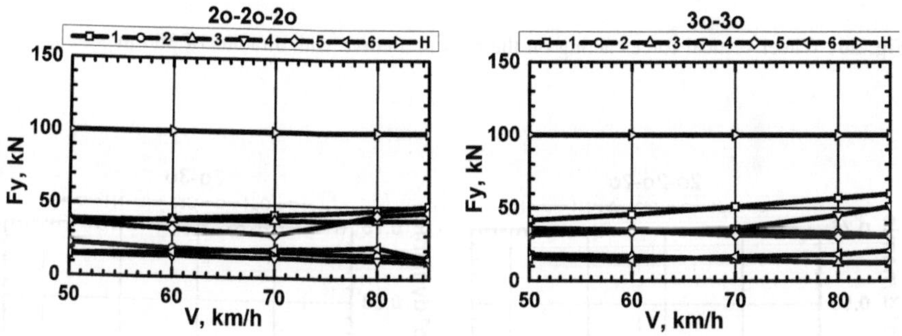

Fig. 34 Lateral forces of the displacement of rails in a curve with a radius of 350 m

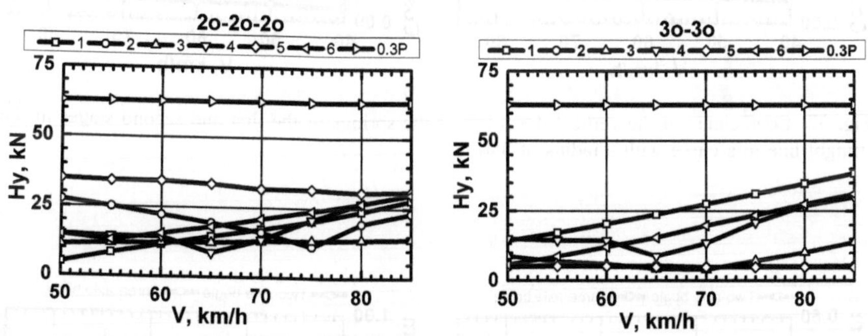

Fig. 35 Frame power in a curve with a radius of 350 m

Table 6 shows the calculated values of speeds corresponding to the unbalanced acceleration of 0.7 m/s^2 for the different curves.

The Simulation Results in a Curve with a Radius of 350 m

The traffic simulation results in a curve with a radius of 350 m are shown in the following figures: Fig. 34, lateral forces of the displacement of rails; Fig. 35, frame power; Fig. 36, wheel vertical forces acting on the rails; Fig. 37, the coefficients of the vertical dynamics of the springs of the first and second stages in the straight line; Fig. 38, specific work of friction forces (per meter of rail) in the contact between the wheels and rails.

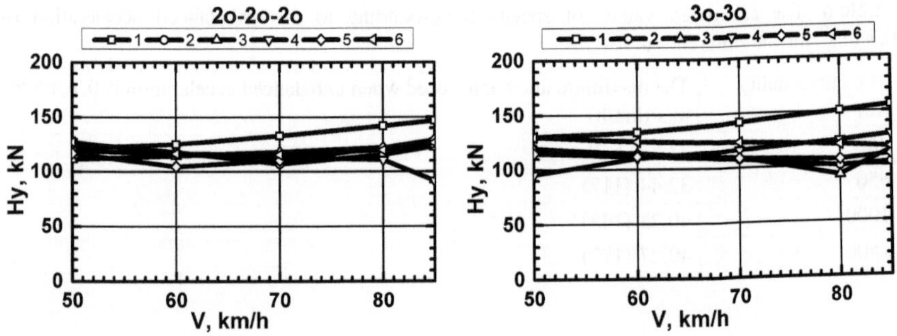

Fig. 36 Wheel vertical forces acting on the rails in a curve with a radius of 350 m

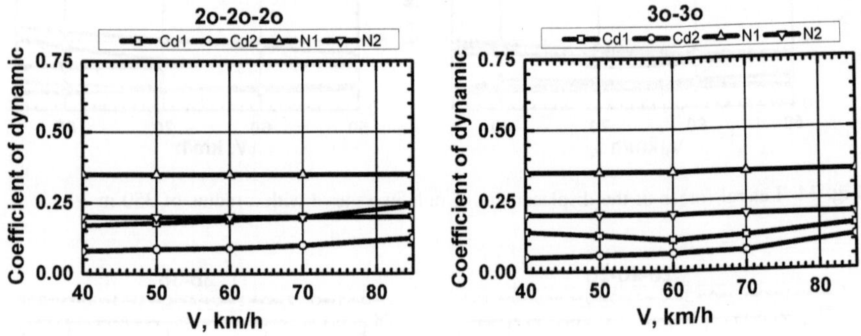

Fig. 37 Coefficients of the vertical dynamics of the springs of the first and second stages in the straight line in a curve with a radius of 350 m

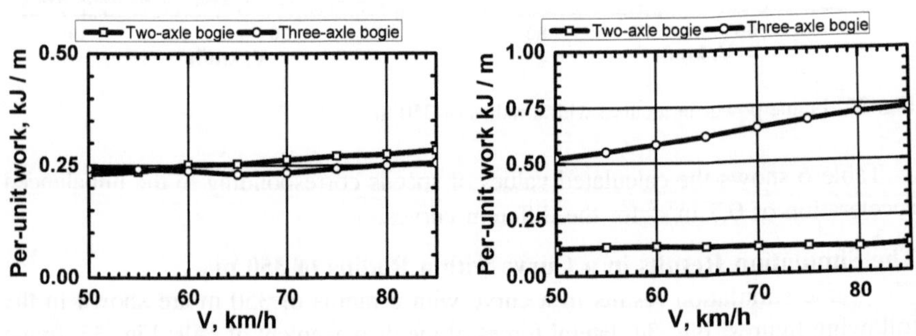

Fig. 38 Specific work of friction forces (per meter of rail) in the contact between the wheels and rails in a curve with a radius of 350 m

The simulation results in a curve with a radius of 350 m showed that the dynamics and impact on the rail were better for the locomotive with a 2o-2o-2o wheel arrangement than the locomotive with a 3o-3o wheel arrangement. Lateral forces of a 2o-2o-2o vehicle do not exceed the standard value of 100 kN in the whole speed range up to 85 km/h and are 12–13% less than in a 3o-3o vehicle in a range of 50–85 km/h. Frame forces of the 2o-2o-2o vehicle at a speed of 85 km/h are 37% less than frame forces of the 3o-3o vehicle and do not exceed the standard value of 63 kN.

The vehicle with a 2o-2o-2o wheel arrangement has a wheel vertical load on the rail 6.1–7.7% lower than the vehicle with a 3o-3o wheel arrangement.

Indicators of specific work of wheel friction forces on the taping line and rail for both vehicles are almost identical. The vehicle with a 2o-2o-2o wheel arrangement has specific work of the frictional forces in contact between wheel flange and rail 4.4–5.8 times smaller than the vehicle with a 3o-3o wheel arrangement (Fig. 38).

The maximum calculated values of the normative parameters are shown in Table 7.

The Simulation Results in a Curve with a Radius of 650 m

The traffic simulation results in a curve with a radius of 650 m are shown in the following figures: Fig. 39, lateral forces of the displacement of rails; Fig. 40, frame forces; Fig. 41, wheel vertical forces acting on the rails; Fig. 42, the coefficients of the vertical dynamics of the springs of the first and second stages in the straight line; Fig. 43, specific work of friction forces (per meter of rail) in the contact between the wheels and rails (Fig. 39).

The simulation results in a curve with a radius of 650 m showed that the lateral forces of the vehicle with a 2o-2o-2o wheel arrangement do not exceed the standard value of 100 kN in the whole speed range up to 126 km/h, and they are 20–22% less than in a 3o-3o vehicle. Frame forces of the vehicle with a 2o-2o-2o wheel arrangement are less than frame forces of the vehicle with a 3o-3o wheel arrangement and have a standard value of 63 kN, which corresponds to the upper

Table 7 The maximum calculated values of the normative parameters for a curve with a radius of 350 m

Parameter	Wheel arrangement	
	2o-2o-2o	3o-3o
The wheel vertical load on the rail (kN)	146.9	155.86
The wheel lateral pressure on the rail (kN)	50.1	56.9
Frame force (kN)	27.9	38.45
The coefficient of vertical dynamics of the first stage	0.24	0.16
The coefficient of vertical dynamics of the second stage	0.12	0.12
Specific work of friction forces (per meter of rail) in the contact between the wheel flange and rail (kJ/m)	0.125	0.724
Specific work of friction forces (per meter of rail) in the contact between the taping line and rail (kJ/m)	0.284	0.255

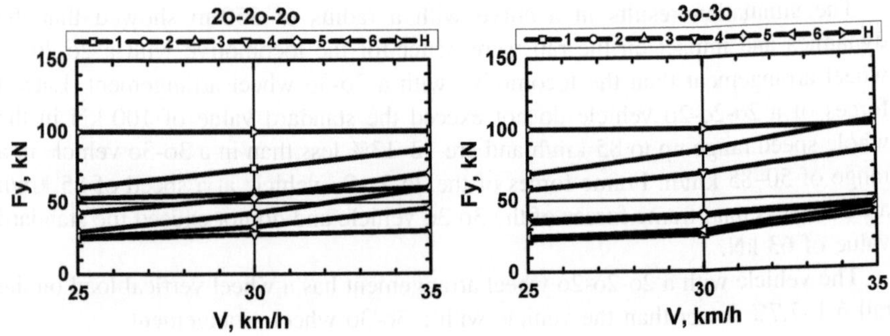

Fig. 39 Lateral forces of the displacement of rails in a curve with a radius of 650 m

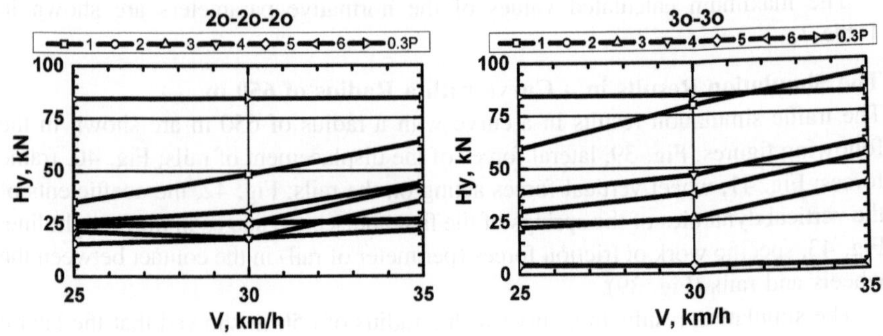

Fig. 40 Frame forces in a curve with a radius of 650 m

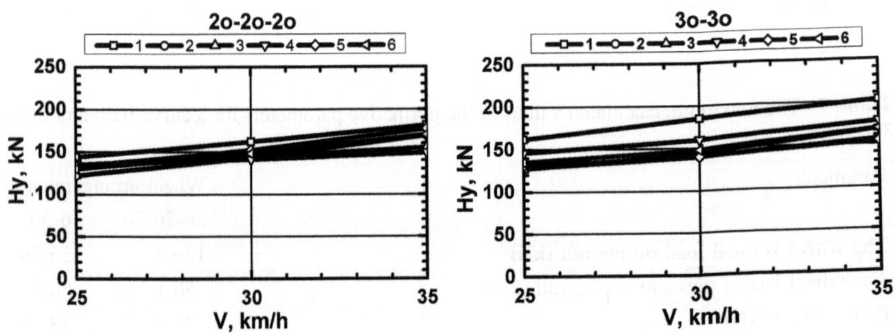

Fig. 41 Wheel vertical forces acting on the rails in a curve with a radius of 650 m

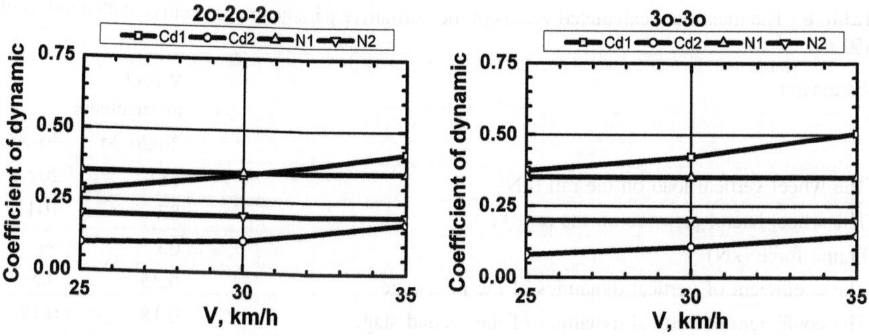

Fig. 42 Coefficients of the vertical dynamics of the springs of the first and second stages in the straight line in a curve with a radius of 650 m

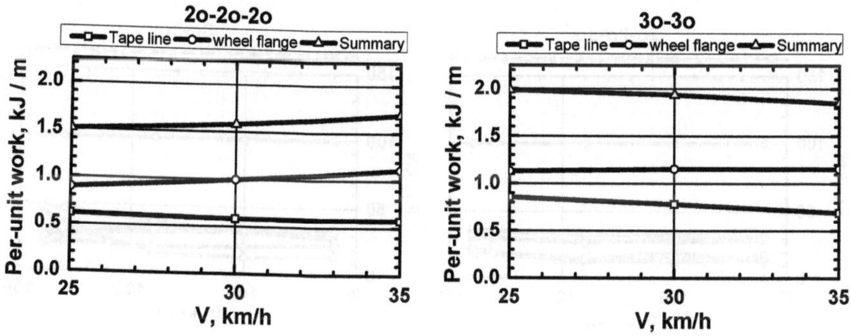

Fig. 43 Specific work of friction forces (per meter of the way) in the contact between the wheels and rails in a curve with a radius of 650 m

value of the permissible frame force. Frame forces of the vehicle with a 3o-3o wheel arrangement reach the standard value of 63 kN at a speed of 94 km/h.

The vehicle with a 2o-2o-2o wheel arrangement has a wheel vertical load on the rail that is 10% lower than the vehicle with a 3o-3o wheel arrangement. The coefficient of vertical dynamics of the box level of the vehicle with a 2o-2o-2o wheel arrangement is 4–29% less than the coefficient of vertical dynamics of the box level of the vehicle with a 3o-3o wheel arrangement, and is dependent on the speed.

The vehicle with a 2o-2o-2o wheel arrangement has 20–39% less specific work in the contact between the wheel flange and rail than the vehicle with a 3o-3o wheel arrangement. The vehicle with a 2o-2o-2o wheel arrangement has 0–26% less specific work in the contact between the taping line and rail than the vehicle with a 3o-3o wheel arrangement (Fig. 43).

The maximum calculated values of the normative parameters are shown in Table 8.

Table 8 The maximum calculated values of the normative parameters for a curve with a radius of 650 m

Parameter	Wheel arrangement	
	2o-2o-2o	3o-3o
The wheel vertical load on the rail (kN)	181	202
The wheel lateral pressure on the rail (kN)	87	104
Frame force (kN)	63	94
The coefficient of vertical dynamics of the first stage	0.35	0.5
The coefficient of vertical dynamics of the second stage	0.18	0.15
Specific work of friction forces (per meter of rail) in the contact between the wheel flange and rail (kJ/m)	1.116	1.118
Specific work of friction forces (per meter of rail) in the contact between the taping line and rail (kJ/m)	0.615	0.856

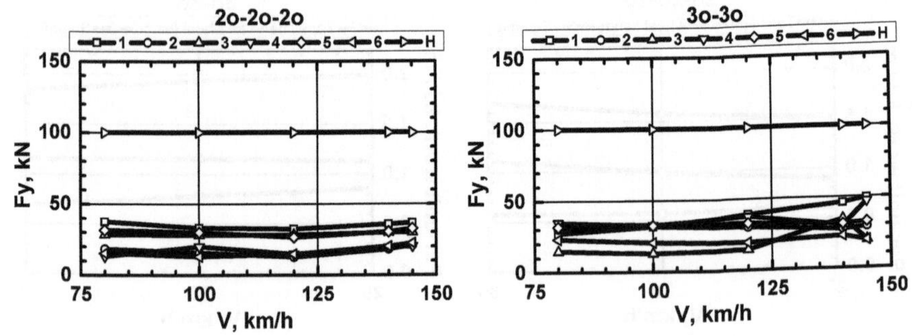

Fig. 44 Lateral forces of the displacement of rails in a curve with a radius of 1000 m

The Simulation Results in a Curve with a Radius of 1000 m

The traffic simulation results in a curve with a radius of 1000 m are shown in the following figures: Fig. 44, lateral forces of the displacement of rails; Fig. 45, frame forces; Fig. 46, wheel vertical forces acting on the rails; Fig. 47, the coefficients o the vertical dynamics of the springs of the first and second stages in the straight line Fig. 48, specific work of friction forces (per meter of rail) in the contact between th wheels and rails.

The simulation results in a curve with a radius of 1000 m showed that the latera forces of a vehicle with a 2o-2o-2o wheel arrangement do not exceed the standar value of 100 kN in the whole speed range up to 145 km/h, and are 34% less tha the lateral forces of the vehicle with a 3o-3o wheel arrangement. Frame forces c the 2o-2o-2o wheel arrangement are 0–15% less than frame forces of the vehicl with a 3o-3o wheel arrangement.

An electric locomotive with a 2o-2o-2o wheel arrangement has 5–14% les vertical impact on the rail than an electric locomotive with a 3o-3o whe arrangement. The vehicle with a 2o-2o-2o wheel arrangement has a coefficient c

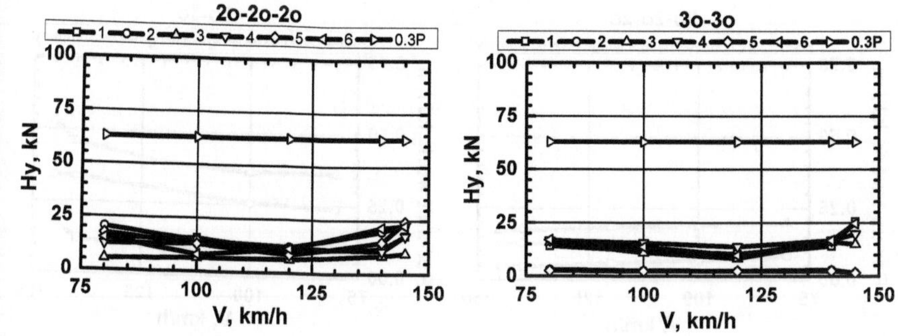

Fig. 45 Frame forces in a curve with a radius of 1000 m

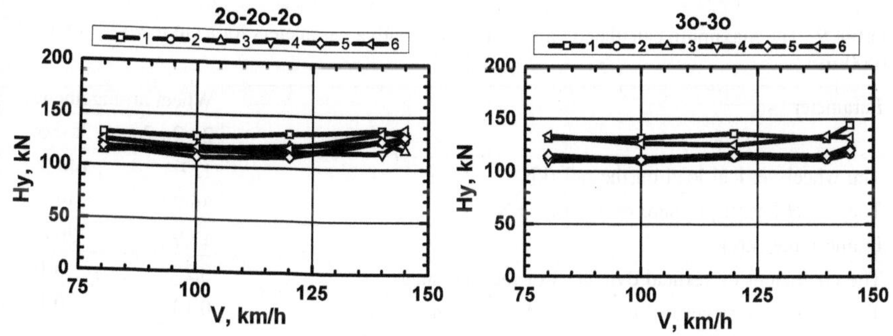

Fig. 46 Wheel vertical forces acting on the rails in a curve with a radius of 1000 m

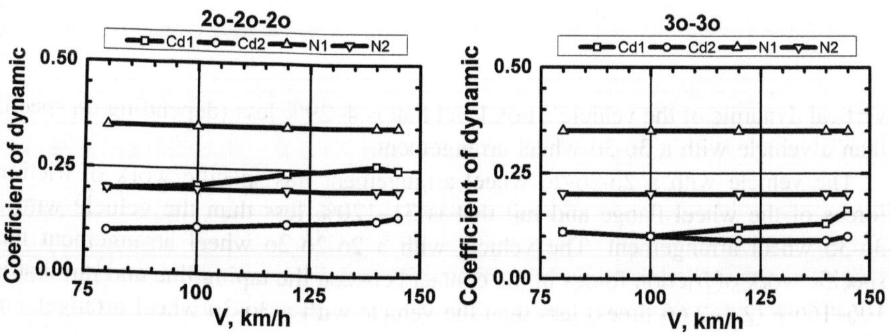

Fig. 47 Coefficients of the vertical dynamics of the springs of the first and second stages in the straight line in a curve with a radius of 1000 m

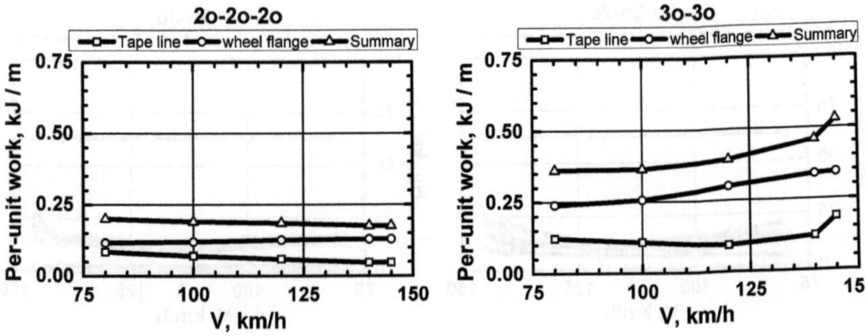

Fig. 48 Specific work of friction forces (per meter of rail) in the contact between the wheels and rails in a curve with a radius of 1000 m

Table 9 The maximum calculated values of the normative parameters for a curve with a radius of 1000 m

Parameter	Wheel arrangement	
	2o-2o-2o	3o-3o
The wheel vertical load on the rail (kN)	127	144.3
The wheel lateral pressure on the rail (kN)	36.7	49.3
Frame force (kN)	17.9	20.6
The coefficient of vertical dynamics of the first stage	0.25	0.16
The coefficient of vertical dynamics of the second stage	0.13	0.11
Specific work of friction forces (per meter the way) in the contact between wheel flange and rail (kJ/m)	0.084	0.187
Specific work of friction forces (per meter the way) in the contact between taping line and rail (kJ/m)	0.129	0.343

vertical dynamic of the vehicle's box level that is 4–29% less (depending on speed) than a vehicle with a 3o-3o wheel arrangement.

The vehicle with a 2o-2o-2o wheel arrangement has specific work of friction forces of the wheel flange and rail that is 31–120% less than the vehicle with a 3o-3o wheel arrangement. The vehicle with a 2o-2o-2o wheel arrangement has specific work of friction forces in the contact between the taping line and rail that is 106–166% (2.1–2.66 times) less than the vehicle with a 3o-3o wheel arrangement (Fig. 56).

The maximum calculated values of the normative parameters are shown in Table 9.

The Simulation Results in a Curve with a Radius of 1500 m

The simulation results in a curve with a radius of 1500 m are shown in the following figures: Fig. 49, lateral forces of the displacement of rails; Fig. 50, frame forces; Fig. 51, wheel vertical forces acting on the rails; Fig. 52, the coefficients of the vertical dynamics of the springs of the first and second stages in the straight line; Fig. 53, specific work of friction forces (per meter of rail) in the contact between the wheels and rails.

The simulation results of a curve with a radius of 1500 m showed that the lateral forces of a vehicle with a 2o-2o-2o wheel arrangement do not exceed the standard value of 100 kN in the whole speed range up to 177 km/h and are 30–66% less than the lateral forces of the vehicle with a 3o-3o wheel arrangement. Frame forces of a vehicle with a 2o-2o-2o wheel arrangement are 2.2 times less than frame forces of a vehicle with a 3o-3o wheel arrangement.

An electric locomotive with a 2o-2o-2o wheel arrangement has 5–14% less vertical impact on the rail than an electric locomotive with a 3o-3o wheel arrangement. The vehicle with a 2o-2o-2o wheel arrangement has a coefficient of vertical dynamics of the vehicle's box level that is 4–29% less (depending on speed) than a vehicle with a 3o-3o wheel arrangement.

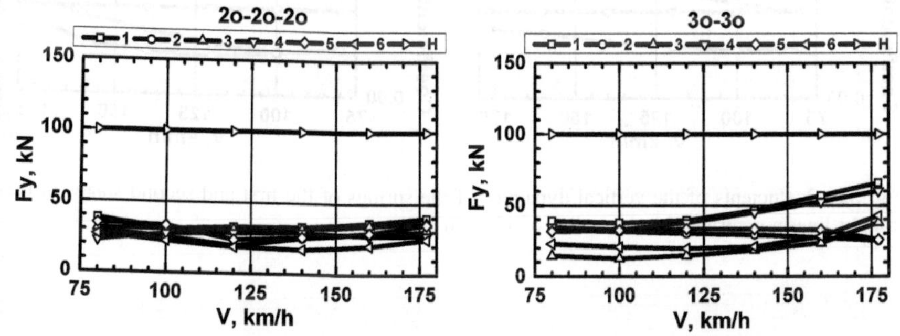

Fig. 49 Lateral forces of the displacement of rails in a curve with a radius of 1500 m

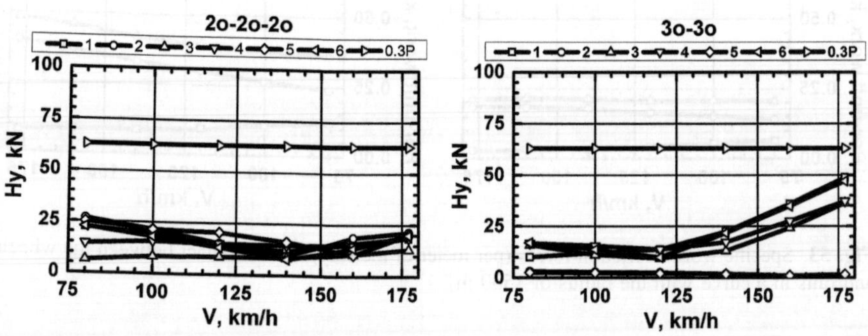

Fig. 50 Frame forces in a curve with a radius of 1500 m

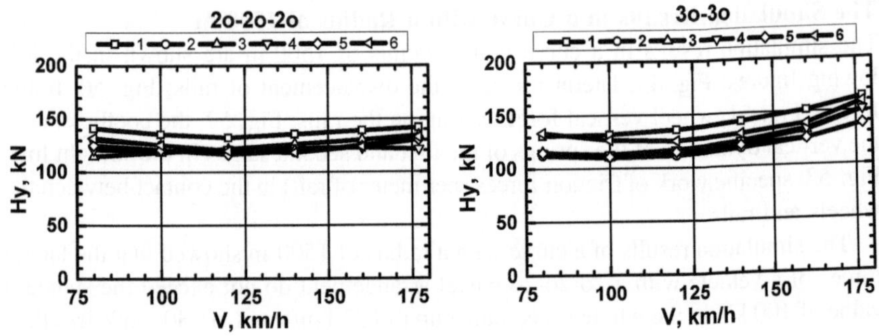

Fig. 51 Wheel vertical forces acting on the rails in a curve with the radius of 1500 m

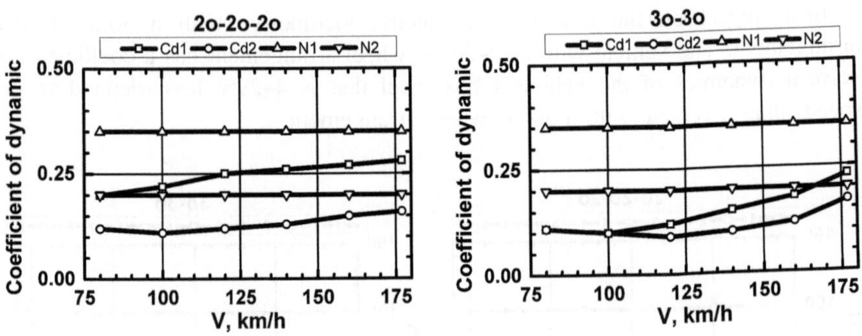

Fig. 52 Coefficients of the vertical dynamics of the springs of the first and second stages in the straight line in a curve with the radius of 1500 m

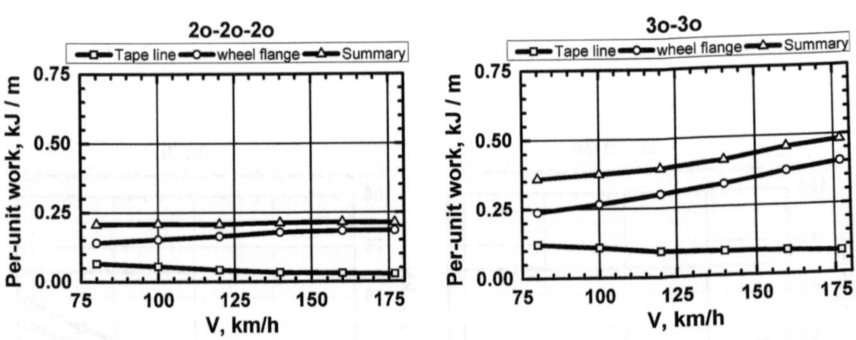

Fig. 53 Specific work of friction forces (per meter of the way) in the contact between the wheels and rails in a curve with the radius of 1500 m

Table 10 The maximum calculated values of the normative parameters for a curve with a radius of 1500 m

Parameter	Wheel arrangement	
	2o-2o-2o	3o-3o
The wheel vertical load on the rail (kN)	131	163.4
The wheel lateral pressure on the rail (kN)	39.7	65.9
Frame force (kN)	18.5	40.9
The coefficient of vertical dynamics of the first stage	0.28	0.23
The coefficient of vertical dynamics of the second stage	0.16	0.17
Specific work of friction forces (per meter the way) in the contact between the wheel flange and rail (kJ/m)	0.185	0.403
Specific work of friction forces (per meter the way) in the contact between the taping line and rail (kJ/m)	0.066	0.121

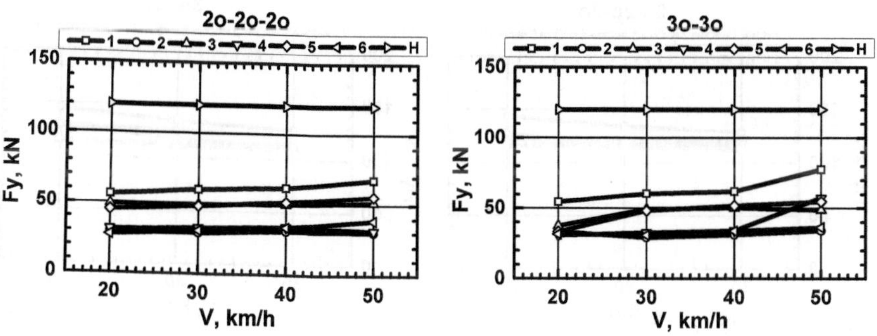

Fig. 54 Lateral forces of the displacement of rails in the turnout R65 1/11

The vehicle with a 2o-2o-2o wheel arrangement has specific work of friction forces of wheel flange and rail that is 51–83% less than the vehicle with a 3o-3o wheel arrangement. The vehicle with a 2o-2o-2o wheel arrangement has specific work of friction forces in the contact between taping line and rail that is 69–118% less than the vehicle with a 3o-3o wheel arrangement (Fig. 53; Table 10).

The Simulation Results in the Turnout R65 1/11

The following graphs show the results of movement simulation on the turnout R65 1/11: Fig. 54, lateral forces of the displacement of rails; Fig. 55, frame forces; Fig. 56, wheel vertical forces acting on the rails; Fig. 57, the coefficients of the vertical dynamics of the springs of the first and second stages in the straight line; Fig. 58, specific work of friction forces (per meter of the way) in the contact between the wheels and rails.

The results of simulation in the turnout R65 1/11 show that the vehicle with a 2o-2o-2o wheel arrangement has lateral forces no more than the standard value of 100 kN in the whole speed range up to 177 km/h and 0–14% less than the vehicle

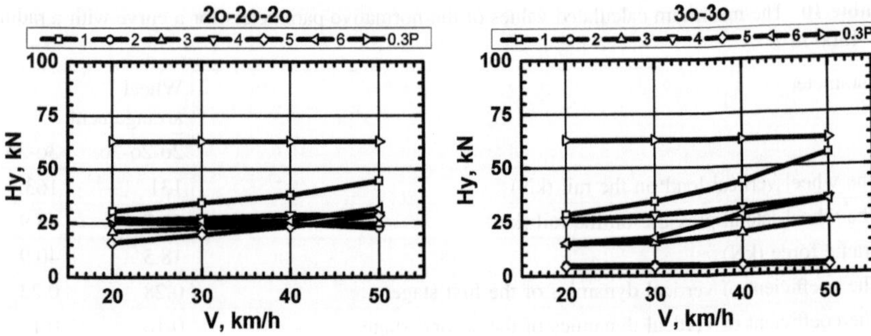

Fig. 55 Frame forces in the turnout R65 1/11

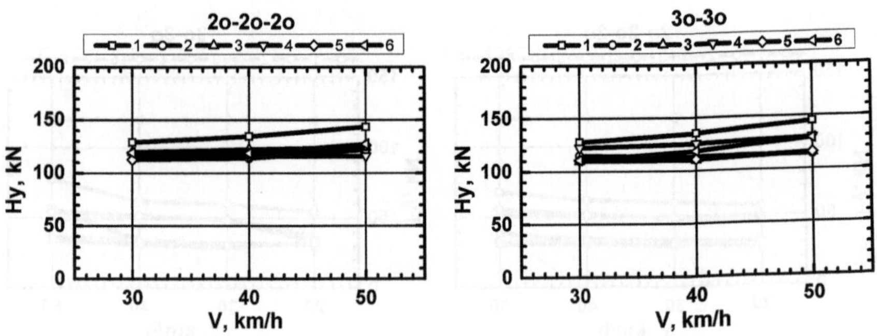

Fig. 56 Wheel vertical forces acting on the rails in the turnout R65 1/11

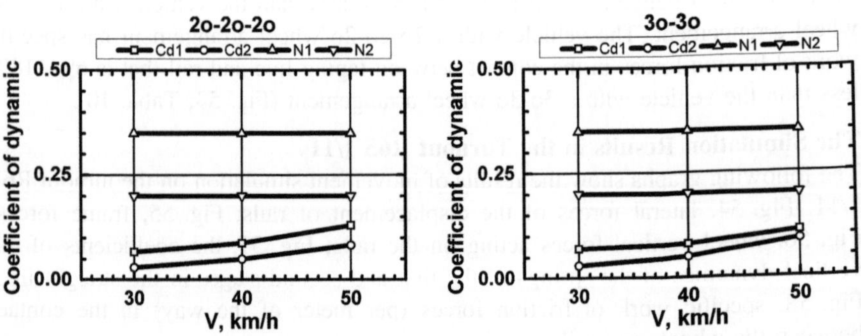

Fig. 57 Coefficients of the vertical dynamics of the springs of the first and second stages in the straight line in the turnout R65 1/11

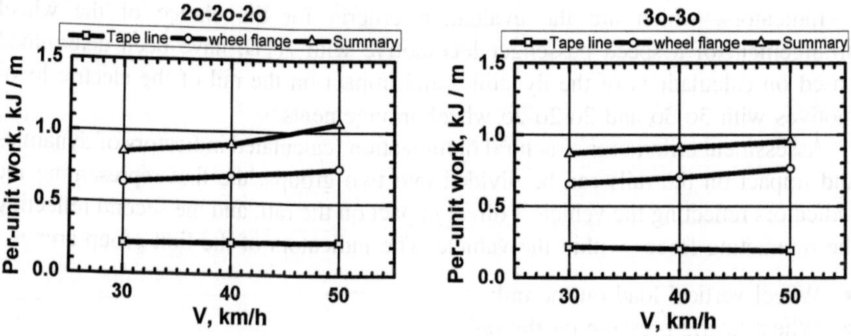

Fig. 58 Specific work of friction forces (per meter of the way) in the contact between the wheels and rails in the turnout R65 1/11

Table 11 The maximum calculated values of the normative parameters for the turnout R65 1/11

Parameter	Wheel arrangement	
	2o-2o-2o	3o-3o
The wheel vertical load on the rail (kN)	122	145
The wheel lateral pressure on the rail (kN)	68	77.6
Frame force (kN)	44.6	56
The coefficient of vertical dynamics of the first stage	0.13	0.11
The coefficient of vertical dynamics of the second stage	0.08	0.09
Specific work of friction forces (per meter of the path) in the contact between the wheel flange and rail (kJ/m)	0.741	0.763
Specific work of friction forces (per meter of the path) in the contact between the taping line and rail (kJ/m)	0.231	0.189

with a 3o-3o wheel arrangement. The vehicle with a 2o-2o-2o wheel arrangement has frame forces 0–26% less than the vehicle with a 3o-3o wheel arrangement.

An electric locomotive with a 2o-2o-2o wheel arrangement has 13–19% less vertical impact on the rail than an electric locomotive with a 3o-3o wheel arrangement. The vehicle with a 2o-2o-2o wheel arrangement has a coefficient of vertical dynamics of the vehicle's box level that is 4–29% less (depending on speed) than a vehicle with a 3o-3o wheel arrangement.

The vehicle with a 3o-3o wheel arrangement has specific work of friction forces of wheel flange and rail 0–22% less than the vehicle with a 2o-2o-2o wheel arrangement at a speed of 50 km/h. The vehicle with a 2o-2o-2o wheel arrangement has 3% (2.1–2.66 times) less specific work of friction forces in the contact between the taping line and rail than the vehicle with a 3o-3o wheel arrangement (Fig. 58).

The maximum calculated values of the normative parameters are shown in Table 11.

Indicators, which are the evaluation criteria for the choice of the wheel arrangement of a speed passenger locomotive with ATD, have been determined based on calculations of the dynamics and impact on the rail of the electric locomotives with 3o-3o and 2o-2o-2o wheel arrangements.

Assessment criteria for structural optimization (calculated indicators of dynamics and impact on the rail) can be divided into two groups: the first representing the indicators reflecting the vehicle's direct impact on the rail, and the second reflecting the interacting forces within the vehicle. The indicators of the first group are:

- Wheel vertical load on the rail;
- Wheel lateral pressure on the rail;
- Specific work of friction forces in the contact between the wheel flange and rail;
- Specific work of friction forces in the contact between the taping line and rail.

The indicators of the second group are:

- Coefficient of vertical dynamics of the first stage;
- Coefficient of vertical dynamics of the second stage;
- Frame force.

The choice of the vehicle part is based on the first group of indicators, as they directly characterize the interaction of the vehicle parts and the rail; the indicators of the second group are taken into account when the indicators of the first group are equal.

The maximum values of the indicators of the first and second groups are shown in Table 12 (marked in bold indicators have best value among the compared).

The calculated values of the forces of interaction between the vehicle parts and the rail, as well as the forces of interaction between the elements of vehicle parts, have shown that the vehicle with a 2o-2o-2o wheel arrangement is better for design of a high-speed passenger locomotive with ATD in terms of the force interaction than the vehicle with a 3o-3o wheel arrangement.

The work of friction forces in the wheel zone (especially the wheel flange) and rail contact on the majority of the investigated part of the path is less for the vehicle with a 2o-2o-2o wheel arrangement. It reduces the flange being worn sharp and the number of wheel turnings while in operation. Currently, the flange being worn sharp is the primary cause of most locomotive's wheels turning.

On the basis of a two-axle bogie, it is possible to create passenger electric locomotives with different wheel arrangements: 2o-2o; 2o-2o-2o; 2(2o-2o). On the basis of a three-axle bogie, it is possible to create only passenger electric locomotives with 3o-3o wheel arrangements. Therefore, the application of wheel arrangement has one more criterion, processability, in addition to the criteria outlined above. A two-axle bogie unifies passenger locomotives with different numbers of axles.

Based on the above, a bogie with a 2o-2o-2o wheel arrangement is recommended for the high-speed passenger six-axle locomotive with asynchronous traction drive.

Table 12 The maximum values of the indicators of the first and second groups

Parameters	Straight track		*R* = 1500 м		*R* = 1000 м		*R* = 650 м		*R* = 350 м		Р 65/11	
	2o-2o-2o	3o-3o	2o-2o-2o	3o-3o	2o-2o-2o	3o-3o	2o-2o-2o	3o-3o	2o-2o-2o	3o-3o	2o-2o-2o	3o-3o
Wheel vertical load on the rail (kN)	**168**	179	**131**	163.4	**127**	144.3	**181**	202	**146.9**	155.86	**122**	145
Wheel lateral pressure on the rail (kN)	**81.78**	85.98	**39.7**	65.9	**36.7**	49.3	**87**	104	**50.1**	56.9	**68**	77.6
Specific work of friction forces (per meter of rail) in the contact between wheel flange and rail (kJ/m)	0.354	**0.269**	**0.185**	0.403	**0.084**	0.187	**1.116**	0.118	**0.125**	0.724	**0.741**	0.763
Specific work of friction forces (per meter of rail) in the contact between taping line and rail (kJ/m)	0.176	**0.138**	**0.066**	0.121	**0.129**	0.343	**0.615**	0.856	0.284	**0.255**	0.231	**0.189**
Frame force (kN)	**46.79**	69.08	**18.5**	40.9	**17.9**	20.6	**63**	94	**27.9**	38.45	**44.6**	56
The coefficient of vertical dynamics of the first stage	**0.35**	0.511	0.28	**0.23**	0.25	**0.16**	**0.35**	0.5	0.24	**0.16**	0.13	**0.11**
The coefficient of vertical dynamics of the second stage	0.073	**0.072**	**0.16**	0.17	0.13	**0.11**	0.18	**0.15**	0.12	0.12	**0.08**	0.09

5 Comparison of Computer Simulation Results and Tests on the Line Results of Electric Locomotive EP20. Comparison of Dynamic Parameters of Electric Locomotives EP10 and EP20

To assess the reliability of the results obtained by computer simulation for the interaction of a passenger locomotive's underframe and rail, we have compared them with the results of tests on the line of the electric locomotive EP20. The locomotive underframe of this locomotive is constructed in line with the applications outlined above.

We have compared the results of tests on the line of electric locomotives EP20 and EP10 to assess the effectiveness of new constructions. These locomotives have a similar construction of the locomotive underframe, they are dual-mode and equipped with ATD, and they have a 2o-2o-2o wheel arrangement. They have almost the same mass distribution between the body and bogies, and the same wheel pair loads on the rails.

5.1 Comparison of Computer Simulation Results and Tests on the Line Results of Electric Locomotive EP20

We have tested electric locomotive EP20-001 h on the test track on the straight sections of the rail up to a speed of 160 km/and on the curved sections of the rail with radii of 350 and 650 m at speeds corresponding to the outstanding acceleration of 0.7 m/s^2.

We have tested electric locomotive EP20-002 on the operational section of the St. Petersburg–Bologoe route at a speed up to 220 km/h.

The electric locomotives were equipped with the necessary sensors for measuring the dynamics results. Figure 59 shows the sensor arrangement.

Figure 59 shows:

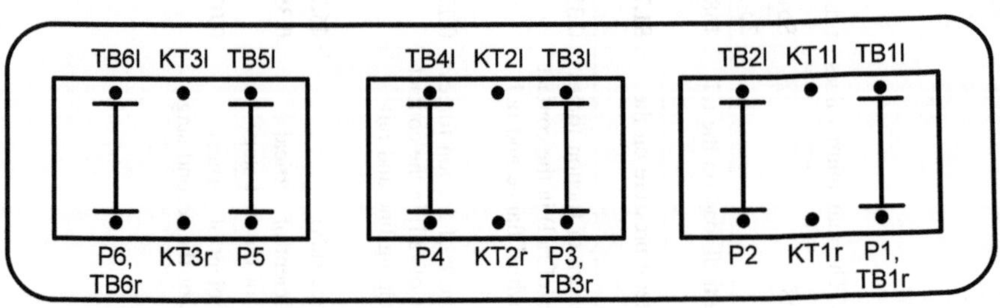

Fig. 59 Sensor arrangement

Sensors for the frame forces measuring:

P1—frame forces of the 1st wheel set
P2—frame forces of the 2nd wheel set
P3—frame forces of the 3rd wheel set
P4—frame forces of the 4th wheel set
P5—frame forces of the 5th wheel set
P6—frame forces of the 6th wheel set

Sensors for measuring the coefficient of vertical dynamics of the first stage of suspension:

TB1l—left box of the 1st wheel set
TB1r—right box of the 1st wheel set
TB2l—left box of the 2nd wheel set
TB3l—left box of the 3rd wheel set
TB3r—right box of the 3rd wheel set
TB4l—left box of the 4th wheel set
TB5l—left box of the 5th wheel set
TB6l—left box of the 6th wheel set
TB6r—right box of the 6th wheel set

Sensors for measuring the coefficient of vertical dynamics of the second stage of suspension:

KT1l—the 1st bogie, left-hand side
KT1r—the 1st bogie, right side
KT2l—the 2nd bogie, left-hand side
KT2r—the 2nd bogie, right side
KT3l—the 3rd bogie, left-hand side
KT3r—the 3rd bogie, right side

To measure the level of the electric locomotive's impact on the rail, the selected areas were equipped with sensors for measuring the load on the rail base edge, forces transmitted from the wheels on the rails. Resistive strain sensors with a base of 10 mm and a nominal resistance of 100 Ω were pasted on the rails to measure the load on the bottom rail base edge and forces transmitted from the wheels on the rails. The signals from sensors arrived by measuring cables in the input of the measuring and computing system and were recorded in the computer.

After testing the electric locomotives, we compared the simulation and test results. Figures 60, 61, 62, 63 and 64 show the comparisons. The electric locomotive EP20 meets the requirements of the impact on the rail and dynamic qualities up to a speed of 220 km/h in the straight and curved sections of the path, with speeds appropriate for outstanding acceleration of 0.7 m/s^2.

In the straight section of the path, the difference between the simulation and the test results is not more than 20% for frame forces, lateral forces or coefficients of the

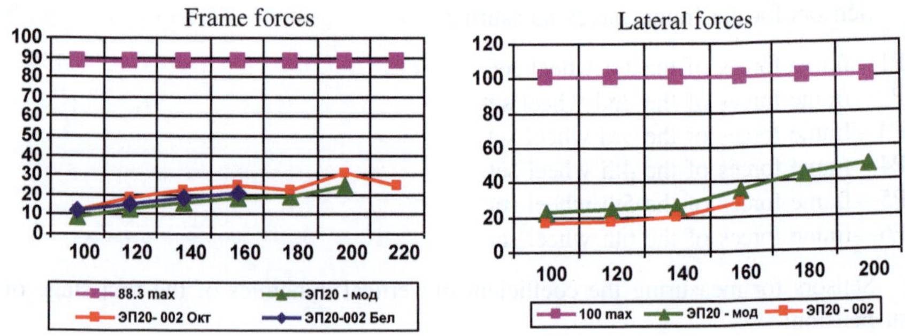

Fig. 60 Forces in the straight section

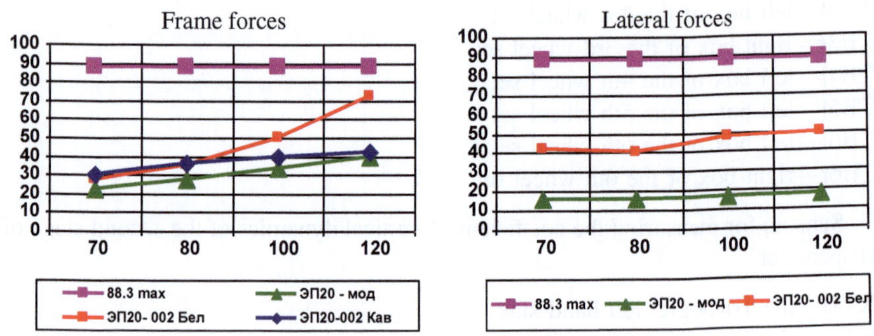

Fig. 61 Forces in the curved section with a radius of 650 m

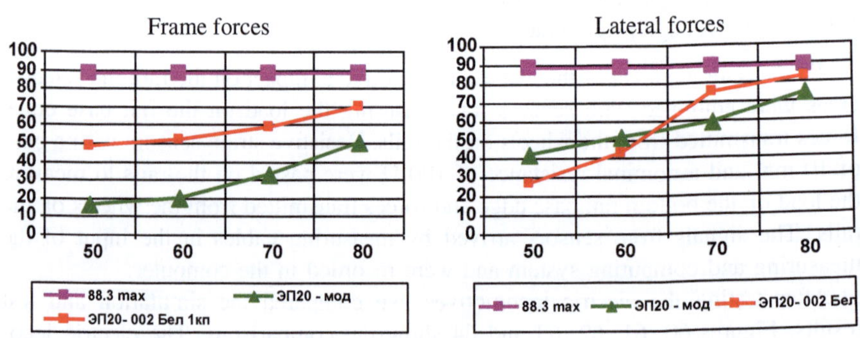

Fig. 62 Forces in the curved section with a radius of 350 m

vertical dynamics of the first and second stage of spring suspension (Figs. 60 and 64). Good convergence is observed for the coefficients of vertical dynamics in curves with the radii of 350 and 650 m (Fig. 63), and when driving in a curve with a radius of 650 m (Fig. 61, EP20-002 Kav).

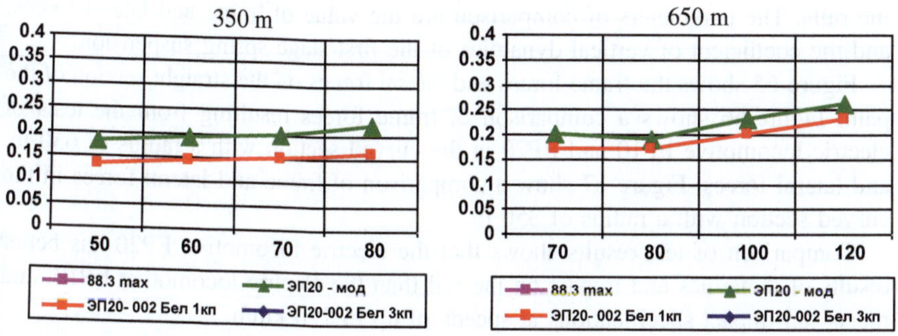

Fig. 63 The coefficient of vertical dynamics of the 1st stage in the curved section

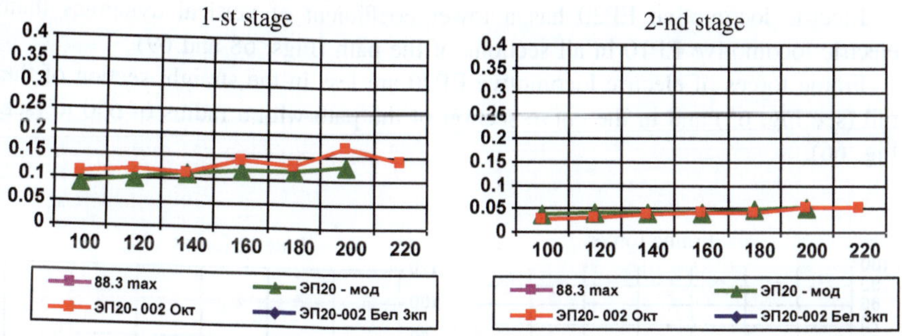

Fig. 64 The coefficient of vertical dynamics in the straight section

Deviation of the test results from the simulation results in a curve with a radius of 350 m for the lateral forces (Fig. 62) does not exceed 35% at a speed of 50 km/h; the convergence improves when speed increases, and it reaches a minimum value of 16% at a speed close to 60 km/h.

Therefore, we have good convergence of computer simulation and test results for most parameters, and we have been able to replace the natural experiment by computer simulation and avoid the need to construct an experimental model of the locomotive underframe. This has significantly reduced the development time for a high-speed electric locomotive with ATD and its associated financial costs.

5.2 *Assessment of the Locomotive Underframe Design Effectiveness*

To assess the effectiveness of the new design solution of the locomotive underframe, we have compared the test results of electric locomotives EP20 and EP10, which have the same wheel arrangement (2o-2o-2o) and the same wheel set load on the rails. The parameters of comparison are the value of frame and lateral forces, and the coefficient of vertical dynamics of the first-stage spring suspension.

Figure 65 shows the frame forces and lateral forces on the straight section of the path. Figure 66 shows a comparison of frame forces resulting from the tests of electric locomotive EP10 and EP20 in the curved section with a radius of 650 m, and lateral forces. Figure 67 show a comparison of frame and lateral forces in the curved section with a radius of 350 m.

Comparison of test results shows that the electric locomotive EP20 has better results of dynamics and impact on the rail than the electric locomotive EP10, and meets rail impact specifications at speeds of up to 220 km/h.

Figures 65, 66 and 67 show that the electric locomotive EP20 has less impact on the rail for the lateral forces than the electric locomotive EP10 in the straight section of the path and in the curved sections of the path with radii of 350 and 650 m.

Electric locomotive EP20 has a lower coefficient of vertical dynamics than electric locomotive EP10 in all sections of the path (Figs. 68 and 69).

Frame forces of electric locomotive EP20 are less in the straight section of the rail (see Fig. 65) and in the curve section of the path with a radius of 650 m (see Fig. 66).

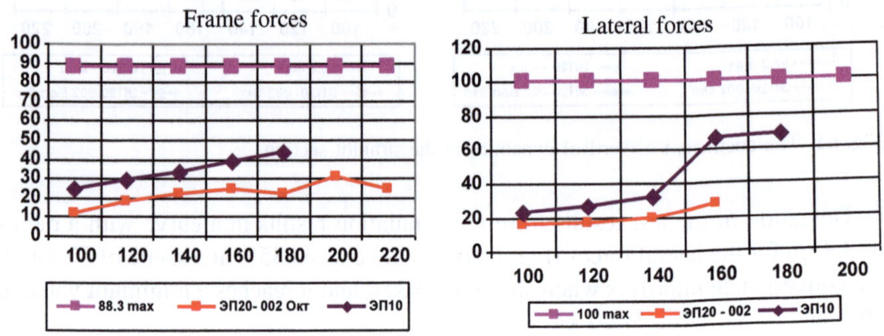

Fig. 65 Forces in the straight section

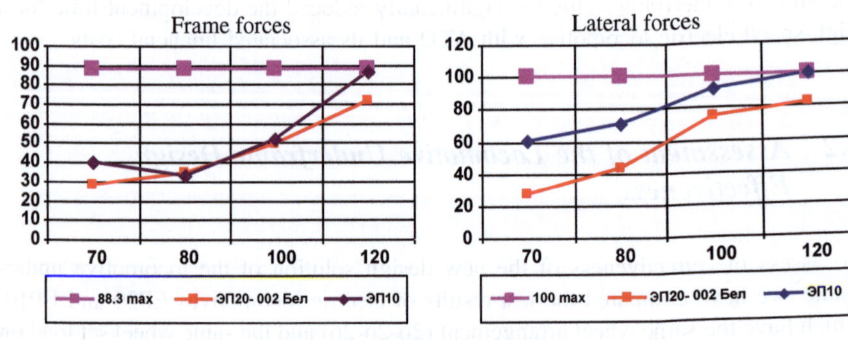

Fig. 66 Forces in the curved section with a radius of 650 m

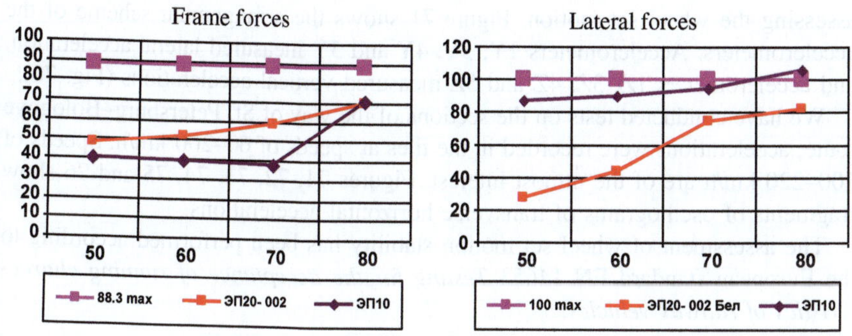

Fig. 67 Forces in the curved section with a radius of 350 m

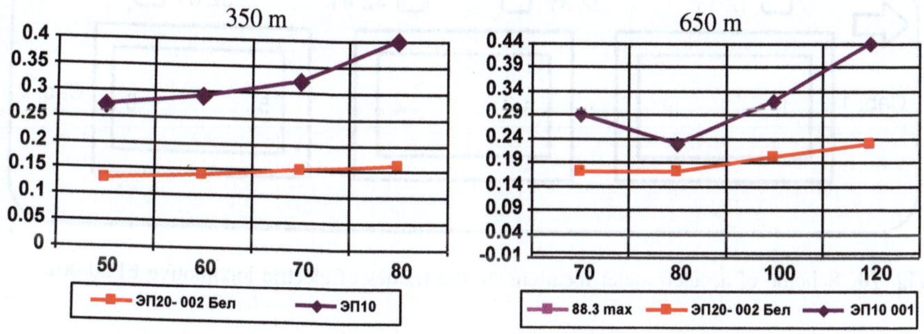

Fig. 68 The coefficient of vertical dynamics of the 1st stage in the curved section

Fig. 69 The coefficient of vertical dynamics of the 1st stage in the straight section

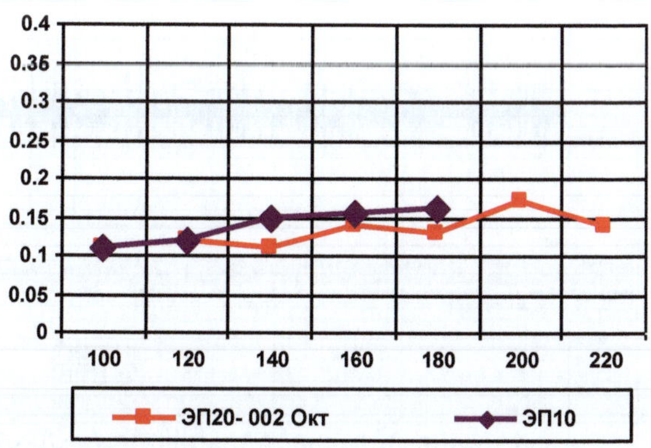

5.3 *The Wheel Set Motion Stability Test at High Speeds*

The assessment of the wheel set motion stability of the vehicle with a 2o-2o-2o wheel arrangement at high speeds is of utmost interest. The values of the bogie frames' horizontal accelerations measured during the test are the initial data for assessing the wheel set motion. Figure 71 shows the arrangement scheme of the accelerometers. Accelerometers 1Y, 3Y, 4Y and 5Y measured lateral acceleration, and accelerometers 1Z, 3Z, 4Z and 5Z measured vertical accelerations (Fig. 70).

We have conducted tests on the sections of the way of St. Petersburg–Bologoye route; accelerations were recorded in the files at speeds of 60–200 km/h. Speeds of 200–220 km/h are of the utmost interest. Figures 71, 72, 73, 74, 75 and 76 show fragments of oscillograms of transverse horizontal accelerations.

The assessment of wheel set motion stability has been performed according to the European standard EN 14653 *Testing for the acceptance of running characteristics of railway vehicles*.

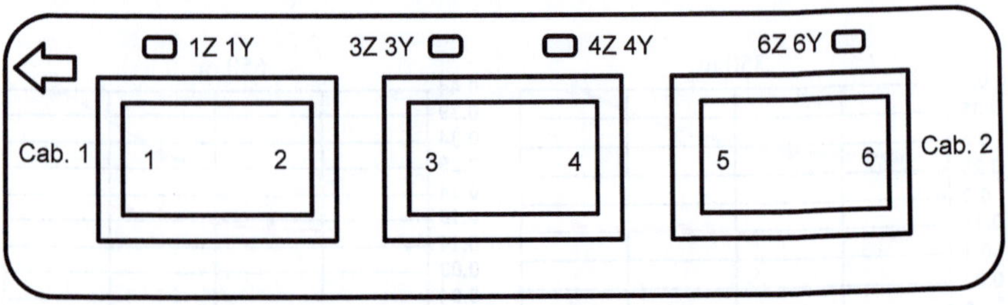

Fig. 70 Scheme of accelerometer locations on the frames of electric locomotive EP20-002

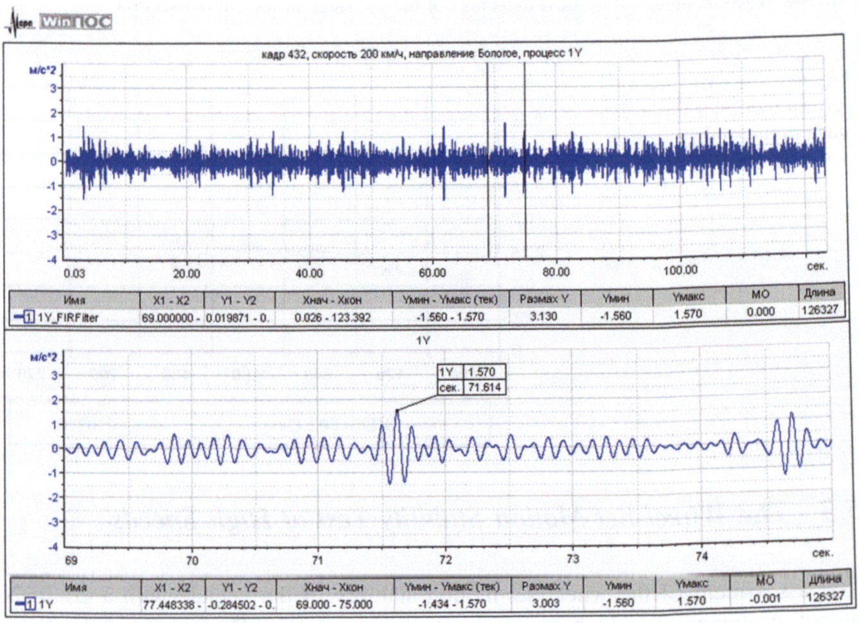

Fig. 71 Accelerations on the frame of the outermost bogie (the first), $V = 200$ km/h

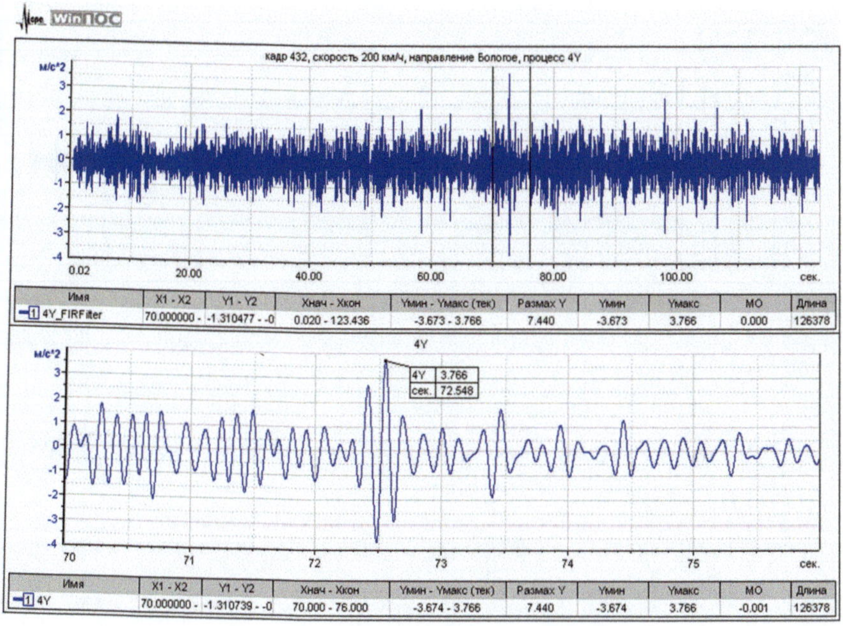

Fig. 72 Accelerations on the frame of the middle bogie, $V = 200$ km/h

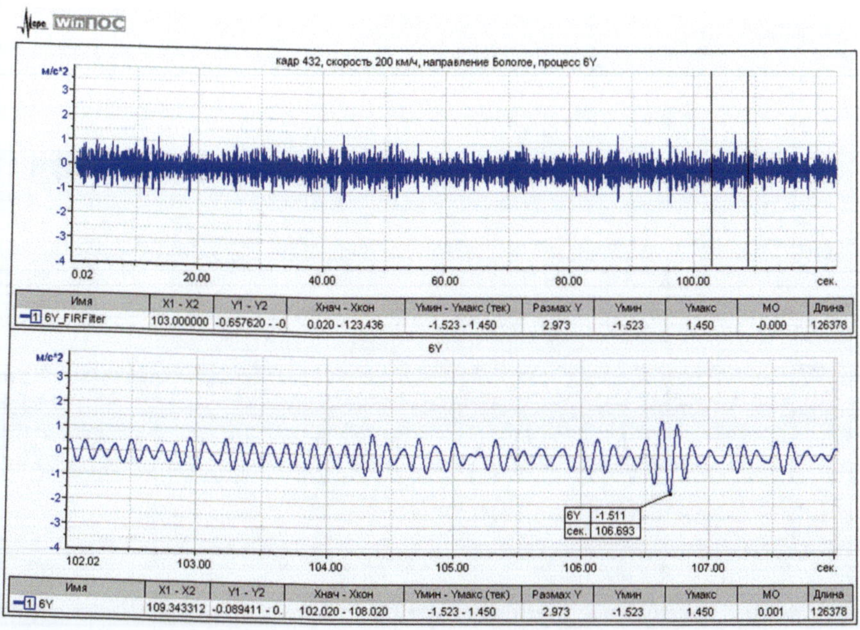

Fig. 73 Accelerations on the frame of the outermost bogie (the third), $V = 200$ km/h

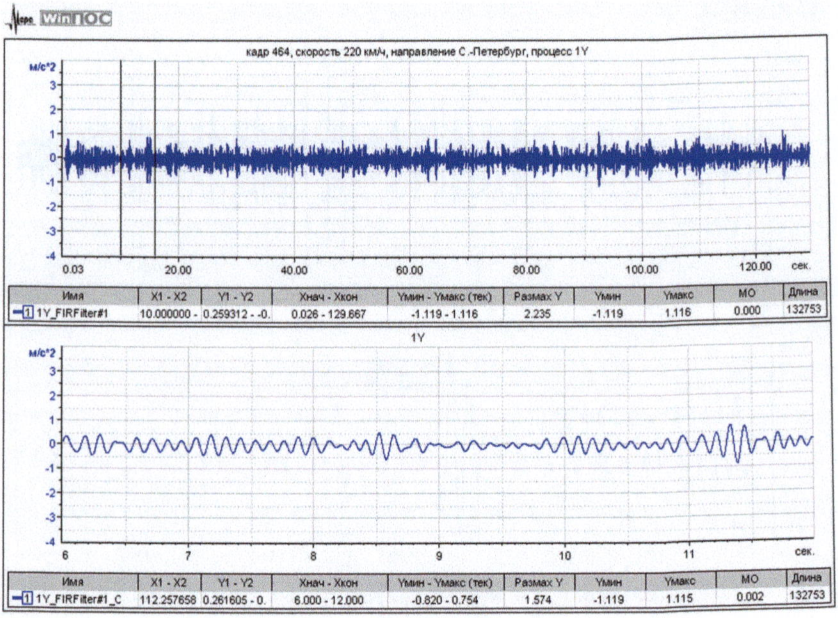

Fig. 74 Accelerations on the frame of the outermost bogie (the first), $V = 220$ km/h

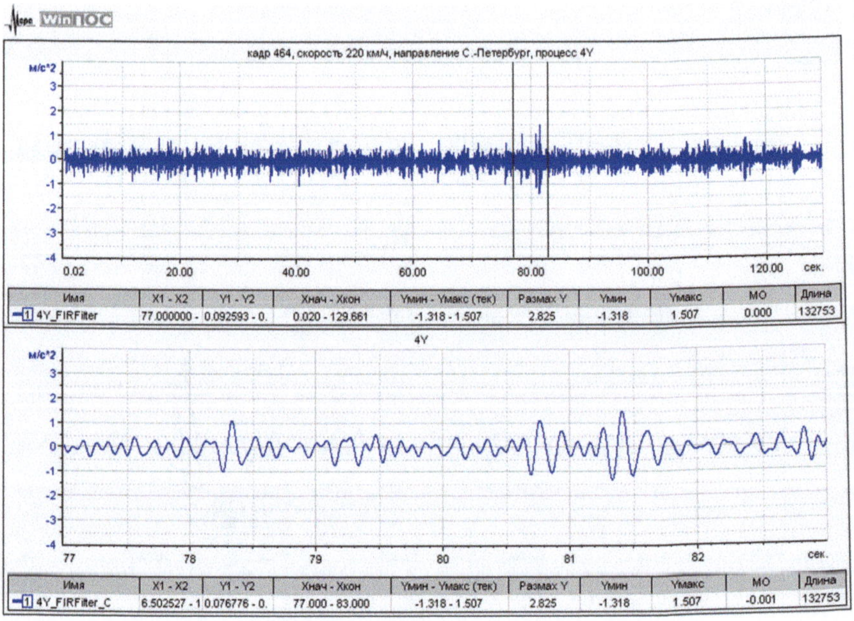

Fig. 75 Accelerations on the frame of the middle bogie, $V = 220$ km/h

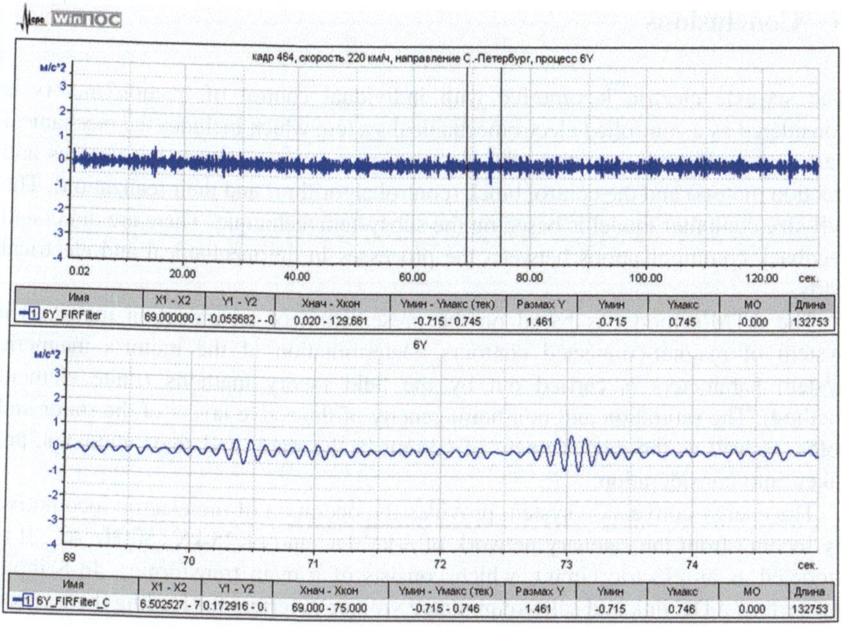

Fig. 76 Accelerations on the frame of outermost bogie (the third), $V = 220$ km/h

Russian regulatory documents do not standardize values of bogies' vertical and horizontal accelerations. In accordance with EN 14653, the allowable value of the ogies' transverse acceleration (in m/s^2) in the frequency range of 0–10 Hz for electric ocomotive EP20 [23] is not more than 0.86 g (8.6 m/s^2). The largest horizontal ransverse acceleration measured on the frame of the middle bogie was 3.77 m/s^2, vhich is well below the standard value, and the horizontal transverse acceleration of he outermost bogies did not exceed 1.77 m/s^2. Noteworthy is the fact that the cceleration amplitude decreased significantly when the speed increased from 200 to 20 km/h. This shows that the locomotive underframe concept and selected parameter alues presented in this work enabled us to satisfy the requirements for electric ocomotives, and can be applied in the future at higher speeds.

The horizontal transverse acceleration of the outermost trucks at speeds of up to 00 km/h does not exceed 0.18 g, and is less than 0.28 g on average overall. At a peed of 160 km/h, the maximum values remained virtually unchanged in the path f the perfect quality, but in the path of a satisfactory quality, they increased to .3 g at the extreme trucks, and up to 0.69 g at the middle trucks.

Thus, the measured horizontal lateral acceleration on the bogies does not exceed e permissible value of 0.86 g, as in the trials in Belorechensk on the test site of VNIIZhT" at speeds of 60–160 km/h, and on the St. Petersburg–Bologoe route ktyabrskaya Railways) at speeds of 160–220 km/h. Consequently, in accordance ith European standard EN 14653, the EP20 electric locomotive meets the regutory requirements for stable behavior of wheel sets.

6 Conclusions

The six-axle electric locomotive with individual control of traction motors is considered as a controlled electromechanical system which includes the mechanical part as a multi-body structure, the electrical part (energy conversion devices and traction motors) and the control block (control algorithms and their realization). The full-size computer model is based on the subsystem technique. There are direct and feedback communications between the processes in the mechanical and electrical parts.

The ATM's model is based on the presentation of an induction motor as a system of magnet-connected contours. Determination of the motor's magnetic system parameters is carried out by the field theory methods (finite element method). The saturation and non-homogeneity of the active layers of the stator and rotor, as well as the displacement of current in the stator and rotor windings, are taken into consideration.

The energy conversion system provides the operation of the electric locomotive by feeding from the catenary network of 3-kV DC and of 25-kV, 50-Hz AC. It is modeled as an electric circuit, which consists of a main transformer, 4q-S input converters, a DC link and self-commutated voltage inverters for feeding the traction motors.

The two-channel automatic control system with independent control of the rotor flux and electromagnetic torque is developed. The stabilization of rotor flux magnitude in all modes eliminates the excessive saturation of the magnetic system.

The mechanical part of the electric locomotive, considered as a multi-body structure, consists of a car body and three two-axle bogies. The model contains 28 rigid bodies. The processes during coasting movement were studied, and the normal reactions and the forces of interaction in the wheel–rail contact are shown. The AFC was built for car body bouncing, pitching and lateral rolling oscillations. Some qualitative features of the dynamic behavior are noted.

Complex evaluation of principles and algorithms of the control system of the ATM and analyses of electromechanical processes in the ATD under various working conditions (locomotive's start and acceleration; movement in traction mode with constant speed on a straight and curve railway sections; wheel-slide protection system with individual regulation of traction forces; control of the voltage inverter at high-speed mode) were executed in Sect. 4.

Implemented approaches and software are used for testing automatic control systems and wheel-slide protection systems for newly designed electric locomotives, as well as for the prediction of performance, including the transient processes in both mechanical and electric parts.

At present, using the developed mathematical model, the analysis and optimization of the energy conversion system's parameters are performed. Based on the developed model, some variants of the control system are tested and compared.

The electrical scheme allows realization of the operational disconnection/connection of one or more axles in the automatic mode, with an account of the actual load. In perspective, this approach would lead to improving the locomotive energy efficiency during partial load work.

Mathematical modeling methods were carried out as part of the research of the dynamic characteristics of electric locomotive EP20 under various conditions: the movement in a straight line, in curves and in the turnouts. The deviation of the test results from the simulation results in a curve for the lateral forces does not exceed 35% at a speed of 50 km/h; the convergence is improved when speed increases and it reaches a minimum value of 16% at a speed close to 60 km/h. In the straight section of the rail, the difference between the simulation and the test results is not more than 20% for frame forces, lateral forces and coefficients of the vertical dynamics of the first and second stages of spring suspension. Good convergence is observed for the coefficients of vertical dynamics in curves. Thus we have good convergence of computer simulation and test results for most parameters, allowing us to replace the natural experiment by computer simulation, and avoiding the need to construct an experimental model of the locomotive underframe. This has significantly reduced the development time for the high-speed electric locomotive with ATD and the associated financial costs.

References

1. Zarifian A, Nikitenko A, Kolpahchyan P, Khomenko B (1997) Computer modeling of dynamic processes in complex electromechanical systems. In: Proceedings of 15th IMACS World Congress, vol 6. Application in modelling and simulation, Berlin, pp 281–286
2. Bakhvalov Yu, Kolpahchyan P, Plokhov E, Yanov V, Zarifian A (2000) Mathematical modelling of electromechanical processes in electric locomotive. In: Proceedings of 16th IMACS World Congress. Book of abstracts, Lausanne, p 331
3. Andrushchenko A et al (2013) Locomotive's asynchronous traction drive. UMC ZhDT, Moscow (in Russian)
4. Bose B (2002) Modern power electronics and AC drives. Prentice Hall PTR, Upper Saddle River
5. Zobory I, Benedek T, Gyoumlrik A, Szaboacute A (1988) Dynamic processes in the drive system of electric traction vehicles. Veh Syst Dyn 17(S1):559–570
6. Vukosavic SN (2013) Electrical machines, vol XXXIII. Springer, New York, 649 pp
7. Chua L, Lin P (1975) Computer—aided analysis of electronic circuits: algorithm and computational techniques. Prentice-Hall, Englewood Cliffs
8. Blaschke F (1972) The principle of field orientation applied to the new trans-vector closed-loop control system for rotating field machine. Siemens Rev 93:217–220
9. Chiasson J (1995) Non linear controllers for induction motors. In: IFAC conference system structure and control, pp 572–583
10. Chiasson J (2005) Modeling and high-performance control of electrical machines. Wiley, New York
11. Bedford BD, Hoft RG (1985) Principles of inverter circuits. Robert E. Krieger Publishing Company, Melbourne, 430 pp
12. Giri F (ed) (2013) AC electric motors control: advanced design techniques and applications. Wiley, Oxford
13. Holmes G, Lipo TA (2003) Pulse width modulation for power converters: principles and practice. Wiley-IEEE Press, 744 pp

14. Glumineau A, de Leon Morales J (2015) Sensorless AC electric motor control: robust advanced design techniques and applications. Springer, Berlin, 244 pp
15. Patel HS, Hoft RG (1973) Generalized harmonic elimination and voltage control in thyristor inverters. Part 1. Harmonic elimination. IEEE Trans Ind Appl 9:310–317
16. Patel HS, Hoft RG (1973) Generalized harmonic elimination and voltage control in thyristor inverters. Part 1I. Voltage control technique. IEEE Trans Ind Appl 10:666–673
17. Patra B, Kumar B, Yadagiri J, Dasgupta A (2012) Estimation of switching angles by using PSO of three phase voltage source inverter. Int J Model Optim 2(4):513–517
18. Chiasson JN, Tolbert LM, McKenzie KJ, Zhong Du (2004) A complete solution to the harmonic elimination problem. IEEE Trans Power Electron 19(2):491–499
19. Madichetty S, Rambabu M, Dasgupta A (2014) Selective harmonic elimination: comparative analysis by different optimization methods. In: Power electronics (IICPE), 2014 IEEE 6th India international conference. Kurukshetra, pp 1–6
20. Kreuzer E (1994) Generation of symbolic equations of motion of multibody systems. Computerized symbolic manipulations in mechanics. Springer, New York, pp 1–67
21. Universal Mechanism/user's manual (2016). URL: http://www.universalmechanism.com/en/pages/index.php?id=3
22. Kolpahchyan P, Pogorelov D (2004) Simulation of electric locomotives as mechatronic systems. In: EUROMECH 452. Colloquium on advances in simulation techniques for applied dynamics. Abstracts. Martin-Luther-University Halle-Wittemberg, Halle, p 19
23. Andryushchenko A (2013) Development of the high-speed passenger electric locomotive with asynchronous traction drive. Abstract of diss., Ph.D., Rostov-on-Don, RSTU (in Russian)
24. Kolpakhchyan P, Zarifyan Jr A (2015) Study of the asynchronous traction drive's operating modes by computer simulation. Part 1: problem formulation and computer model. Transp Prob 10(2):125–136
25. Kolpakhchyan P, Zarifyan Jr A (2015) Study of the asynchronous traction drive's operating modes by computer simulation. Part 1I: simulation results and analysis. Transp Prob 10(3):5–15
26. Kalker JJ (2000) Rolling contact phenomena: linear elasticity. Reports of the Department of Applied Mathematical Analysis. Report 00-09, Delft, 90 pp
27. Kolpahchyan P (2015) Animation of the asynchronous traction motor magnetic field. URL: https://youtu.be/Fm0_an2OPj0
28. Standards for evaluating the strength of load-bearing elements, dynamic qualities and impacts on the way the underframe part of RUSSIAN railways locomotives of gauge 1520 mm (1998) VNIIZhT, 145 pp (in Russian)
29. Garg VK, Dukkipati RV (1984) Dynamics of railway vehicle systems. Academic Press, New York, p 407
30. EN 14363:2016 Testing for the acceptance of running characteristics of railway vehicles
31. Cherkashin YM, Shestakov AA (1982) On the sustainability movement of railway rolling stock. VNIIZhT Proc 649:42–49 (in Russian)
32. Kogan AY (1997) Dynamics of ways and its interaction with the rolling stock. Transport, Moscow (in Russian)
33. Wickens AH (1993) Dynamic stability of articulated and steered railway vehicles guided by lateral displacement feedback. the dynamics of vehicles on roads and tracks. in: Proceedings of 13th IAVSD symposium. Supplement to vehicle system dynamics, vol 23, pp 541–553
34. Instructions for maintenance and operation of buildings, devices and rolling stock and traffic on sections treatment of high-speed passenger trains (1996) CRB-393. RF Ministry of Railways (in Russian)

The Aspects of Modernization of Diesel-Electric Locomotives and Platform for Transportation of Railway Switches in Lithuanian Railways

Lionginas Liudvinavičius and Stasys Dailydka

Abstract Electrical locomotives made in Russia, Czechoslovakia and Ukraine have been used primarily in the railways of the former Soviet Union. Russian companies have manufactured the TEM1 and TEM2 diesel-electric shunting locomotives and the TEP-60 and TEP-70 passenger locomotives, while the 2M62 freight locomotives have been manufactured in Ukraine. The ČME2 and ČME3 shunting diesel-electric locomotives, manufactured in Czechoslovakia, were manufactured with analogous control systems of the entire powertrain and electric drive, which have many deficiencies, the most important of which is high fuel consumption. Reducing power transmission losses from the primary power source—the diesel engine—to the wheel sets is critical. JSC Lietuvos geležinkeliai, who owns a fleet of TEM2 and ČME3 typical shunting locomotives, 2M62 freight locomotive and TEP-70 passenger locomotive, made the decision to modernize them. To this end, JSC Lietuvos geležinkeliai established a subsidiary company, Vilniaus lokomotyvų remonto depas UAB, where locomotives were modernized and new locomotives were manufactured for JSC Lietuvos geležinkeliai and railways abroad during the period 2005–2015. Modernization was performed together with scientists from Vilnius Gediminas Technical University (VGTU). Companies participating in the modernization effort included Vilniaus lokomotyvų remonto depas UAB, Czech company CZ Loko a.s., CJSC TMHB Transmashholding, the Briansk machine building plant (Russia), Transmashholding, Caterpillar, MTU, and the Hungarian company Woodward-Mega Kft, among others.

Keywords Diesel-electric locomotive · Traction generator · Static converter · Traction generator load characteristic · Microprocessor (computer) control

---

L. Liudvinavičius (✉) · S. Dailydka
Department of Railway Transport, Vilnius Gediminas Technical University, J. Basanavičiaus g. 28, Vilnius, Lithuania
e-mail: lionginas.liudvinavicius@vgtu.lt

S. Dailydka
e-mail: stasys.dailydka@vgtu.lt

A. Sładkowski (ed.), *Rail Transport—Systems Approach*, Studies in Systems, Decision and Control 87, DOI 10.1007/978-3-319-51502-1_4

1 Introduction

In order to improve the energy characteristics of diesel-electric locomotives 2M62, TEM1, TEM2, ČME2, ČME3 and TEP-70, analysis of their automatic control systems was performed. Based on this analysis, the systems which were proposed by the author and were subsequently implemented enabled significant reductions in fuel consumption and improvements in driver working conditions and traffic safety. The major technical tasks were to improve the general reliability of the locomotives and to reduce operating costs. The author proposed that special electric machines not be used for excitation of traction generators, and in their place using contactless semiconductor static converters. They also proposed replacing the DC machines in auxiliary drive systems, which due to the collector brush unit alone, feature low reliability. An artificial three-phase 3 × 400-V network is created for AC auxiliary electric machines in all modernized and new locomotives manufactured by JSC Lietuvos geležinkeliai. Engineering solutions adopted during modernization of diesel-electric locomotives have not only significantly improved the energy characteristics of typical diesel-electric locomotives, but have also improved driver working conditions and traffic safety through implementation of microprocessor-based control systems for the entire powertrain and electric drive, as well as diagnostic systems for the main parameters. Also very important is the implementation of kinetic energy control systems in locomotives. Typical TEM2 and ČME3 diesel-electric locomotives were not equipped with electric braking systems, and electric braking systems in TEP-70 passenger locomotives were not used due to their limited reliability. Furthermore, typical TEP-70 passenger locomotive did not have reliable heating systems for passenger cars. Vilniaus lokomotyvų remonto depas UAB, in cooperation with companies abroad and VGTU scientists, has not only improved the technical characteristics of modernized (and manufactured new) locomotives, but has also designed and manufactured MPP-61 platforms for the company Voestalpine VAE Legetecha UAB, intended for transportation of assembled railway switches and their components. Tilt equipment manufactured for these platforms has enabled the assembly of switches and their components (oversized cargo) at the site of Voestalpine VAE Legetecha UAB and safe transportation by rail to the installation point. This has led to a considerable reduction in rail traffic delays and the use of fewer service vehicles, as well as a reduction in the installation costs of switches and their components.

2 Aspects of Shunting Locomotive Modernization

In traction vehicles (diesel-electric locomotives, diesel trains, motrices, automotive rail maintenance machines), energy from the internal combustion engine (ICE) is transferred through a powertrain (mechanical, hydraulic, electric) to wheel sets, where it is converted to a locomotive traction force. The primary energy source in traction vehicles is typically a diesel engine. Control of this energy using mechanical power drives is possible only in a forward direction from a diesel engine to the wheel sets. Kinetic energy in traction vehicles using mechanical power drives is not exploited. Energy control using hydraulic and electric power drives is possible in forward and backward directions. In this case, kinetic energy is used, and braking of the traction vehicle is achieved without the use of traditional friction brake pads. Kinetic energy in traction vehicles with an electric power drive is used to create an electric braking mode. Kinetic energy can be converted to heat energy in braking resistors or, in hybrid locomotives, accumulated in an energy accumulator and storage batteries. The primary energy consumers in traction vehicles with electric drives and in electric traction vehicles are the same: DC or AC traction motors. The locomotive traction force $F_k = f(v)$ must be inversely proportional to the movement speed v. This characteristic is called hyperbole, and it enables optimal control of the locomotive across the entire speed range, i.e., using the diesel engine power to its full extent. Once such locomotive traction characteristics are created, the diesel engine capacity in the hyperbole characteristics is constant. To solve this task the traction characteristics $F_k = f(v)$ and traction generator characteristics $U_g = f(I_g)$ must be hyperbolic shape. Diesel engine characteristics $M_D = f(n_D)$ and $N_D = f(n_D)$ are not in line with the theoretical traction requirements. Figure 1 illustrates the structural scheme and characteristics of the autonomous traction rolling stock with DC/DC, AC/DC or AC/AC electric drive structures. To create traction force, three energy transformers are used: diesel engine, power drive and moving axle-wheels. The dependence of the diesel engine

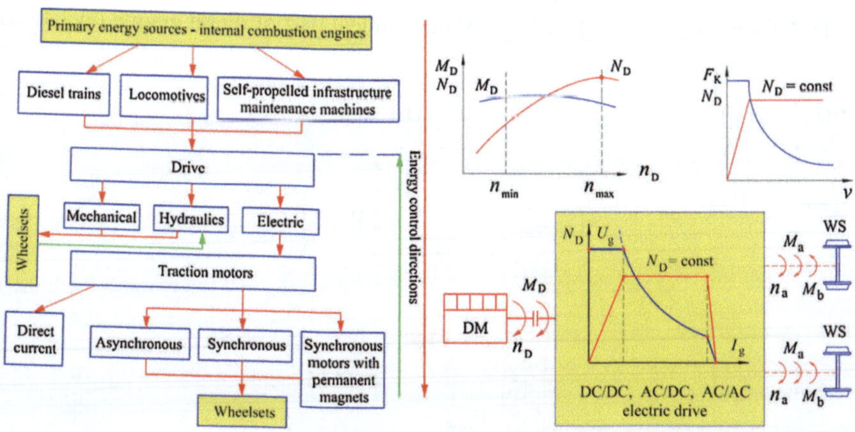

Fig. 1 Traction rolling stock and kinetic energy control of electric drive structural schemes: DM—diesel engine; M_D, n_D—diesel engine crankshaft parameters; M_a, n_a—electric drive transformed axle-wheel parameters; M_D—diesel engine rotation momentum; N_D—diesel engine power; M_b—dynamic braking momentum; U_g, I_g—traction generator parameters; F_K—traction force; v—autonomous traction rolling stock speed; WS—wheel sets

crankshaft capacity $N_D = f(n_D)$ and torque $M_D = f(n_D)$ on the crankshaft speed shows that the torque of the diesel engine has little to do with the speed of the crankshaft, and the capacity of the diesel engine is the greatest with maximum rotation of the crankshaft. The traction force of the diesel locomotive with electric drive depends on the main generator voltage and current, the drive efficiency factor and the moving speed. As the diesel locomotive moves, its speed changes, as does the traction motor current and the traction force in the axle-wheel tangential. Alongside this, traction generator current I_g changes.

To ensure traction generator capacity in the entire locomotive, self-propelled railway track maintenance machines and other traction rolling stock speed range would be constant $P_g = I_g U_g$ = const, and the diesel engine capacity N_D = const would be constant, the shift of the traction generator voltage must be inversely proportional to the current I_g—hyperbole law, i.e., the artificial characteristics of the direct current traction generator $U_g = f(I_g)$ would be related by points BC (see Fig. 2).

An electric drive creates a diesel-electric locomotive traction characteristic which is close to "ideal". To reach this target, the nature of the DC or AC traction generator load characteristic is modified by automatic control systems. To ensure optimal use of diesel engine power across the entire speed range, the traction generator load characteristic $U_g = f(I_g)$ must be shaped so that it is the same as the traction characteristic $F = (f)v$ in terms of its nature (see Fig. 2) [1].

Ideally, the traction generator's load characteristic should be hyperbolic. However, in real diesel-electric locomotive systems, the nature of the generator's external load characteristics is not hyperbolic. The major deficiency of the system of artificial shaping of a traction generator's load characteristics is that the part of hyperbole where N_D = const (the most economical part of diesel engine operation)

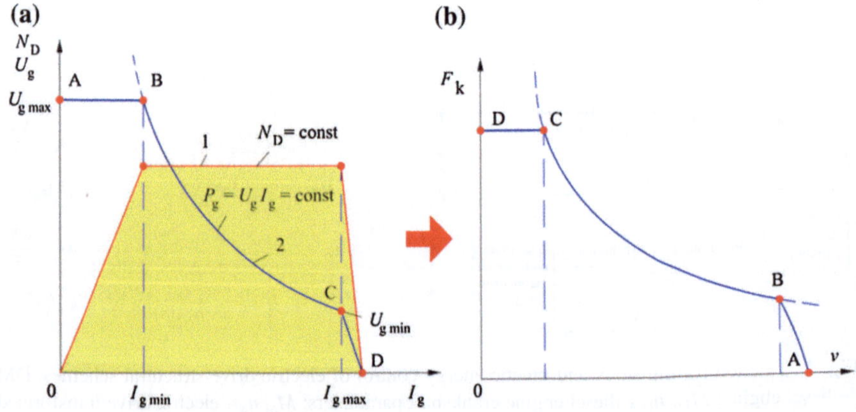

Fig. 2 Diesel-electric locomotive traction generator artificial load (points ABCD), diesel engine power (**a**) and traction characteristics (**b**)

is not shaped precisely. "Distorted" hyperbole shows that the "free" power of such a locomotive diesel engine intended for wheel set rotation is not fully used across the entire locomotive speed range, which leads to increased fuel consumption and affects the economical parameters of a diesel-electric locomotive. Traction generator control systems are modernized by applying microprocessor (computer) control systems. This enables precise shaping of the traction generator load characteristic of a required nature and improves the energy characteristics of diesel-electric locomotives.

2.1 Automatic Control System (ACS) of a Diesel-Electric Locomotive Traction Generator

Theoretical aspects. The automatic control system (ACS) of a diesel-electric locomotive traction generator includes automatic control based on load current [2]. The main factor causing a change in traction generator voltage U_g is its load current I_g. ACS application in diesel-electric locomotives enables the traction generator's load characteristic $U_g = f(I_g)$ to approach a hyperbolic nature [2, 3]. Diesel-electric locomotive traction generator artificial load (points ABCD), diesel engine power (a) and traction characteristics (b) are provided in Fig. 2.

When a diesel-electric locomotive starts moving, the traction generator current is restricted by load characteristic $U_g = f(I_g)$ CD points. The maximum voltage of a traction generator of a DC/DC diesel-electric locomotive (e.g., TE-3, TEM1, TEM2, ČME-3) is restricted by saturation of the exciter (specially designed DC electric generator EG) magnetic system [4]. Artificial load characteristics of a traction generator: (a) when startup currents are limited in all positions of locomotive drive controller LD; (b) when startup currents are not limited in the initial positions of locomotive drive controller LD; (c) when startup current (line AB) is limited regardless of the position of the locomotive drive controller LD (see Fig. 3).

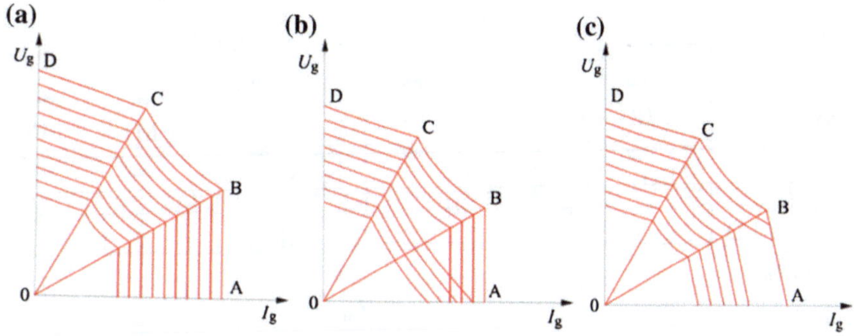

Fig. 3 Artificial load characteristics of a traction generator

2.1.1 DC/DC Locomotive Traction Generator's ACS with a Specially Designed Excitation Generator

ACS of a traction generator based on load current (interference). Such a control principle is applied in DC/DC diesel-electric locomotives such as the TE-3, TEM1, TEM2 and ČME3. The target of automatic control in these locomotives is a separately excited DC traction generator TG [3]. An electric machine ACS is used to control the traction generator TG [5]. The main interference for the traction generator is load current. The load current signal is sent to the exciter's differential winding. A signal summing unit is the excitation system of the electric generator EG. The control setting is given by a driver who changes the positions of the locomotive drive controller LD.

Features of the ACS of a shunting diesel-electric locomotive traction generator. A wiring diagram of DC/DC diesel-electric locomotive TEM2 is provided in Fig. 4a. Traction generator ACS based on load current is used in electric drives of shunting diesel-electric locomotives TEM2 and ČME3. The following electric machines are used to control the traction generator (section C of Fig. 4a): the auxiliary generator BG and the specially designed exciter EG (generator), running in generator mode. The specially designed exciter EG (with divided poles; section

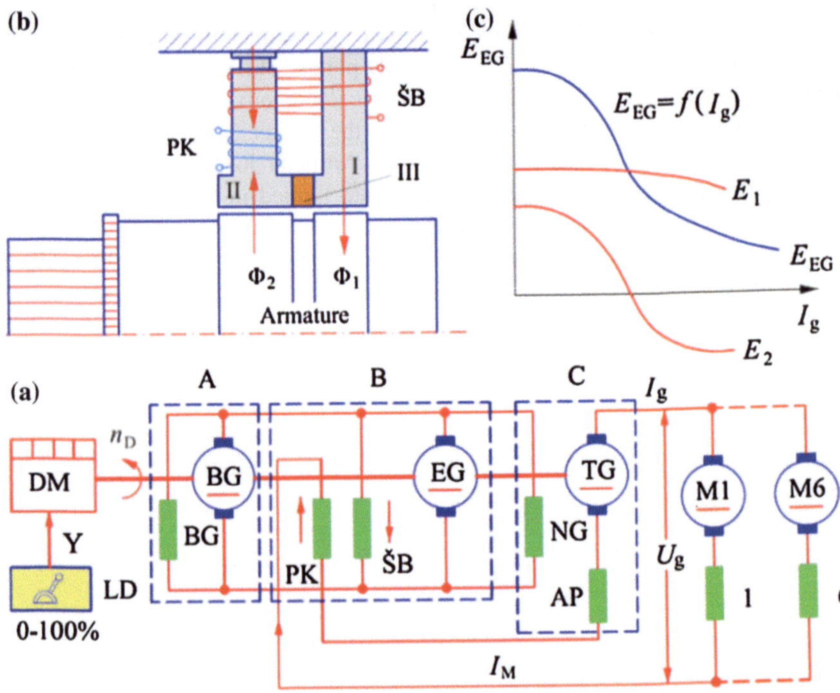

Fig. 4 Wiring diagram of diesel-electric locomotive TEM2 traction generator ACS (**a**), exciter EG (**b**) and load characteristic (**c**)

A of Fig. 4a) and the auxiliary generator BG (section B of Fig. 4a) compose the single assembly of two electrical machines. The traction generator (GP-300B type) independent excitation winding NG is supplied from the EG [3, 6–8]. The exciter is designed with lengthwise-divided poles. The exciter is a special-purpose quadrupole DC generator. Each pole is divided into two separate parts (Fig. 4b): unsaturated I and saturated II. The pole is equipped with differential PK and parallel (main) ŠB excitation windings. The pole parts are divided by a brass plate (III). The exciter's differential PK winding and winding AP of additional poles of the traction generator are connected serially. Figure 4c shows the nature of variation of electromotive forces E_1, E_2 and E_{EG}. The directions of electromotive forces of winding PK and winding ŠB are opposite one another (see Fig. 4a). For exciter EG, it is determined by the equation:

$$E_{\mathrm{EG}} = C_{\mathrm{EG}}\Phi_{\mathrm{EG}}n_{\mathrm{D}}, \tag{1}$$

where E_{EG} is the electromotive force of the exciter, V; C_{EG} is the constant of the exciter; Φ_{EG} is the magnetic flux of the exciter, Wb; and n_{D} is the diesel engine crankshaft speed (rpm). The total magnetic flux Φ_{EG} of exciter EG can be expressed as an algebraic sum of the magnetic fluxes of the saturated and unsaturated parts:

$$\Phi_{\mathrm{EG}} = \Phi_1 \pm \Phi_2. \tag{2}$$

The electromotive force of exciter EG is expressed by the equation:

$$E_{\mathrm{EG}} = C_{\mathrm{EG}}\Phi_1 n_{\mathrm{D}} \pm C_{\mathrm{EG}}\Phi_2 n_{\mathrm{D}} = E_1 \pm E_2 \tag{3}$$

where E_1 is the induced electromotive force of the unsaturated part of the pole, and E_2 is the induced electromotive force of the saturated part of the pole. Load characteristics of exciter EG of (a) diesel-electric locomotive TEM2 (MBT-25/9 type) and (b) traction generator (GP-300B type) at different positions of locomotive controller LD [9] are shown in Fig. 5.

Conclusion: The load characteristics of the TEM2 shunting diesel-electric locomotive traction generator $U_g = f(I_g)$ are similar to the load characteristics of the exciter $U_{\mathrm{EG}} = f(I_g)$ [10].

Analysis of the ACS of the traction generators of the TEM2 and ČME3 shunting diesel-electric locomotives shows that the ACS has many deficiencies: the influence of the traction generator and traction motor winding heating, the influence of hysteresis (flux reversal) of electric machines, and the influence of the auxiliary generator on the traction generator output voltage are not estimated [11–13]. Using an electric machine ACS alone, it is not possible to restrict the generator's maximum voltage and current values, and it is also difficult to coordinate ACS components. The load characteristics of the TEM2 diesel-electric locomotive traction generator (GP- 300B type) are not of a hyperbolic nature (see Fig. 5b). Thus, the power of the diesel engine is not fully exploited. The load characteristics of the traction generator depend on the characteristics of the exciter and the auxiliary generator, which can change during repair due to violation of technology, magnetic

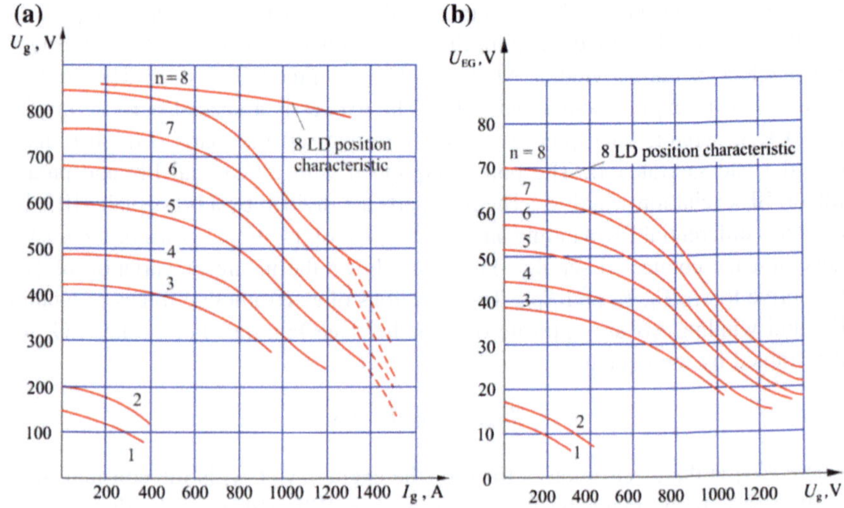

Fig. 5 Load characteristics of (**a**) exciter EG of diesel-electric locomotive TEM2 (MBT-25/9 type) and (**b**) traction generator (GP- 300B type), at different positions of locomotive controller LD

system assembly inaccuracy and other reasons. All these deficiencies cause high fuel consumption in TEM2 and ČME3 shunting diesel-electric locomotives.

2.1.2 DC/DC Locomotive Traction Generator's ACS with Magnetic Amplifier

Analysis of the ACS of a DC/DC diesel-electric traction generator. Control of the traction generator based on loading current and coordination of the control according to interference and deviation is applied to traction generators of DC/DC diesel-electric locomotives.

A DC traction generator ACS with magnetic amplifiers is more advanced than an ACS where special electric machines (exciters) are used. An ACS with a magnetic amplifier is applied for excitation of traction generators of diesel-electric locomotives including TE10, 3TE10M, TE10M, 2TE10B, 2TE10L, TEP-60, M62 and 2M62. A diagram of an ACS with a magnetic amplifier applied for electric drive of a DC/DC diesel-electric locomotive (Fig. 6) shows how the traction generator is automatically controlled in a real locomotive and what kind of feedback exists. The DC traction generator has independent excitation. Its excitation current is changed using a low-power DC generator with a traditional design (i.e., EG).

The EG excitation current is controlled by the output current of a magnetic amplifier (MA), the value of which depends on feedback values of I_g, U_g and diesel engine power N_D. The control object is the DC traction generator and regulator–magnetic amplifier. The main interference of the traction generator is its load

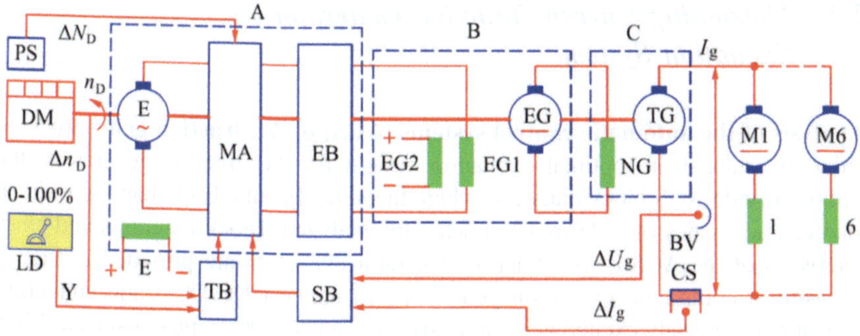

Fig. 6 ACS of DC/DC diesel-electric locomotive traction generator when a low-power single-phase AC generator E is used and magnetic amplifier (MA) is used for processing of control signals: A—low-power single-phase generator part; B—exciter (EG) part; C—traction generator (TG) part

current, whose signal is sent from the direct current sensor (CS). Control windings of the MA signal the summing unit in diesel-electric locomotives. The controlled variable is the traction generator voltage U_g. Voltage sensor BV is used to obtain the feedback signal of generator G voltage U_g. This signal is the main feedback signal. The signal of the traction generator load current from the CS is sent to the selector block (SB), where the selective sampling (of current and voltage) is carried out. Signals are compared to the setpoint signal and signals from the contactless tachometric block (TB) and power sensor (PS). The influence of "free" diesel engine power variation on the ACS is assessed according to diesel engine crankshaft speed n_D, traction generator voltage U_g and I_g deviation signals, correspondingly acting to regulator—MA control windings. Figure 6 shows the ACS of a DC/DC diesel-electric locomotive traction generator when a low-power single-phase AC generator (E) is used and an MA is used for control signal processing. Two low-power electric power machines, AC auxiliary generator E and exciter EG operating in generator mode are used to control the traction generator.

New designations of Fig. 6: LD—locomotive drive controller; n_D—diesel engine crankshaft speed; TG—DC traction generator; EG—exciter; M1–M6—direct current traction motors; 1–6—direct current traction motors excitation winding; BV—DC traction generator voltage sensor; SB—selective block; TB—tacho block; NG—DC traction generator separately excited winding; PS—diesel engine power sensor; CS—current sensor; Δn_D—diesel engine crankshaft speed deviation signal; ΔU_g—DC traction generator voltage U_g deviation signals; ΔI_g—DC traction generator current I_g deviation signal; EB—excitation control block; MA—magnetic amplifier; EG1, EG2—exciter EG excitation windings; ΔN_D—diesel engine power deviations signal; Y—control signal.

2.2 *Alternating Current Traction Generator Excitation System*

Analysis of the automatic control systems (ACS) of AC traction generators. In order to establish an optimal excitation system for AC traction generators, the author provide ACS block diagrams where theoretical methods of shaping artificial load characteristics of a desired character are analyzed. Block diagrams and characteristics of the ACS of AC traction generator excitation are provided in Fig. 7: (a) using a compound excitation generator, (b) using a DC/DC static converter, (c) using a separately excited generator, (d) using an AC/DC static converter. When DC/DC or AC/DC static semiconductor converters are used for excitation of AC traction generators, artificial load characteristics of a desired character can be shaped (see Fig. 7b, d). When AC traction generators are excited using the diagram of the separately excited generator provided in Fig. 7c, the load characteristics can be shaped from natural to artificial by varying the excitation current I_E according to the function provided by the green line (see Fig. 7c). The ACS diagram of an AC/DC diesel-electric locomotive traction generator when a controlled rectifier (CR) is used (a), and time diagrams applying a TIR are provided in Figs. 7a, b and 8. Using a compound excitation generator, artificial load characteristics are shaped by magnetic fluxes of serial D1–D2 and parallel winding F1–F2 of natural characteristics. ACS control signals Y of the AC traction generator excitation control ON (t_1) and OFF (t_2) durations of the TIR (IGBT transistor) connected to the circuit

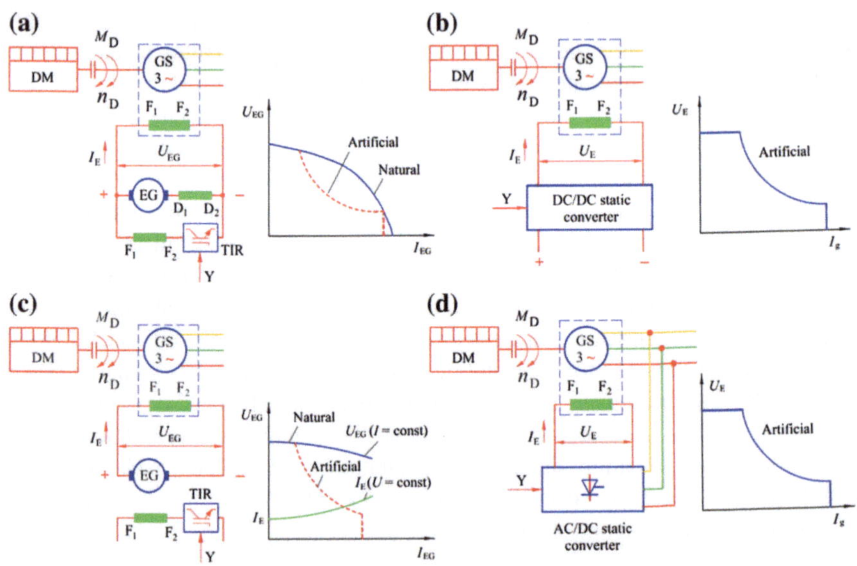

Fig. 7 Block diagrams and characteristics of the ACS of AC traction generator excitation **a** using a compound excitation generator, **b** using a DC/DC static converter, **c** using a separately excited generator, and **d** using an AC/DC static converter

of parallel winding F1–F2 (see Figs. 7a and 8b). When the pulse width modulation principle is applied by varying ON and OFF durations of the IGBT transistor, excitation current I_E and magnetic flux Φ_M of the EG generator change, and consequently the EG generator output voltage U_{EG} changes as well. By changing ON and OFF durations of excitation winding F1–F2, the average value of the excitation current is regulated, and load characteristics of required character are shaped. Time diagrams of AC traction generator excitation current I_E control, when the TIR is used, are provided in Fig. 8b. The U_{EG} diagram versus ON (t_1) and OFF (t_2) durations of the IGBT transistor clearly shows that in order to increase the voltage of the locomotive's traction generator, the IGBT transistor must be in ON status longer (pulse width and average voltage U_{EG} are higher in this case), and conversely, in order to reduce the voltage of the traction generator, the IGBT transistor must be in ON status for a shorter period (in this case, pulse width and average voltage U_{EG} are lower). ON (t_1) and OFF (t_2) durations of the IGBT transistor are controlled by sending (from the control unit) corresponding pulses to the IGBT transistor gate. Figure 8a shows the ACS diagram of an AC/DC diesel-electric locomotive traction generator when a low-power single-phase AC generator (E) and controlled rectifier (CR) are used.

A single-phase controlled rectifier consists of thyristors T1 and T2 and diodes D1 and D2. Such an analogue ACS system of an AC traction generator is used in typical diesel-electric locomotives TEP-70 produced in Russia. This system is more advanced than the ACS used in the TEM2 and ČME-3 shunting locomotives and the ACS with an MA applied in the control systems of diesel-electric locomotives TE10, 3TE10M, TE10M, 2TE10B, 2TE10L, TEP-60, M62 and 2M62.

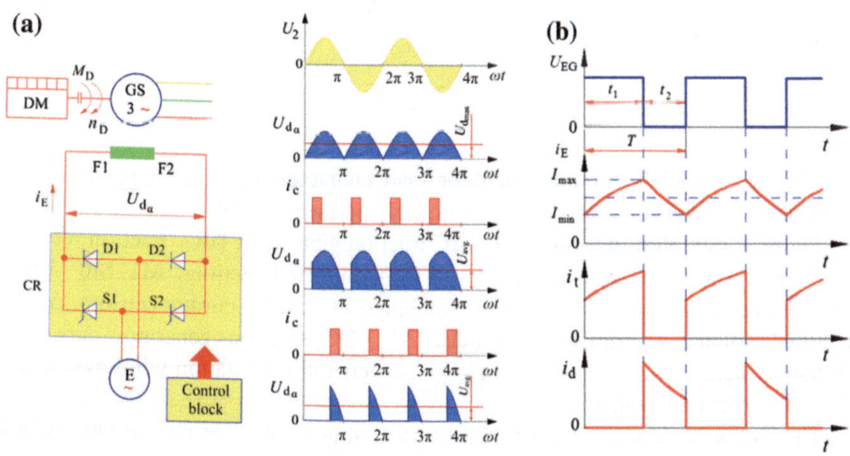

Fig. 8 ACS diagram of AC/DC diesel-electric locomotive traction generator **a** when a controlled rectifier (CR) is used, and **b** time diagrams applying transistor impulse voltage regulator (TIR): GS—synchronous traction generator; E—low-power single-phase generator; I_E—traction generator excitation current; CR—controlled rectifier; D1, D2—diodes, S1, S2—thyristors; $U_{d\alpha}$—thyristor rectified voltage; i_c—thyristor control signal; T—pulse period; i_t—IGBT transistor current

Only one low-power electric machine is used to control the traction generator: a single-phase AC generator. Load characteristics of the traction generator $U_g = (f)I_g$ are shaped by changing the thyristor control angle α in the CR. By changing the thyristoric control angle α [14], the thyristors may be opened by corresponding time moments, and the total length of time during which the thyristor releases the current may be changed and in this way is possible a smooth regulation of AC generator's existation current value. By increasing the opening angle α of the thyristor, the thyristor is opened later and later; therefore, the average rectified voltage U_{avrg} will be smaller. By increasing the control angle α of the thyristors, the curves of the rectified voltage are reduced, and the abscissa load is restricted as well as the average rectified voltage U_{avrg}. The greatest value of the rectified average voltage U_{avrg} is reached when $\alpha = 0$, and equals zero when $\alpha = 180°$. With an active load, the value of the rectified voltage is calculated as follows [15]:

$$U_{d\alpha} = \frac{1}{\pi}\int\limits_{0}^{\alpha} \sqrt{2}U_2 \sin\omega t d\omega t = \frac{2\sqrt{2}}{\pi}U_2\frac{1+\cos\alpha}{2}, \quad (4)$$

Here, U_2 is the effective value of the source voltage. The effective value of the load current is calculated as follows:

$$I_{d\alpha} = I_{da}\sqrt{\frac{\pi - \alpha}{\pi}}, \quad (5)$$

Here, $I_{d\alpha}$ is the effective value of the load current, where $\alpha = 0$. The average current values of the thyristors S1 and S2 and the diodes D1 and D2 are calculated as follows.

$$I_T = \frac{\pi - \alpha}{2\pi} I_{d\alpha}. \quad (6)$$

Shaped (artificial) traction generator load characteristics $U_g = (f)I_g$, using the controlled rectifier approach the hyperbole shape (see Fig. 9b).

Figure 9 **new designations**: TG—AC traction generator; UR—traction rectifier; E—low-power single-phase generator; CR—controlled rectifier; M1–M6—direct current traction motors; BV—rectified voltage sensor; CS—current sensors; ΔU_d—rectified voltage U_d deviation signals; ΔI_g—DC traction generator current I_g deviation signal; E—low-power single-phase generator excitation windings; $U_{d\alpha}$—thyristors rectified voltage.

The author have analyzed the ACS of traction generators of typical DC/DC and AC/DC diesel electric generators and have established that shaped load characteristics of the traction generators are not of a hyperbolic shape. Due to this, a significant portion of the transformed diesel engine power is lost. Better results can be obtained by using microprocessor (computer) control systems for the traction

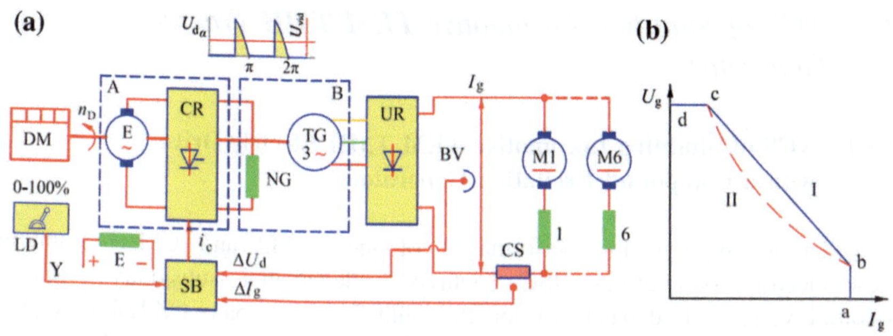

Fig. 9 ACS of DC/DC diesel-electric locomotive traction generator, when low power single-phase AC generator E is used and magnetic amplifier MA is used for control signal processing: A—low-power single-phase generator part; B—traction generator (TG) part

generator ACS. All conclusions obtained from the analysis of DC/DC and AC/DC diesel-electric locomotive traction generator ACS were assessed in modernizing and manufacturing new JSC Lietuvos geležinkeliai locomotives.

Modernization of diesel-electric locomotive:

1. **Modernization of diesel-electric locomotive series 2M62 (ST 44)**. Diesel-electric locomotive series 2M62 is for heavy cargo transportation. Available for 1435/1520-mm gauge with a maximum speed up to 100 km/h. Year of beginning of modernization: 2005. Number delivered: 58 units. Delivery countries: Lithuania and Latvia.
2. **Production of shunting locomotive series ČME-3M**. Shunting locomotive series ČME-3M is for heavy cargo transfers. Available for 1435/1520-mm gauge with a maximum speed of 90 km/h. Year first produced: 2008. Number delivered: 22 units. Delivery country: Lithuania.
3. **Production of shunting locomotive series TEM TMH**. Shunting locomotive series TEM TMH is for heavy transfer and main train use. Available for 1435/1520-mm gauge with a maximum speed of 100 km/h. Year first produced: 2008. Number delivered: 75 units. Delivery countries: Lithuania, Estonia and Russia.
4. **Production of shunting locomotive series TEM 33**. Shunting locomotive series TEM 33 is for heavy transfer. Available for 1435/1520-mm gauge with a maximum speed of 100 km/h. Year first produced: 2013. Number delivered: 1 unit. Delivery country: Russia.
5. **Production of shunting locomotive series TEM 35 (hybrid)**. Shunting locomotive series TEM 35 is for mid–heavy transfer. Available for 1435/1520-mm gauge with a maximum speed of 100 km/h. Year of first produced: 2013. Number delivered r: 1 unit. Delivery country: Russia.
6. **Modernization of passenger locomotive series TEP-70M**. Modernized passenger locomotive series TEP-70M is for main line transfers. Available for 1520-mm gauge with a maximum speed of 160 km/h. Year of beginning of modernization: 2015. Number delivered: 1 unit. Delivery country: Lithuania.

2.3 ACS of Shunting Locomotive TEM TMH Traction Generator

2.3.1 ACS of Shunting Locomotive TEM TMH Electric Drive with a Compound Excitation Generator

In order to improve the performance of the TEM2 and ČME3 shunting diesel-electric locomotives, JSC Lietuvos geležinkeliai subsidiary Vilniaus lokomotyvų remonto depas UAB, together with Czech company CZ Loko a.s. [16] and the Briansk machine building plant (Russia), completed their modernization (new diesel-electric locomotives were produced): the diesel engine was replaced, the traction generator was changed from DC to AC, an artificial three-phase network for supplying asynchronous motors of auxiliary equipment was created, and computer control and a diagnostic system of the traction or electrodynamic braking was initiated for the entire power assembly. DC traction engines of electric drive diesel-electric locomotive TEM TMH are supplied with AC traction generator rectified voltage. The speed and torque of the traction motors are varied according to the setpoint (position) of the traction controller, and by changing the value of rectified voltage U_d in the terminals of the traction motor armature windings. A block diagram of the electric drive of diesel-electric locomotive TEM TMH, when a compound excitation generator EG is used for excitation of the traction generator, is provided in Fig. 10. The excitation current value of the synchronous traction generator is changed by changing the output voltage value of the compound excitation generator.

Compound excitation of DC machines is provided by two field windings: a series winding and a shunt (or separate) winding F_1–F_2. The shunt winding provides a magnetic excitation flux in the machine that corresponds to the nominal voltage under no load conditions. The series winding D_1–D_2 is intended for

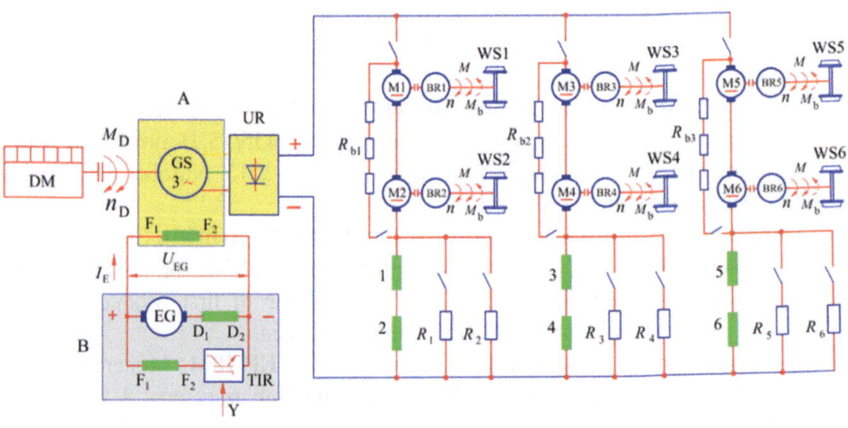

Fig. 10 Block diagram of the electric drive of the TEM TMH diesel-electric locomotive

automatic voltage regulation of the machine voltage depending on the load. Machines of this type are called compound excitation machines.

The output voltage value of the compound excitation generator when the ICE speed changes is controlled by a TIR, connected in series with excitation windings F_1–F_2. The TIR processes information from the traction generator current I_g, voltage U_g, crankshaft n_D, and traction engine armature speed sensors, and corrects the output voltage value of the compound excitation generator accordingly. Thus, the excitation current values of the synchronous traction generator (GS) and traction generator output voltage value also change. By increasing the excitation current of the synchronous traction generator or diesel engine crankshaft speed, the output voltage (rectified) of the traction generator–rectifier—and simultaneously the locomotive speed and traction force—is increased.

New designations of Fig. 10: GS—synchronous traction generator (A); M1–M6—DC series excited traction motors; UR—three-phase rectifier; EG—compound excitation generator (B); D1–D2—series winding; F1–F2—shunt winding; R_1–R_6—resistors; 1–6—traction motor excitation windings; I_E—synchronous traction generator excitation current; U_{EG}—compound excitation generator output voltage; TIR—transistor-switching synchronous traction generator excitation current regulator; Y—TIR control signal; BR1–BR6—wheel set speed sensor; WS1–WS6—wheel sets 1–6.

According to the block diagram of the electric drive of diesel-electric locomotive TEM TMH shown in Fig. 10, new diesel-electric locomotives up to No. 036 were produced where compound excitation generators were used for excitation of the traction generator. In order to further improve the energy characteristics of the TEM TMH, future TEM TMH locomotives were manufactured using a static semiconductor converter proposed by the author for excitation of the traction generator. Examples of the shunting diesel-electric locomotives are shown in Fig. 11.

Fig. 11 Examples of shunting diesel-electric locomotives: **a** six axles; **b** eight axles

The decision was made to manufacture new TEM TMH series diesel-electric locomotives based on the diagram in Fig. 11a. The main purpose of the TEM TMH series with electric drive is to distribute freight cars in distribution stations. The maximum speed of the TEM TMH series is 100 km/h. The driver's cabin is provided in the back of the locomotive. A computer (microprocessor) control system is provided for automatic control of the diesel engine generator and regulation of the powertrain parameters.

The TEM TMH diesel-electric locomotive is a 126-ton (t) locomotive with six drive axes with the main axis formula Co'–Co', and an axial rail load of 21 t. The locomotives are equipped with 1455-kW (CAT 3512B) or 1550-kW (CAT 3512C HD) Caterpillar diesel engines. These TEM TMH diesel-electric locomotives are equipped with typical diesel-electric locomotive bogies (TEM18) with six traction motors (324 kW of DC serial excitation, type ED 118AU1). The main power source in the TEM TMH is the powertrain, which includes the diesel engine-generator Locat 3512/631. The CAT 3512B diesel engine (CAT 3512C HD) and the Siemens 1FC2 631-6BO29T AC traction generator (Siemens 1FC2 631-6T in the case of CAT 3512C HD) are rigidly connected by elastic coupling.

Other important parts of the diesel engine generator include the auxiliary 120-kVA AC generators such as the Siemens 1FC2 631-6BO29P or Siemens 1FC2 631-6P connected with traction generators into one electric machine structure. Independently from the main locomotive computer control system, diesel engine parameters are monitored and controlled by an autonomous Caterpillar electronic control system.

This system controls the startup of the diesel engine, prevents power increase up to critical values and limits it in the case of overloading. All diesel engine parameters are controlled; the diesel engine is stopped in case of overloading up to critical values. The arrangement of TEM TMH assemblies is shown in Fig. 12.

Assemblies of diesel electric locomotive TEM TMH are arranged so that axial rail loads would not differ more than 3 %. Air unit and air system control elements are arranged at the front hood of the TEM TMH.

Further installations include the ICE cooling unit, the compressor and its cooler for auxiliary equipment, cabinet of electric equipment with frequency converters, cooling fan of first bogie traction engines, coolant reservoir and the ICE. The diesel engine installed in the common frame of the TEM TMH diesel-electric locomotive is rigidly connected to the AC traction generator and exciter. At the back side, a high-voltage chamber is installed with power circuit electrical devices: startup circuit contactors, traction motor reversing equipment and two units of electric braking resistors.

The driver's cabin is provided for locomotive control. The main technical characteristics of the TEM TMH locomotive are provided in Table 1.

The TEM TMH diesel-electric locomotive is equipped with a flange lubrication system, ensuring lubrication of flanges of the first wheel set in the driving direction. A diagram of a TEM TMH bogie is shown in Fig. 13.

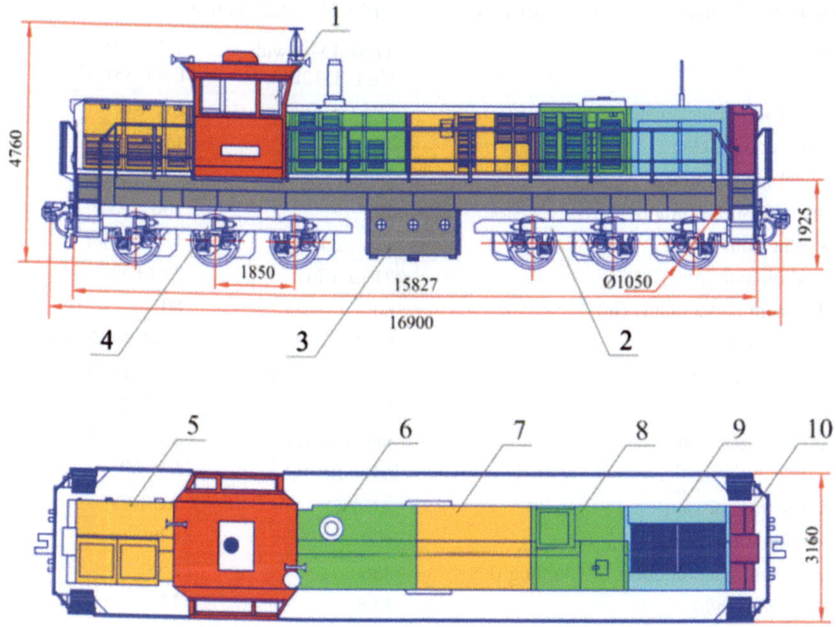

Fig. 12 Arrangement of TEM TMH diesel-electric locomotive assemblies: 1—cabin; 2, 4—bogies; 3—fuel tank; 5—electric power equipment (contactors, etc.); 6—exciter connected by a shaft to the AC traction generator; 7—diesel engine; 8—electric equipment; 9—cooling sections; 10—air equipment

2.3.2 Shunting Locomotive TEM TMH Electric Drive with a Static Converter

The author proposed the use of a semiconductor static converter supplied from the generator of auxiliary equipment for excitation of AC traction generators (see Fig. 14) [17, 18]. A block diagram of the computer control and diagnostics of power assembly of AC/DC diesel-electric locomotive TEM TMH is provided in Fig. 14. Automatic control of electric drive parameters of the TEM TMH is performed by the computer control system, which receives diagnostic signals from sensors supplying continuous information about deviations of diesel engine crankshaft speed, pressure in the main tanks, temperatures in the main cooling circuit of the ICE and traction generator bearings, as well as braking resistor unit RB1, currents of the traction motors of groups I, II and III, rectified current I_d, traction motor excitation currents (in the mode of electrodynamic braking), traction generator rectified voltage U_d and wheel set speed. The onboard computer processes data of the rectified current I_d, voltage U_d, crankshaft n_D, and speed sensors of traction motor armature and adjusts, respectively, output voltage of the static

Table 1 The main technical characteristics of the TEM TMH locomotive

Title	TEM TMH with CAT 3512B	TEM TMH with CAT 3512C HD
Wheel set arrangement	Co'-Co'	Co'-Co'
Rated power (kW)	1455	1550
Wheel gauge (mm)	1435/1520	1435/1520
Maximal speed (km/h)	100	100
Weight (tons)	126 (±3%)	126 (±3%)
Load on axle (tons)	21 (±3%)	21 (±3%)
Minimum curve radius (m)	80	80
Diesel engine type	CAT 3512B	CAT 3512C HD
Traction alternator	Siemens 1FC2	Siemens 1FC2
	631-6B029T	631-6T
Auxiliary alternator	Siemens 1FC2	Siemens 1FC2
	631-6B029P	631-6P
Power transmission	AC/DC	AC/DC
Traction effort of coupling (kN)	400	436
Permanent traction effort of coupling (kN)	266	259
At speed (km/h)	13.5	14.3
Control circuit voltage (V)	24	24

converter by signal Y_2. An increase in the synchronous traction generator's excitation [19–21] current or speed of the diesel engine crankshaft also increases the output voltage (rectified) of the traction generator-rectifier and, simultaneously, the locomotive speed and traction force.

New designations of Fig. 14: A—locomotive driver signal converter; G1–G2—brushless synchronous traction generator; M1–M2—I group of DC traction motors; Δn_1–Δn_2 traction motor M1 and M2 speed deviation signals; BR1—traction motor's M1 speed sensor; BR2—traction motor's M2 speed sensor; BV—rectified voltage sensor; CS1–CS2—current sensors; ΔU_d—rectified voltage deviation signals; ΔI_d—traction motors current I_d deviation signal; A/D—analog-digital converters; BT1–BT3—temperature sensors; R_{T1}–R_{T3}—temperature sensors/thermistors; ΔT_1—traction generator winding temperature deviation signal; ΔT_2—traction rectifier temperature deviation signal; ΔT_3—braking resistor RB1 temperature deviation signal; MV—braking resistor cooling fan; Y1–Y3—control signals; WS1–WS2 wheel sets 1 and 2.

Traction generator load and diesel engine power characteristics of the new diesel-electric locomotive TEM TMH are provided in Fig. 15, which shows how the traction generator's voltage U_g changes at traction power of 50% (P_e = 558 kW, 1372 rpm) and 100% (P_e = 1200 kW, 1800 rpm) and current I_g.

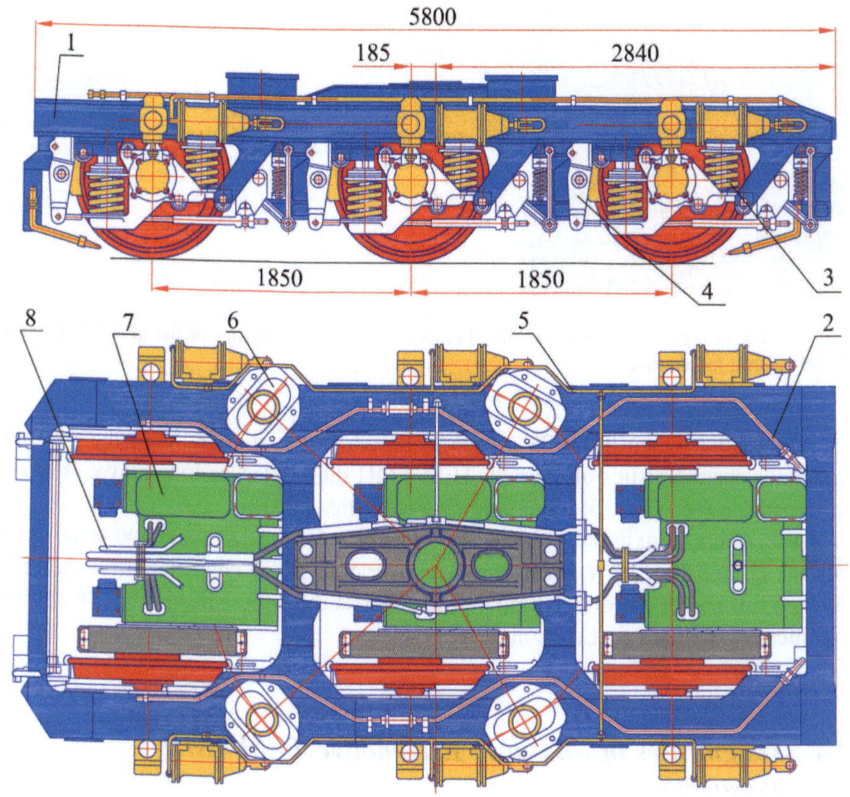

Fig. 13 Diagram of a bogie of diesel-electric locomotive TEM TMH: 1—lateral longitudinal beams; 2, 5—tubes; 3—external springs; 4—brake system pull rod; 6—support; 7—DC traction motor; 8—traction motor winding connection cables

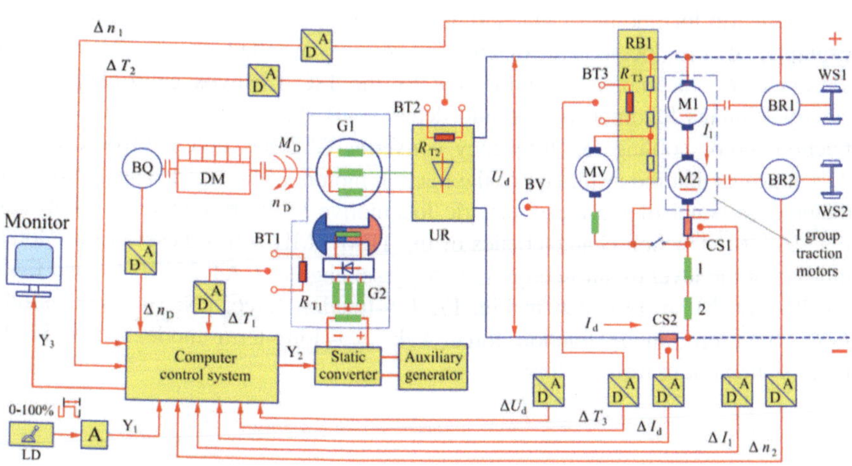

Fig. 14 Block diagram of computer control and diagnostics of power assembly of AC/DC diesel-electric locomotive TEM TMH

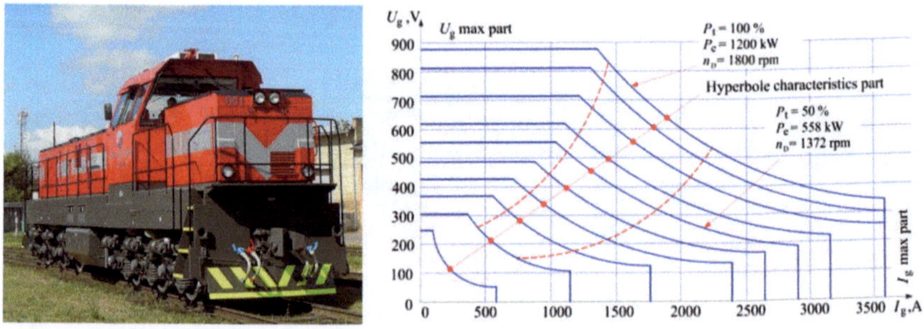

Fig. 15 Traction generator load and diesel engine power characteristics of the new TEM TMH diesel-electric locomotive

We see that when the ACS uses a static converter, the nature of the traction generator characteristics of the new TEM TMH diesel-electric locomotive is close to hyperbole (see Fig. 15; hyperbole characteristics part).

Furthermore, maximum voltage U_g (see Fig. 15; U_g max part) and maximum current I_g (see Fig. 15; I_g max part) are restricted in all characteristics. All this enables full use of the diesel engine power across the entire speed range, and thus significantly reduces fuel consumption.

2.4 Diagnostics of Shunting Locomotive TEM TMH Parameters

Controlled parameters of the technical state of an AC/DC diesel-electric locomotive. A "locomotive diagnostics" device is installed in the AC/DC diesel-electric locomotives, which signals the error codes in case of failures. For example, if the locomotive diagnostics device installed in the TEM TMH diesel-electric locomotive manufactured by the JSC Lietuvos geležinkeliai subsidiary Vilniaus lokomotyvų remonto depas UAB shows error code no. 846, the traction motor cooling fan frequency converter GS2 has failed. Error code no. 227 indicates that the insulation of the traction motors is damaged. The control and diagnostics panel of the diesel-electric locomotive with error codes is shown in Fig. 16 [16]. Traction characteristics of the TEM TMH series diesel-electric locomotives at the level of auto-coupling with diesel engines CAT 3512B (a) and CAT 3512C HD (b) are provided in Fig. 17. F_t—traction force, kN; v—speed, km/h; $F_t\infty$—long-term mode traction force, kN; v_∞—long-term mode speed, km/h; F_{tpmax}—maximum traction force.

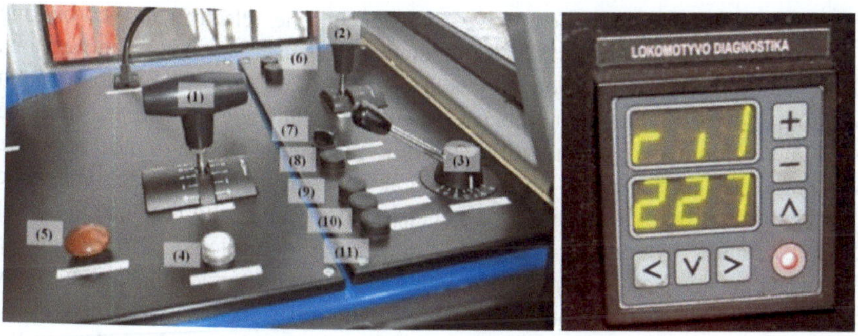

Fig. 16 Diesel-electric locomotive TEM TMH control and diagnostics panel with error codes

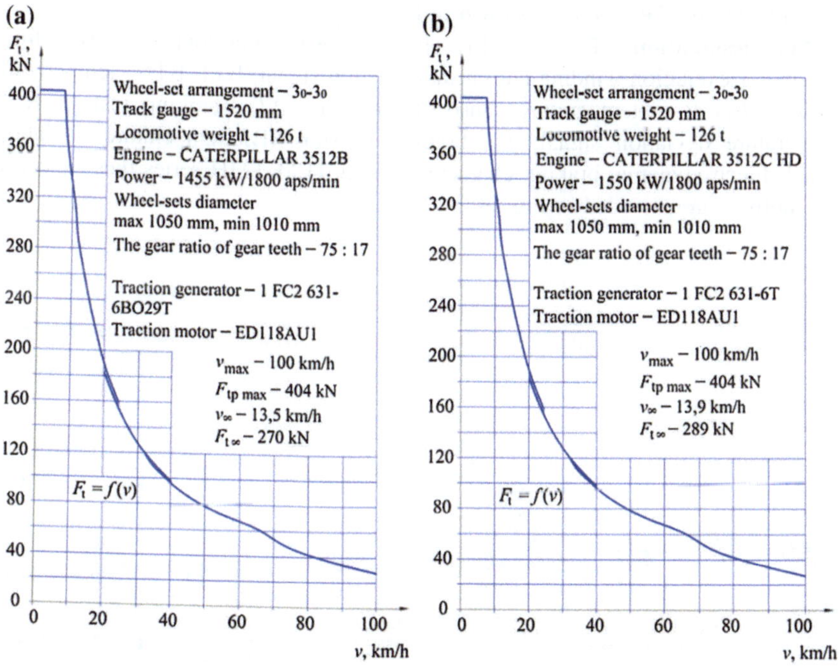

Fig. 17 Traction characteristics of TEM TMH series diesel-electric locomotives at the level of auto-coupling with diesel engines CAT 3512B (**a**) and CAT 3512C HD (**b**)

2.5 Electrodynamic Braking System of Shunting Locomotive TEM TMH

The locomotive braking controller (LB) provided in both driver control panels is intended to control driving and electrodynamic brakes. Based on the position of this controller, the computer control system sets the ICE speed, strength of AC excitation current, and locomotive power. In the electrodynamic braking (EDS) mode, traction motors M1–M6 operate as separately excited DC generators. The traction circuit is switched over with the help of contactors so that excitation windings 1–6 of all traction motors are connected serially and are supplied from the AC traction generator through the traction rectifier. Traction motor armature windings are connected to individual braking resistors (R_b). Electromagnetic braking torque (braking force) may be changed (1) by changing the braking circuit current I_b value (by splitting braking resistor into separate steps), (2) by changing the magnetic flux value of traction engines running in generator mode, or (3) by changing both parameters simultaneously. Braking resistors are intensively cooled by an axial fan driven by a DC motor, which is supplied by a voltage drop in the braking resistor. A diagram of computer control of a TEM TMH AC/DC diesel-electric locomotive electric drive in EDS mode, which was developed and implemented by the author, is shown in Fig. 18.

New designations of Fig. 18: LB—locomotive braking controller; ΔI_E—traction motor excitation winding current I_E deviation signal; BT1–BT5—temperature sensors; R_{T1}–R_{T5}—temperature sensors/thermistors; ΔT_4—braking resistor RB2 temperature deviation signal; T_5—braking resistor RB3 temperature deviation signal. Electrodynamic braking characteristics of TEM TMH series diesel-electric locomotives are provided in Fig. 20.

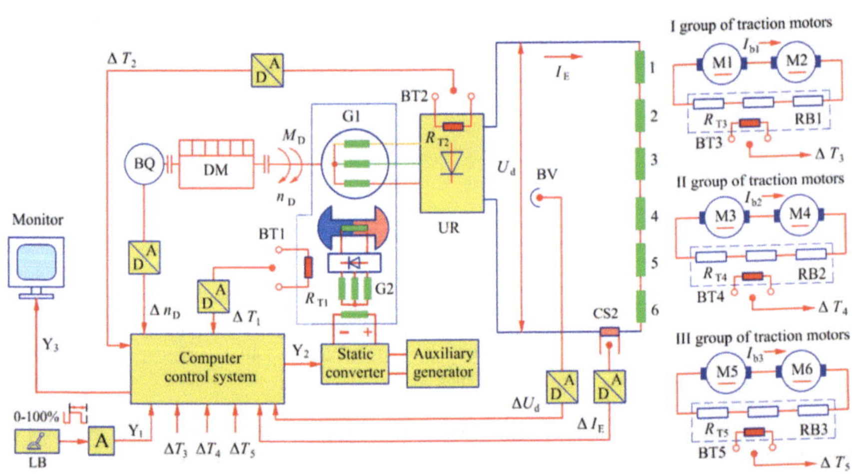

Fig. 18 Diagram of computer control of AC/DC diesel-electric locomotive TEM TMH electric drive in electrodynamic braking mode

2.6 *Auxiliary Equipment Power Supply System of Shunting Locomotive TEM TMH*

Auxiliary equipment of the diesel-electric locomotive is supplied by 120-kVA AC generators (Siemens 1FC2 631-6BO29P or 1FC2 631-6P) connected with traction generators to one electric machine structure. An auxiliary AC generator is intended to create an onboard locomotive three-phase network of 3 × 400 V. A block diagram of the supply of auxiliary equipment of the TEM TMH diesel-electric locomotive created by the author is provided in Fig. 19. The use of industrial asynchronous engines to drive auxiliary equipment (compressors, fans, etc.) is enabled by the system where the rotor of auxiliary synchronous generator G3–G4 is driven by the diesel engine. The value of voltage induced in windings of the synchronous generator stator U_1 depends on the ICE crankshaft speed and generator excitation flux. Varying the diesel engine crankshaft speed also varies the voltage and frequency values of the synchronous generator G3-G4 network, i.e., U_1 = var, f_1 = var.

Purpose of frequency converters. Frequency converter 1 is intended for supply of asynchronous motors M1 and M2 of cooling fans of traction motors of the first and the second bogies. Frequency converter 2 is intended for supply of asynchronous motors M3 and M4 of the cooling fans of the main and auxiliary ICE cooling circuits. Frequency converter 3 is intended for supply of compressor drive asynchronous motor M5. Frequency converter 4 is intended for supply of compressor cooler asynchronous motor M6. An artificial three-phase supply system must be created for stabilization of voltage and frequency values of the synchronous generator G3–G4 network in the TEM TMH diesel-electric locomotive, which would correspond to industrial equipment asynchronous motor supply system parameters (400-V AC, 50 Hz). With that intention, the supply system for the

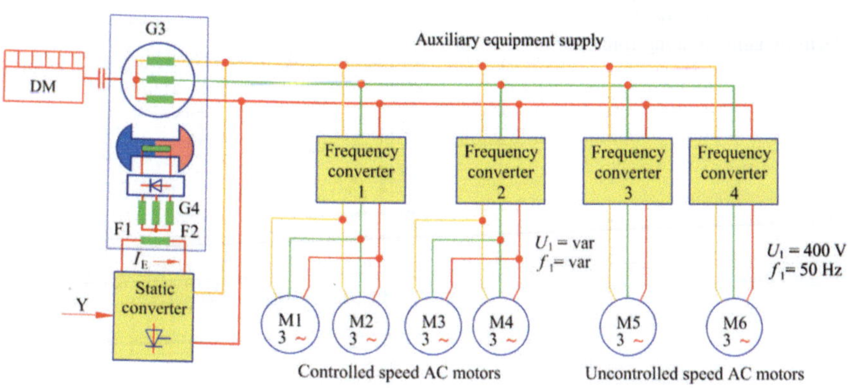

Fig. 19 Block diagram of supply of auxiliary equipment of TEM TMH diesel-electric locomotive

asynchronous motors of auxiliary equipment in the diesel-electric locomotive is developed with semiconductor frequency converters 1–5 connected to the synchronous generator output voltage network (Fig. 20).

The frequency converters of the auxiliary equipment supply of the TEM TMH diesel-electric locomotive are shown in Fig. 21. Semiconductor frequency converters can be divided into two groups. The first group includes frequency converters with variable power supply voltage and frequency. These are frequency converters 1 and 2, and they supply asynchronous motors M1 and M2 of the cooling fans of traction motors of the first and second bogies, as well as asynchronous motors M3 and M4 of the cooling fans of the main and auxiliary ICE cooling circuits. Their speed is varied by adjusting the output parameters of the respective frequency converter—voltage and frequency of the power supply. The second group includes frequency converters with constant power supply voltage and frequency. These are converters 3 and 4, and the speed of their supplied asynchronous motors M5 and M6 is not regulated. The speed of asynchronous fan motors M1, M2, M3 and M4 is adjusted based on information supplied from temperature sensors of the controlled objects. If, for example, windings of the first or second group of traction motors are overheating due to increased armature circuit currents, temperature sensors send data to the onboard computer, which in turn adjusts frequency converter output parameters (power supply voltage and frequency), and at the same time the speed of asynchronous motors M1, M2, M3 and M4 increases. The increased speed of M1 and M2 means a higher flow of cooling air to the traction motors. Increased speed of M3 and M4 means a higher flow of

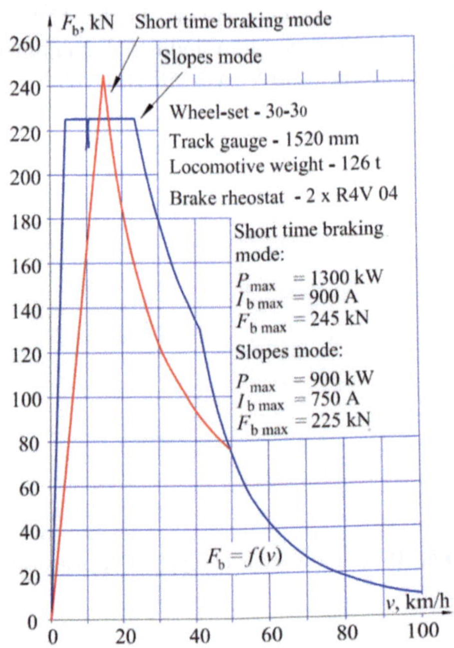

Fig. 20 Electrodynamic braking characteristics of TEM TMH series diesel-electric locomotives: F_b—braking force, v—speed, km/h; F_{bmax}—maximum electrodynamic braking force, kN

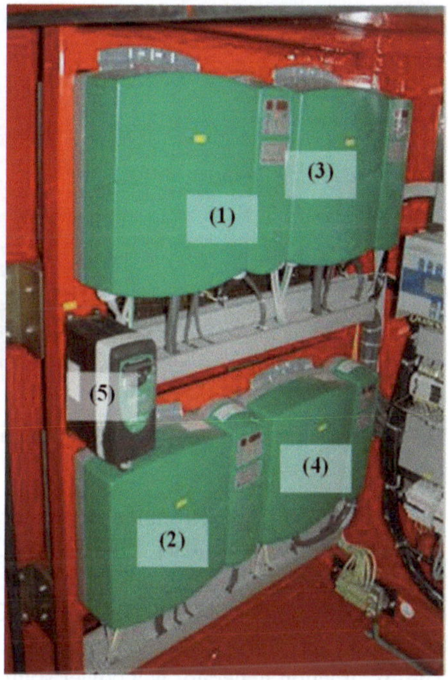

Fig. 21 Frequency converters of auxiliary equipment supply of TEM TMH diesel-electric locomotive

cooling air to the ICE cooling radiators. Cooling of the traction motors of the first and the second bogies in EDS is equivalent to "traction" mode. If at least one cooling fan of traction motors of the first or second bogie fails, then the locomotive power is limited. In such a case, only emergency travel to the repair depot is allowed. The onboard computer in this case limits traction engine currents and diesel-electric locomotive power.

2.7 *Description of the AC Diesel-Electric Locomotive TEM LTH*

The main purpose of the TEM LTH diesel-electric locomotive (with CAT C27 709-kW diesel ICE) with electric drive is the distribution of freight cars in distribution stations. The maximum speed of the TEM LTH series locomotives is 90 km/h. A computer (microprocessor) control system is provided for automatic control of the diesel engine generator and regulation of the powertrain parameters. The TEM LTH is an 86-t locomotive with four drive axles, axle formula B'o–B'o, and an axial rail load of 22.5 t. The locomotive is equipped with the Caterpillar C27 diesel engine. The TEM LTH uses typical diesel-electric locomotive bogies (TEM103) made in Ukraine, with four AC traction motors (AD 917) installed. The main power source in the TEM LTH diesel-electric locomotive is the powertrain,

Fig. 22 General view of TEM LTH diesel-electric locomotive

which includes a diesel engine–generator. The CAT C27 diesel engine and the Siemens 1FC2 631-6BO29T 1350-kVA AC traction generator are rigidly connected via elastic couplings. An important part of the diesel engine-generator is the Siemens 1FC2 631-6BO29P 120-kVA AC auxiliary generator. The traction generator and auxiliary generator are joined to a single electric machine structure. The AC traction generator is a power source intended to supply the traction electric system, and the auxiliary AC generator is intended to create an onboard locomotive three-phase network of 3 × 400 V. The static converter is intended for excitation of the GA1 AC traction generator. Another part of the powertrain is the generators for battery charging and starters to start the diesel engine. Control of the locomotive's ICE and the entire powertrain is performed by an automatic speed control system (ASCS) manufactured by MSV elektronika, which controls startup of the diesel engine, prevents power increase up to critical values and limits power in case of overloading. All diesel engine parameters are signaled; the diesel engine is stopped in case of overloading up to critical values (Fig. 22).

2.7.1 Technical Characteristics of TEM LTH Shunting Locomotive

The main technical characteristics of the TEM LTH locomotive are provided in Table 2.

The composition of shunting locomotive TEM LTH 001 is presented in Fig. 23.

The arrangement of the TEM LTH shunting locomotive is presented in Fig. 24, and consists of the following main equipment: 1—air unit; 2—traction rectifier; 3—fuel discharge path dampener; 4—fuel discharge pipe; 5—auxiliary AC generator; 6—coarse cleaning filter; 7—the main AC generator; 8—filter for diesel engine air inlet to cylinders; 9—internal combustion engine (ICE); 10—ICE cooling unit; 11—traction converters—individual inverters; 12—cooling unit of electrodynamic

Table 2 The main technical characteristics of the TEM LTH locomotive

Title	TEM LTH with Caterpillar C27
Wheel set arrangement	B'o–B'o
Rated power (kW)	709
Wheel gauge (mm)	1435/1520
Maximal speed (km/h)	90
Weight (tons)	90 (+3%, −1%)
Load on axle (tons)	22.5 t (±3%)
Minimum curve radius (m)	80
Diesel engine type	CAT C27 ACERT
Traction alternator	Siemens 1FC2 631-6B029T
Auxiliary alternator	Siemens 1FC2 631-6B029P
Power transmission	AC/AC
Traction effort of coupling (kN)	320
Permanent traction effort of coupling (kN)	139
At speed (km/h)	14
Control circuit voltage (V)	24

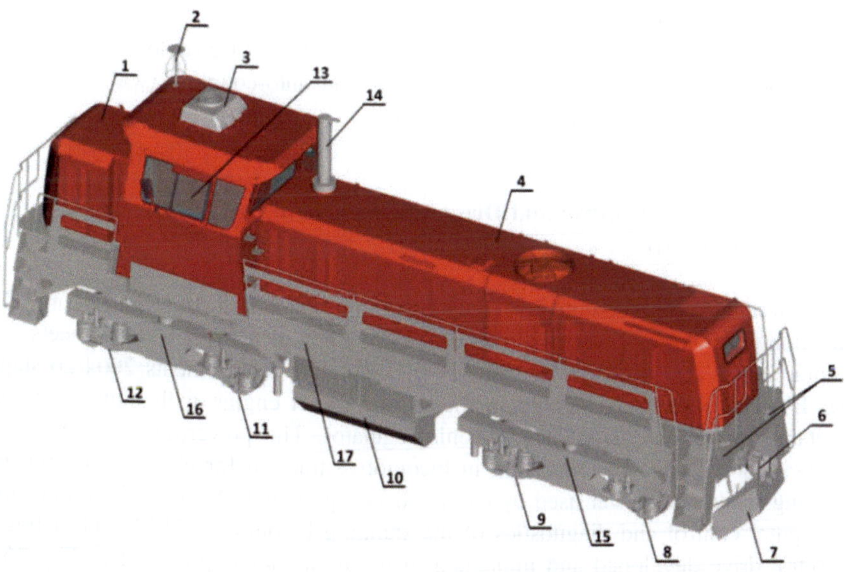

Fig. 23 Composition of the TEM LTH 001 shunting locomotive: 1—back hood; 2—KLUB-U system antenna; 3—climate control equipment; 4—front hood; 5—signal lights; 6—automatic coupling SA-3; 7—guard; 8—first wheel set; 9—second wheel set; 10—fuel tank; 11—third wheel set; 12—fourth wheel set; 13—driver's cabin; 14—fuel discharge pipe; 15—first bogie; 16—second bogie; 17—main frame

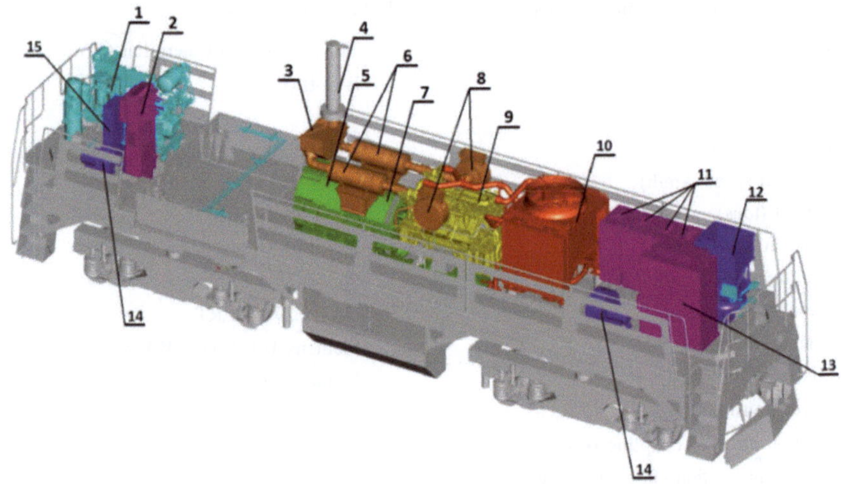

Fig. 24 Arrangement of main equipment of TEM LTH shunting locomotive

braking resistors; 13—power distribution cabinet; 14—traction motor cooling fan; 15—electrical cabinet.

Traction characteristics of the TEM LTH series diesel-electric locomotive at an auto-coupling level and traction characteristics at auto-coupling (with limitation based on adhesion) are provided in Fig. 25.

2.7.2 Automatic Control and Diagnostic System of the TEM LTH Locomotive

The diesel-electric locomotive is equipped with a powertrain consisting of an ICE, traction generator and electric drive. Emission, when diesel fuel is used, meets EU requirements for such type of vehicles (EU directive requirements 2004/26 stage III B). The powertrain includes a four-stroke diesel engine with liquid cooling, suitable for rail use with an electronic regulator. The powertrain ensures automatic, smooth, step-less variation in locomotive traction force depending on the driving speed and power used by the auxiliary equipment. A block diagram of the computer control and diagnostics of the shunting locomotive TEM LTH I bogie electric drive developed and implemented by the author is provided in Fig. 26.

Figure 26 **New designations**: G1–G2—synchronous traction generator 1FC2 631-6B029T; I_1–I_2—inverters; RB1–RB2—electrodynamic braking resistors; BR1–BR2—I bogie wheel set speed sensors; M1–M2—asynchronous traction motors; CS—shunt DC circuit current sensor; Y_1—driving task signals; Y_2—static converter control signals; Y_3—diagnostic signals; Y_4, Y_5—inverter control signals

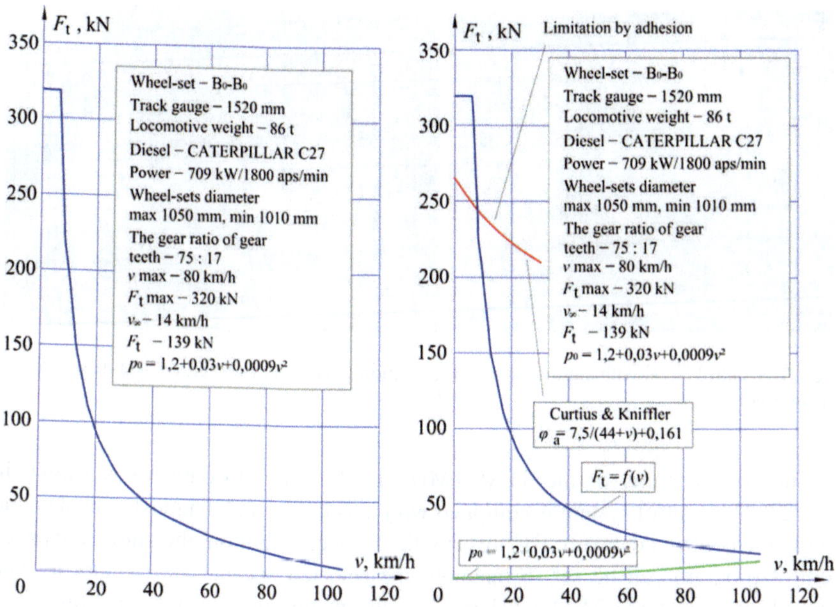

Fig. 25 TEM LTH Shunting locomotive traction characteristics and traction characteristics with limitation based on adhesion: F_t—traction force, kN; v—speed, km/h; $F_t\infty$—long-term mode traction force, kN; v_∞—long-term mode speed, km/h; $F_t\infty$—long-term mode traction force, kN; F_{tpmax}—maximum traction force, kN

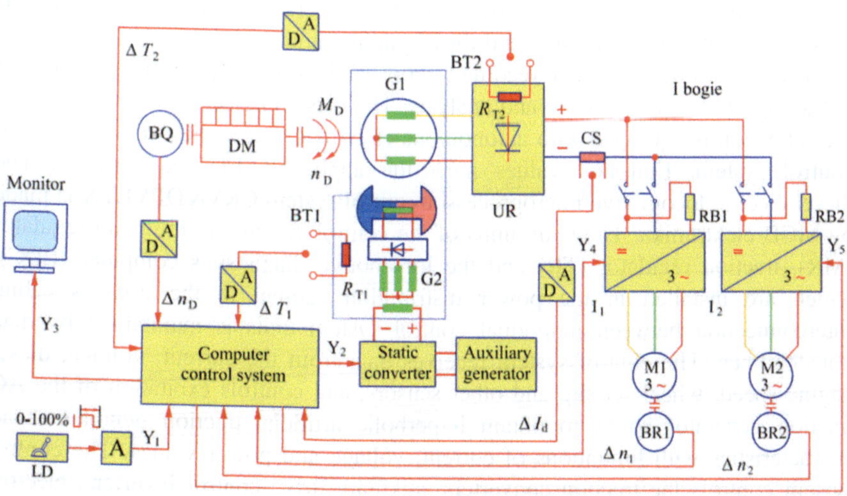

Fig. 26 Block diagram of computer control and diagnostics of shunting locomotive TEM LTH I bogie electric drive

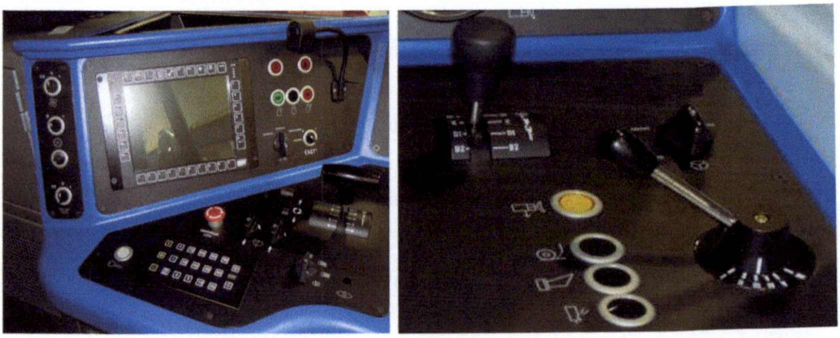

Fig. 27 Control panel with automatic speed regulation keyboard of the TEM LTH diesel-electric locomotive

The control panel of the TEM TMH diesel-electric locomotive is shown in Fig. 27. The control system is equipped with a technical and diagnostic display and automatic control keyboard. By pressing the respective button, the automatic speed control program can be set to maintain the selected speed. The monitoring and control system of the TEM LTH powertrain ensures optimal traction force in a tangential direction and prevents wheel slip. It is integrated into the onboard control and monitoring system of the entire diesel-electric locomotive, which also ensures that all required parameters are displayed for the driver. The onboard computer has all required interfaces with the safety system and brakes. It registers moto-hours of the powertrain from the vehicle commissioning and displays them on a screen in the driver's cabin. The diagnostic system ensures that errors are recorded, with indication of error code, date, time and diesel engine moto-hours.

Data about moto-hours and recorded errors can be transferred using a laptop into a desktop PC for data storage and analysis. The diesel engine CAT 27 with direct fuel injection is equipped with autonomous electronic diagnostics, monitoring and control system. Emission values meet the applicable EU requirements. The diesel-electric locomotive microprocessor control system CRV&DPVLTX is made by MCB elektronika. The main units of the control system are the main regulator (MR), traction regulator (TR) and the locomotive diagnostics computer (LDC), which are installed in the power distribution cabinet in the driver's cabin. Interconnection between individual control system units is maintained by data transfer lines. The control system receives data from the current, voltage, diesel engine speed, wheel set slip and other sensors, and controls excitation of the AC traction generator so as to obtain hyperbolic artificial traction generator load characteristics with limitations of current, voltage and power values. The control system controls the traction converters, auxiliary drive control inverters, electrodynamic brake, ICE parameters, auxiliary loops (sand pouring, opening of electrodynamic brake resistor cooling louvers, etc.) and diesel-electric locomotive diagnostics. The electric drive AC traction motors M1–M4 are supplied three-phase voltage generated by inverters I_1–I_4. The speed and torque of the TEM LTH traction engines are adjusted based on the traction controller LD setpoint (position), by changing inverter output voltage U_1 and frequency f_1 values. The 1350-kVA traction generator 1FC2 631-6B029T and 120-kVA auxiliary AC generator 1FC2 631-6B029P form a single electric machine whose rotor is rotated by the diesel engine. Excitation of the 1FC2 631-6 AC traction generator is controlled from a static converter. The traction converter consists of the common three-phase rectifier UR with DC-rated voltage of 940 V and individual inverters I_1–I_4, whose output third-phase-rated AC voltage range is 0–660 V and frequency range is 0.1–200 Hz.

The auxiliary generator 1FC2 631-6B029P with 120-kVA power supplies a three-phase network whose voltage is stabilized 3 × 400 V and frequency ranges within 30–90 Hz. The three-phase 3 × 400-V network of the 120-kVA auxiliary generator 1FC2 631-6 B029P is created by static converters composed of a rectifier with an output voltage range of 560 V (10–13%) and three-phase inverters for supply of electric drives of auxiliary equipment.

Energy efficiency of modernized locomotives. Diesel fuel consumption charts of shunting locomotives are shown in Fig. 28, which reveal that consumption decreased significantly after modernization of the TEM2 and ČME-3 typical diesel-electric locomotives and when the fleet of shunting locomotives was supplemented with new TEM TMH diesel-electric locomotives.

Diesel fuel consumption in freight traction traffic during 2012–2016 is provided in Fig. 29. A significant decrease in consumption is observed after modernization of the 2M62 typical diesel-electric locomotive.

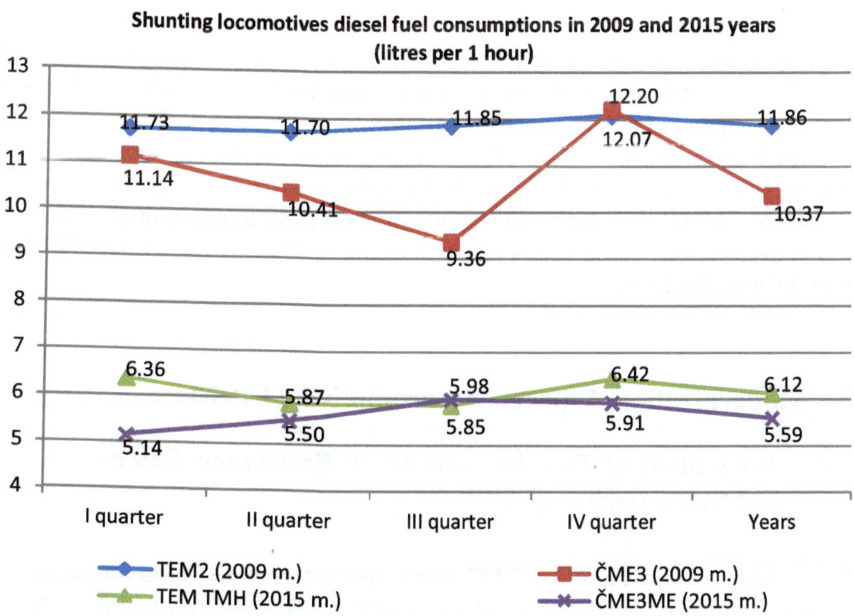

Fig. 28 Shunting locomotives diesel fuel consumption 2009 and 2015 (l/h)

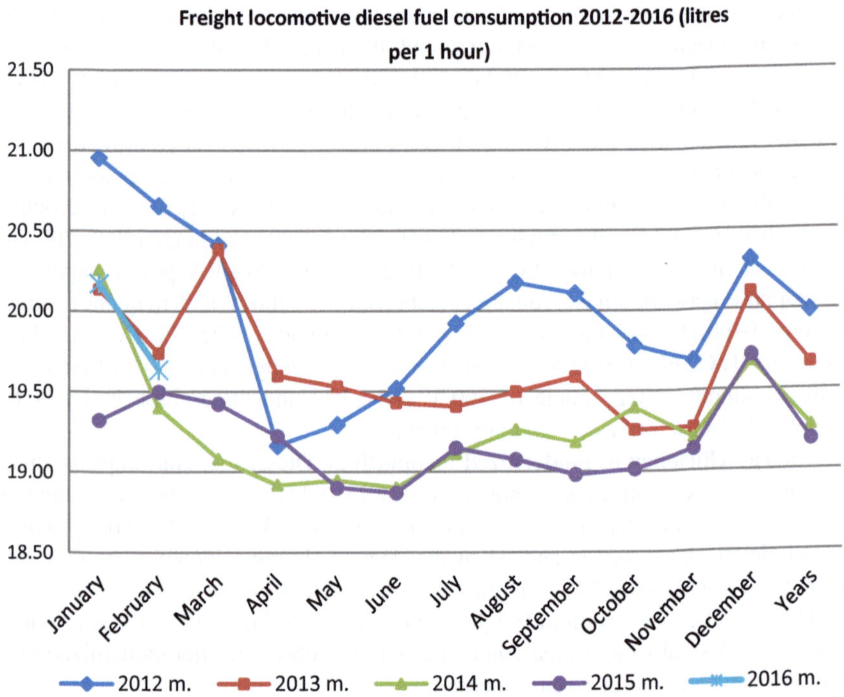

Fig. 29 Freight locomotive diesel fuel consumption 2012–2016 (l/h)

Conclusions: In 2009, before modernization, the average fuel consumption of the ČME3 shunting diesel-electric locomotive was 10.37 l/h; in 2015, the average fuel consumption of the modernized ČME3ME was 5.59 l/h, a 46% reduction. The average fuel consumption of the TEM2 shunting diesel-electric locomotive in 2009, before modernization, was 11.86 l/h. In 2015, the average fuel consumption of the new TEM TMH diesel-electric locomotive was 6.12 l/h, a reduction of 48%.

On average, fuel consumption of the modernized 2M62M locomotive (l/h), has been reduced by 25%.

3 Modernization Aspects of Locomotive TEP-70

3.1 Assessment of Traction and Train Resistance Forces in Modernization of TEP-70

Modernization of the TEP-70 AC/DC diesel-electric locomotive used for traction of passenger cars was completed in 2015 by VLRD UAB, subsidiary of JSC Lietuvos geležinkeliai, with participation of CJSC TMHB Transmashholding and Hungarian company Woodward-Mega Kft. During the design phase of modernization, the author assessed how locomotive dynamic properties are influenced by axial load variation of drive wheel sets.

In order to increase locomotive–rail adhesion properties, the author, during the design phase, proposed to model the layout of the main equipment with different weight values using SolidWorks software. Modeling data then should be checked by real weighing of modernized diesel-electric locomotives using Schenck V039905 equipment. The diagram of redistribution of rail axial loads of drive wheel sets of the modernized diesel-electric locomotive TEP-70 in traction mode is provided Fig. 30. The diagram shows "overturn" torques M_T created in the diesel-electric locomotive TEP-70 in traction mode, which reduce the rail axial load of the first bogie wheel sets by force ΔP (lifts up) and increase rail axial load of the second bogie wheel sets by force ΔP (presses down). These torques are caused by the difference between traction forces F created by locomotive drive wheel sets acting in a direction tangential to the wheel set, and train movement resistance force W acting at the level of autocoupling at height h.

Figure 30 **designations**: 1—autocoupling; 2—diesel engine cooling equipment; 3—braking resistors; 4—battery and fuel tank; 5—traction motor; F_{a1}, F_{a2}, F_{a3}, F_{a4}, F_{a5}, F_{a6}—traction forces of drive wheel sets; W_j—movement resistance forces that act on the rolling stock; P_1—the first wheel set–rail static load; P_2—the second wheel set–rail static load; P_3—the third wheel set–rail static load; P_4—the fourth wheel set–rail static load; P_5—the fifth wheel set–rail static load; P_6—the sixth wheel set–rail static load; h—height from the railhead to automatic coupling point; r_k—wheel set radius; B—distance between bogie centers (basis).

The traction force acting at the railhead level and movement resistance force at the autocoupling level together create a pair of forces with a shoulder equal to the coupling height from the railhead surface, or the torque which reduces load to the front wheel sets and increases load to the end wheel sets.

Therefore, the first least-loaded wheel set will first lose rail adhesion. In most cases, the first wheel set slips more frequently. Locomotive traction force is also

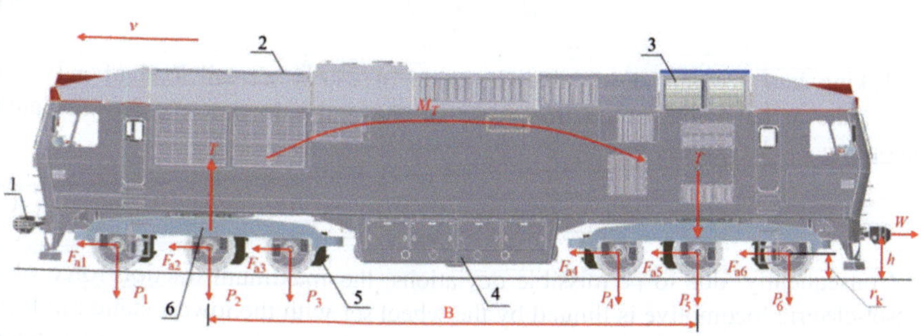

Fig. 30 Equipment and acting forces of the TEP-70 modernized diesel-electric locomotive: 1—autocoupling; 2—diesel engine cooling equipment; 3—braking resistors; 4—battery and fuel tank; 5—traction motor

influenced by the irregularity of traction forces of each wheel set. If a less-loaded wheel set creates lower traction force, then its slippage will begin at a lower traction force of the entire locomotive.

3.2 Assessment of Adhesion Weight in Modernization of TEP-70

Traction vehicles such as electric locomotives, diesel-electric locomotives, and electric or diesel trains should be designed with equal load to the wheel sets. In order to realize these theoretical requirements, centers of the equally spaced bogies should be equally loaded so that characteristics of the elastic elements of suspensions coincide. When these conditions are implemented, locomotive weight is distributed uniformly and wheel set–rail load is equal. However, in practice, realization of such conditions is complicated, because locomotive equipment of different weight values must be arranged uniformly. Therefore, typically, according to technical requirements, the permissible deviation of wheel set–rail axial load for diesel-electric locomotives with static rail load up to 225 kN (23 tf) is ±3%, and ±2% in the case of a static axial rail load of 245 kN (25 tf). The traction force F_{ki} developed by each wheel set, limited by wheel set–rail load force $6P_i$ based on rail adhesion conditions, is calculated as follows:

$$F_{ki} \leq \psi 6P_i \tag{7}$$

Entire locomotive traction force F_k:

$$F_k \leq \psi P_k \tag{8}$$

where P_k is the locomotive adhesion weight force, which is calculated as follows:

$$P_k = \sum 6P_i \tag{9}$$

Taking into account permissible deviation of wheel set–rail load of ±3% or ±2%, the actual value of traction force of the entire locomotive is calculated according to (10) and is less by 2–3%.

$$F_{k\min} = 0.97\psi 6P \tag{10}$$

Consequently, due to permissible deviations, the maximum traction force of a diesel-electric locomotive is limited by the wheel set with the lowest static rail load and may be less by 2–3% compared to the design value. In this case, it can be stated that adhesion weight of the modernized diesel-electric locomotive TEP-70 is not used optimally, resulting in reduced traction force and higher fuel losses.

3.3 Assessment of Influence of Load Changes in Modernization of TEP-70

The bogie-to-rail force redistribution scheme (in traction mode) of the six-axle locomotive is provided in Fig. 31. Frames of bogies are affected by drive wheel set traction forces F_{a1}, …, F_{a6}, and the automatic coupling point of rolling stock is affected by movement resistance forces W_j that are equal: $W_j = n \cdot F_a$. Due to the operation height difference of the traction force F_T generated by locomotive drive wheel sets and train movement resistance force W_j that acts at the coupling point, there is torque M_T:

$$'M_T = n \cdot F_a \cdot (h - r_r) \quad (11)$$

where n is the wheel set number (units), F_a is the traction force of one drive wheel set (N), h is the coupling point height (distance from railhead to coupling point, longitudinal axis; m) and r_r is the wheel set rolling radius (m).

Torque M_T tilts the locomotive body at axis Y that crosses locomotive weight center C, and it changes the value of vertical force T acting on the bogies, which is calculated as follows:

$$T = M_T / B \quad (12)$$

where B is the distance between bogies (basis; m). The change in wheel set force ΔP_i is calculated as follows:

$$\Delta P_i = |R_T| = \frac{T}{m} = \frac{M_T}{m \cdot B} \quad (13)$$

where m is the bogie wheel set number. At torque M_T, the first bogie is less loaded and the second is more loaded.

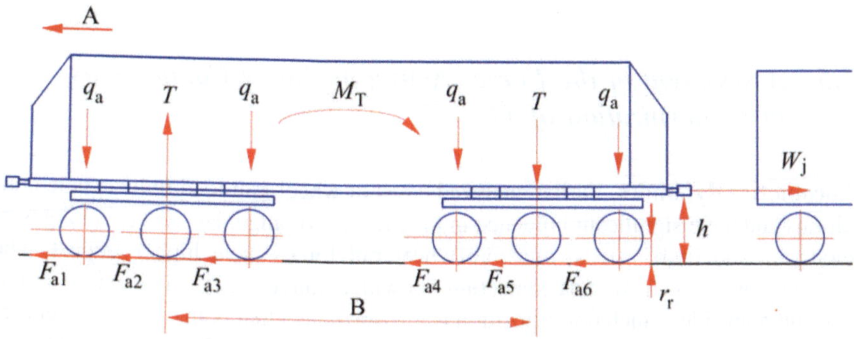

Fig. 31 Redistribution scheme of locomotive bogie to rail force: F_{a1}, F_{a2}, F_{a3}, F_{a4}, F_{a5}, F_{a6}—traction forces of drive wheel sets; W_j—movement resistance forces that act on the rolling stock; q_a—static wheel to rail force of a wheel set; h—height from the railhead to automatic coupling point; B—distance between bogie centers (basis). A—direction of locomotive movement

3.4 Assessment of Influence of Adhesion Factor Changes in Modernization of TEP-70

The traction force of the modernized diesel-electric locomotive is affected by non-uniformity of traction forces of each wheel set. If a less-loaded wheel set creates higher traction force, then its slippage will begin at lower traction force of the entire locomotive. The adhesion factor of the modernized diesel-electric locomotive TEP-70 is also affected by fretting of wheel rims, rail wear and differences in wheel diameter. In low-radius track curves, wheel set–rail adhesion is affected by wheel set slippage caused by a difference in running distances. The adhesion factor is also influenced by the type of locomotive drive (DC/DC, AC/DC, AC/AC), traction motor connection methods and the nature of traction motor mechanical characteristics. Locomotive traction force due to redistribution of wheel set–rail load may be reduced by the maximum force calculated as follows: $\Delta P = \Delta P_1 + \Delta P_2 + \Delta P_3$. The force of the least-loaded wheel set is calculated according to the formula $6P_{\min} = 6P(1 - 0.03) - \Delta P$; here, 0.03 is the deviation of static wheel set–rail load permitted by the technical conditions of the locomotive when force $6P$ is below 225 kN. The load ratio of the least-loaded wheel set and design load is referred to as locomotive adhesion utilization factor β_k and is expressed as follows:

$$\beta_k = \frac{6P_{\min}}{6P} = \frac{6P(1 - 0.03) - \Delta P}{6P} = 0.97 - \frac{\Delta P}{6P}. \tag{14}$$

The adhesion utilization factor shows what proportion of traction force the locomotive actually develops. Based on analysis, we can state that the adhesion utilization factor of the modernized diesel-electric locomotive TEP-70 depends on the layout of the wheel sets, traction motors, type of traction motor suspension and traction force transmission equipment.

3.5 Assessment of the Forces Acting on Bogie Components in Modernization of TEP-70

Locomotive dynamics. Earlier analyzed cases of wheel–rail load reflect static conditions and have significant influence in the low-speed range. However, in ranges of moderate and high speeds, dynamic wheel–rail forces have higher impact. The “passing” wheel in a curve is contacting rail with a flange, and the last wheel set has gaps on both sides. Such bogie position is intermediate when both end wheel sets are pressed against the rail during the highest leaning in a curve. The diagram of forces acting on the bogie of the modernized diesel-electric locomotive TEP-70 is provided in Fig. 32. The bogie of the modernized diesel-electric locomotive TEP-70 runs on a

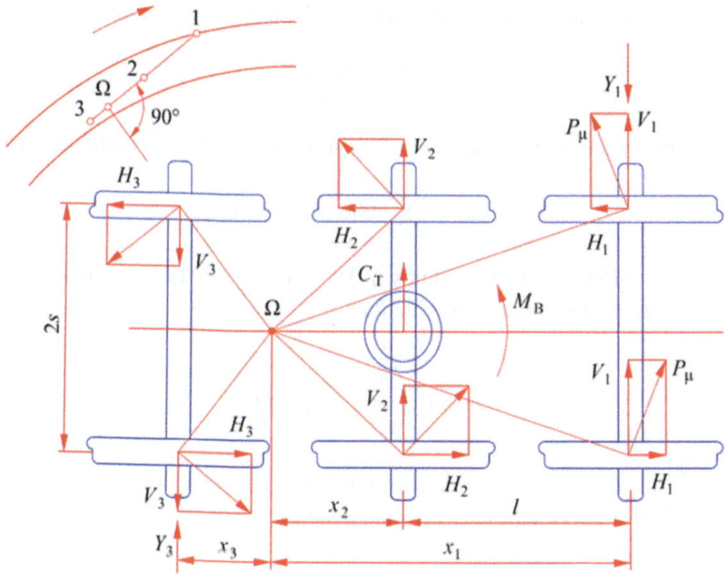

Fig. 32 The diagram of forces acting on the bogie of the modernized TEP-70 diesel-electric locomotive

track by a complex path close to sinusoid (meandering). Let us determine its momentary center of the turn **Ω** (see Fig. 32) by laying a perpendicular line from the curve center to the wheel set axis. Bogie turning and cross-movement against the track axis is influenced by wheel set friction forces P_μ acting in the opposite direction (μ, friction factor). These forces in each wheel set are perpendicular to the radii, connecting the force action point with the turn angle **Ω**. These forces consist of longitudinal force H_i and transverse force V_i components.

They can be determined graphically or analytically. The longitudinal force H_i and transverse force V_i components are calculated as follows:

$$H_i = P_\mu \frac{s}{\sqrt{x_i^2 + s^2}}, \tag{15}$$

$$V_i = P_\mu \frac{x_i}{\sqrt{x_i^2 + s^2}}. \tag{16}$$

A bogie is exposed to a part of centrifugal force C_T caused by the outside rail elevated by h, whose value depends on the locomotive speed:

$$C_T = 3mP\left(\frac{v^2}{gR} - \frac{h}{3s}\right). \tag{17}$$

Let us compose an equilibrium equation of forces acting against a bogie:

$$-Y_1 + C_T + \sum V_i = 0. \quad (18)$$

Taking into account that the sum of all transverse forces is equal to zero:

$$C_T x_2 + \sum M_{TP} - M_B = 0. \quad (19)$$

Using Eqs. (18) and (19), unknown Y_x and C_T can be determined, and the permissible bogie speed v can be calculated:

$$v = \sqrt{\left(\frac{C_T}{3mP} - \frac{h}{3s}\right) gR}, \quad (20)$$

where v is the bogie speed, m is the number of bogie axles, h is the height of elevated rail in a curve and R is the curve radius.

3.6 Assessment of the Influence of Traction Motor Layout on Loads of TEP-70

The author present an analysis of how the adhesion factor is influenced by the layout of the traction motors. Results of this analysis are applied to improve practical construction of a locomotive. The traction motor of the TEP-70 diesel-electric locomotive develops torque M, which rotates a wheel set. A shaft of the traction motor is connected to a wheel set through a gearbox. The gear wheel connected to the traction motor has fewer teeth than the gear wheel pressed on the wheel set axle. The diagrams showing the assessment of the influence of the traction motor on the static rail load (a) and practical traction motor layout in bogies (b) are provided in Fig. 33. When the locomotive runs in a given direction, force Z is directed upwards. If we add two equal opposite forces Z^{I} and Z^{II} in point B (on the wheel set axle) with value equal to force Z, then we can see that Z and Z^{II} create a force pair, forcing rotation of the wheel set, and force Z^{I} reduces the wheel set load. After movement, the direction of the modernized TEP-70 is changed, and force Z^{I} increases the load of the locomotive wheel set. When the locomotive traction motor suspension is of the axial frame type, and traction motors in the first and second bogies are arranged symmetrically in relation to their wheel sets, the number of wheel sets with increased and reduced loads in a locomotive is equal, because locomotive weight does not change.

In the locomotive with three-axle bogies, rail loads of the three wheel sets are increased, and those of the other three are reduced. However, the reduction in wheel set loads limits the maximum force of the diesel-electric locomotive even more. Reaction Z_1 is acting on a gear wheel tooth in point C according to Newton's third

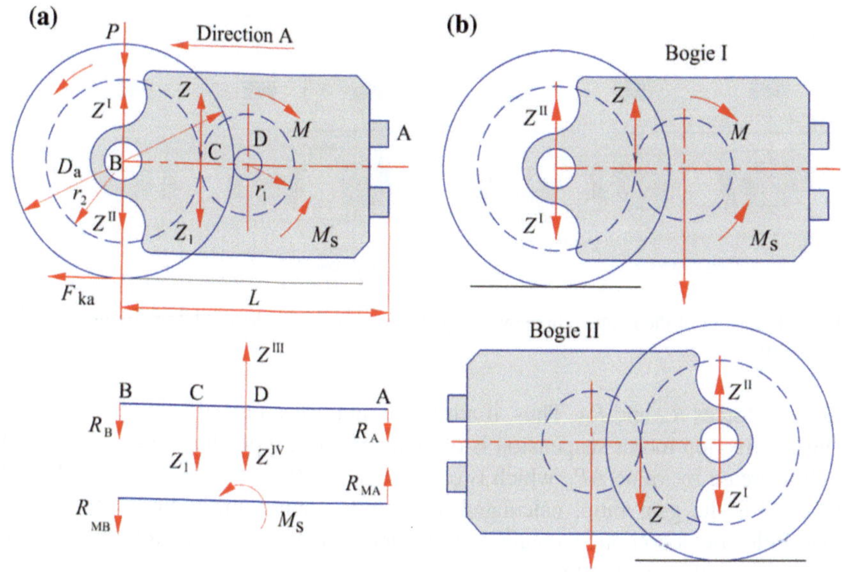

Fig. 33 The diagrams of assessment of traction motor influence on static rail load (**a**) and practical traction motor layout in bogies (**b**). *P*—locomotive wheel set–rail static force; M_s—static resistance momentum added to a traction motor shaft; D_a—wheel set diameter; r_1, r_2—diameters of gear wheels

law, which is equal to force Z in value, but in an opposite direction. If we add two equal and opposite forces Z^{III} and Z^{IV}, equal in value $Z = Z_1$, to point D (on traction motor axis), we can see that forces Z_1 and Z^{III} create reactive momentum, $M_p = M$, equal in value to the motor's developed torque *M*. The force Z^{IV} through the traction motor rotor shaft bearings acts on a body and supports (designated as A in the figurc). Let us determine the forces which change the wheel set–rail load of the modernized TEP-70. The force R_B at point B, increasing wheel set–rail load, when the locomotive movement direction is A, is calculated as follows:

$$R_B = Z\frac{L-(r_1-r_2)}{L}. \tag{21}$$

The force R_A at point A increases the load of the bogie frame when the locomotive movement direction is A. When the locomotive movement direction changes, forces R_A and R_B are directed upwards. In addition, the motor stator is subjected to the momentum $M_s = M$, which is added to motor supports by the force R_{MB}, calculated as follows:

$$R_{MB} = |R_{MA}| = M/L. \tag{22}$$

The force R_{MB} increases the wheel set–rail load, and the force R_{MA} reduces the wheel set–rail load. The absolute values of forces R_{MA} and R_{MB} are lower than

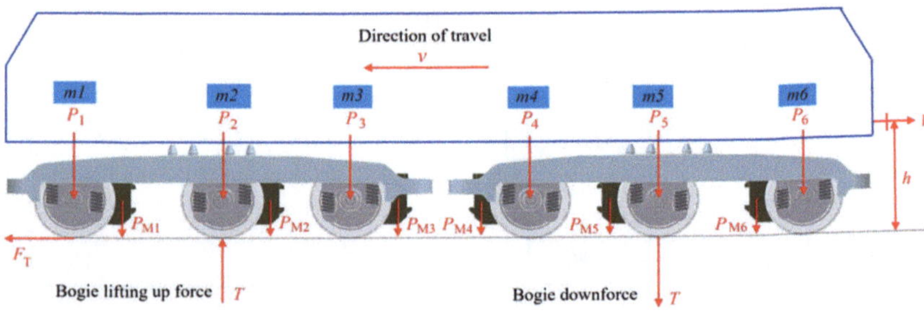

Fig. 34 Layout of the main equipment of equal weight in the modernized TEP-70 diesel-electric locomotive

those of forces R_A and R_B. Thus, if a traction motor is arranged behind a wheel set, and the traction motor suspension is of an axial frame type, the wheel set–rail load will decrease by value ΔP_i, which is calculated as $\Delta P_i = Z - R_B - R_{MB} = M \cdot i/L$ where i is the gear ratio, calculated as $i = r_2/r_1$. When the modernized TEP-70 diesel-electric locomotive changes movement direction, the wheel set–rail load increases by the same value.

During design, the main equipment should be arranged so as to ensure minimal dispersion of wheel set loads. This is a fairly complex task. In practice, there is no possibility of distributing equipment weight uniformly (see Fig. 34). Consequently, different static forces will act on the first and second wheel sets of a bogie. This leads to different adhesion factors, and slipping of the locomotive will occur. In order to reduce this influence, traction motors are specially arranged according to the diagram in Fig. 33b. The layout of the main equipment of equal weight in the modernized TEP-70 is provided in Fig. 34. According to the layout of the TEP-70 traction motors, in traction mode, due to the specific arrangement of the traction motors, the first bogie lifting force T will be reduced. This will lead to an increase in the adhesion factor of the first bogie, all of which will reduce slipping and skidding.

3.7 Weighing Tests of Modernized Locomotive TEP-70M

The modernized TEP-70M 0249 diesel-electric locomotive was weighed on February 2, 2015, at the site of the Vilnius locomotive repair depot, using Schenck V039905 equipment according to weighing procedures used by UAB VLRD. The purpose of the measurement of wheel set–rail load is to determine what static force is acting on rails, and what the dispersion of static loads of individual wheel sets is (Figs. 35, 36, 37 and 38).

Weighing conclusion: The deviation in diesel-electric locomotive wheel-set loads comparing the left and the right sides is 1.89%. Percentage deviation of the diesel-electric locomotive wheel sets in the transverse direction is 1.5%. This meets the permissible deviation of wheel set loads of 3%. The total locomotive mass is 133,335 kg (Table 3).

Fig. 35 Diesel-electric locomotive weighing equipment used by UAB VLRD

Fig. 36 Weighing process of the modernized diesel-electric locomotive TEP-70M

Fig. 37 Weighing process of the modernized diesel electric locomotive TEP-70M

Fig. 38 Recording of the weighing process results of the modernized diesel-electric locomotive TEP-70M

Table 3 Weighing results for the modernized diesel-electric locomotive TEP-70M

Parameter	Wheel set					
	1	2	3	4	5	6
Wheel set load, left side. F_{Rk} (tf)	10.645	11.076	10.992	10.788	11.380	10.543
Wheel set load, right side. F_{Rd} (tf)	12.012	11.991	10.972	10.931	11.155	10.890
Total wheel set load $F_R = F_{Rk} + F_{Rd}$ (tf)	22.657	23.067	21.964	21.719	22.535	21.433
Bogie load $F_V = \sum F_R$ (tf)	67.688 (1st bogie)			65.687 (2nd bogie)		
Total left side locomotive load $F_{BRk} = \sum F_{Rk}$ (tf)	65.424					
Total right side locomotive load $F_{BRd} = \sum F_{Rd}$ (tf)	67.951					
Total diesel-electric locomotive mass m (t)	133.375					
Average wheel set load in a bogie $F_{VAv} = F_V/3$ (tf)	22.563			21.896		
Average wheel set load in a locomotive $m/6$ (t)	22.229					
Deviation of locomotive wheel set loads in a longitudinal direction $(F_{BRk} - F_{BRd})/(F_{BRk} + F_{BRd}) \times 100$	−1.89% Deviation of locomotive loads in longitudinal direction					
Deviation of locomotive wheel set loads in a transversal direction $(F_{v1} - F_{v2})/(F_{v1} + F_{v2}) \times 100$	1.5% Deviation of locomotive loads in transversal direction					

3.8 *Main and Control Equipment of the Modernized Locomotive TEP-70*

The maximum design speed of the modernized diesel-electric locomotive TEP-70M 0249 is 160 km/h. The main technical data for the modernized diesel-electric locomotive TEP-70M 0249 is provided in Table 4.

After analysis of the influence on locomotive dynamics, assemblies of the modernized diesel-electric locomotive TEP-70 are arranged so that axial rail loads would not deviate more than 3%. The diesel engine installed in the center of the common frame of the diesel-electric locomotive TEP-70 is rigidly connected to AC traction, auxiliary equipment supply and passenger car heating generators.

Generators for traction, auxiliary equipment supply and passenger car heating and their exciters are installed in a single electric machine. The driver's cabin is provided for locomotive control. Layout of the main equipment of the modernized diesel-electric locomotive TEP-70 is shown in Fig. 39.

The following assemblies and equipment are arranged in the modernized diesel-electric locomotive TEP-70: Central cabinet of equipment, auxiliary equipment control cabinet, passenger car heating rectifier, control panel VZA-A, control panel VZA-B, electrical cabinet; machinery compartment cabinet; battery compartment, MTU ICE, battery charging generator and two starters, AC traction generator G1 and AC exciting G2 generator, AC auxiliary equipment and passenger car heating system supply generator G3 and AC exciting generator G4, compressor; cooling fan of traction motors of the first bogie, cooling fan of traction motors of the second bogie, three-phase bridge rectifier (UR).

Table 4 The main technical data of the modernized diesel-electric locomotive TEP-70M 0249

Locomotive TEP-70M 0249 technical data	Value
Maximal speed (km/h)	160
Wheel gauge (mm)	1520
Weight (tons)	133 t (±3%)
Load on axle (tons)	22.2 t (±3%)
Wheel-set arrangement	Co‘-Co‘
Diesel engine type	MTU20V4000R43L
Rated engine power (kW)	3000
Traction alternator	1TAG-2900/700-TEP-70
Traction alternator rated power (kW)	2900
Minimum curve radius (m)	125
Power transmission	AC/DC
Wagon heating power (kW)	400

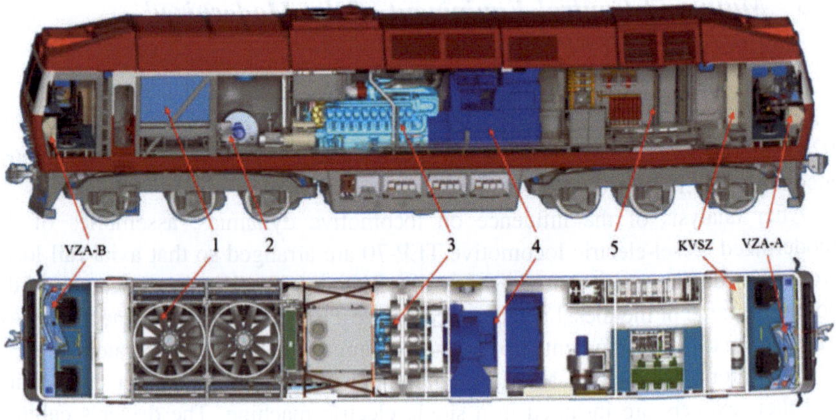

Fig. 39 Layout of the main equipment of the modernized diesel-electric locomotive TEP-70: 1—diesel engine cooling equipment; 2—compressor; 3—diesel engine; 4—assembly of traction, auxiliary equipment generators and traction rectifier; 5—passenger car heating equipment; VZA-A and VZA-B—control panels; KVZ70—central cabinet of equipment

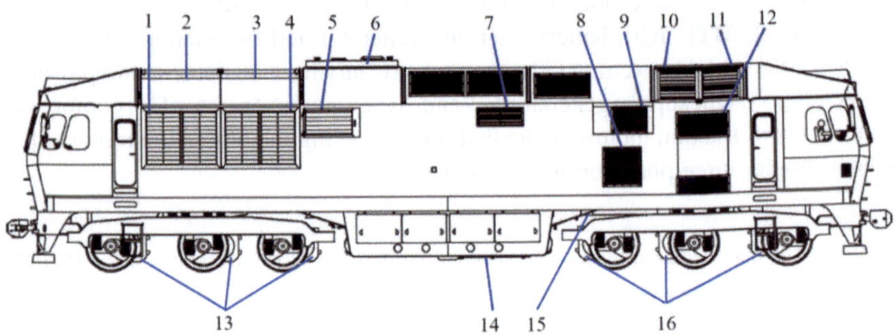

Fig. 40 Layout of the main equipment of the modernized diesel-electric locomotive TEP-70M 0249

Figure 40 **designations**: 1, 4—cooling compartment; 2, 3—cooling fans; 5—compressor louver; 6—roof fan louver; 7—generator cooling fan; 8—cooling fan; 9—machinery compartment 1 cooling fan; 10, 11—electrodynamic braking resistor cooling fans; 12—transformer compartment fans; 13, 16—traction motors (M1–M6) 14—battery unit AKK2; 15—external supply socket.

The structure of the bogie of the modernized diesel-electric locomotive TEP-70 is shown in Fig. 41.

The driver's cabin is arranged so as to ensure good visibility; all outside dimensions would be visible. Front and back windows are made of multilayered protection glass, equipped with electric wipers and washers. To enable observation of both locomotive sides for the driver, electrically heated rear-view mirrors are installed outside the cabin. The driver's seat rotating 280° ensures good visibility of

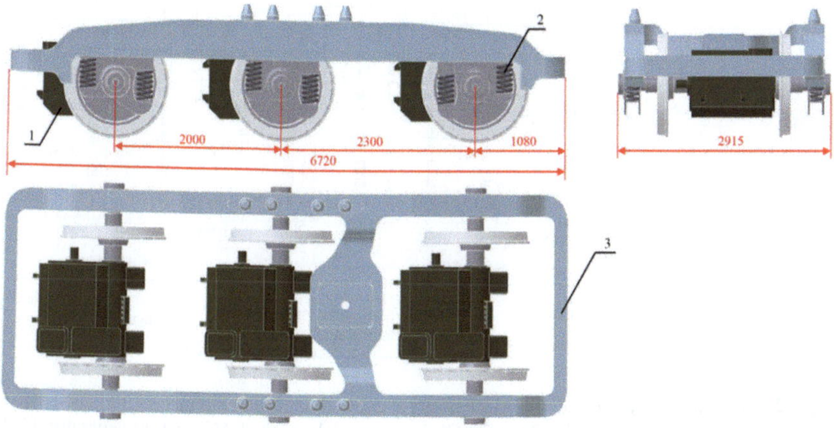

Fig. 41 Structure of the bogie of the modernized diesel-electric locomotive TEP-70: 1—traction motor; 2—spring; 3—frame

the track and signals. The control panel contains all important indicators and controls, including the speedometer, kilometer meter, tachometer, work hour meter, battery charging indicator, braking system pressure gauges, fuel indicator, oil pressure and engine temperature. Each driver's cabin is equipped with a traction and braking controller at the right side of the panel; the driver's and assistant's seats meet requirements of ergonomics, hygiene and fire safety rules. Controllers and activation devices of the main and auxiliary systems of the diesel-electric locomotive's safety, communication and measuring devices, and their arrangement and mounting in the driver's cabin and on the panel meet the requirements of UIC 651. Cabin lighting installation meets the requirements of UIC 555 for lighting intensity and emergency lighting. Heating and air conditioning equipment ensures a cabin temperature of 21–28 °C when the outside temperature is −40 to +40 °C, according to HN 69:2003.

Diagnostics of electric drive parameters of the modernized diesel-electric locomotive TEP-70. The block diagram of the electric drive with diagnostics signals of the modernized TEP-70 AC/DC diesel-electric locomotive is provided in Fig. 42. The block diagram shows electric drive control signals sent by sensors. Information of these signals is used for diagnostics.

The information window in the display of control panel A of the TEP-70 is shown in Fig. 43. This display window shows the following parameters in real time: crankshaft speed (1798 rpm) and rated speed (1800 rpm), coolant temperature (96 °C), coolant temperature in the main circuit (96 °C), traction generator load current (3624 A), voltage (657 V) and output power (2380 kW).

The diagnostics of the control system of the modernized diesel-electric locomotive TEP-70. The first window of the control system parameters of the modernized TEP-70 is shown in Fig. 44. This display window shows the following parameters in real time: crankshaft speed (1798 rpm), oil pressure (0.92 bar),

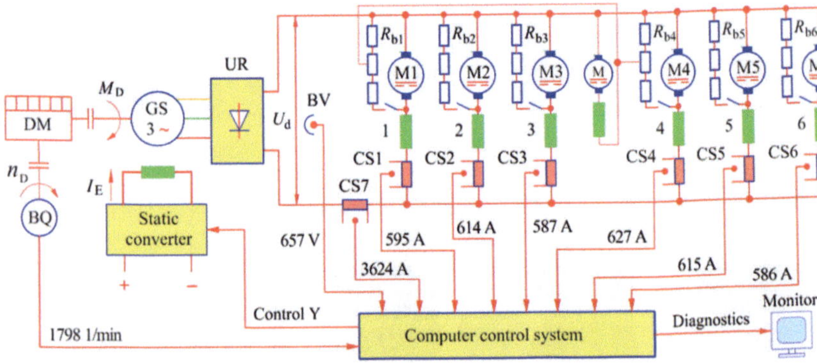

Fig. 42 The block diagram of the electric drive with diagnostics signals of the modernized AC/DC diesel-electric locomotive TEP-70. GS is a separately excited AC traction generator; CS1–CS6 are armature current sensors of DC traction motors; CS7 is the traction generator's rectifier current I_d sensor; M1–M6 are series wound DC traction motors; BQ is a diesel speed and crankshaft position sensor; BV is the traction generator's rectifier voltage U_d sensor; 1–6 are the DC traction motor excitation windings; Y is the excitation regulator control signals; BQ is the diesel engine speed and crankshaft position sensor

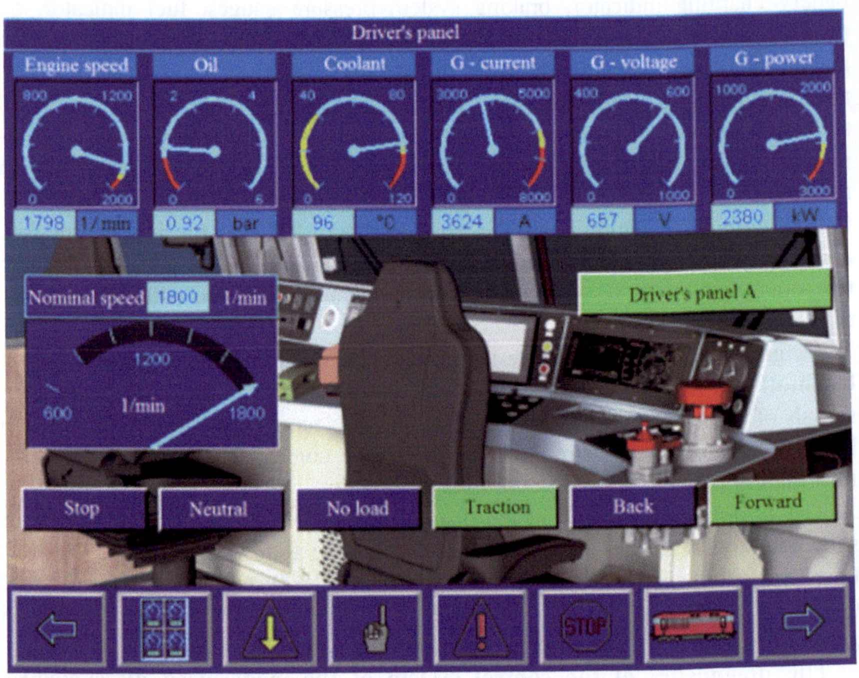

Fig. 43 Information window in the diagnostics display of control panel A in the diesel-electric locomotive TEP-70

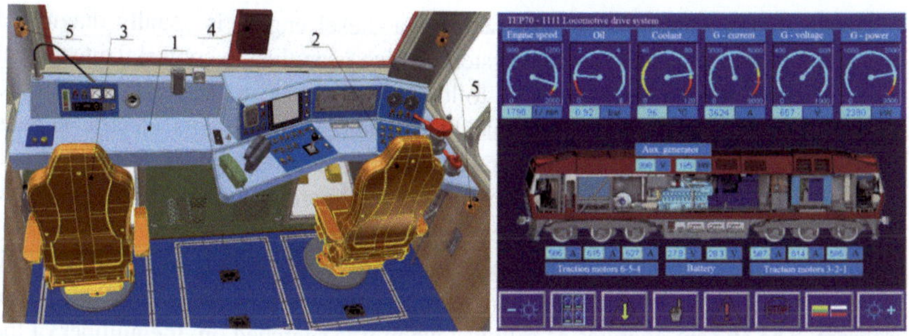

Fig. 44 Driver's cabin equipment and control system parameters window 1 in the diagnostics display of the diesel-electric locomotive TEP-70: 1—panel; 2, 3—driver's seats; 4—ТСКБ panel; 5—alertness monitoring buttons

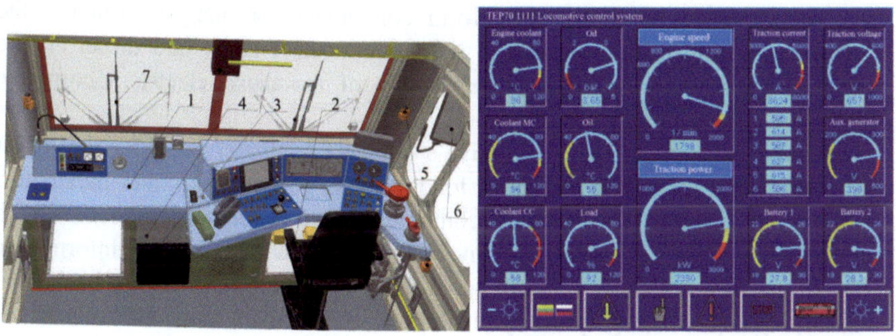

Fig. 45 Equipment in driver's cabin control station 1 and control system parameters window 2 in the diagnostics display of the diesel-electric locomotive TEP-70: 1—panel; 2—driver's seat; 3—microwave oven; 4—refrigerator; 5—alertness monitoring buttons; 6—mirrors; 7—wipers

coolant temperature (96 °C), auxiliary battery charge voltage (27.8 V), battery charge voltage (28.3 V), power of the auxiliary three-phase generator (195 kW, voltage 398 V), DC traction motor armature circuit currents: sixth—586A, fifth—615A, fourth—627A, third—587A, second—614A, first—595A. In addition to the diesel engine parameters, the diagnostics window also shows traction generator load current (3624 A), voltage (657 V) and developed power (2380 kW).

The second window of the control system parameters of the TEP-70 is shown in Fig. 45.

This display window shows the following parameters in real time: crankshaft speed (1798 rpm), oil pressure (3.85 bar), oil temperature (92 °C), coolant temperature (96 °C), coolant temperature in the main circuit MC (96 °C), coolant temperature in the auxiliary circuit CC (58 °C), traction generator load current (3624 A), voltage (657 V) and output power (2380 kW). Auxiliary battery (battery 1) voltage (27.8 V), main battery (battery 2) voltage (28.3 V), DC traction motor armature circuit currents: sixth—565A, fifth—615A, fourth—627A, third—587A, second—614A, first—595A.

Diesel engine diagnostics. The state of a diesel engine is usually diagnosed using a diagnostics and forecasting system, which is applied to control the technical state of many machines. It has the following functions: to convert non-electric sensor signals to analogue or digital electric signals, to process electric signals based on the defined algorithm, to compare results with limit values, to convert to a digital format and to store it in a memory of a board computer, to forecast the remaining lifetime of the monitored assemblies and to continuously monitor parameter values determining the operational mode of the diagnosed object. The process of diagnosis is carried out in an automatic mode according to the preset program. The system operates as follows: when it is switched on, the number of the typed diagnostic parameter is displayed in the control unit panel. This number is read from the memory unit and written to the random access memory of the control unit.

This information is transmitted via the router to the primary and secondary transmitters, which present it in a form convenient for analysis. Finally, the information is sent to a display. The right side control panel and diesel engine parameter window in the diagnostics display of the diesel-electric locomotive TEP-70 are shown in Fig. 46.

Onboard computer. The locomotive's onboard computer provides the driver with control and diagnostic functions for the entire rolling stock and parameters necessary for driving. The onboard computer also has required interfaces with traffic safety system and brakes. A driver is provided with necessary information about all the locomotive's operating parameters. Operating information in the driver's display is provided graphically (analogous scales, histograms, etc.), combined with digital information; however, presentation of information exclusively in digital form is avoided. The onboard computer provides the ability to transfer system log data to external memory storage (USB memory stick, SD card, etc.) for saving and analysis on a PC.

The ability to correct or cancel diagnostic system data is possible only with special software or password permission. The engine electronic regulation and control system with diagnostics is integrated into the common control system of the vehicle powertrain.

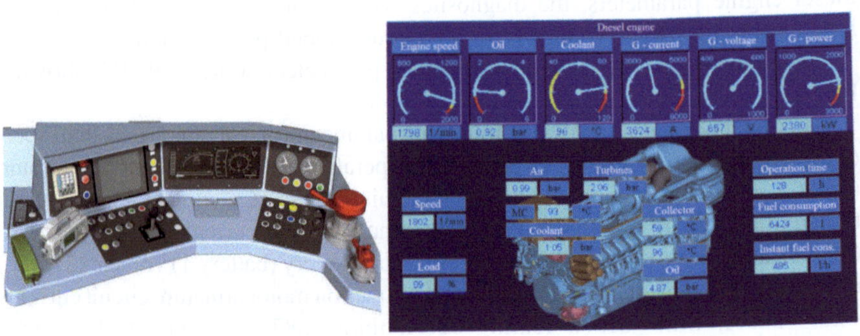

Fig. 46 Right side control panel and diesel engine parameter window in the diagnostics display of the modernized TEP-70 diesel-electric locomotive

The system ensures: automatic engine starting by pushing a button in the driver's panel, i.e., checking operations before start are performed automatically, including automatic disconnection of the starting system after the engine starts running, automatic monitoring of engine operating parameters (crankshaft speed, load, work hours, fuel level, fuel consumption), monitoring of oil, coolant temperature and level, and activation of the dialogue terminal during engine adjustment for indication of engine parameters.

3.9 Electric Drive of the Modernized Locomotive TEP-70

The main purpose of the modernized diesel-electric locomotive TEP-70 series with electric drive is the traction of passenger cars. The maximum speed of the TEP-70 is 160 km/h. Driver cabins are installed in the back part of the locomotive. The TEP-70 is equipped with a computer (microprocessor) control system for automated control of the diesel engine generator and power assembly. The locomotive weighs 126 t and has six driving axles, with the axis formula C_0'–C_0'. The locomotive is equipped with the MTU 20-cylinder 4000 AG capacity diesel engine. Typical Russian-produced diesel-electric locomotive trolleys are used, in which six DC serially excited traction motors (ED 121AU1) with 324-kW capacity are installed. The main power source in the TEP-70 is the power assembly consisting of the diesel engine and two AC generators: traction and auxiliary. Diesel engines and AC generators are rigidly connected with elastic coupling. Another important part of the diesel engine generator is the auxiliary AC generator with 700-kVA capacity, intended for the locomotive's onboard three-phase network with 3 × 400 V and 50 Hz, as well as for heating of passenger cars. The AC traction generator with 2900-kVA capacity is the power source supplying the electrical traction system. Static converter-regulators are used for excitation of generators. Independently of the main locomotive computer control system, diesel engine parameters are monitored and controlled by the autonomous electronic control system designed by the company MTU, which controls startup of the diesel engine, prevents power increase up to critical values and limits the power in the case of overload. Parameters of all engine operation modes are signalled, and the diesel engine is stopped in the case of overload up to the critical values. DC traction motors of the electric drive of the modernized diesel-electric locomotive TEP-70 are supplied by the rectified voltage U_d of the AC traction generator. The author analyzed the ACS of traction generators of typical diesel-electric locomotives 2M62, TEM1, TEM2, ČME3 and TEP-70, and proposed using semiconductor static converters for excitation of traction generators in the modernized JSC Lietuvos geležinkeliai diesel-electric locomotives [17, 18]. Their connection diagrams are provided in different chapters. The block diagram of computer control and diagnostics of the modernized TEP-70M diesel-electric locomotive developed and implemented by the author is provided in Fig. 47.

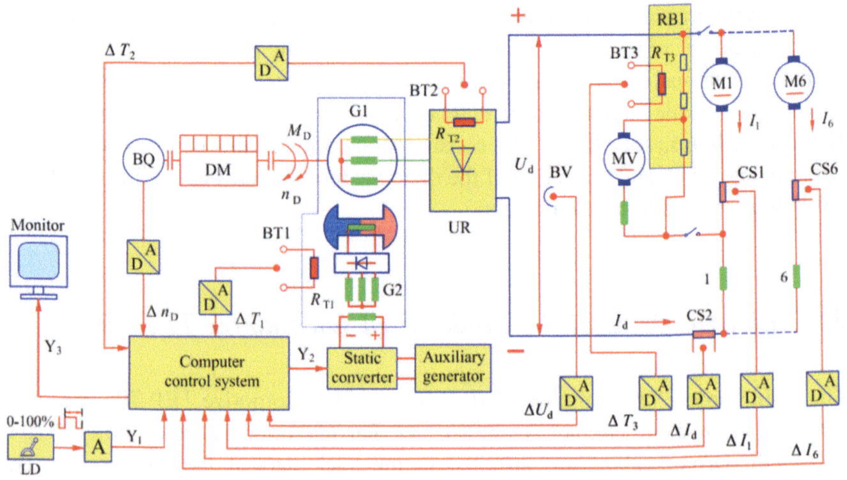

Fig. 47 The block diagram of computer control and diagnostics of the modernized diesel-electric locomotive TEP-70M

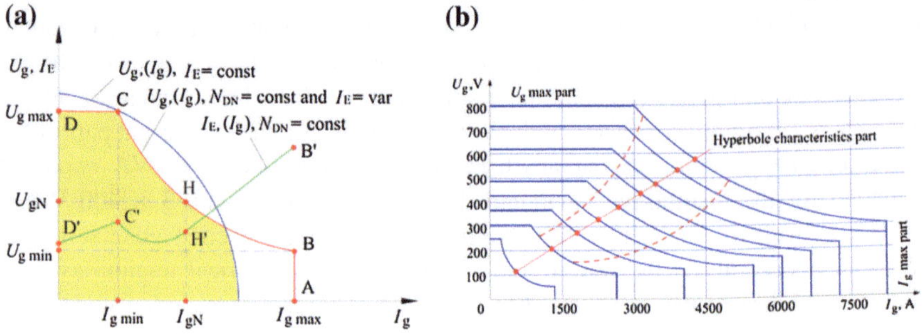

Fig. 48 AC traction generator's theoretical characteristics (**a**) and practical artificial load characteristics of the modernized diesel-electric locomotive TEP-70M (**b**)

To ensure optimal use of diesel engine capacity across the entire speed range, the shape of the traction generator load characteristics $U_g = (f)I_g$ must be the same as that of traction characteristics $F = (f)v$ [1]. Ideally, load characteristics of the traction generator should be of a hyperbolic shape. In order to shape characteristics marked in red (see Fig. 48a) from the natural (blue), excitation current of the synchronous traction generator has to be changed according to function D′C′H′B (green). Characteristics obtained using the synchronous traction generator excitation method proposed by the author are provided in Fig. 48b. The AC traction generator's theoretical characteristics (a) and practical artificial load characteristics of the modernized diesel-electric locomotive TEP-70M (b) are provided in Fig. 48.

3.10 Electrodynamic Braking System of the Modernized Locomotive TEP-70

The electrodynamic braking mode of the modernized TEP-70 is set by an integrated braking (three-position) controller installed in the locomotive's control panel. Braking force is set by the controller's position within a range of 0–100%. If the locomotive is braked in an electrodynamic braking mode, then the braking effect can be increased by using an automated braking (pneumatic) system. For that purpose, an automatic braking valve installed in the locomotive's control panel must be used.

Electrodynamic braking conditions of the modernized diesel-electric locomotive TEP-70. The following actions in the system are taking place when the driver switches the braking controller into electrodynamic braking mode: DC serially excitation traction motors M1–M6 are disconnected from the power supply (terminals of synchronous traction generator-rectifier), and traction motors M1–M6 are connected to the individual braking resistors R_{b1}–R_{b6}. Closed dynamic braking circuits of each traction motor are created where dynamic braking currents I_{b1}–I_{b6} are flowing. At the same time, serially connected traction motor excitation windings 1–6 are connected to the terminals of the synchronous traction generator-rectifier.

Control of electrodynamic braking modes. Depending on the braking switch position, the onboard computer LocoControl receives automatic control information. Many sensors are connected to the onboard computer LocoControl which supply information on the general state of the locomotive: changes in the compressor's oil, pressures in the main tanks, temperature in the main ICE cooling loop, temperature of the front bearing of the traction generator, temperature of the rear bearing of the traction generator and temperature of the braking resistors. Sensors provide information about changes in currents of traction motors 1–6, traction generator current after rectification, traction motor excitation current (in EDB mode) and traction generator rectified voltage U_d. The electromagnetic braking power circuit of the modernized diesel-electric locomotive TEP-70 includes braking resistors R_{b1}–R_{b6} and DC serially excited traction motors M1–M6. The temperature of the traction motor dynamic braking resistors is monitored by thermo-sensors BT3–BT8 (see Fig. 49). Sensor BV is used to obtain deviation signals of rectified voltage U_d. Sensors BT1–BT2 are used to obtain deviation signals of traction generator winding temperature and traction rectifier temperature. Sensor CS7 is used to measure excitation current deviation, which generates deviation signals of excitation circuit current ΔI_E. Output voltage of the synchronous traction generator is proportional to diesel engine crankshaft speed n_D and magnetic flux Φ_G. Electrodynamic braking resistors of the modernized diesel-electric locomotive TEP-70 are cooled by fans (MV). Fan speed is directly proportional to the values of braking currents I_{b1}–I_{b6}. When braking current values increase, at the same time, braking voltage U_b and the speed of the DC fan motor rotors increases.

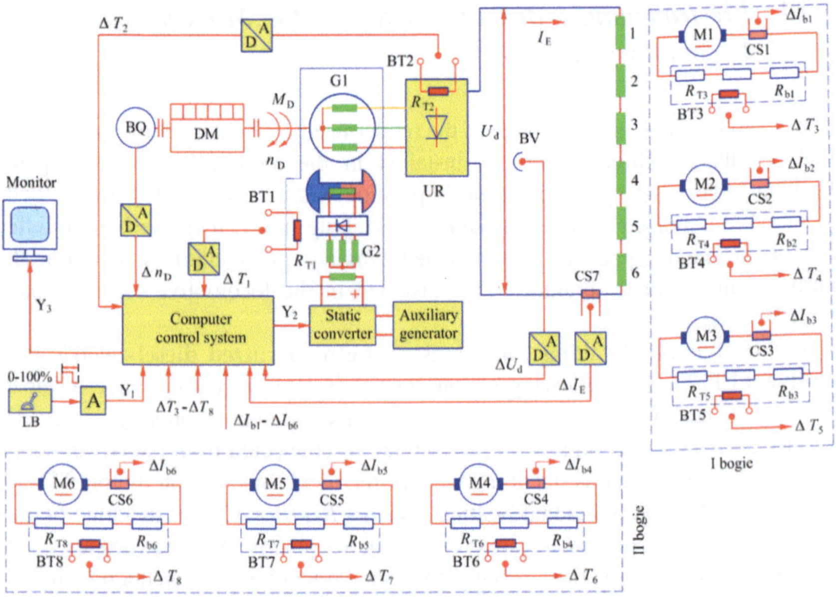

Fig. 49 Diagram of computer control of the modernized AC/DC diesel-electric locomotive TEP-70M electric drive in electrodynamic braking mode. CS1–CS6—current sensors; BT1–BT6 – temperature sensors; R_{T1}–R_{T6}—temperature sensors/thermistors; ΔT_3–ΔT_8—braking resistor temperature deviation signals; I_{b1}–I_{b6}—braking currents 1–6

The automatic braking resistor cooling system operates so that the speed of the DC fan motor rotors increases in proportion to the increase in braking currents and temperature of the braking resistors. Thus, automatic heating and cooling of the braking resistor is ensured.

Electrodynamic braking force regulation. A diagram of computer control of the AC/DC diesel-electric locomotive TEP-70 electric drive in electrodynamic braking mode is provided in Fig. 49. When excitation circuit current I_E of traction motors M1–M6 operating in generator mode changes, the magnetic flux Φ_M also changes. Braking force characteristics are obtained when the braking resistor value is constant and the magnetic flux of the traction motors operating in generator mode is changed.

3.11 The Diagnostics of Electrodynamic Braking of Modernized Diesel-Electric Locomotive TEP-70

The diagnostics window of electrodynamic braking parameters of the modernized diesel-electric locomotive TEP-70 is shown in Fig. 50. In the red background of the display, connection diagrams of the individual traction motor armature windings to

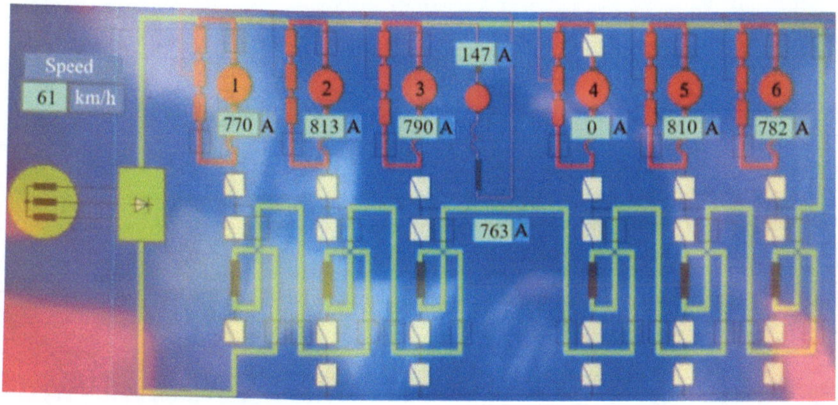

Fig. 50 The window of electrodynamic braking parameters of the modernized diesel-electric locomotive TEP-70M

the individual resistors and values of currents flowing in the respective circuits are shown. The current in the winding of the first traction motor armature is 770 A. The current in the winding of the second traction motor armature is 813 A. The current in the winding of the third traction motor armature is 790 A. The current in the winding of the fourth traction motor armature is 0 A (traction motor is disconnected). The current in the winding of the fifth traction motor armature is 810 A. The current in the winding of the sixth traction motor armature is 782 A. In the green background of the display, the diagrams of connection of the individual motor excitation windings the traction rectifier output voltage. All electrodynamic braking parameters of the modernized diesel-electric locomotive TEP-70 are given at a speed of 61 km/h: coolant temperature (96 °C), coolant pressure (1.05 bar), traction generator load current (3624 A), voltage (657 V) and output power (2380 kW), power, current and three-phase network frequency of the diesel engine cooling roof fans 1 and 2 (58.6 kW, frequency 48.7 Hz, current 92 A); coolant temperature in the main circuit MC (96 °C) and coolant temperature in the auxiliary circuit CC (58 °C). Electrodynamic braking resistors of the modernized AC/DC diesel-electric locomotive TEP-70 are cooled by the fan M. The speed of the fan M is directly proportional to the values of braking currents I_{b1}–I_{b6}. With increasing braking currents, braking voltage U_b increases, and the DC fan motor rotor speed increases simultaneously. The automatic cooling system of the braking resistors operates so that when braking currents increase, the temperature in the braking resistors also increases, and the DC fan motor rotor speed increases proportionally. Thus, automatic heating and cooling of the braking resistors is ensured.

Figure 51 shows electrodynamic braking circuits and values of their currents. In the traction mode, the traction motor M4 is disconnected due to insulation damage. Therefore, current does not flow in the traction motor M4 during electrodynamic braking, i.e., it is equal to 0 A. Current of 147 A flows in the circuit of the fan motor.

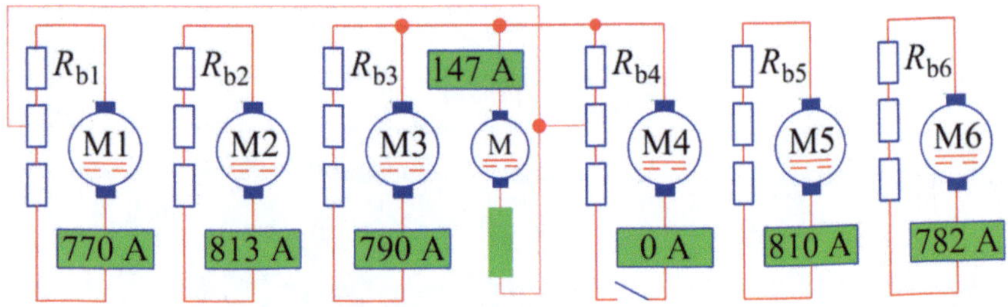

Fig. 51 Electrodynamic braking circuits of the modernized diesel-electric locomotive TEP-70M

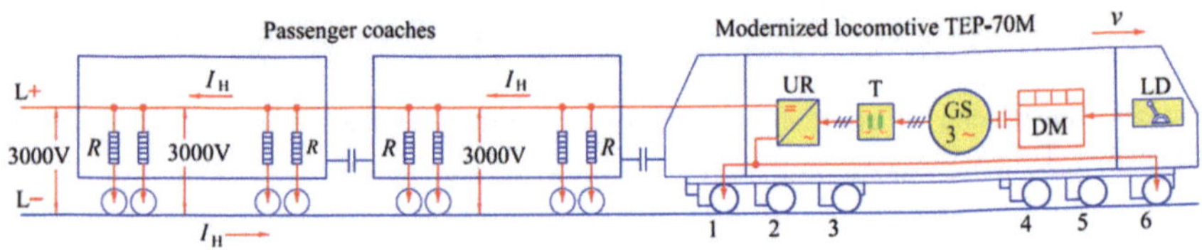

Fig. 52 Diagram of the passenger car heating system of the modernized diesel-electric locomotive TEP-70M: GS—synchronous heating generator; T—dividing transformer; LD—locomotive motor; UR—three-phase rectifier; *R*—electricity consumers/heaters; I_H—current of the heating circuit; *v*—running speed

3.12 *Passenger Car Heating System of Modernized Locomotive TEP-70*

Generator for high-voltage supply to passenger cars. The locomotive is equipped with a generator for high-voltage supply to passenger cars. Trailing passenger cars are heated with a voltage of 3000 V, the output power of which is sufficient for heating 8–10 passenger cars. A diagram of the passenger car heating system of the modernized diesel-electric locomotive TEP-70M is provided in Fig. 52. The power supply source for heating passenger cars is the auxiliary three-phase generator G3–G4 with brushless excitation systems, with total power of 700 kVA. The auxiliary three-phase generator G3–G4 with brushless excitation system is an electric machine in the common housing. The passenger car heating system of the TEP-70 includes auxiliary three-phase generator G3–G4, whose voltage is maintained at a level of 3 × 400 V. This voltage is increased by step-up transformer T with two secondary windings. This voltage is rectified by three-phase rectifiers UR1 and UR2. In order to obtain DC voltage rated at 3000 V for passenger car heating, the outputs of three-phase rectifiers UR1 and UR2 are connected serially. The capacity of the power source for heating of trailing passenger cars is 400 kW.

Automatic control of passenger car heating system parameters of the modernized diesel-electric locomotive TEP-70. The automatic control of passenger car heating system parameters of the TEP-70 is implemented according to the ACS developed by the author using current sensor CS and voltage sensor BV. With increasing numbers of trailing passenger cars, heating circuit current increases and heating circuit voltage decreases. In this case, a voltage of 3000 V is stabilized by increasing the value of the G4 generator's excitation current I_E. This value is adjusted by the output voltage of the static converter when the number of trailing passenger cars increases. The output voltage of the static converter is adjusted by the onboard computer after processing signals of the heating circuit current, heating circuit voltage and passenger car temperature V1 and V*i* sent by sensors CS, BV and BT. The passenger car heating computer control system of the modernized diesel-electric locomotive TEP-70M, which is shown in Fig. 53, is designed for 50 and 100% modes. In 50% mode, cars are supplied with 1500 V from one rectifier (UR1 or UR2). In 100% mode, cars are supplied with 3000 V from serially connected rectifiers.

In this case, the current sensor (bridge) will supply continuous information to the onboard computer upon deviation of passenger car heating circuit current I_H, and the voltage sensor will supply information upon deviation of the 3000-V passenger car heating circuit (Fig. 54).

Conclusions: In 2015, the average fuel consumption of the TEP-70 and TEP-70 BS passenger diesel-electric locomotives (before modernization) was 2.21 l/km. In 2016, the average fuel consumption of the modernized TEP-70M passenger diesel-electric locomotive was 1.7 l/km. Thus the fuel consumption of modernized locomotive TEP-70M (l/km) has been reduced by 23%.

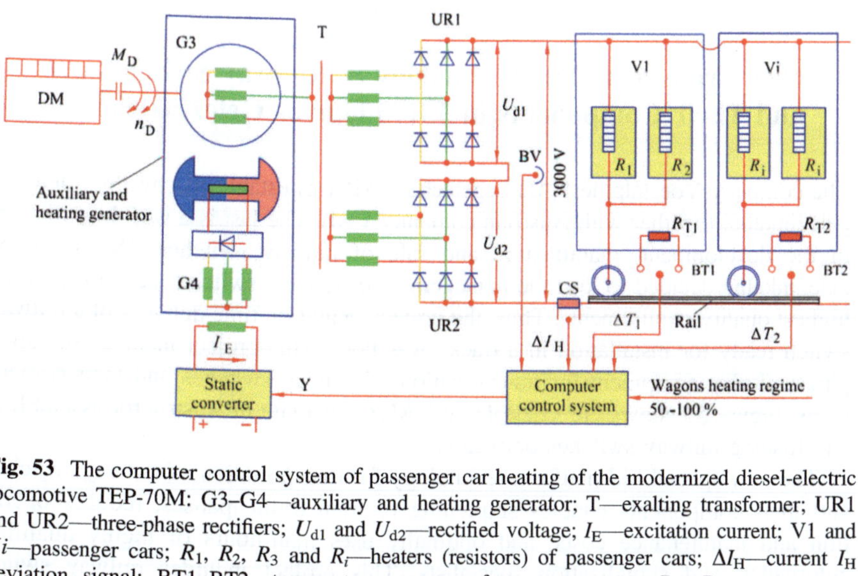

Fig. 53 The computer control system of passenger car heating of the modernized diesel-electric locomotive TEP-70M: G3–G4—auxiliary and heating generator; T—exalting transformer; UR1 and UR2—three-phase rectifiers; U_{d1} and U_{d2}—rectified voltage; I_E—excitation current; V1 and V*i*—passenger cars; R_1, R_2, R_3 and R_i—heaters (resistors) of passenger cars; ΔI_H—current I_H deviation signal; BT1–BT2—temperature sensors of passenger cars; R_{T1}–R_{T2}—temperature sensors/thermistors, ΔT_1–ΔT_2—temperature deviation signal of passenger cars

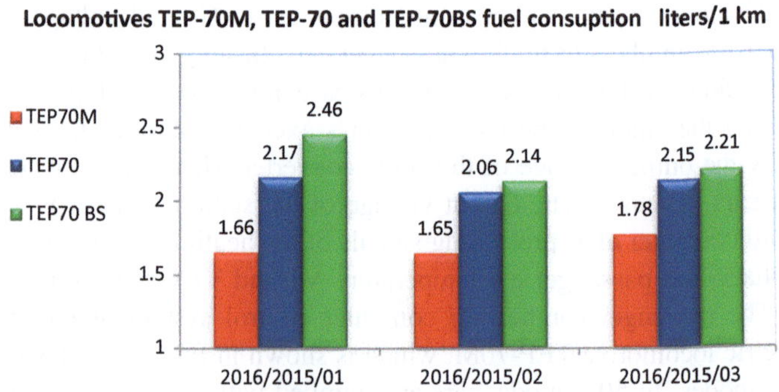

Fig. 54 Modernized passenger locomotive TEP-70 M, TEP-70 and TEP-70BS diesel fuel consumption (l/km)

Fig. 55 Voestalpine VAE GmbH site in Austria for assembling and loading railway switches to cars

4 New Transportation Technology of Railway Switches and Their Components in JSC Lietuvos Geležinkeliai

The company Voestalpine VAE Legetecha UAB was established by JSC Lietuvos geležinkeliai together with Austrian company VAE. The last is a worldwide leader in the development, manufacture and sale of railway switches. JSC Lietuvos geležinkeliai decided to lead the management of railway switches according to the highest quality requirements. Thus, the project "Right on time delivery of a railway switch ready for installation in a track" was born. This project includes so-called plug-and-play equipment for transportation of railway switches and their components. Figure 55 shows the Voestalpine VAE GmbH site in Austria for assembling and loading railway switches onto cars.

The new specialized platform technology for transportation of railway switches and their components enables shortening of train traffic pauses, reduces operation and maintenance costs, and optimally uses man-hours of highly qualified automation and signalization specialists. This project includes railway switch development and manufacturing, site delivery, installation and operation phases.

The project concept is technology according to which a railway switch that is unsuitable for operation is disassembled from a track, loaded to the special platforms for oversize cargo and safely transported to a railway switch repair and manufacturing site, where its assemblies and parts are inspected and refurbished. The repaired or new railway switch is then transported to the installation site on the new specialized platform. Following the implementation plan of the program *Site delivery and installation technology of a railway switch ready for installation in a track*, based on the order of Voestalpine VAE Legetecha, associated company of JSC Lietuvos geležinkeliai, TMHB UAB (representative of Transmashholding ZAO for the Baltic states) delivered the typical platform, and Vilniaus locomotyvų remonto depas UAB, together with Palfinger UAB, modernized the typical platform made in Russia and adapted LST EN 13232 for transportation of railway switches.

4.1 Design Aspects of Platform for Transportation of Railway Switches and Their Components

The technological equipment manufactured by Voestalpine VAE Legetecha UAB for the transportation of railway switches and their components is shown in the transportation position in Fig. 56. The equipment consists of the following: A—MPP-61 platform (modernized tilted platform, no. 00000001) with three-component angle lifting equipment intended for transportation of railway switches and their components; B—technological platform; C—technological platform.

The MPP-61 platform is used for transportation of switches and their components. During the modernization project, transportation of the following switches and their components manufactured by Voestalpine VAE Legetecha UAB was planned on the special platform: intermediate crossing assembly (switch R65-300-1:11, drawing no. E14-006/1, point assembly (switch R65-300-1:11, drawing no. E14-006/2), prepoint assembly (switch R65-300-1:11, drawing no. E14-006/3). The intermediate crossing assembly is shown in Fig. 57.

Technical data of the intermediate crossing assembly P&P switch R65-300-1: weight m = 28,720 kg, length l = 21,475 mm, width a = 4903 mm.

The point assembly (switch R65-300-1:11, drawing no. E14-006/2, Voestalpine VAE Legetecha UAB) is shown in Fig. 58.

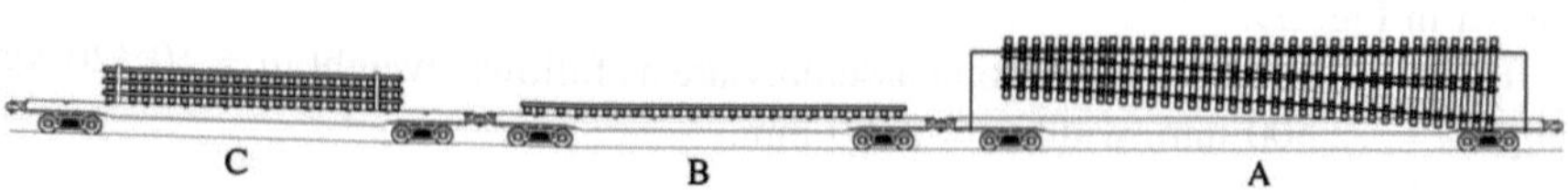

Fig. 56 Technological equipment for transportation of railway switches and their components made by Voestalpine VAE Legetecha UAB

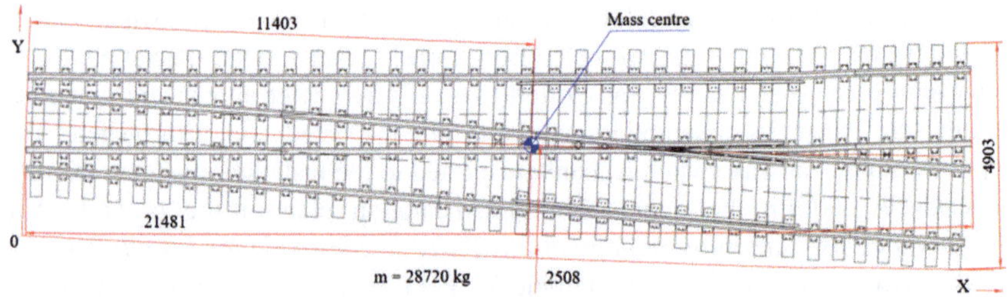

Fig. 57 Intermediate crossing assembly (switch R65-300-1:11, drawing no. E14-006/1, Voestalpine VAE Legetecha UAB)

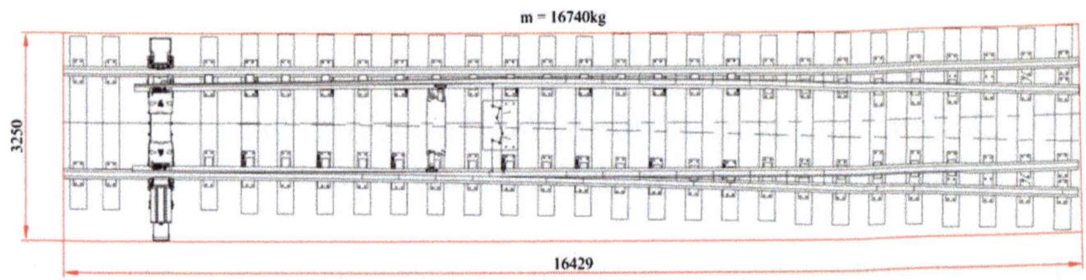

Fig. 58 Point assembly (switch R65-300-1:11, drawing no. E14-006/2, Voestalpine VAE Legetecha UAB)

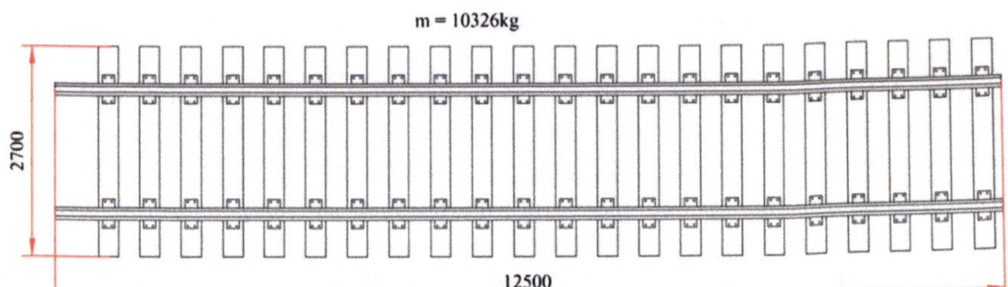

Fig. 59 Pre-point assembly (switch R65-300-1:11, drawing no. E14-006/3, Voestalpine VAE Legetecha UAB)

Technical data for the point assembly are as follows: weight m = 16,740 kg, length l = 16,429 mm, width a = 3250 mm. The pre-point assembly (switch R65-300-1:11, drawing no. E14-006/3, Voestalpine VAE Legetecha UAB) is shown in Fig. 59.

Technical data for the pre-point assembly are as follows: weight m = 10,326 kg, length l = 12,500 mm, width a = 2700 mm.

4.2 Evaluation of Axial Loads of MPP-61 Platform with Three-Component Tilt Cargo Lifting Equipment Without Cargo

A diagram of axial load (forces acting in transverse direction) distribution of the MPP-61 platform with three-component tilted cargo lifting equipment is provided in Fig. 60. A, B and C—newly produced tilted components for fixing and transportation of railway switches and their components.

The main components of the MPP-61 platform with three-component tilted cargo lifting equipment are shown in Fig. 61.

Calculation of static forces. Static forces acting from the empty platform 1–2 wheel sets against a rail are calculated as follows:

$$P\Sigma_{1-2} = 5608\,\text{kgf} + 6770\,\text{kgf} + 6016\,\text{kgf} + 7199\,\text{kgf} = 25,543\,\text{kgf};$$

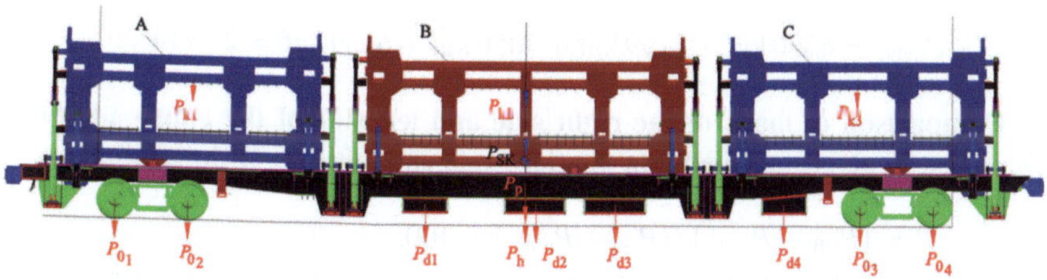

Fig. 60 Diagram of axial load (forces acting in transverse direction) distribution of platform MPP-61 with three-component tilted cargo lifting equipment P_{01}–P_{04}—wheel set axial loads; P_h—hydraulic station load; P_{d1}–P_{d4}—toolbox load; P_{k1}–P_{k3}—load of cargo lifting-landing equipment

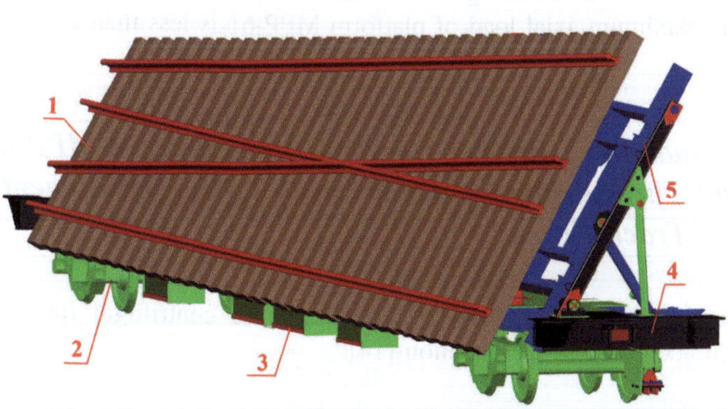

Fig. 61 The main components of platform MPP-61 with three-component tilted cargo lifting equipment: 1—intermediate crossing assembly; 2—bogies of typical platform; 3—diesel engine, hydraulic pump and electrical equipment; 4—frame of typical platform; 5—tilted cargo lifting component

Static forces acting from the empty platform 3–4 wheel sets against a rail are calculated as follows:

$$P\Sigma_{3-4} = 5873\,\text{kgf} + 7056\,\text{kgf} + 5934\,\text{kgf} + 6689\,\text{kgf} = 25,552\,\text{kgf}.$$

Comparison of loads of the empty platform wheel sets 1–2 and 3–4. Load deviation in percentage between bogies I and II:

$$\begin{aligned}\Delta P &= \left(P_{1\text{bogie}} - P_{2\text{bogie.}}\right)/\left(P_{1\text{bogie.}} + P_{2\text{bogie}}\right) \times 100\\ &= (25{,}543\,\text{kgf} - 25{,}552\,\text{kgf})/(25{,}543\,\text{kgf} + 25{,}552\,\text{kgf}) \times 100\\ &= -0.02\%.\end{aligned}$$

Total sum of static forces of the empty platform left side acting against a rail:

$$P_{\text{left.}} = 5608\,\text{kgf} + 6016\,\text{kgf} + 5873\,\text{kgf} + 5934\,\text{kgf} = 23{,}431\,\text{kgf}.$$

Total sum of static forces of the empty platform right side acting against a rail:

$$P_{\text{right.}} = 6770\,\text{kgf} + 7199\,\text{kgf} + 7056\,\text{kgf} + 6689\,\text{kgf} = 27{,}714\,\text{kgf}.$$

Comparison of loads of the right side and left side of the empty platform. Deviation in percentage between the right side and left side of the platform:

$$\begin{aligned}P &= \left(P_{\text{left}} - P_{\text{right.}}\right)/\left(P_{\text{left}} + P_{\text{right.}}\right) \times 100\\ &= (23{,}431\,\text{kgf} - 27{,}714\,\text{kgf})/(23{,}431\,\text{kgf} + 27{,}714\,\text{kgf}) \times 100\\ &= -8.4\%.\end{aligned}$$

Conclusion: Calculations show that the axial loads of platform MPP-61 meet the requirements of technical design and technical specifications for interoperability (TSI). The maximum axial load of platform MPP-61 is less than 22.5 tf.

4.3 Evaluation of Driving Conditions of Platform MPP-61 with Three-Component Tilt Cargo Lifting Equipment in a Track Curve

Evaluation of centrifugal force *F*. The value of centrifugal force F (kN) is determined according to the equation [14]:

$$F = \frac{mv^2}{R}; \tag{23}$$

where m is the platform weight with cargo (kg; P&P switch R65-300-1:11; 1520-mm intermediate crossing assembly drawing no. E12-036/5 Voestalpine VAE Legetecha UAB), v is the speed of the platform moving in a curve (km/h) and R is the radius of the flat curve (m).

The main components of the track plan are straight track sections and curved track sections. Each curved track section consists of round and two transitional curves and is usually referred to as a flat curve. In some cases, transitional curves are not accounted for (due to high radii of circular curves) such that a flat curve corresponds to a circular curve. The sizes of curve radius are determined by the standards according the category of railway track. Adjacent curves must be separated by straight inserts of a defined length. Forces acting in the platform MPP-61 No. 00000001 curve are provided in Fig. 62.

We can state that the train running speed in the railway track plan is limited by small radii of flat curves, elevation of the outside rail, and the lengths of transitional curve inserts and straight inserts. When the platform is running in a curve, it is exposed to centrifugal and centripetal forces, which cause additional wheel pressure against the outside and inside rails. This leads to faster rail wear, distortions appear in the railway track structure, tension in rails and sleepers increases, and train traffic safety is compromised. forces acting on the platform running in a curved track section in a general case are shown in Fig. 62.

Fig. 62 Forces acting in the platform MPP-61 No. 00000001 curve

Figure 62 **designations**: F—centrifugal force, kN; F_h—centripetal force, kN; h—elevation of outside rail, mm; G—platform weight force, kN; S—distance between axes of railheads, mm; α—car tilting angle to a horizon, degrees.

Evaluation of influence of centripetal force F_h: The platform is also exposed to a gravity force caused by outside rail elevation and acts on the inside rail of a curve

$$G = mg,$$

where g is the free-fall acceleration. Centripetal force F_h (kN) is calculated according to the equation:

$$F_h = G \cdot tg\alpha, \tag{24}$$

when

$$tg\alpha = \frac{h}{S}; \tag{25}$$

where h is the elevation of the outside rail (mm) and S is the distance between the axes of railheads (mm).

In order to equalize influence of the both acting forces, the condition of equal wear of outside and inside rails must be met: $F = F_h$. Such outside rail elevation must be installed, which meets the following condition: by inserting the expressions (23)–(25) into this condition, the following equation will be obtained:

$$\frac{mv^2}{R} = mg\frac{h}{S} \tag{26}$$

4.4 *Evaluation of Influence of Outside Rail Elevation in a Track Curve on the Railway Switch Transportation Platform*

In order to reduce the action of side wheels to outside rails, to ensure equal wear of both rails, and to reduce tension in the rails, outside rail elevation h (in mm) is installed in a curve, as shown in Fig. 62. In the general case, the elevation of the outside rail is calculated from Eq. (28) by expressing h (in mm) as:

$$h = \frac{v^2 \cdot S}{g \cdot R}. \tag{27}$$

By inserting S and g values into this equation and matching units, we determine that the elevation of the outside rail in a curve in mm is calculated according to

Eq. (28). The elevation of the outside rail in a curve is calculated when the speed of the platform MPP-61 (modernized tilted platform, No. 00000001) with three-component tilted cargo lifting equipment:

$$h = 12.5\frac{v^2}{R}, \tag{28}$$

4.5 *Evaluation of Influence of Running Speed of Platform MPP-61 with Three-Component Tilt Cargo Lifting Equipment*

The calculation scheme of deviation of the gravity center of the platform MPP-61 with three-component tilted cargo lifting equipment in relation to the longitudinal axial line of the platform is provided in Fig. 63.

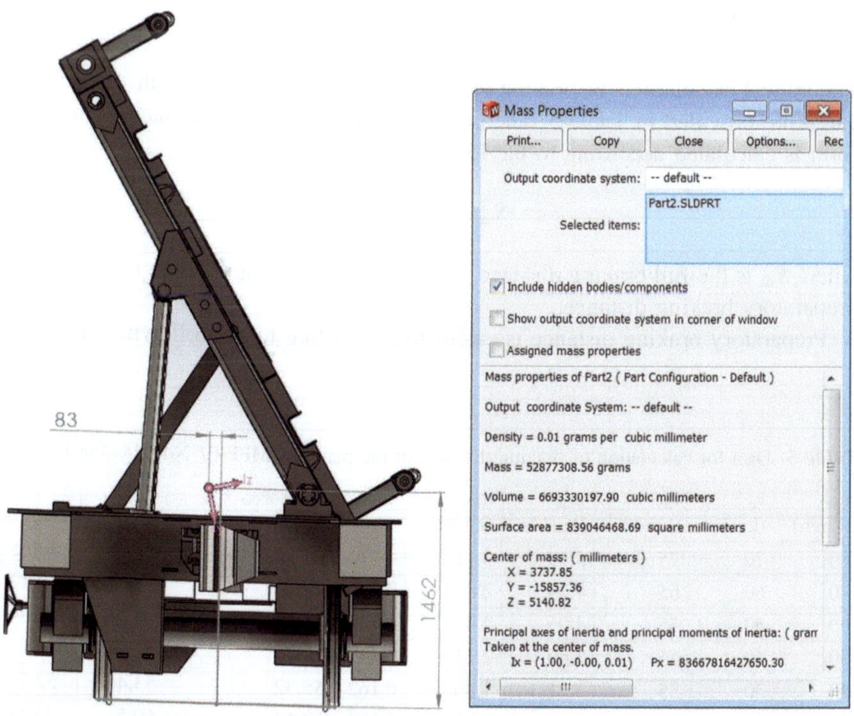

Fig. 63 Calculation scheme of deviation of the gravity center of the platform MPP-61 with three-component tilted cargo lifting equipment in relation to the longitudinal axial line of the platform

By expressing from Eq. (26) the speed v (in km/h) of platform MPP-61, we obtain the following expression for calculation of platform speed:

$$v = \sqrt{\frac{R \cdot h}{12.5}}. \tag{29}$$

Calculations of the deviation of the gravity center of platform MPP-61 in relation to the longitudinal axial line of the platform are made using SOLIDWORKS software.

Conclusion: Evaluation of the influence of running conditions of the platform MPP-61 with three-component tilted cargo lifting equipment in a railway track curve shows that the centrifugal and centripetal forces acting on the platform and the deviation of the gravity center from the platform longitudinal axial line do not exceed permissible values.

4.6 Theoretical Braking Distance Calculations for Platform MPP-61 with Three-Component Tilt Cargo Lifting Equipment

Theoretical calculations of braking distance of platform MPP-61 with the heaviest cargo are provided in Table 5. Theoretically, the full braking distance of the platform is calculated according to the following equation:

$$S_{su} = S_{st.} + S_{p.} \tag{30}$$

where S_{su} is the full braking distance, $S_{st.}$ is the real braking distance and $S_{st.}$ is the preparatory braking distance.

Preparatory braking distance is calculated according to the following equation:

Table 5 Data for calculation of braking distance of the platform MPP-61 No. 00000001

v_p (km/h)	v_g (km/h)	$v_{av.}$ (km/h)	w'_{te} (kgf/t)	w_{0te} (kgf/t)	$\varphi_{tr.}$	$b_{st} + w_{0te} + i_r$ (kgf/t)	b_{st} (kgf/t)	$S_{st.}$ (m)
80	70	75	5.19	3.11	0.122	43.23	40.12	144.65
70	60	65	4.59	2.82	0.128	45.10	42.28	120.12
60	50	55	4.06	2.46	0.136	47.47	45.01	96.54
50	40	45	3.60	2.14	0.147	50.73	48.59	73.97
40	30	35	3.21	1.86	0.162	55.32	53.46	52.71
30	20	25	2.89	1.62	0.183	62.12	60.5	33.52
20	10	15	2.64	1.42	0.217	72.98	71.56	17.18
10	0	5	2.46	1.26	0.277	92.74	91.48	4.45

$$S_{p.} = 0.278 v_p t_p \tag{31}$$

where v_p is the vehicle speed at the beginning of braking and t_p is the braking preparation time.

Real braking distance is calculated according to the following equation:

$$S_{st.} = \sum \frac{500 \cdot (v_p^2 - v_g^2)}{\zeta \cdot (1000 \cdot \varphi_{tr.} \cdot \vartheta_{sk.} + w_{0te} + i_r)} \tag{32}$$

where i_r is the slope of the braking section, $v_{p.}$ is the initial vehicle speed, $v_{g.}$ is the vehicle speed at the end of braking, $\varphi_{tr.}$ is the main resistance to vehicle movement, w_{0te} is the additional resistance to vehicle movement, ζ is vehicle deceleration caused by braking activity and ϑ_{sk}. is the braking factor. Formulas and factors required for calculations are provided in [22].

The theoretical braking distance up to the full stop, calculated using formulas and including preparatory braking distance, is as follows:

(1) 696.36 m, when the initial speed of the platform MPP-61 No. 00000001 is 80 km/h;
(2) 393.59 m, when the initial speed of the platform MPP-61 No. 00000001 is 60 km/h;
(3) 184.54 m, when the initial speed of the platform MPP-61 No. 00000001 is 40 km/h.

4.7 *Dynamic Testing of a Loaded Platform MPP-61 with Three-Component Tilt Cargo Lifting Equipment*

Target of the dynamic tests: To assess the dynamic properties of the modernized platform MPP-61 with cargo, an evaluation was performed on November 7, 2014, in a real railway section with modernized platform MPP-61 transporting the heaviest cargo. The tests included acceleration and braking modes. A braking mode experiment was conducted to determine braking distance and to compare it with theoretical calculations.

Results of the dynamic tests: Analysis of the dynamic tests was performed on the basis of data for safety system KLUB-U of the locomotive pulling the platform. This system includes electronic storage which can be decoded, and all locomotive movement, driver's actions and other data can be determined. The results of the braking tests with cargo at speeds of 60–0 and 80–0 km/h are provided in Figs. 64, 65, 66 and 67. These figures show speed data from the panels of locomotive safety system KLUB-U and decoded data of KLUB-U electronic storage. Hard braking

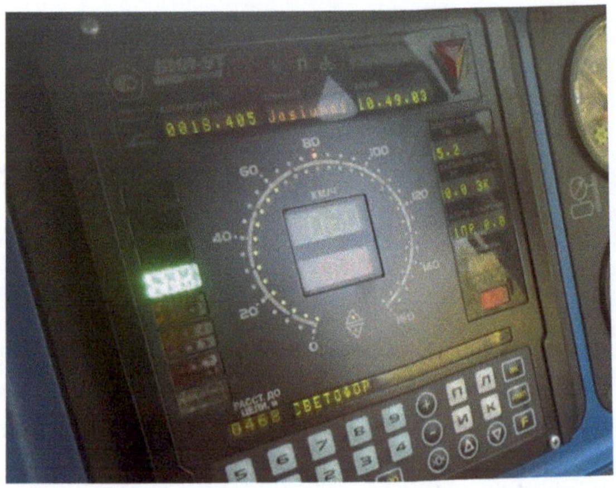

Fig. 64 Driver's display data during braking system tests of the platform MPP-61 No. 00000001 at a speed of 60 km/h

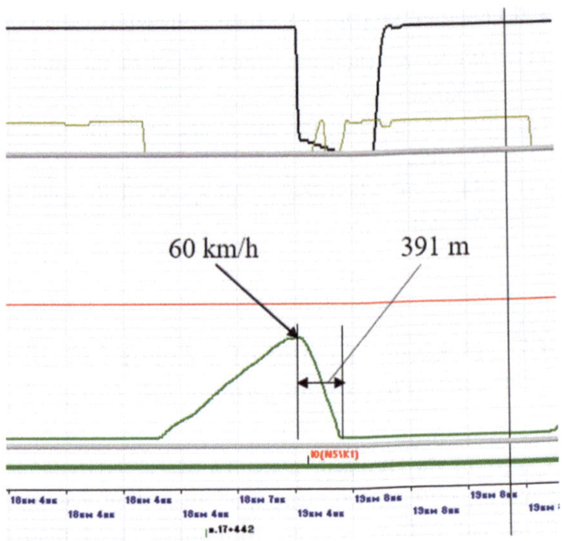

Fig. 65 Braking of the platform MPP-61 No. 00000001 from 60 to 0 km/h (*green* speed trend; *red* permissible speed in the section)

distance was determined during analysis of braking tests of platform MPP-61 with cargo at speeds of 60–0 km/h and 80–0 km/h.

Braking test data of platform MPP-61 with cargo at speeds of 60–0 and 80–0 km/h are provided in Table 6.

Hard braking distances for platform MPP-61 from a maximum speed of 80–0 km/h and 60–0 km/h determined by decoding the electronic storage data of traffic safety system KLUB-U of the locomotive TEM TMH are provided in Figs. 65 and 67.

Fig. 66 Driver's display data during braking system tests of the platform MPP-61 No. 00000001 at a speed of 80 km/h

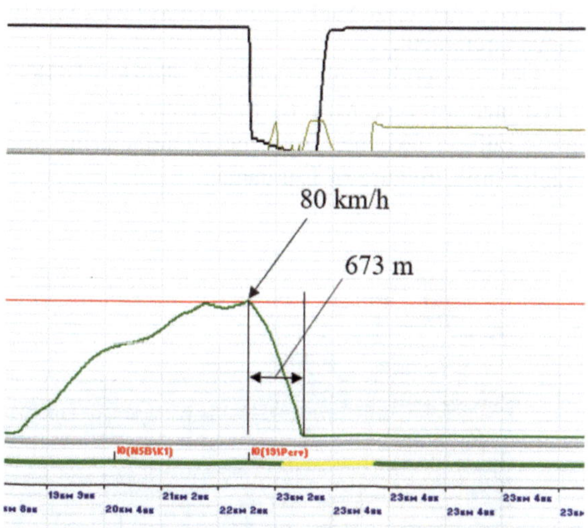

Fig. 67 Hard braking of the platform MPP-61 No. 00000001 from 80 to 0 km/h (*green* speed trend; *red* permissible speed in the section)

Table 6 Results of hard braking of the platform MPP-61 No. 00000001 from 60 to 0 km/h, and from 80 to 0 km/h (direction **Valčiūnai-Jašiūnai** station)

Initial braking speed (km/h)	Braking start/end	Braking distance (m)	Other information
60–0	19 km 4 pk 71 m/19 km 8 pk 62 m	391	Braking cylinder pressure 5.2 kg/cm^2
80–0	22 km 8 pk 11 m/23 km 4 pk 84 m	673	Braking cylinder pressure 5.2 kg/cm^2

Conclusion: The braking distance of platform MPP-61 with heaviest cargo measured during dynamic tests was shorter than the theoretical braking distance down to a full stop. The measured braking distance of the platform with the heaviest cargo at the maximum speed of 80 km/h in the direction of Valčiūnai-Jašiūnai station during dynamic braking tests was 673 m, and the theoretically calculated value was 696.36 m. According to technical requirements, braking distance at a speed of 80 km/h must not exceed 830 m.

4.8 Main Parameters of Platform MPP-61 with Three-Component Tilt Cargo Lifting Equipment

A photo of platform MPP-61 with cargo is shown in Fig. 68.

Wheel sets, bogies, the main frame, the brake system, lifted platforms and autocouplings with dampeners of machinery platforms ППК-3Д made in Russia remain unmodified during modernization. Hydraulic equipment, including main cylinders, hydraulic station and their control, is replaced by the advanced autonomous equipment.

After modernization, machinery platform ППК-3Д is renamed the modernized tilted platform type MPP-61 (hereinafter referred to as the platform), which transports railway switches from the manufacturing site to the installation site in the public infrastructure of Lithuania.

Fig. 68 Photo of platform MPP-61 with three-component tilted cargo lifting equipment with cargo

Transportation of loaded and unloaded platform MPP-61 is carried out with lifted and fixed, tilted load lifting equipment (consisting of three sections).

Modernized platform MPP-61 is intended for transportation: one switch unit of width not exceeding 5.25 m, length 22.5 m.

The platform is autonomous, transported by locomotive, for transportation of new switches from the manufacturing site to a switch installation site.

Used switches are transported from their dismantlement place to the manufacturing site.

The platform is designed for operation in open air at ambient temperature −10 to +40 °C. The chassis consists of two two-axle bogies with wheel sets.

The platform is equipped with a standard brake system. A manual brake is also installed. The three-component tilt cargo lifting equipment is installed on the main frame.

For tilt lift, hydraulic equipment including six pieces of equal main hydraulic cylinders is used. Autocouplings of type SA-3 with dampeners are installed on the main frame.

The hydraulic equipment includes: six equal hydraulic cylinders designed for cargo tilt lift and lowering, six equal double-stroke cylinders for locking of cargo lifted at 61°, hydraulic station with diesel engine and its control equipment (remote controller with cable, radio remote controller).

The diesel engine is designed to transmit torque to the hydraulic pump through a UIC coupling.

The diesel engine is equipped with assemblies and equipment: silencer, fuel tank, starter and battery.

The main parameters of the modernized platform are provided in Table 7.

The modernized platform MPP-61 is for transporting cargo with width not exceeding 5.25 m, length up to 22.5 m and weight up to 36 tons. The main frame is made from welded constructions. It is installed on two bogies. The equipment which is intended to lift the cargo at an angle of 61° and consists of three components is mounted on the main frame. Hydraulic equipment consisting of equal basic hydraulic cylinders is used to lift the load. Year of first production: 2014. Delivered number: 4 units. Delivery country: Lithuania.

4.9 Functional Performance Testing of Platform MPP-61 with Three-Component Tilt Cargo Lifting Equipment

Functional performance testing of the lifting-lowering process of platform MPP-61 was executed in order to assess the performance of all process equipment under real conditions. Functional performance testing of the lifting-lowering process of platform MPP-61 with cargo (intermediate crossing assembly, switch R65-300-1:11,

Table 7 The main parameters of the modernized platform MPP-61 of VLRD UAB

Title	Parameter
Wheel gauge (mm)	1520
Maximum load (tons)	36.0
Maximum dimension of the transported unit (m)	
– Length	22.5
– Width	5.25
The maximum tilt angle of the platform block during transportation, in degrees	61
The duration of transporting the platform from operating position to transporting position, in minutes, no more than	10
Speed of transporting (km/h) no more than	
– In unloaded position	90
– In loaded position	80
Actuator of the executive mechanism	Hydraulic
Gauge of the platform in the transporting positions, when the platform is turned into an angle of 61° when the width of the switch element does not exceed 4.95 m – In loaded position – In unloaded position	The third upper gauge category
Gauge	1-T GOST 9238
Maximum weight of the unloaded platform	52 ± 3%
Overall dimensions in transport position (mm)	
– Length based on the axis	26220 ± 50
– Width	2800 ± 5
– Maximum height	4660 ± 10
Engine	DEUTZD 2011 L021
Engine power (kW)	23

drawing no. E14-006/1, Voestalpine VAE Legetecha UAB) was performed using a remote controller. Cargo lifting-lowering times in normal and emergency modes were tested with the platform. A photo of the functional performance testing of the lifting-lowering of platform MPP-61 with fixed cargo (switch component) is shown in Fig. 70.

Control equipment of platform MPP-61 is shown in Fig. 69.

Times of the tilt lifting process of platform MPP-61 with fixed cargo (the heaviest switch component) by controlling with a remote controller: (1) time of cargo lifting with hydraulic drive determined during tilt lift testing of cargo (the heaviest switch component) is 2–6 min; (2) time of cargo lifting with hydraulic drive determined during emergency tilt lift testing of cargo (the heaviest switch component) using batteries (in case of failure of the main diesel engine) is 12 min.

Fig. 69 Control equipment of platform MPP-61. **a** Designations: 1—remote controller of platform MPP-61; 2—transmitter of the remote control system; 3—manual control panel. **b** Designations: 1—electrical cabinet; 2—diesel engine; 3—hydraulic pump. **c** Designations: 1—hydraulic cylinder; 2—support; 3—part A of tilt platform; 4—part B of tilt platform

Conclusion: After completion of functional performance testing of the lifting-lowering process of platform MPP-61 with the heaviest cargo intermediate crossing assembly (switch R65-300-1:11, drawing No. E14-006/1, Voestalpine VAE Legetecha UAB) in the site, the following was determined: the cargo fixing process and cargo lifting and lowering processes using a remote controller meet the process requirements provided in the modernization design of VLRD UAB (Fig. 70).

Fig. 70 Photo of the functional performance testing of lifting-lowering of platform MPP-61 No. 00000001 with fixed cargo

5 Other UAB Vilnius Locomotive Repair Depot Activities

History. The depot was founded on December 29, 1884, at which time it was called Vilnius Locomotive Return Depot. In 1888, the depot already had its own fleet of locomotives and was called Vilnius Steam Engine Depot. In 1970, with an increase in the number of locomotives, the depot was renamed Vilnius Locomotive Depot. On October 1, 2003, UAB Vilnius Locomotive Repair Depot started its operations. The depot is a subsidiary of JSC Lietuvos geležinkeliai, a Lithuanian railway company, established on October 1, 2003, by restructuring JSC Lietuvos geležinkeliai Vilnius Locomotive Depot into two divisions. The restructuring was carried out according to the requirements of the European Parliament and Council Directive No. 2001/12EB, and sought to reform the rolling stock facilities, separating operational activities from the rolling stock repairs, which could speed up the introduction of new technology and develop modern rolling stock repair facilities. Our company was the first in the Baltic region to engage in complex and responsible modernization of locomotives, and is now implementing a traction fleet renewal program which is of vital importance to the railways. In 2007, our company was nominated by JSC Lietuvos geležinkeliai as the Best Subsidiary of 2007. In 2009, we assembled the TEM TMH locomotive, which was nominated by the Confederation of Lithuanian Industrialists as the Lithuanian Product of the Year and awarded the gold medal.

Certificates. The implementation of quality and environmental management systems conforming to the International Organization for Standardization (ISO) 9001 and ISO 14001 management system standards began in April 2005.

After intense, coherent and concerted efforts by the company's employees, in June 2006, Vilnius Locomotive Repair Depot was granted international certificates certifying that the company had implemented a quality management system in accordance with standards BS EN ISO 9001:2001 and Environmental Management System BS EN ISO 14001:2005.

New locomotive production. Since 2009, our company has cooperated in a joint project with the companies Transmashholding (Russia) and CZ Loko a.s. (Czech Republic) in the production of a series of new shunting locomotives, TEM TMX.

Remotorization. In April 2004, a cooperation agreement was signed with the Kolomna Energy Service OU Company (Estonia), which developed the remotorization project. The diesel locomotive 2M62 underwent extensive repairs simultaneously with ICE replacement; the old 14D40 diesel was replaced with the modern four-stroke diesel 5-26DG.01, without changing the locomotive control systems or auxiliary equipment. In July 2005, the remotorization of M62 diesel locomotives was begun. During 2005–2007, the company focused its efforts on the remotorization of 25 2M62 diesel locomotives and 16 M62 diesel locomotives; in total, JSC Lietuvos geležinkeliai commissioned 66 diesel locomotive sections to be remotorized. The remotorization program was successfully completed. In terms of operating characteristics, compared to the old-style locomotive 14D40 engines and diesel locomotives, the 5-26DG.01 type four-stroke diesel engines achieved a 16–18% reduction in diesel fuel consumption, 4–5% reduction in motor oil consumption and significantly lower diesel locomotive servicing and repair costs.

Modernization. This international project was carried out in 2004: 2M62 locomotives were overhauled, and KR-2 with a Caterpillar engine was installed. For this purpose, a contract was signed with the Hungarian company MAVSzolnoki Jarmujavito Kft, under which the company carried out the modernization of the 2M62 freight locomotives. Modernization included the modification of the locomotive body to adapt it for the installation of new modern units and assemblies.

All mechanical transmission reducer drives were replaced with hydraulic actuators from the German Rexroth company, which ensured the reliability and reduced maintenance and repair costs. During 2004–2008, the company streamlined 58 sections of 2M62M locomotives. The locomotive now has a modern, cost-efficient, USA-made Caterpillar internal combustion engine (CAT 3512B-HD-LC), the German AVK asynchronous DSG86M1 three-phase generator with a Mega Techno DH-6x 1000/2 current rectifier.

Modernization of diesel-electric locomotive series 2M62 (ST 44). Diesel-electric locomotive series 2M62 is for heavy cargo transportation. It is available for 1435/1520-mm gauge with a maximum speed up to 100 km/h. Year modernization began: 2005. Delivered number: 58 units. Delivery countries: Lithuania and Latvia. Diesel-electric locomotive 2M62 is presented in Fig. 71.

The main technical characteristics of the 2M62 locomotive are provided in Table 8.

Fig. 71 Diesel-electric locomotive 2M62

Table 8 The main technical characteristics of locomotive 2M62

Title	2M62	Title	2M62
Wheel set arrangement	Co'-Co'	Diesel engine type	CAT 3512
Rated power (kW)	1700	Traction alternator	DSG 86 M1-4/AVK SEG
Maximal speed (km/h)	100	Power transmission	AC/DC
Weight (tons)	120 (±3%)	Auxiliary transmission	Hydrostatic
Load on axle (tons)	20 (±3%)	Traction effort of coupling (kN)	379.5
Minimum curve radius (m)	80	Control circuit voltage (V)	24

ČME3 locomotive modernization to CAT 3512 and CAT 3508 engines. Beginning in the second half of 2007, the company, in cooperation with the Czech company CZ Loko, undertook implementation of the modernization project for CME-3 M shunting locomotives. The first phase included the modernization of locomotives with modern Caterpillar ICEs made in the USA, the Caterpillar 3512B DITA (1450 kW) and 3508B DITA (940 kW). For the first time in the practice of diesel locomotive modernizations, modular units were applied; each of the modules was produced and completed in the shop and then moved onto the main frame.

Production of shunting locomotive series ČME-3M. Shunting locomotive series ČME-3M is for heavy cargo transfers. It is available for 1435/1520-mm gauge with a maximum speed up to 90 km/h. Year production began: 2008.

Delivered number: 22 units. Delivery country: Lithuania. The shunting locomotive series ČME-3 M is presented in Fig. 72.

The main technical characteristics of the ČME-3M shunting locomotive are provided in Table 9.

The locomotive was assembled from several main modules: the electric control module, the driver's compartment module, the power module, the auxiliary equipment module, the cooling module and the pneumatic module. Single locomotive use can reduce fuel consumption by up to 35% and maintenance and repair costs by up to 70%.

Fig. 72 Shunting locomotive series ČME-3M

Table 9 The main technical characteristics of the ČME-3M series shunting locomotive

Title	ČME-3M	Title	ČME-3M
Wheel set arrangement	Co'-Co'	Diesel engine type	CAT 3512C/ CAT 3508C
Rated power (kW)	1550/970	Traction alternator	Siemens 1FC2 631-6B029T
Wheel gauge (mm)	1435/1520		
Maximal speed (km/h)	90	Power transmission	AC/DC
Weight (tons)	120 (±3%)	Auxiliary transmission	Electric
Load on axle (tons)	20 (±3%)	Traction effort of coupling (kN)	436
Minimum curve	80	Permanent traction effort of coupling (kN)	259
		At speed (km/h)	14.3
Control circuit voltage (V)	24		

New projects. The main technical characteristics of the AGRc 1200 catenary system maintenance machine are provided in Table 10.

In 2012, the company turned a new page in their business history by starting production of railway infrastructure machinery. The KTD motor trolley for catenary system maintenance was created and manufactured together with SVI S.p.A., an Italian rail transport production company.

At the same time, another machine, the AGRc-1200, operating on 1520 and 1435-mm gauge railway tracks, was manufactured. This motor trolley will be used for railway track inspection and repair and for laying of new track. The AGRc 1200 catenary system maintenance machine is presented in Fig. 73.

Table 10 The main technical characteristics of the AGRc 1200 catenary system maintenance machine

Title	AGRc 1200	Title	AGRc 1200
Wheel gauge (mm)	1435/1520	Diesel engine type	CAT 3512C/CAT 3508C
Maximal speed (km/h)	80	DEUTZ engine power (kW)	2 × 440
Weight (tons)	50	Transmission type	Hydro SAUER-DANFOSS
Motorized axles	4	Fuel tank capacity (l)	800
Maximum length of frame (mm)	18500	Air compressed system power (m^3/h)	60
Maximum high (mm)	3800	Environment temperature (° C)	−32/+40
Loading capacity (t)	12	Control circuit voltage (V)	24
Minimum curve radius (m)	150		

Fig. 73 Catenary system maintenance machine AGRc 1200

In 2013, a new project for our company was started. In cooperation with Ukrainian company Lugansteplovoz, Russian company CJSC Bryansk Machine Building Plant and Czech company CZLoko, we began production of a new TEM LTH four-axle shunting diesel locomotive. This locomotive represents our first product equipped with an induction motor drive and a traction motor.

Production of shunting locomotive series TEM 33. Shunting locomotive series TEM 33 is designed for heavy transfer and is available for 1435/1520-mm gauge with a maximum speed up to 100 km/h. Year production began: 2013. Delivered number: 1 unit. Delivery country: Russia. Shunting locomotive series TEM 33 is presented in Fig. 74. The main technical characteristics of shunting locomotive series TEM 33 are provided in Table 11.

Production of hybrid drive shunting locomotive series TEM 35. In the fourth quarter of 2012, in cooperation with UAB Transmashholding and CZLoko Company, the production of a new TEM TMX 35 hybrid drive shunting diesel locomotive began. This locomotive is intended for the Russian market. Innovation: an additional energy storage unit would allow the effective use of diesel motor power as well as reduce fuel consumption to the minimum. Hybrid drive shunting

Fig. 74 Shunting locomotive series TEM 33

Table 11 The main technical characteristics of the TEM 33 series shunting locomotive

Title	TEM 33	Title	TEM 33
Wheel set arrangement	Co‘-Co‘	Diesel engine type	Caterpillar C18
Rated power (kW)	2 × 571	Traction alternator	2xGS523
Wheel gauge (mm)	1435/1520	Power transmission	AC/AC
Maximal speed (km/h)	100	Traction effort of coupling (kN)	351
Weight (tons)	123 (±3%)	Permanent traction effort of coupling (kN)	206
		At speed (km/h)	7.8
Load on axle (tons)	20.5	Control circuit voltage (V)	110
Minimum curve radius (m)	80		

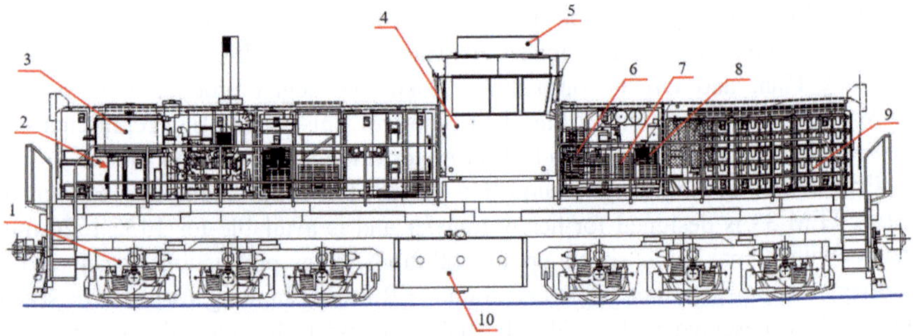

Fig. 75 Hybrid drive shunting locomotive TEM 35: 1—bogie with AC traction motor; 2, 3—diesel engine equipment; 4—cab; 5—conditioner; 6—air purifier; 7—compressor; 8—traction motor cooling fan; 9—energy storage batteries (supercapacitors); 10—fuel tank

Table 12 The main technical characteristics of hybrid drive shunting locomotive TEM 35

Title	TEM 35
Wheel-set arrangement	Co‘-Co‘
Rated power (kW)	571
Rated energy of power accumulators (MJ)	22.7
Wheel gauge (mm)	1435/1520
Maximal speed (km/h)	100
Weight (tons)	120 (±3%)
Load on axle (tons)	20
Minimum curve radius (m)	80
Diesel engine type	Caterpillar C18
Traction alternator	2xGS523
Power transmission	AC/AC
Power on the shaft of traction motors in continuous operation (when powered by engine) (kW), not less than	324
Power on the shaft of traction motors in continuous operation (when powered by engine and storage device, with taken 410 kW power from storage device) (kW), not less than	683
Speed of continuous operation when traction motors are powered from diesel generator (km/h)	7.8
Control circuit voltage (V)	110

locomotive TEM 35 is presented in Fig. 75. The main technical characteristics of hybrid drive shunting locomotive TEM 35 are provided in Table 12.

TGM4 locomotive modernization with the CAT engine. Our company carries out modernization for TGM4 shunting locomotives with the Caterpillar CAT 3508B engine. The shunting locomotive TGM-4 is presented in Fig. 76, and its main technical characteristics are provided in Table 13.

Fig. 76 Shunting locomotive TGM-4

Table 13 The main technical characteristics of shunting locomotive TGM-4

Title	TGM-4	Title	TGM-4
Wheel set arrangement	B'o-B'o	Diesel engine type	Caterpillar 3508
Rated power (kW)	621	Power transmission	Hydrodynamic
Wheel gauge (mm)	1520	Transmission type	UGP 750
Maximal speed (km/h)	55	Traction effort of coupling (kN)	259
Weight (tons)	80 (±3%)	Permanent traction effort of coupling (kN)	176.5
		At speed (km/h)	5
Load on axle (tons)	20	Control circuit voltage (V)	24
Minimum curve radius (m)	40	Fuel consumption saving (in comparison with base model)	30%

It is a leap into the new era in the modernization of 600-kW class shunting locomotives. The four-axle shunting locomotive TGM-4 is equipped with an electric AC/DC drive and the Caterpillar 3508B ICE (640 kW), without altering the long-term traction and speed of the diesel locomotive. The diesel locomotive control computer system is adapted for one-person operation, so the driver no longer needs an assistant.

Modernization includes the KR-2, with major repairs of the locomotive frame, bodywork, driver's car, carriages, hydraulic drive, auxiliary power plant equipment, compressors and pneumatic system, sander system, road safety and radio communication systems, in accordance with the TGM-4 locomotive overhaul instructions.

TGK-2 locomotive modernization with the Volvo-Penta engine. In 2010, our company offered modernization projects for the TGK-2 low-power shunting

Fig. 77 Shunting locomotive TGK-2

Table 14 The main technical characteristics of shunting locomotive TGK-2

Title	TGK-2	Title	TGK-2
Wheel set arrangement	0-2-0	Diesel engine type	Volvo-Penta TAD95851VE
Rated power (kW)	224	Hydraulic transmission type	UGP 230
Wheel gauge (mm)	1520	Traction effort of coupling (kN)	100.5
Maximal speed (km/h)	60	Permanent traction effort of coupling (kN)	70
Maximal speed in shunting mode (km/h)	30	Clearance	03-T
Weight (tons)	30 (±3%)	Minimum curve radius (m)	50
Maximal train weight (tons)	700		

locomotive. The first modernized locomotive was equipped with a new cab and a new hood, a new Volvo-Penta ICE and diagnostic controls. The TGK-2 shunting locomotive is presented in Fig. 77, and its main technical characteristics are provided in Table 14.

6 Conclusions

1. After modernization of JSC Lietuvos geležinkeliai shunting locomotives, fuel consumption was reduced up to 35%.
2. The traction generator excitation control system proposed by the author enabled the artificial load characteristics to approach a hyperbolic nature; ICE power is fully used across the entire speed range.

3. The traction generator's artificial load characteristics are stable. Parts of characteristics of maximum current, current restriction and its hyperbolic nature were shaped in the ACS.
4. The traction generator excitation system (contactless, without moving parts) has become more reliable; it does not contain the electric machine intended for excitation, which is driven by a diesel engine in typical arrangements.
5. Contactless components are used in the static converter—controlled semi-conductors—thyristors or IGBT transistors.
6. The traction generator's excitation system is long-lived.
7. The static converter has enabled a significant reduction in the traction generator's control power.
8. After modernization of JSC Lietuvos geležinkeliai freight locomotives, fuel consumption was reduced up to 25%.
9. Fuel consumption of the modernized diesel-electric locomotive TEP-70M decreased by 23%.
10. The modernized diesel-electric locomotive TEP-70M can reliably heat 8–10 passenger cars.
11. The implemented diagnostic systems enable prompt elimination of complex faults.
12. The time required for replacement of railway switches and their components has been reduced fourfold after implementation of the new transportation technology for switches and their components.
13. The time required for transportation of railway switches and their components was reduced after manufacturing of the modernized platform MPP-61.
14. The modernized platform MPP-61 has reduced the time required for loading/unloading of railway switches and their components.
15. The modernized platform MPP-61 has reduced the number of special track repair machines used for installation of railway switches and their components.

References

1. Kossov EE, Sukhoparov SI (1999) Optimization of operating modes of diesel generators. Intekst, Moscow (in Russian)
2. Kalinin VK (1991) Electric locomotives and electric trains. Transport, Moscow (in Russian)
3. Liudvinavičius L, Dailydka S, Vaičiūnas G (2015) Traukos energetinės sistemos ir jų valdymas. Vadovėlis. Vilnius: Technika. [In Lithuanian: Traction energy system and their control. Coursebook]
4. Volkov NI, Milovzorov VP (1978) Electric machinery automation. Higher School, Moscow (in Russian)
5. Yuferov FM (1976) Electric machines automatic devices. Higher School, Moscow (in Russian)
6. Östlund S (2011) Electric railway traction. KTH, Stockholm
7. Strekopytov V, Grishchenko A, Kruchek A (2003) Electric drives of the locomotives. Marshrut, Moscow

8. Fuest K, Döring P (2000) Elektrische Maschine und Antriebe. Lehr- und Arbeitsbuch. Wiesbaden: Vieweg [In German: An electrical machine and drive. Teaching and workbook. Vieweg]
9. Filonov CP et al (1996) Diesel locomotive 2TE116. Transport, Moscow (in Russian)
10. Kossov EE et al (2011) Diesel generator sets rates using the energy storage device in the electric the power transmission. The Science and Business, Development Ways, vol 6, pp 95–100
11. Slemon GR (1992) Electric machines and drives. Addison-Wesley Publishing Company
12. Jolevski D (2009) Excitation system of synchronous generator. University of Split, Faculty of Electrical Engineering, Mechanical Engineering and Naval Architecture
13. Liudvinavičius L, Lingaitis LP (2009) Locomotive energy savings possibilities. Transp Prob 4 (3–1):35–41
14. Lingaitis LP, Liudvinavičius L, Butkevicius J, Podagėlis I, Sakalauskas K, Vaičiūnas G, Bureika G, Gailienė I, Petrenko V, Subačius R (2009) Geležinkeliai. Bendrasis kursas: vadovėlis. Vilnius: Technika [In Lithuanian: Railways. General course: text book]
15. Liudvinavičius L, Lingaitis LP, Dailydka S (2010) Traukos riedmenų elektros pavaros ir jų valdymas: bendrasis aukštųjų mokyklų vadovėlis. Vilnius: Technika [In Lithuanian: Traction rolling stock power drives and their control: common high school text book]
16. CZ LOCO a.s. (2009) Lokomotiv TMX series, technical description (in Russian)
17. Machowski J, Bialek JW, Robak S, Bumby JR (1998) Excitation control system for use with synchronous generators. IEE Proc Gener Transm Distrib 145(5):537–546
18. Sumina D, Erceg G, Idžotić T (2005) Excitation control of a synchronous generator using fuzzy logic stabilizing controller. In: European conference on power electronics and applications. Dresden 6 pp
19. IEEE Power Engineering Society IEEE (2006) Recommended practice for excitation system models for power system stability studies. IEEE 421-5-2005, New York, pp 1–93
20. Miskovic M, Mirosevic M, Milkovic M (2009) Analysis of synchronous generator angular stability depending on the choice of the excitation system. Energija 58(4):430–445
21. Bodefeld T, Sequenz H (1971) Elektrische Maschinen, 8th edn. Springer, Berlin, Heidelberg, New York [In German: Electrical machinery]
22. Bureika G (2002) Riedmenų traukos teorija: paskaitų tekstas. Vilnius: Technika [In Lithuanian: Rolling stock traction theory. Instructional book]

Systems Approach to the Organization of Locomotive Maintenance on Ukraine Railways

Eduard Tartakovskyi, Oleksander Ustenko, Volodymyr Puzyr and Yurii Datsun

Abstract A basic criterion of locomotive maintenance system optimization in rail transport is the ability to provide the required volume of transportation and safety at the lowest cost. Early investigations into locomotive maintenance did not touch on problems encountered in modern rail engineering. Within the context of railway reform efforts in Ukraine, this has highlighted the practical necessity of actually making optimization of the locomotive maintenance system the top priority. The reforms carried out on Ukraine's rail transport system include considerable changes, both qualitative and quantitative, in locomotive repair plants, their rearrangement by certain criteria to distinguish between repair and maintenance, and the formation of repair clusters oriented to service a certain type of locomotive operating on a certain rail site. The current study deals with the development of a virtual maintenance, service and repair system for rolling stock which takes into account the type of locomotive, state of repair depots and plants, and a method for assessing the organizational and technical level of a locomotive repair plant.

Keywords Maintenance system · Rail transport · Optimization · Organizational and technical level · Locomotive repair plant · Rolling stock

1 Introduction

In rail transport, the system of locomotive maintenance has a whole range of peculiarities of complex technical systems, including a common objective, manageability, interrelation of components and hierarchical structure. The main criterion for optimization of such a system is the ability to provide the required traffic volume and safety at the lowest cost.

E. Tartakovskyi (✉) · O. Ustenko · V. Puzyr · Y. Datsun
Ukrainian State University of Railway Transport,
Feiierbakh sq. 7, Kharkiv, Ukraine
e-mail: cvntt@rambler.ru

A. Sładkowski (ed.), *Rail Transport—Systems Approach*, Studies in Systems, Decision and Control 87, DOI 10.1007/978-3-319-51502-1_5

Early investigations into locomotive maintenance carried out by a great number of scientific institutions did not touch on problems occurring nowadays in rail engineering. In particular, some researchers [1–7] have devoted much attention to determining the rational characteristics of modern locomotives, while ignoring the current state and potential of locomotive repair plants.

Some works [8–11], in determining the parameters for improved locomotive performance, have taken into account changes in their state over the whole life-cycle, but have considered their maintenance system as constant.

A great number of studies [12–22] dealing with locomotive maintenance systems have embraced issues such as structure, service region, modes and operational conditions, but have ignored the organizational and technical state of repair facilities, repairability and repair quality at locomotive repair plants.

Some studies [8, 14, 23] have provided detailed assessments of railway repair plants and determination of their main characteristics. Other works [24, 25] have defined an organizational and technical level of locomotive repair plants as a formal procedure—notably, certification—the main disadvantage of which is the subjective nature of the conclusions and variation among expert commissions certifying repair plants.

Problem statement: The analysis that has been carried out has shown the theoretical and practical need for solving problems in order to optimize the locomotive maintenance system on Ukraine's railways in the period of reforms.

Rail reform in Ukraine has been under way for several years, and the process has involved various organizational stages, with the current stage stipulating considerable change, both quantitative and qualitative, at locomotive repair plants, their rearrangement by certain criteria differentiating between repair and maintenance, and the organization of repair clusters focused on handling a certain type of locomotive running within a certain rail site.

One of the most important issues in the reform process is developing methods for assessing the organizational and technical level of locomotive repair plants. A number of criteria contributing to optimization must be determined. In practice, this will allow a reduction in operating expenses due to the elimination of small repair shops at ineffective sections, with repair services concentrated at large automated plants with highly qualified personnel, which will be included in a single information space of the industry.

In choosing solutions for the various problems that arise, the authors used their long-term experience in the assessment of locomotive repair plants. The scientific foundation of the research involves methods for simulating the behaviour of a virtual repair plant, and application of genetic algorithms, expert methods and fuzzy logic techniques.

The research tasks include the following:

- Developing a concept of the virtual maintenance, service and repair system for rolling stock based on its type and the condition of repair depots and plants

- Developing a model of the virtual maintenance, service and repair system for locomotives
- Developing a method for assessing the organization and technical level of a locomotive repair plant.

2 The Concept of a Virtual Control System of Tractive Rolling Stock Maintenance

Ukraine's railways handle a great volume of various types of traffic; therefore, there are many types and series of locomotives and multiple units. The whole locomotive fleet is operated and repaired at more than 60 locomotive depots, classified by purpose, type of locomotive and service capacity. Production spaces, repair equipment, and the state and quality of service and repair are characteristics that specify a certain depot for certain types of repair and certain types of locomotives [26, 27]. As noted in the Concept and Restructuring Plan on Railway Transport [28], locomotive depots will be divided into maintenance and repair types, which will contribute to higher-quality maintenance and increased reliability.

To maintain locomotives in good operational condition, the railways apply a planned preventive maintenance and repair system, which may fail during periods of unstable transport flows, irregular spare parts supply and lack of funds. Owing to modernization and procurement of new-generation locomotives, with their respective maintenance and repair systems, as well as the rearrangement of locomotive depots, a system more suited to the current situation is needed which will improve the quality of service of new, modernized and existing locomotives, and reduce operating expenses.

A reasonable solution can be reached not only by considering foreign experience, but also by simulating and studying the virtual locomotive maintenance control system (VLMCS), meaning that the locomotive maintenance system can be amended on the basis of a systems approach considering structural peculiarities, maintenance and repair system, condition and capacity. This can be realized in four stages. The first stage includes the analysis of the locomotive's structure and reliability, so the following should be fulfilled and determined:

(1) Collection of technical/economic factors for a certain type of locomotive
(2) A list of design reliability factors for a certain type of locomotive, and their values
(3) A list of repairability factors for a certain type of locomotive, and their values

The second stage includes assessment of the technical level of the locomotive depots involved in the maintenance, service and repair of a certain type of locomotive, either currently or planned for the future. The following should be performed and determined:

(1) Assessment of the general state of existing depots which maintain, service and repair a certain type of locomotive
(2) Production capacity of each depot
(3) Operational factors of production lines in a depot
(4) Mechanization and automation levels of production processes in the depot
(5) A depot's technical level
(6) A list of additional factors characterizing a repair plant

The third stage deals with methods for assessing the state of locomotives during their operation and maintenance, in which the following should be defined:

(1) A list of factors characterizing locomotives during operation
(2) Methods for defining the state of a locomotive
(3) The service and repair system for a given locomotive
(4) A list of factors characterizing a given locomotive maintenance system, and their values

The fourth stage deals with the development of the virtual locomotive maintenance control system, and requires the following:

(1) Defining approaches to the development of virtual business systems
(2) Establishing dependences between factors of design, reliability, operational and technical service, repair of locomotives, and factors characterizing the state and technological potential of repair plants
(3) Designing a model of the virtual locomotive maintenance system (designing processes in the system, binding them to resources and resource search by contracts)
(4) Operating the virtual locomotive maintenance control system (monitoring and control)

The VLMCS is a network computer organizational structure consisting of heterogeneous components located in different places. The adjective "virtual" means "artificial, imaginary, unreal", or "extended due to joint resources". The VLMCS is designed by selecting needed organizational and technological resources from different locomotive operational and repair plants, scientific institutions and various enterprises engaged in a repair cycle, and their integration through the computer network. This will lead to the formation of a flexible and dynamic organizational system most suitable for maintaining locomotives in operational condition and with the most effective usage. It should be noted that the virtual locomotive maintenance control system cannot be completely virtual, i.e. without basic physical structures. This involves intensive cooperation between specialists and divisions of locomotive depots, scientific institutions and virtual enterprises on the basis of state-of-the-art information and communications technologies. This cooperation is aimed at intensifying interrelation and coordination of depots, scientific institutions and enterprises, and eventually higher competitiveness of production, reliability, and lower operational expenses at the second stage of the locomotive's lifecycle. Furthermore, the VLMCS is completely oriented to fulfilling the main objective of rail transport, i.e. effective, safe and timely passenger and freight transportation.

Among the advantages of the VLMCS are the following:

- Speed of locomotive service and repair
- Possibility of decreasing rolling stock maintenance, service and repair expenses
- Flexible adaptation of the maintenance and repair system to changes in traffic volume and quality of repair performed by locomotive repair plants

However, there are disadvantages:

- Economic dependency on partners due to narrow specialization of the network's members
- Lack of social and material support of partners due to refusal to use classic long-range contractual forms and conventional working relations
- Possible loss of simplicity, particularly due to the diverse nature of the members, unclear membership procedures, openness of networks, dynamics of self-organization, and uncertain planning for members of the virtual enterprise

Appropriate managerial approaches are also important for planning, organizing and coordinating the VLMCS. The main general management functions of the VLMCS are as follows:

(1) Setting objectives and goals for the virtual locomotive maintenance system
(2) Identifying and assessing possible maintenance and repair depots, scientific institutions and enterprises participating in operational and repair locomotive cycles
(3) Ranking executives which match certain tasks
(4) Continuously monitoring and realigning partners and resources by task
(5) Dispatching virtual locomotive maintenance

An analysis of studies [8, 29] and expert methods helped define the major characteristics of the VLMCS:

- Open distributed structure
- Flexibility
- Priority of horizontal links
- Autonomy and narrow specialization among network members
- High status of information and human resources engaged in integration
- Integration of the advanced technologies, means of production and experience within strategically smart formations
- Concentration of resources around key business processes
- Formation of autonomous working groups, locally distant
- Temporary nature of relations among partners, flexibility, possibility of realignment
- Combination of centralization and decentralization in management, with priority of decentralization

- Priority character of coordination bonds
- Interrelation of specialists on the basis of computers
- Development of heterogeneous computer networks and environments, application of different software tools

The key advantage of virtual forms is the ability to choose and use the best resources, knowledge and capacities in the most effective way. The following are necessary for successful operation of the VLMCS:

- Choice of criteria and assessment of efficiency of locomotive depots, scientific institutions and different enterprises engaged in locomotive maintenance and repair during the process of selecting them for the virtual locomotive maintenance system
- Choice of an optimal design solution for routing technological processes of locomotive maintenance, service and repair
- Determination of an optimal production structure of the virtual locomotive maintenance, service and repair system

From a practical perspective, the VLMCS is a network of agents cooperating freely and located remotely. These agents develop one or more joint projects and serve as partners who cooperate, collaborate and coordinate the project. Multi-agent systems are one of the most rapidly growing spheres of artificial intelligence. The development of the VLMCS deals with intelligent simulation of how complex, heterogeneous, distant agents interrelate. Creation of multi-level friendly interfaces among agents is of primary importance. Therefore, the first step entails building models of relations between producer, customer, supplier and contractor that encourage greater flexibility and responsiveness of the enterprise. An important role is given to partners' models based on multi-purpose cooperation logics among some enterprises and their divisions (for example, three-digit logic, i.e. cooperation, competitiveness, indifference), based on such factors as the compatibility of the enterprise's goals, mutual trust and responsibility among partners in fulfilling commitments, the necessity of aligning resources and lack of production experience for certain enterprises. The VLMCS can be regarded as a metasystem uniting the goals, resources, traditions and experience of several enterprises while developing complex innovation projects or maintaining and repairing world-class locomotives. Planning, organization and coordination of the VLMCS require appropriate managerial approaches, namely the ability to identify and integrate the necessary co-participants, and organization of production and distribution.

As a network of partners, the main functions of the VLMCS management are to define the project (design) requirements, identify and assess potential partners (co-participants), define and assign participants most efficiently according to set objectives, and continuously monitor and, if needed, realign executives and resources by production stage.

3 A Model of the Virtual Locomotive Maintenance Control System According to a Minimum Cost Criterion

A network of locomotive depots forms a subsystem of Ukraine's rail transport system, the functions of which are to provide timely and quality repair and maintenance of the locomotive fleet. The rail network operates in a competitive market environment, where prices for energy and material resources are escalating. Therefore, one of the most important objectives for the system to be functional is to ensure its economic efficiency. The economic efficiency of the subsystem of maintenance and repair depots is influenced by many negative factors: continuous wear of locomotives, shortage of new locomotives, lower freight volume due to the global economic crisis, and instability and higher prices for material resources. All these factors are objective, so the only solution for increasing the operational efficiency of the locomotive repair subsystem is the rearrangement of its internal structure, abolishment of obsolete work techniques, and transformation to logistic management principles providing flexibility to the maintenance and repair system [30–32]. The main task of such management is effecting optimal work distribution among depots included in the rail network to minimize total costs for repair and maintenance of the locomotive fleet. The major factors influencing the cost and reliable operation of locomotives are the cost of repair, railway losses due to unused locomotives, losses due to unplanned repairs, and expenses for the transfer of locomotives to repair sites. Therefore, an urgent need in the current environment is the formation of a flexible maintenance and repair system based on virtual enterprises. The formal approach requires setting optimization goals regarding an effective operational distribution scheme among locomotive depots. The target function of an optimization model may generally be presented as follows:

$$E = C_{\mathrm{pr}} + C_{\mathrm{ur}} + C_{\mathrm{d}} + C_{\mathrm{nr}} \to \min, \tag{3.1}$$

where C_{pr} is the total input for planned repairs for all locomotives of the railway during a planning period, C_{ur} is the cost of unplanned repairs for all locomotives of the railway during a planning period, C_{d} is the revenue deficit due to withdrawal of locomotives from the operating fleet for repair, and C_{nr} is the total expense due to non-production locomotive runs while transferring to repair depots.

Minimization of the target function should be executed by redistribution of repair work among locomotive depots. Therefore, the vector of variables of the model must reflect information on locomotives being assigned to depots for next planned repair. Let the vector X be the vector of variables of a model of the maintenance and repair system. The number of components of the vector must be equal to k, where k is the number of locomotives assigned to all depots of the railway. Let the vector d contain information on assigned depots for locomotives. The vector d also contains k components describing the depot of assignment. The number of the element of the vector d corresponds to the number of a locomotive, with the value of the element of the vector indicating the number of the main locomotive depot. Each number of the element of the vector X also corresponds to the number of a locomotive, with the value of the element of the vector indicating the number of the depot where the next planned repair of the locomotive must be executed.

Let the vector C^{r} contain m elements, where m is the number of repair depots at a railway site. The number of an element corresponds to the number of the depot, and the value of the element corresponds to the cost of repair in the depot.

Thus the first summand of the target function reflects the general cost of planned repairs and can be expressed as follows:

$$C_{\mathrm{pr}} = \sum_{i=1}^{k} C^{\mathrm{r}}_{X_i}. \tag{3.2}$$

The second summand of the target function corresponds to the costs of unplanned repairs, which are random in nature. However, one of the most important factors influencing the numerical characteristics of the process is the repair quality factor. Lower repair quality leads to a greater number of unplanned repairs. Each depot can be characterized by the repair quality factor indicating the level of reliability restored in the repair process. The repair quality factor can be determined as the probability of no unplanned repair occurring at inter-repair periods, its value being determined for each depot on the basis of statistical data. The costs of unplanned repair, unlike those of planned repair, can vary according to amount of work. Therefore, each depot can determine an average cost of unplanned repair based on statistical data. Thus, the average cost of unplanned repair is the cost of repair fulfilled over and above the plan and repair needed afterwards. Therefore, the average cost of unplanned repair is a factor which, together with a repair quality factor, also characterizes the quality of repair in a given depot. On this assumption, one can determine the total costs for unplanned repairs:

$$C_{\mathrm{ur}} = \sum_{i=1}^{k} C^{\mathrm{ur}}_{X_i} \cdot \left(1 - \psi_{X_i}\right), \tag{3.3}$$

where $C^{\mathrm{ur}}_{X_i}$ is the average cost of unplanned repair for a depot where planned repair of the ith locomotive will be made, and ψ_{X_i} is the repair quality factor for a depot where planned repair of the ith locomotive will be made.

The third summand of the target function represents revenue deficits due to withdrawal of locomotives from operation during repair and lost revenue from railway fares suffered by the railways. The period during which a locomotive is withdrawn from operation is calculated as the sum of the period when the locomotive is waiting for repair and the repair period. The value of the revenue deficit is the cost of lost runs multiplied by a level of profitability. Thus, one can note:

$$C_{\mathrm{d}} = \sum_{i=1}^{k} \left(R_{\mathrm{d}_i} \cdot C_{\mathrm{d}_i}^{n} \cdot \frac{W_{\mathrm{d}_i}^{1}}{24} \cdot \left(t_{X_i}^{p} + t_{X_i}^{\mathrm{wf}} \right) \right), \tag{3.4}$$

where R_{d_i} is the level of production profitability of a depot as an enterprise engaged in freight transportation to which the *i*th locomotive is assigned, profitability is determined as the net profit-to-cost price ratio, $C_{\mathrm{d}_i}^{n}$ is the cost price of runs in the depot to which the *i*th locomotive is assigned, $W_{\mathrm{d}_i}^{1}$ is the daily average efficiency of a locomotive in the depot to which the *i*th locomotive is assigned, $t_{X_i}^{\mathrm{r}}$ is the duration of repair works in the depot where the *i*th locomotive is repaired, and $t_{X_i}^{\mathrm{wf}}$ is the average downtime for the locomotive waiting for repair in the depot where the *i*th locomotive is repaired.

An average downtime for the locomotive waiting for repair in the depot can be determined by the queuing theory mathematical apparatus. The locomotive depot is the single- or multi-channel queuing system with a line of an unlimited number of places. A complex of depots within a rail site or Ukraine's railway system is a queuing network. The input flow of such a system can be regarded as simple, i.e. Markovian, as it consists of independent events. But service flows in such a system are not Markovian, since the repair time consists of times of sequential operations, the duration of which is standardized. Therefore, service intervals in such a system can be regarded as distributed according to the normal law, or deterministic. An average queuing time for a locomotive before repair begins in the depot can be divided into two parts:

$$t_{\mathrm{wf}} = t_{\mathrm{wf},1} + t_{\mathrm{wf},2}. \tag{3.5}$$

The first part is a time interval until the end of the repair of the locomotive being repaired at the moment when the next locomotive arrives for repair in the depot. Thus, that is the remaining repair time, which can be defined as follows:

$$t_{\mathrm{wf},1} = \frac{\bar{t}_{\mathrm{r}}}{2} \cdot \sigma_{\mathrm{r}}, \tag{3.6}$$

where $\bar{t}_{\mathrm{r}}$ is the average repair time (mathematical expectation for repair as a random quantity), and σ_{r} is the shape parameter of the repair time distribution law, i.e. an average deficiency in the normal repair time law as a random variable.

The second part of a queuing time deals with actual waiting time. If $\bar{n}_{\mathrm{q}}$ is an average queue length, according to Little's Law:

$$\bar{n}_{\mathrm{q}} = \lambda \cdot \bar{t}_{\mathrm{wf}}, \tag{3.7}$$

where λ is the average number of locomotives arriving in the depot per time unit (intensity of an input flow), and $\bar{t}_{\mathrm{wf}}$ is the average queuing time for a locomotive until the beginning of repair, the value of which should be determined. Each locomotive in queue will be repaired within the period $\bar{t}_{\mathrm{r}}$. Thus the second part of the time corresponds to the actual queuing of the locomotive and can be determined as follows:

$$t_{\mathrm{wf},2} = \bar{n}_{\mathrm{q}} \cdot \bar{t}_{\mathrm{r}}. \tag{3.8}$$

Therefore, according to Little's Law, one can note:

$$t_{\mathrm{wf},2} = \lambda \cdot \bar{t}_{\mathrm{wf}} \cdot \bar{t}_{\mathrm{r}}. \tag{3.9}$$

Taking into account that $\bar{t}_{\mathrm{r}} = \frac{1}{\mu}$ and $\frac{\lambda}{\mu} = \rho$, where μ is repair intensity, i.e. an average number of repaired locomotives per time unit (24 h), and ρ is the relative intensity, which for a single-channel queuing system is the probability that the system appears to be occupied at any random point of time, one can determine the total queuing time:

$$t_{\mathrm{wf}} = t_{\mathrm{wf},1} + \rho \cdot t_{\mathrm{wf}}. \tag{3.10}$$

The total average queuing time is then:

$$t_{\mathrm{wf}} = \frac{t_{\mathrm{wf},1}}{1-\rho}, \tag{3.11}$$

$$t_{\mathrm{wf}} = \frac{\rho \cdot \bar{t}_{\mathrm{r}} \cdot \sigma_{\mathrm{r}}}{2(1-\rho)}. \tag{3.12}$$

This notation corresponds to the Pollaczek–Khinchine formula for a single-channel queuing system. However, some repair depots are equipped with two repair bays. The repair time for a locomotive according the TO-3 Technical Maintenance is determined by a rated interval of 16 h. Thus, we can take a repair in such a queuing system as close to deterministic, and to define an average queuing time for a locomotive, one may use the approximation Molien's formula:

$$t_{\mathrm{wf}} \approx \frac{n}{n+1} \cdot E2 \frac{t_p}{n-\rho} \cdot \frac{1-(\rho/n)^{n+1}}{1-(\rho/n)^n}, \tag{3.13}$$

where $E2$ is the Erlang-C formula which determines queuing probability and may be presented as:

$$P_{\mathrm{wf}} = \frac{\frac{\rho^n}{n!} \cdot \frac{n}{n-\rho}}{\sum_{i=1}^{n-1} \frac{\rho^i}{i!} + \frac{\rho^n}{n!} \cdot \frac{n}{n-\rho}}. \tag{3.14}$$

Thus an average queuing time for a locomotive is:

$$t_{\mathrm{wf}} \approx \frac{n}{n+1} \cdot \frac{\frac{\rho^n}{n!} \cdot \frac{n}{n-\rho}}{\sum_{i=1}^{n-1} \frac{\rho^i}{i!} + \frac{\rho^n}{n!} \cdot \frac{n}{n-\rho}} \cdot \frac{t_p}{n-\rho} \cdot \frac{1-\left(\frac{\rho}{n}\right)^{n+1}}{1-\left(\frac{\rho}{n}\right)^n}. \tag{3.15}$$

Therefore, the average queuing time for a locomotive waiting for repair in a depot can be presented as a function of the repair time t^{r} and the relative intensity of flows ρ and number of repair bays in the depot n:

$$t^{\mathrm{wf}} = f(t^{\mathrm{r}}, \rho, n). \tag{3.16}$$

The summand of the target function is the deficit in transportation revenue, and can be presented as:

$$C_{\mathrm{d}} = \left(R_{\mathrm{d}_i} \cdot C^{n}_{\mathrm{d}_i} \cdot \frac{W^{1}_{\mathrm{d}_i}}{24} \cdot \left(t^{p}_{X_i} + \frac{n_{X_i}}{n_{X_i}+1} \cdot \frac{\frac{\rho_{X_i}^{n_{X_i}}}{n_{X_i}!} \cdot \frac{n_{X_i}}{n_{X_i}-\rho_{X_i}}}{\sum_{i=1}^{n-1} \frac{\rho^{i}_{X_i}}{i!} + \frac{\rho^{n}}{n_{X_i}!} \cdot \frac{n_{X_i}}{n_{X_i}-\rho_{X_i}}} \cdot \frac{t^{p}_{X_i}}{n_{X_i} - \rho_{X_i}} \cdot \frac{1 - \left(\rho_{X_i}/n_{X_i}\right)^{n_{X_i}+1}}{1 - \left(\rho_{X_i}/n_{X_i}\right)^{n_{X_i}}}\right)\right), \tag{3.17}$$

where ρ_{X_i} is the relative intensity of flows in the depot to which the ith locomotive is assigned, and n_{X_i} is the number of repair bays in the depot to which the ith locomotive is assigned.

The second summand of the target function is the costs of non-productive locomotive runs during their transfer to repair depots. These costs can be determined as the duration of non-productive runs multiplied by the cost of locomotive-hours (Clh) related to locomotive operation (fuel consumption, wages of locomotive crews, etc.):

$$C_{\mathrm{nr}} = C^{\mathrm{lh}} \cdot \sum_{i=1}^{k} \frac{L(d_i, X_i)}{v^{x}}, \tag{3.18}$$

where $L(d_i, X_i)$ is the distance between the assigned depot of the ith locomotive and the depot where its next repair will be carried out, and v^{x} is the running speed at which locomotives transfer at station-to-station hauls of the rail site.

Thus, if a locomotive is repaired in the assigned depot, this element of the sum is equal to zero, as the distance function equals zero. Therefore, the target function is presented as:

$$\begin{aligned} C(X) = {} & \sum_{i=1}^{k} C^{\mathrm{r}}_{X_i} + \sum_{i=1}^{k} C^{\mathrm{ur}}_{X_i} \cdot (1 - \psi_{X_i}) \\ & + \sum_{i=1}^{k} \left(R_{d_i} \cdot C^{n}_{d_i} \cdot \frac{W^{1}_{d_i}}{24} \cdot t^{\mathrm{r}}_{X_i} \cdot \left(1 + \frac{\frac{\rho_{X_i}^{n_{X_i}} \cdot n^{2}_{X_i}}{n_{X_i}! \cdot (n_{X_i} - \rho_{X_i})} \cdot \left(1 - \left(\frac{\rho_{X_i}}{n_{X_i}}\right)^{n_{X_i}+1}\right)}{(n_{X_i}+1) \cdot \left(\sum_{i=1}^{n_{X_i}-1} \frac{\rho^{i}_{X_i}}{i_{X_i}!} + \frac{\rho_{X_i}^{n_{X_i}} \cdot n_{X_i}}{n_{X_i}! \cdot (n_{X_i} - \rho_{X_i})}\right) \cdot (n_{X_i} - \rho_{X_i}) \cdot \left(1 - \left(\frac{\rho_{X_i}}{n_{X_i}}\right)^{n_{X_i}}\right)}\right)\right) \\ & + C^{\mathrm{lh}} \cdot \sum_{i-1}^{k} \frac{L(d_i, X_i)}{v^{x}} \to \min. \end{aligned} \tag{3.19}$$

The target function requires minimization at the following limitations:

$$\begin{cases} \sum_{i=1}^{k} (t^{\mathrm{sr}} \cdot (1 - |\mathrm{sgn}(j - X_i)|)) \leq n_j \cdot s_j \cdot t_j^{\mathrm{ws}} \cdot t^{\mathrm{w}}, \quad j = 1, 2 \ldots m \\ v^x \leq v^N \end{cases}, \tag{3.20}$$

where m is the number of repair depots at a rail site, sr is the standard repair time, n_j is the number of repair bays in the jth depot, s_j is the number of working shifts in the jth depot, t_j^{ws} is the duration of working shifts in the jth depot and t^{w} is the number of working days during a planning period.

The first limitation involves the fact that the total volume of work in each depot should not exceed this depot's capacity, i.e. the duration of repairs planned at each depot of a rail site should not exceed the duration of working time by each depot based on the number of bays. The second limitation is in maintaining the speed level in transferring locomotives to repair sites and technical maintenance to guarantee safe operation. The maximum speed at station-to-station blocks in transferring locomotives to the depot should not exceed the speed indicated in operating rules for Ukraine railways or the speed indicated in operating precautions for a section. The above-mentioned mathematical model is intended for compiling an optimal repair plan for locomotives used in freight transportation at a rail site. Among the basic requirements for choosing a mathematical apparatus and a method for realizing a given model, the following should be included: low sensibility to dimension of the task, ease in transferring a mathematical model of a classic form to a standard solution, and computer orientation. These requirements can be met by methods for solving complex tasks of functional optimization relating to evolution calculation, which is a sphere of artificial intelligence, and particularly by genetic algorithms. One of the peculiarities of genetic algorithms is that the vector of the model's variables is presented as a chromosome, and some variables are presented as genes of a chromosome. Genetic algorithms relate to heuristic search methods, during which they generate mechanisms similar, by basic principles, to biological evolution mechanisms. An important element when solving tasks using a mathematical apparatus is creating a fitness function. A genetic algorithm does not operate with one set of variables presenting a possible solution, but simultaneously with many such sets, called a population. A fitness function, also called a suitability function, is a target function of a special nature; it generates information used by a generic algorithm in scaling, crossing, mutation and selection for vitality assessment of individuals of a new generation. An ideal fitness function must be correlated with operating principles and objectives of the selected genetic algorithm in the best possible way. Such an approach provides algorithm convergence within a practically appropriate time. A fitness function can also include a mathematical model limitation system. While solving the task, the limitation system is included in fitness functions, limitations being observed by a system of penalty points of the fitness function. To realize the model, software was created using MATLAB, a language of technical computing, which is the de facto standard in mathematical computation. The given optimization model is of a universal nature and can be used at sites of Ukrainian railway locomotive depots.

4 Organizational and Technical Assessments of Locomotive Repair Plants

Organizational and technical level of production (OTLP) is a complex qualitative factor presenting the conformance of technologies and production processes to production system requirements. This factor indicates the efficiency level of an enterprise's innovation policy.

The complex nature of OTLP requires the use of a systematic approach, taking into account and distinguishing the most important factors influencing OTLP.

A common method for determining and structuring the factors influencing the subject of inquiry is the use of Ishikawa diagrams, which expose the key relation among different factors and establish cause-and-effect relations between them [33].

Ishikawa diagrams were drawn for the organizational and technical level of locomotive repair plants, taking into account the "6M's + E" principle, which stands for *man, machine, method, materials, measurement, management and environment.* In addition, the experience of an expert group testing locomotive industry is considered.

As shown in Fig. 1, the organization and technical level of locomotive repair production is influenced by a number of factors which cannot be explicitly

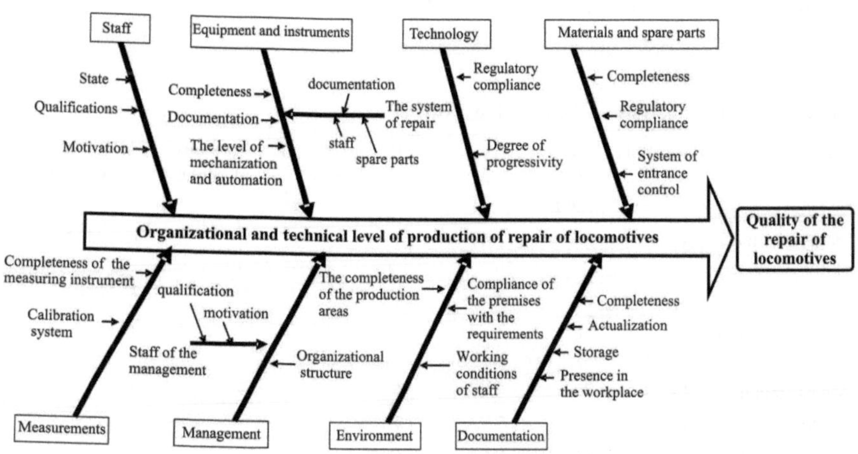

Fig. 1 Cause-and-effect diagram of the factors' influence on the organizational and technical level of locomotive production and quality of repair

estimated and measured, and the level of their influence cannot be measured using rigorous methods. Therefore, the assigned task can be solved with expert assessment, which combines mathematical methods with the experience of experts.

The most common method for presenting expert information on the comparative advantages of alternative options is through ranking [34]. This method is characterized by procedural simplicity and short and easy expert training.

When ranking, an expert must allocate N factors in descending order by significance. The ranks are designated by numbers from 1 to n, where n is the number of ranks. Here, the sum of ranks will be equal to the sum of natural scale numbers [35].

To carry out the procedure, a group of experts was formed, comprising 15 participants with administrative roles in locomotive depots. Participant requirements included higher education and more than 15 years of experience.

To analyze the data obtained by ranking, a method of average arithmetic ranks is used, which is not absolutely correct, as the expert assessments are measured on an ordinal scale. A more reasonable approach is the method of medians. And according to [36], it is reasonable to use the two methods simultaneously to comply with the general scientific concept of robustness (distinguishing conclusions obtained using different methods). The results presented in Table 1 show that the calculation of the resulting ranks using the two methods produces consistent results, and according to [37], they can be regarded as reliable.

The consensus of experts is determined in accordance with the Kendall rank correlation coefficient (Kendall's tau coefficient) (Table 2):

$$W = \frac{12D}{m^2(n^2 - n)}, \tag{4.1}$$

where D is the sum of squares of ranks, m is the number of experts, and n is the number of factors in the analysis.

Table 1 The data analysis of the expert assessment

Factors	Symbols	Sum of ranks	Average rank	Resulting rank by average	Medians of ranks	Resulting rank by medians
Materials and spare parts	A_1	37	2.47	1	2	1
Staff	A_2	65	4.33	4	3.5	3
Equipment and instruments	A_3	58	3.87	3	3	2
Measurements	A_4	73	4.87	5	4	4
Technology	A_5	86	5.73	7	5.5	5
Documentation	A_6	85	5.67	6	6.5	6
Environment	A_7	98	6.53	8	7.5	7
Management	A_8	38	2.53	2	2	1

Table 2 Characteristics of linguistic variables

Indicator level	Marking	Characteristics	Membership function
Very high	A	Full compliance with requirement, innovation implementation	$\mu_1(x) = \begin{cases} 0, 0 \le x < 0.75 \\ 10(x - 0.75), 0.75 \le x < 0.85 \\ 1, 0.85 \le x \le 1 \end{cases}$
High	B	Full compliance with requirements	$\mu_2(x) = \begin{cases} 0, 0 \le x < 0.55 \\ 10(x - 0.55), 0.55 \le x < 0.65 \\ 1, 0.65 \le x < 0.75 \\ 10(0.85 - x), 0.75 \le x < 0.85 \\ 0, 0.85 \le x < = 1 \end{cases}$
Middle	C	Compliance with the requirement with minor remarks	$\mu_3(x) = \begin{cases} 0, 0 \le x < 0.35 \\ 10(x - 0.35), 0.35 \le x < 0.45 \\ 1, 0.45 \le x < 0.55 \\ 10(0.65 - x), 0.55 \le x < 0.65 \\ 0, 0.65 \le x < = 1 \end{cases}$
Low	D	Non-compliance	$\mu_4(x) = \begin{cases} 0, 0 \le x < 0.15 \\ 10(x - 0.25), 0.15 \le x < 0.25 \\ 1, 0.25 \le x < 0.35 \\ 10(0.45 - x), 0.35 \le x < 0.45 \\ 0, 0.45 \le x < = 1 \end{cases}$
Very low	E	Absence of basic assessment elements	$\mu_5(x) = \begin{cases} 1, 0 \le x < 0.15 \\ 10(0.25 - x), 0.15 \le x < 0.25 \\ 0, 0.25 \le x \le 1 \end{cases}$

To determine the statistical significance of ranking the χ, distribution with $(n - 1)$ degrees of freedom is used.

$$\chi_p^2 = m(n - 1)W. \tag{4.2}$$

The calculated Kendall's tau coefficient of a given ranking is 0.37, which indicates a low degree of expert consensus. However, based on $\chi_p^2 = 38.85$ greatly exceeding tabulated values for seven degrees of freedom and significance level ($\alpha = 0.01$, $\chi_p^2 = 18.5$), the hypothesis regarding the statistical significance of ranking is accepted.

The results of expert assessment of the basic factors influencing the organizational and technical level of locomotive repair plants made it possible to establish their rank succession by severity of exposure:

$$\{A_3., A_8.\} < A_4 < A_6 < A_2 < A_1 < A_7 < A_4 \tag{4.3}$$

According to the expert assessment carried out, the factors A_3 (Materials and spare parts) and A_8 (Management) have the greatest influence on the organizational and technical level of locomotive repair plants. The influence of the other factors decreases in accordance with (4.3). Investigations into the impact of factors on the locomotive production level and repair quality [38, 39] using statistical methods demonstrate a high level of consensus with the results obtained. The fact that the "Materials and spare parts" factor has a decisive influence can be explained by the difficult economic situation in the industry and the insufficient supply and poor quality of materials and spare parts. The large influence of the "Management" factor can be attributed to an outdated management structure and managerial methods. Most enterprises are managed "manually", i.e. under continuous managerial pressure. A low consensus level among experts, determined according to Kendall's tau coefficient, cannot cover the entire situation at locomotive repair plants. Actual expert groups included representatives of various enterprises (depots). The impact of factors on the organizational and technical level of locomotive repair plants may vary between enterprises and within an enterprise but at different time points. Consequently, obtaining individual characteristics for a specific plant requires a separate expert assessment.

The additive convolution is used to obtain the generalized criterion according to (4.3):

$$K = \sum_{i=1}^{8} r_i A_i, \tag{4.4}$$

where r_i is the weight coefficient of the ith index, and A_i is the value of the ith index.

In the space formed by dedicated indices, one can draw a hyperplane separating the enterprises which meet the requirements from those of a low organizational and technical level. By transferring the plane (4.4) in a parallel manner, one can observe how successful and unsuccessful enterprises rearrange themselves to the subareas formed by the given plane. Accordingly, one can establish the threshold standard Kn: when $K < Kn$, the enterprise's level does not meet the requirements; when $K > Kn$, the enterprise's level is sufficient to carry out locomotive repair. To determine the K value of the enterprise, one should take the values and their weight coefficients. According to the ranking (4.3) and Fishburn's rule [40], their value is:

$$r = \frac{2(N - i + 1)}{(N + 1)N}. \tag{4.5}$$

Then, marking these indices as A_1, A_2, ... A_8, the expression (4.4) can be presented as

$$\begin{aligned} K = {} & 0.08A_6 + 0.11A_4 + 0.22A_1 + 0.17A_3 \\ & + 0.03A_7 + 0.14A_2 + 0.05A_5 + 0.19A_8. \end{aligned} \tag{4.6}$$

From the expression (4.6), one can estimate the impact of certain factors A_i on the total K. However, at this stage it is impossible to determine the total index of the enterprise because, practically, the value of A_i can be characterized just by mismatches identified by the experts. To formalize such type of data, the most suitable are methods of fuzzy logic, where assessment of each indicator is

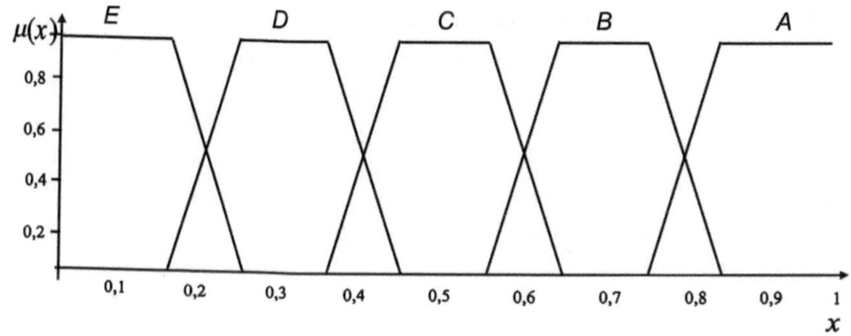

Fig. 2 Location of trapezoidal membership functions of a linguistic variable

regarded as linguistic variables (Fig. 2). For simplification, all linguistic variables were assigned the same number of terms with trapezoidal membership functions (Fig. 2) [41].

By applying membership functions of terms, the parameters can be specified by the results of experiments. The initial membership functions are constructed with symmetrical arrangement of nodes in classification (0.1, 0.3, 0.5, 0.7, 0.9). In these nodes, the value of a corresponding membership function is equal to 1, while all other functions equal zero.

Each expert assessment corresponds to a vector of five membership functions:

$$S_*(x) = \{\mu_{*1}(x), \mu_{*2}(x), \mu_{*3}(x), \mu_{*4}(x), \mu_{*5}(x)\}. \quad (4.7)$$

According to [42], the sweep of all the vectors $S_*(x)$ with the weight coefficients r is carried out according to the formula:

$$\sum_{i=1}^{N} r_i \times \{\mu_{i.1}, \mu_{i.2}, \mu_{i.3}, \mu_{i.4}, \mu_{i.5}\} = \left\{\sum_{i=1}^{N} r_i \times \mu_{i.1}, \sum_{i=1}^{N} r_i \times \mu_{i.2}, \sum_{i=1}^{N} r_i \times \mu_{i.3}, \sum_{i=1}^{N} r_i \times \mu_{i.4}, \sum_{i=1}^{N} r_i \times \mu_{i.5}\right\} \quad (4.8)$$

The resulting indicator of production status is determined as a vector of five membership functions $S_0 = \{\mu_{0i}\}$, the sum of which is equal to 1. According to Nedosekin, the scalar vector characterizing the status of the enterprise is:

$$K = \sum_{i=1}^{5} (0.2i - 0.1) \times \mu_{0i}, \quad (4.9)$$

where $(0.2i - 0.1) = (0.1, 0.3, 0.5, 0.7, 0.9)$ are nodal points of a standard five-tier fuzzy classifier.

Table 3 Expert assessment of a locomotive repair plant

Characteristics	Weight coefficients	Terms of linguistic variables				
		A	B	C	D	E
		λ_{i1}	λ_{i2}	λ_{i3}	λ_{i4}	λ_{i5}
Documentation	0.08	0	1	0	0	0
Measurements	0.11	0	0	1	0	0
Materials and spare parts	0.22	0	0	1	0	0
Equipment and instruments	0.17	0	0	1	0	0
Environment	0.03	0	1	0	0	0
Staff	0.14	0	0	1	0	0
Technology	0.05	0	0	1	0	0
Management	0.19	0	0	1	0	0
$\sum r_i \lambda_{ij}$		0	0.11	0.88	0	0

According to the above-mentioned method, it is possible to determine the level of locomotive repair production in the depot *L*. The expert assessments are summarized in Table 3.

A single-phase sweep (4.8) gives the vector $S_0 = \{0, 0.11, 0.88, 0, 0\}$. One can see that the result is close to the average and high values of the factors. From (4.9) we can obtain $K = 0.473$. If we recognize the value by membership functions (Table 2), we can obtain "Middle level" of the membership degree 1, i.e. the general level of production is "Comply with minor remarks".

5 Conclusions

1. In assessing the efficiency of a locomotive maintenance system, it is advisable to apply a model of the virtual rolling stock maintenance system which links locomotive characteristics with the state and capacities of repair plants.
2. The optimal repair plan for locomotives used in freight transportation can be compiled with a mathematical model which takes into account cost minimization criteria. The number of repair depots for the Southern Railway cluster has been optimized by means of calculation.
3. The organizational and technical levels of locomotive repair plants depend on quantitative impact factors (ranked in descending order)—"Materials and Spare Parts", "Management", "Equipment and Instruments", "Stuff", "Measurements", "Documentation", "Technology" and "Environment"—the value of which can by defined by expert assessments.
4. To formalize expert assessments while estimating the organizational and technical level of a locomotive repair plant, it is advisable to use an algorithm based on fuzzy logic.

References

1. Basov GG (2000) The method of assessment of the technical level of diesel trains. Collection of scientific articles of the Kharkiv State Academy of Railway Transport, vol 41, pp 5–8 (in Russian)
2. Basov GG (2007) Improving of the modeling of the technical service of the motor-car rolling stock. Collection of scientific articles of the Ukrainian State Academy of Railway Transport, vol 81, pp 26–31 (in Russian)
3. Basov GG, Mishchenko KP (2003) Developing of the type of modern rolling stock for Ukrainian Railways. Vestnik of East Ukrainian National University named after V. Dahl, vol 9(67), pp 90–95 (in Russian)
4. Zimmermann G (2011) Mehrkriterielle Konfigurationsoptimierung dieselelektrischer Lokomotiven mit Hilfe evolutionärer Algorithmen. Technische Universität Dresden, Diplomarbeit
5. Tartakovskyi E, Falendysh A, Zinkivskyi A, Mikheev S (2015) Refining the models of performing service tests of upgraded locomotives. East-Eur J Enterp Technol 3(74):26–31
6. Kim J, Shim K, Kim J-H (2007) Load modeling of electric locomotive using parameter identification. J Electr Eng Technol 2(2):145–151
7. Sz Fischer (2015) Traction energy consumption of electric locomotives and electric multiple units at speed restrictions. Acta Technica Jaurinensis 8(3):240–256
8. Tartakovskyi ED, But'ko TV, Ostapchuk VN, Matiash BA (1996) Modeling and optimization of the system of technical service and repair of locomotives on total unit costs. Collection of scientific articles of Dnipropetrovsk National University of Railway Transport, vol 300, pp 87–91 (in Russian)
9. Dan'ko NI, Lomot'ko DV, Tartakovskyi ED, Kalabukhin YE, Falendysh AP (2011) Updating of rolling stock taking into account the life cycle. Railway Transp 12:42–44 (in Russian)
10. Tartakovskyi ED, Grishchenko SG, Kalabukhin YE, Falendysh AP (2011) Methods for assessing the life cycle of traction rolling stock of railways: monograph. Knowledge, Lugansk, 174 p. (in Russian)
11. Dae-Sung Bae, Jong-Duk Chung (2005) Life-cost-cycle evaluation analysis of the shunting locomotive. J Korean Soc Railway 8(3):260–266
12. Bodnar EB (2001) Methods for determining the periods between repairs of locomotives. Collection of scientific articles of Dnipropetrovsk State University of Railway Transport, vol 9, pp 33–87 (in Russian)
13. Bodnar EB, Ochkasov OB, Shepotenko AP (2004) Assessment of economic efficiency of introduction of a rational system of repair of locomotives. Vestnik of East Ukrainian National University named after V. Dahl, vol 8(78), pp 25–28 (in Russian)
14. But'ko TV (1996) Improvement of methods of calculation of parameters of system of technical service of locomotives. Dissertation. Kharkiv State Academy of Railway Transport (in Russian)
15. Gorski AB, Vorobev AA (1994) Optimization of the system of repair of locomotives. Transport, Moscow, 209 p. (in Russian)
16. Zhang Z, Gao W, Zhou Y, Zhang Z (2012) Reliability modeling and maintenance optimization of the diesel system in locomotives. Maintenance and Reliability, vol 14(4), pp 302–311
17. Huang J, Miller CR, Okogbaa OG (1995) Optimal preventive-replacement intervals for the Weibull life distribution: solutions and applications. In: IEEE proceedings annual reliability and maintainability symposium. Washington, DC, pp 370–373
18. Gao W, Zhang Z, Zhou Y, Wang R, Jiang L (2013) Optimal combinatorial replacement policy under a given maintenance interval for the combined governor in diesel locomotives. Maintenance and Reliability, vol 15(2), pp 89–98
19. Candfield J (2011) New trains for old? Rolling stock maintenance and replacement in the UK. Eur Railway Rev 17(2):73–78

20. Bolotin MM, Osinovsky LL (1989) Automation of production processes in the manufacture and repair of cars. Moscow, Transport, 206 p. (in Russian)
21. Chen L, Li C (2013) Overall objectives and principles of optimizing adjustments in locomotive affair productivity layout. ICTE, pp 1309–1315
22. Oliver D (1999) Concept analysis: modeling and organization transformation. INCOSE international symposium. Brighton, England, vol 9(1), pp 1207–1213
23. Palo M, Lin J, Larsson-Kraik P (2013) Maintenance performance improvement for rolling stock wheels. Chem Eng Trans 33:727–732
24. Puzyr VG, Zarubin VM, Datsun YM (2006) The certification of the locomotive depots is a necessary stage of the technical re-equipment of the industry. Locomotive-inform, vol 4, pp 28–31 (in Russian)
25. Sergienko MI, Puzyr VG, Datsun YM (2007) The certification of the locomotive production as a part of ensuring of a traffic safety. Collection of scientific works Ukrainian State Academy of Railway Transport, Kharkiv, vol 82, pp 5–8 (in Ukranian)
26. Osyaev AT (1996) A comprehensive system of technical service and current repair of mainline locomotives. Basics. Rostov-on-Don, VNIIZhT, 27 p. (in Russian)
27. Khomich AZ, Zhalkin SG, Simson AE, Tartakovsky ED (1977) Diagnosis and adjustment of diesel locomotives. Moscow, Transport, 222 p. (in Russian)
28. The Concept of the State program of reforming of railway transport. Magistral, vol 1(1179), pp 2–3 (In Russian)
29. Bondar BE, Ochkasov AB, Gilewicz OI (2000) To optimization of information-diagnostic system of locomotive. Information management systems in railway transport, vol 4, pp 118–119 (in Russian)
30. Ayzinbud SY, Kelperis PI (1990) Operation of locomotives. Transport, Moscow, 261 p. (in Russian)
31. Lashko AD (2007) The reform of the railway transport as one of the steps on the way of Ukraine to the world community. Collection of scientific articles of the Kiev University of Transport Economics and Technology, vol 1, pp 49–51 (in Russian)
32. Ustenko OV (2010) Improving the system of technical operation of rolling stock of Railways of Ukraine in the period of their reform. Vestnik of East Ukrainian National University named after V. Dahl, vol 4(147), pp 90–95 (in Russian)
33. Ishikawa K (1985) What is total quality control?. The Japanese Way, London, Prentice Hall
34. Lytvak BG (1996) Expert evaluation and decision making. Patent, Moscow, 271 p. (in Russian)
35. Trofimova LA, Trofimov VV. Management decisions (making and implementation): training manual. Ed. SPSUE, 190 p. (in Russian)
36. Orlov AI (2011) Organizational and economic modeling: P.2: expert evaluation. Ed. Bauman MSTU, Moscow, 486 p. (in Russian)
37. Morozova LA, Bortnyk OA, Kravchuk IS (2009) Expert methods and technologies of the complex evaluation of economic and innovative potential of the enterprises. Training manual. MSURE, Moscow, 81 p. (in Russian)
38. Korytov AY (2010) Organizational and methodical maintenance of the quality of a capital repair of a rolling stock. Dissertation. Moscow State University of Railway Engineering (in Russian)
39. Kornylov SN (2010) The formation and development of the system of repair of railway rolling stock of industrial enterprises on the base of logistic principles. Dissertation. Petersburg State Transport University (in Russian)
40. Fishburn PC (1970) Utility theory for decision making. Wiley, New York
41. Nedosekin A (2003) Fuzzy financial management. AFA Library, Russia, Moscow 184 p.
42. Nedosekin A (2004) Risk assessment of business based on fuzzy data. Monograph. St. Petersburg, 100 p. (in Russian)

Part II
Infrastructure and Management in Rail Transport

Systematic and Customer Driven Approach to Cost-Efficiently Improving Interlocking and Signaling in Train Transport

Jörn Schlingensiepen, Florin Nemtanu and Marin Marinov

Abstract A big challenge for European train companies in the next year will be upgrading large parts of interlocking and signaling to enable Europe-wide train transport. Additional transitions having heavy impact on infrastructure development, including the transition from conventional to high-speed trains and the application of new operation patterns in rail freight aimed at increasing the share of rail in the freight market. The main aspect in accomplishing this transformation is the development and deployment of new technology, but focusing only on technology will fail, since many other organizational components are involved in this transformation. This chapter proposes a holistic approach, starting with gathering customers' requirements, while also respecting development of the organization as a whole and the personal development of employees. This chapter extends various earlier works of the authors and shows how extending the principles of systems engineering to other domains can help to manage this transformation successfully. The conclusion presents a reference process for service development that can be used as a blue print for participating companies.

Keywords ERTMS · Signaling · Interlocking · Rail transport · Technical and organizational transition · Service engineering · Safety

---

J. Schlingensiepen (✉)
I3CM Fulda, Fulda, Germany
e-mail: joern@schlingensiepen.com
http://www.i3cm.de; http://www.thi.de

J. Schlingensiepen
TH Ingolstadt, Ingolstadt, Germany

F. Nemtanu
"POLITEHNICA" University of Bucharest, Bucharest, Romania
e-mail: fnemtanu@yahoo.com
http://www.upb.ro/

M. Marinov
Newcastle University, Newcastle upon Tyne, UK
e-mail: marin.marinov@ncl.ac.uk
http://www.ncl.ac.uk/

A. Sładkowski (ed.), *Rail Transport—Systems Approach*, Studies in Systems, Decision and Control 87, DOI 10.1007/978-3-319-51502-1_6

1 Introduction

1.1 Requirements of Infrastructure Changes

A big challenge for European train companies in the next year will be upgrading large parts of interlocking and signaling to enable Europe-wide train transport as described in ERTMS—operational principles and rules [1] and ERTMS—Work Plan of the European Coordinator [2]. Additional transitions having heavy impact on infrastructure development, include the transition from conventional to high-speed trains, as required in the U.S. High-Speed Rail Strategic Plan [3] and in the EU paper High-speed Europe [4], and the application of new operation patterns in rail freight aimed at increasing the share of rail in the freight market. In fact all of today's challenges to railway industry are related to improvements in interlocking and signaling. For example, the objective of interoperability implies running trains across borders without any technical, regulatory and operational issues; this can be achieved only by investing in current infrastructure to make it interoperable.

The main aspect in this transformation is developing and deploying new technology, which makes engineering (mechanical, electrical and civil engineering) the leading domain in this process. Systems engineering is a well-known and approved approach to combining the strength of the engineering domains with product and manufacturing process development. But focusing only on technology will fail, since many organizational components are involved in this transformation, and the process relies also on these as well as on engineering. Ultimately, the use of new technologies makes sense only if the customer rewards these investments by asking for and paying for new or better services. That is why we propose to follow a general approach, starting with gathering customers' demands, while also respecting development of the organization as a whole and personal development of employees. This chapter extends various earlier works of the authors, and shows how extending the principles of systems engineering to other domains can help to manage this transformation successfully. Due to the fact that each organization and each transformation project has specific characteristics and conditions, we cannot show a single set of overall instructions to manage these kinds of projects, but we can offer a checklist allowing structured planning of specific projects.

Figure 1 shows the well-known triangle of people, infrastructure and services. These three main aspects in Providing Train Transport Service are extended by the introduction of new technologies. While customers and their (new) demands are a driving force for investing in new technologies, the transport service providers must identify those new demands and define services that can be sold as products on the market. Those new services lead to requirements on technologies that allow companies to decide which technologies shall be introduced in which time frame.

As already stated, technology may be the driver, but it is not the only aspect in this transition. New technologies always introduce new processes that require training or education for employees. Regarding time frame, investment to support these changes must be planned and managed. As an example of how failing in this

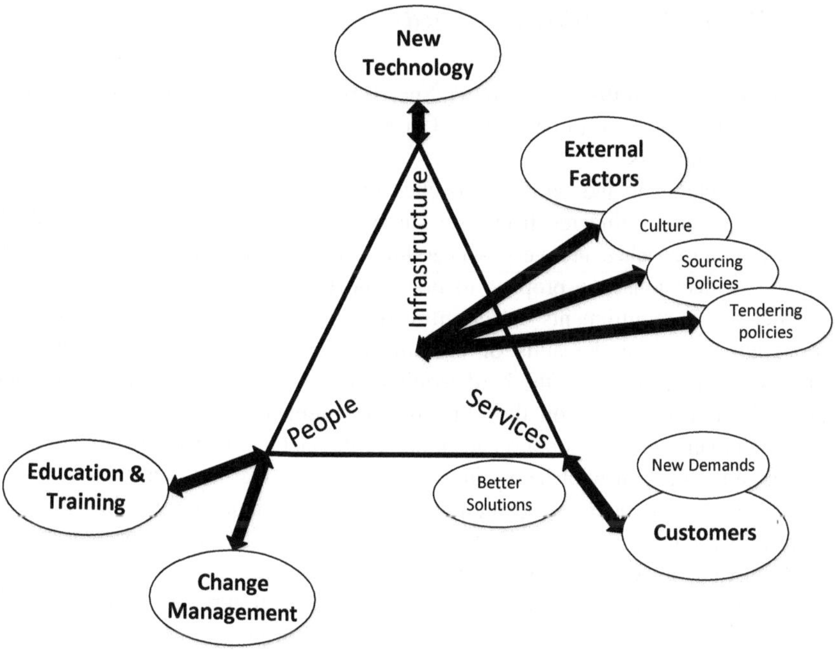

Fig. 1 Aspects of infrastructure transition

aspect causes major issues, we refer to the trouble in August 2013 in Mainz, a city with 200,000 inhabitants in the south of Germany. Deutsche Bahn, though running railway control centers using the latest technologies, had to shut down the railway service around the city for some days and could provide only limited service for over two months [5, 6]. This problem caused major issues in the national railway system, since Mainz is an important hub not only for local transport, but also for long distance trains and freight trains. A later investigation showed that due to a deficiency in forecasting, there were not enough apprenticeships offered to young people, so the number of traffic controllers were not enough to keep up the service [7, 8].

To illustrate the general approach, this chapter gives an overview following the described steps. After starting with a short introduction on how to determine customers' needs and define services with related payment schemes, an overview is given about how the principles of interlocking and signaling are related to new technologies and how these principles support optimal decision-making with regard to technology adoption. After that, an overview of external factors on the decision process is given and finally, some training and education forms are presented, showing their advantages and disadvantages in terms of employee development and change management.

1.2 Demands—Customers' Needs

As already shown in [9–11], customers' needs may vary, thus groups of customers with similar needs have to be identified in order to create unique selling points for groups of customers.

In both classical and agile approaches to product and service development, requirements are engineered from user surveys, user stories, use cases, scenarios, etc. These approaches are very successful. To obtain an even better basis for investment decisions, we propose to first collect all customer demands and then later derive the requirements on the infrastructure, processes and competencies of employees from the demands of customers. Here, we introduce a distinction between, on the one hand, the hard requirements on infrastructure, processes and skills of employees, and on the other the requirements collected from the customers' demands. This distinction accounts for the fact that in transport logistics, on the one side, we deal with heavy industrial structures like long-term investments, use of heavy equipment and technology-driven innovations, and on the other, the customer's focus on acquiring a service, not a technology. Thus, the presentation to the market has to follow the rules of the service market.

This chapter provides a brief overview of the general approach, using methods and best practices in the development of service products. For further details, references are given for additional information.

2 General Process for Designing a Service

There are two basic and well-known trends in the development of services: New Service Development (details can be found in [12]) and Service Engineering (details are to be found in [13, 14]).

In contrast to the strong marketing oriented New Service Development, the so-called Service Engineering is aimed at a meaningful transfer of existing engineering expertise from classical product development and engineering to the development of services.

We propose to follow the general reference model introduced in [15] that is based on the models from [16] and [13], as well as the agile methods introduced in software engineering (e.g. [17, 18]) and adapt these to the needs of the transport industry and the companies of which it is comprised. The main advantage of this model is that it takes into account that services are not entirely new, but always developed on the basis of existing services. The next section gives a short overview on the reference model; further details can be found in [15] and related works.

2.1 Reference Model

As seen in Fig. 2, there are two main processes, *Service Engineering* and *Service Management*. The processes are connected through the specification, usually in the form of documents. This is in line with the division of [16] between processes that design the newly developed services (Service Engineering), and those deploy and support them (Service Management). In addition there is a third process, *Curating Service Engineering*, a metaprocess in which the procedure for Service Engineering is defined and developed. Curating Service Engineering is a company-specific process which ensures that the organization learns about Service Engineering and keeps the knowledge over time.

As seen in Fig. 2, Curating Service Engineering defines the way services are developed in Service Engineering. Service Engineering defines the description of services in specifications and defines the way those services are maintained during Service Management.

2.2 Curating Service Engineering

Curating Service Engineering describes the activities that a company needs in order to develop new services in a systematic way. The organization must provide a common understanding of tools and methods to be used in developing new services or modifying existing ones. Documentation must be provided describing the general process of Service Engineering, as well as details of how this documentation and tools are provided. Processes must be defined for activities such as investment decisions, definition of prerequisites, and roles and tasks of involved people and

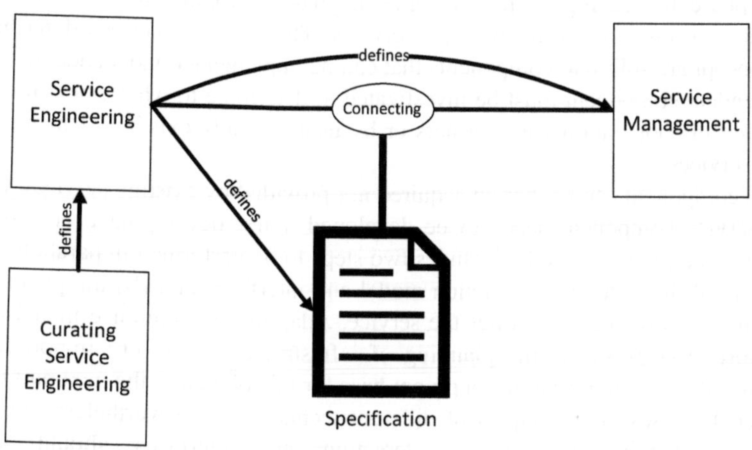

Fig. 2 General processes in reference model adapted from [15]

departments. These processes should follow the principles of the company's quality management process and use its evaluation methods and procedures.

2.3 Service Engineering

Service Engineering should be an iterative process. We propose the meta model from [19] that can be found in [14]. This model is described in three phases:

I. Define the Service
II. Conceive the Service
III. Implement the Service

Phases I and II are functions of Service Engineering, and phase III describes the hand-over to Service Management. The model describes an iterative process, where jumping forward and backward on the process steps is allowed.

As shown in Fig. 3, the most important results of the definition phase are the description of the new service and the planning of the necessary steps for development. To create these, customer requirements and employee suggestions are collected, ordered and clarified in order to show the resulting benefits for the customer. To ensure a well-documented result is generated, well-known methods such as "Quality Function Deployment (QFD)" should be used during this step [20]. Structuring the result in this way will help to define independent service components in phase II. In the next step an integrated project plan is set up, which includes all other activities and documents that are required as part of the development or the provision of the service. At the end of this phase the achieved results are reviewed. If the documentation of service definition and the project plan can be released, phase II is started; otherwise, previous steps are repeated as needed to complete the documentation.

In phase II (see Fig. 4) the outcome of phase I is converted into an implementable design concept for the new services. The service functions described in the concept are split into components that can be implemented independently. Each independent component must be investigated to determine to what extent it can be provided in cooperation with partners or by using already existing elements from other services.

If a component can neither be acquired nor provided by existing services, then a new service component needs to be developed. After developing a new service function, the reference model includes two steps to be performed in parallel: (I) the planning of the customer interaction model and interfaces and (II) the planning of the infrastructure needed to offer the service. Adapting the reference model to rail transport services shows the planning of infrastructure is a very important and complex step. In order establish a proper base for this planning, the next part of this chapter discusses the principles of different technologies. Nevertheless, the other step—the planning of customer interactions and interfaces—should not be neglected. Transport services have not changed their general approach to customer

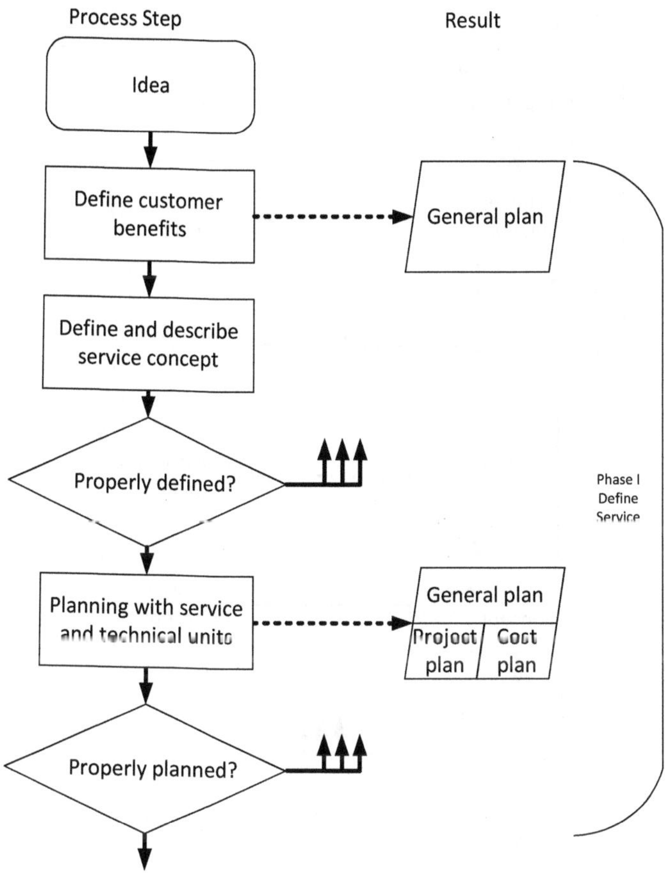

Fig. 3 Jaschinski model: phase I—define service [19]

interactions over the last 150 years. Even today, many transport service providers underestimate the impact of developing new customer interaction protocols and therefore lose market share to innovative competitors or other transport modes. At the end of phase II, the service concept is summarized from the outcome of the previous steps. Again, there may be deficiencies in the plans or documentation, so a review of the results should be performed before starting phase III.

2.4 *Service Implementation*

As shown in Fig. 5, the implementation phase starts with specification of implementation and introduction of the new designed services, followed by planning the way in which the company must be organized to provide the planned services. Once

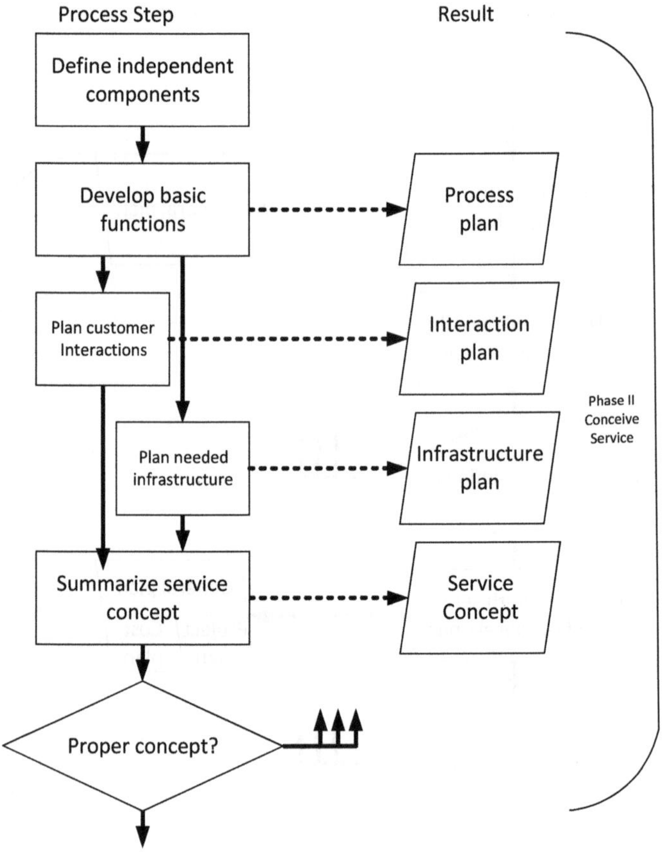

Fig. 4 Jaschinski model: phase II—conceive service [19]

organization is defined, the technical implementation can be specified. Since the infrastructure in railway industry is always a massive and long-term investment, the implementation plan has to take in account the advantages, disadvantages and specific constraints of the underlying technology. Therefore the principles and selection attributes discussed in the following sections have to be considered. After planning the technical infrastructure, that is, the needed improvements in current infrastructure, a distribution concept has to be defined. In contrast to the marketing concept (to be defined later), the distribution concept describes the way the service is brought to the customer. In conventional industry process, this means planning transportation, distribution channels, dealer networks, etc. When providing services, this means planning how services may be offered—as a single service product or bundled with other products—and the way customers will order services and interact with staff, dealers or online interfaces. A distribution concept must include an understanding of situations, customers' demands and individual services or service bundles.

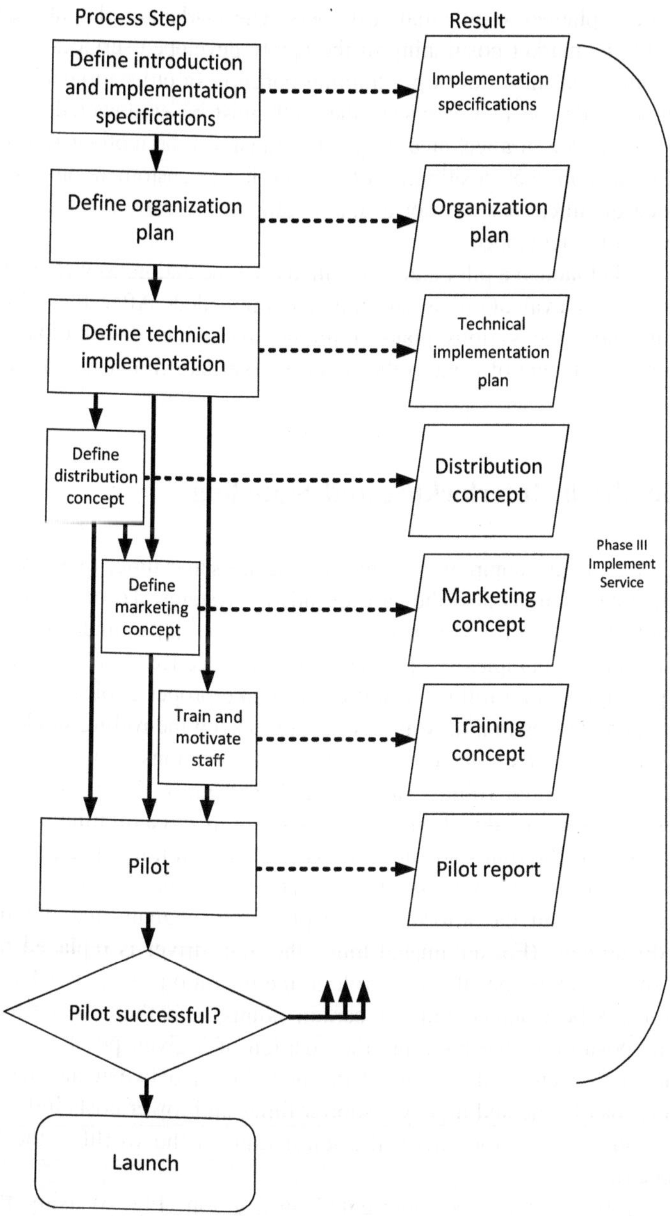

Fig. 5 Jaschinski model: phase III—implement service [19]

Based on the distribution concept, the marketing concept can be defined. The marketing concept covers the one-to-one customer interactions that occur while ordering services as well as the way the services are presented and, if needed, the way they are explained to potential customers. The marketing plan also addresses branding, that is, market positioning of the newly developed services.

The organization must develop a training concept. In order to enable the organization to provide the new services, the staff must be empowered and trained, providing the service in a technical way. Staff must also be motivated to represent the branding and market positioning of the service to customers and partners. If there needed qualifications are not available among the current staff, a recruiting concept has to be developed.

Before market launch a pilot and its detailed documentation, as well as its careful and conscientious evaluation, is strongly recommended. After the pilot is successfully finished, the various plans (training, marketing, distribution, technical implementation, change of organization) and market launch can be implemented.

3 Principles in Interlocking and Signaling

Transport systems are complex systems but, at the same time, they are based on simple principles. This is also the case in railway systems. To ensure selection of appropriate technology during new service implementation, an understanding of the technologies and the complexity of these systems are needed. This section describes the main principles to be followed in order to make proper choices.

The main challenges in a railway system are to keep the rolling stock stable at a given speed and to steer the trains in different directions. To resume this, the movement of trains and rolling stocks could be simply described through two characteristics: speed and direction. The speed of trains has a minimum at 0 km/h or mph, which means the trains are stopped. The maximum train speed is determined by the rails, rolling stock, and other factors. The actual speed is determined by the train driver, based on the information displayed by signals and the movement authority documents. (For automated trains the train driver is replaced by a computer, but this computer has the same role in the operation of trains.) The direction of the train can be changed only at certain points, and there is a time relation between the position of the train and the position of a given point.

The railway system is the result of the negotiation between the need for efficiency (in terms of time and money—shorter times and lower costs indicate greater efficiency) and the need for safety (the safest train in the world is the train with 0 km/h speed).

The safety function can be represented on a graph (Fig. 6) using the mathematical function $f(x) = ax/x$ (as an example of this principle). The main idea is to have the lowest level of safety, if the speed is at the highest level and to have the highest level of the safety when the speed is at the lowest level. This safety function could take different shapes based on the level of technological investment and new technologies installed. Based on technological investment the curve could be close to or far from the origin. The function of efficiency is represented in the diagram

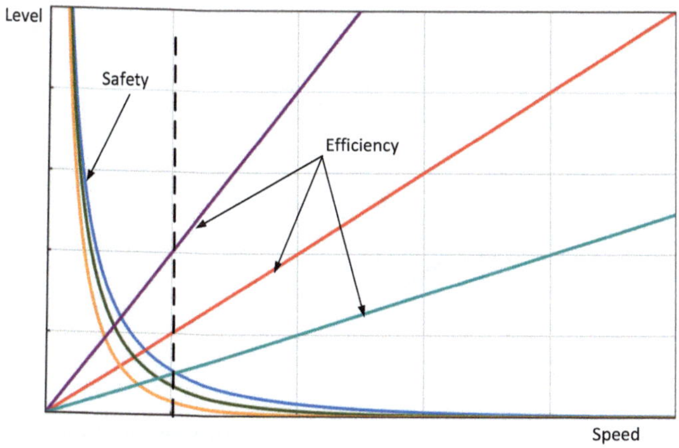

Fig. 6 The relation between speed, safety and efficiency

based on the elementary mathematical function $f(x) = ax$. (The authors' assumption is that the dependency is linear between efficiency and speed. This assumption was taken only to explain the principles; in reality one could use any linear or nonlinear function). The principle is that when the speed has the highest value the efficiency is also at the highest level. (For this principle the authors used travel time as the main indicator for efficiency.) When the speed is at its lowest level, then the efficiency is also at its lowest level.

For a given speed, efficiency can be improved if the system is more advanced in terms of technological solutions, and if other logistics were integrated into the process. For instance, during the transport process the freight might undergo transformations that increase the value of the products. This would indicate that the efficiency of the transport is increased, while maintaining the same level of speed.

Starting with the relation between speed and safety, the system has to be designed to control the speed of the train during the entire trip and to transmit this information at the control board of the train. For this reason, it is important to set a level of safety on the railway network and, based on this level of safety, to then adapt the speed to external conditions.

Every day, new technologies are coming, and technological progress is difficult to predict. For this reason it is important to analyze the role of technological investment in the railway system. This analysis has two variables as main indicators: the efficiency of the railway system and its safety.

Figure 7 presents a simplified model of the relationships between the investment in technologies (in terms of money or costs), the efficiency of railway system and its level of safety. For efficiency, the model illustrates the following aspects:

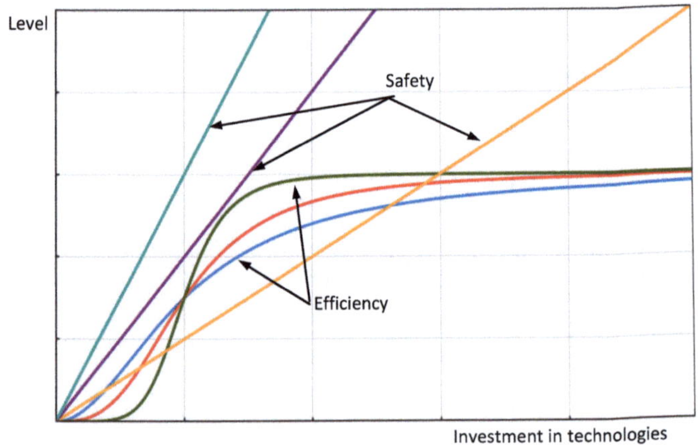

Fig. 7 The relationships between investments, safety, and efficiency

- With zero investment in technology, the level of efficiency is close to zero (in the lowest position of the graphical representation).
- Based on the adopted technical solution the system could be efficient after a low investment or could be efficient after a high level of investment.
- There is a limitation on the efficiency, no matter how much investment is involved in the railway system. There is a point at which efficiency grows only slightly with additional investment, after which additional investment is not profitable.

The analysis was done also for the relation between safety and the level of investment in technologies and the following conclusions emerged:

- With zero investment in technology and new technologies (so-called advanced technologies), the level of safety is in the lowest part of its trend.
- Based on the adopted technical solution the system could have a lower or higher level of safety for a given level of investment in technologies.
- There is also a limitation for safety level, but this limitation effect is much more pronounced for efficiency than it is for safety.

The radar presented in Fig. 8 could be drawn with three variables: investment, efficiency and safety, but the speed variable is important to emphasize this characteristic of efficiency and to link efficiency with time saved. A comparison of three different train technologies is presented in terms of highlighting the link between the speed of the train (as the main characteristic of efficiency) and the technological investment and safety.

In the following sections, we provide a set of main principles used in railway design, especially in interlocking and signaling; this approach is based on the system analysis and system engineering.

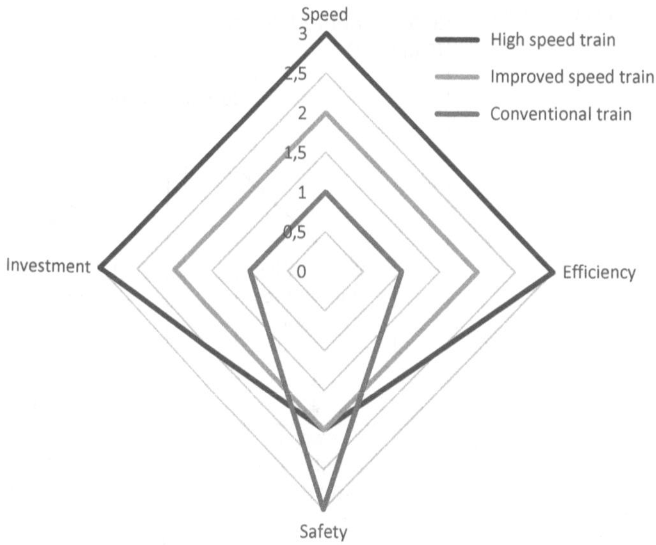

Fig. 8 The radar: investment-safety-efficiency-speed

3.1 *The "Fail-Safe" Principle*

A fail-safe system is one that, in case of a failure, causes a minimum of harm to the rest of the system's components, especially to the people involved in operation and other transport related activities. The definition could be converted into a day-by-day or operational definition: a fail-safe system is a system that, in the case of failure, will cost money (a minimum amount of money) only, and not property damage or lives. The fail-safe approach has to be adopted at the beginning, in the phase of planning and design of a railway system. The main idea is to introduce during the design phase some intermediate states of the system in case of failure. In case of a failure the system will not go directly to the damage state, but will be driven to an intermediate state that includes a lower level of risk than the damage state. The fail-safe design of the systems could generate three states of the system:

- Normal or True
- Failure with False positive response
- Failure with False negative response

The ideal function is to generate only the normal state of the system but, in reality, this supposition is always true and sometimes a failure appears. The main goal of the system's designer is to manage this failure state and to design the system in a manner to switch the false negative response into a false positive one. This approach is specific to a safety oriented design or fail-safe design. It assumes that the false negative state has a probability of appearance, based on experience.

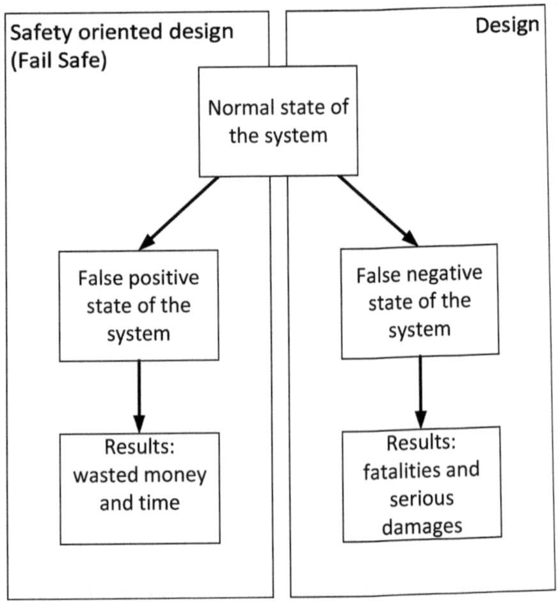

Fig. 9 Safety oriented design

Increasing the complexity of the system increases the probability of a false positive. Figure 9 shows this principle of safety oriented design.

Real-time systems, including railway systems, require both logical correctness of the result (which could be translated into the logical function of the system) and correctness of its timing (which is the moment of processing or acting) [21]. It is important to logical processing at the proper moment in time.

It is important to develop tools to measure the effect of introducing fail-safe principles as well as the effect of faults on the system. Methods to control the effects of faults are important; in [22] the authors summarized the major methods to control the effects of single random fault: Composite Fail Safe, Reactive Fail Safe, and Inherent Fail Safe.

3.2 The "Redundancy" Principle

One of the most important measures in the fail-safe approach is to use redundancy in the design phase, especially when building new railway systems. Redundancy is the inclusion of extra components which are not strictly necessary to functioning but, in case of failure, will take over the function of the damaged components. Redundancy is an additional resource supporting parallel computation of the same process [21]. Redundancy costs additional money, but it increases the level of safety. It is important to analyze the real effect of including extra components in terms of safety; if there is a real increase in safety the investment in redundancy is justified, if not, the redundancy is not needed.

The main principle in evaluating redundancy is to use a majority voting system based on majority voting logic, as well as additional components which are able to take over the functionality, if a failure appears.

Figure 10 shows a sample for the principle of redundancy. It also shows that redundancy could be cascaded by combining redundant subsystems with redundant components in subsystems and redundant communication systems. On the control and supervision level, redundancy is realized by so-called decision systems using a majority voting system. Independent controllers or predictors are implemented to calculate controller signals. In case these independent systems put out different signals, the signal calculated by the majority of the systems is used. Therefore, the number of independent calculations has to be odd. In practice, the overall system is brought to a safe state if calculations are different over a predefined period of time.

Another example of redundancy is AVTMR (all voting triple modular redundancy); an application of this was developed in [23].

3.3 *The "Speed Control" Principle*

Speed control is one of the most important activities in railway systems. Speed is a key performance indicator for efficiency and it is a technical parameter of the infrastructure. In controlling speed, there is a negotiation between the need for efficiency and the need for safety, which is infrastructure related.

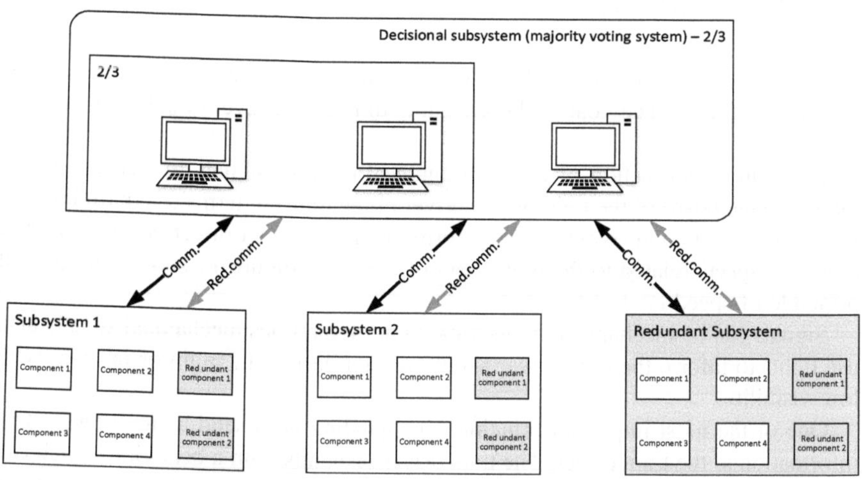

Fig. 10 Principles of redundancy

A two layer system was developed to control the speed of a train:

- The first layer is to send information about normal, reduced and zero speeds. This has been done by a signaling system.
- The second layer delivers information about the exact value of the speed in a specific area. This has been done by using physical signs along the railroad. The role of these physical signs is to delimit the zone within which a speed rule applies.

This principle can be applied to new systems. One example is ERTMS (especially the ETCS part of this system). In ETCS level 3, the signals and signs will be removed from the trackside and will be installed on the board of the train. For this to work, it is mandatory to have a link between the real position of the train and the information displayed on all signals and signs. More details about the levels of ETCS as part of ERTMS could be found in [24].

A train protection system was created to maintain the level of safety that incorporates a railway technical installation to ensure safe operations in the event of human failure. A better step in the development of this train protection system is automatic train control (ATC), which is a general class of train protection systems for railways that involves a speed control mechanisms designed using external factors and their inputs. ATC paved the way for ETCS (European Train Control System), which has the following elements: a regional center, a mobile network (based on GSM-R), an onboard computer, devices embedded on the track and a train protection system [25].

3.4 The "Signaling" Principle

The main components of a railway yard are track segments or sections, points, signals and routes. The goal of the system is to keep the level of safety as high as possible [26].

Signaling is the action of using mechanical or light signals to transmit information to the board of the train (to the driver of the train) in terms of controlling the speed of the train and sometimes to inform the driver about direction or additional relevant aspects related to the movement of the train. Figure 11 shows the general principle of speed control by signals.

Signals can be mechanical or electrical light. Sometimes mechanical signals also use lights to inform the train driver about position during the night or in cases with low visibility.

One of the most important principles in signaling is to send a small amount of information to the train driver. The train driver responds with a behavior, which can be modeled using the AIDA model (Attention-Interest-Decision-Action) originally developed for car drivers or the Perceive-Decide-Act-Model for train drivers introduced in [27]. For every stage of this model a time can be defined; total time between attention and action is the sum of the times of each stage. This time is

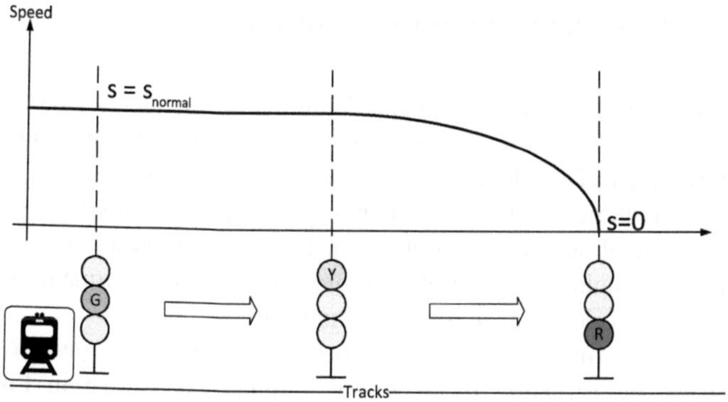

Fig. 11 Speed control-signaling system

influenced by the skill and ability of the train driver as well as the signaling system used. If the signaling scheme is complicated, the total time between attention and decision will be increased; this time can be reduced by keeping information to a necessary minimum and by training the train drivers.

3.5 The "Blocks" Principle

A block is a defined section of track. Using this concept, the distance between two adjacent railway stations can be split into several blocks. Absolute block signaling is a signaling scheme planned and designed to ensure a given level of safety for the operation of a railway by allowing only one train to occupy a defined section of track at any time. This system is used on single, double or multiple lines, where use of each line is assigned a direction of movement.

Once the isolated section is defined as a fixed block, the physical dimension of the block is unmodified.

A railway network faces huge demands in capacity in certain areas. Engineers have to design solutions to increase the capacity of these areas. One solution is to use moving blocks. A moving block respects all principles of a fixed block, but the boundaries of the section are moving, so in fact it is a virtual block. Application of this solution is creates more moving blocks on a given track section, which means more trains can pass in the same time, with the requirement that all trains have to occupy a single moving block.

The new approach in terms of railway systems is ERTMS (European Railway Traffic Management System), which is defined by two main components: ETCS (European Train Control System) and GSM-R (Global System for Mobile Communication-Railway). In ETCS level 3, the concept of the moving block is highlighted to increase the capacity of the railway network.

3.6 *The "Movement Authority" Principle*

The European Railway Agency's official definition of Movement Authority is "the permission for a train to run to a specific location within the constraints of the infrastructure" [28]. This definition underscores an important aspect of railway system, which is the existence of an authority giving or restraining permission to run trains in general, at specific locations or links or with an allowed speed. This authority may be implicit, by a set of laws and a number of departments, or explicit, by a transport authority or even a railway authority.

Movement authority is the main link between the railway operator, the train driver and the infrastructure (or technical system). As shown in Fig. 12, the principle is as follows: the railway operator wants to move a train from one location to another using a dedicated path and based on information collected from the railway system. To begin, the information has to be sent from the operator to the train driver and the train driver has to guide the train based on this information. The ETCS system together with its components, the signaling system and the interlocking system, has the task to transmit the information from the railway operator to the train driver in the form of a movement authority. The virtualization or digitalization of the movement authority is made in ETCS level 3. This approach generates more ICT and cyber security issues than previous solutions. An example is the new standards published by CENELEC (CENELEC-EN 50159) [29].

Fig. 12 The movement authorities

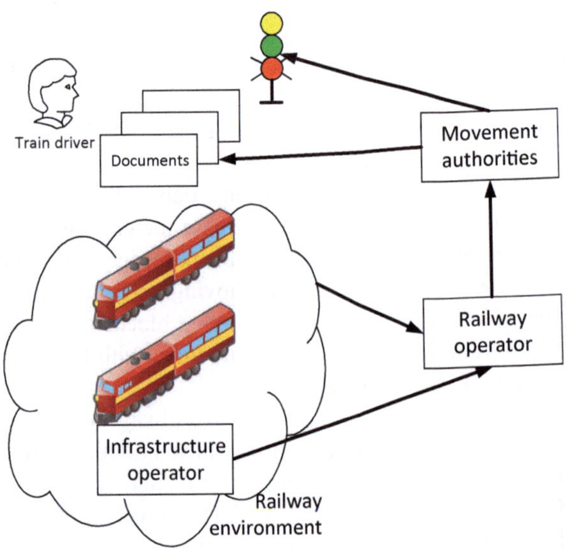

3.7 The "Resilience" Principle

Another important principle applied in railway systems is the resilience of the subsystems or components in terms of assuring the continuity of railway services. Resilience (also called reliability and risk management) refers to a system's ability to accommodate variable and unexpected conditions without catastrophic failure [30]. The railway system is critical infrastructure, meaning that the interruption of functionality—damage or failure—of this infrastructure could affect other components of the societal system with a huge risk and negative effects. Resilience could be covered using operational procedures, but now, based on the new and advanced technologies, a set of measures to improve the resilience of the system can also be developed.

In [31] a resilience model for a railway system is proposed consisting of three boundaries putting pressure on the operating state: Safety, Performance and Workload. A simulation of the resilience of the system can be done using a modeling tool.

Resilience can be defined as internal resilience or as external or integrated resilience. The second approach is oriented on multimodal transport systems, where the main idea is to link different modes of transport to provide transport services in the case of the failure of one transport mode.

3.8 The "Real Time Acquisition of Data" Principle

The acquisition of real time information by real-time sensors is an important property of today's transport system. This means collecting real-time information about the vehicles and the state of infrastructure to provide information and management services for a higher level of transport safety. The track circuit is one type of sensor which is able to collect information about the presence of train on a given track section and can also provide information about the integrity of the rails on that section. The analysis of the best sensor involves the definition of several key performance indicators. For instance, if this analysis is based only on the detection of the train and the detection of broken rails, the solution based on track circuit is much better than a solution based on an axle counter. A detailed analysis was done in [32] and this comparative assessment could be extended to all types of sensors. Another approach is proposed in [33], where a reference model for a sensor is proposed and all new sensors will be assessed relative to this reference.

The main principles for a classical track circuit, as seen in Fig. 13, are as follows:

- If the train is present on the controlled track, the receiver is not able to receive any electrical signal from the emitter and the section will be considered occupied. This is the case on the left side of the figure.

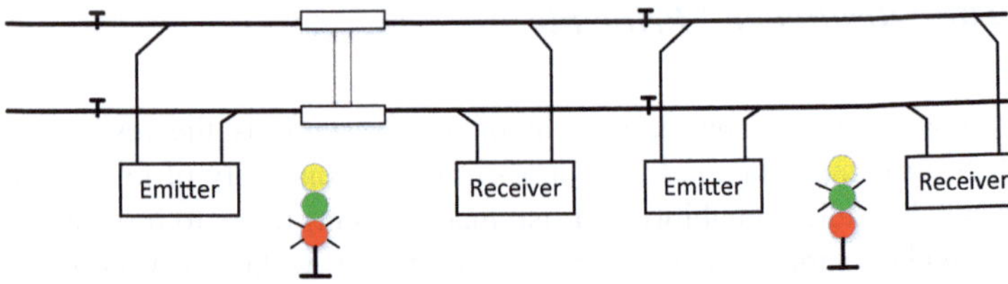

Fig. 13 The principle of a track circuit

- If the rail is broken, the wires between the emitter and receiver are interrupted and the receiver is not able to receive any signal. As in the preceding case, the section will be considered occupied.
- If either the emitter and receiver, or both, has failed, the section will be considered occupied.
- Each track circuit has to be isolated from other track circuits. In the figure this isolation is done in a physical manner, but it is possible to use different codes or frequencies for adjacent track circuits.
- The system has to generate, in case of a failure, a false positive response instead a false negative one.
- The coverage area of the sensor is defined based on some safety conditions; one is to ensure a sufficient braking distance for the train.

Based on these principles, different solutions can be developed using new technologies. One solution is to create a virtual section within a moving block. (In this example, the principle of separation between different sections is the same.)

Another type of sensor for train detection is the axle counter. The main principle of this sensor or detector is to count the number of axles passing a point A, (which is the input in the section A–B) and after that to count the number of axles in point B. If the number is the same, which means the section A–B is free, no train or rolling stock occupies this section. The rest of the principles are similar to the track circuit but the integrity of the rail is not detected.

The following principles are important for designing a sensor which is able to detect a train:

- The safety aspect: in terms of fail-safe principles, a false positive response instead of a false negative response
- The sensor has to be energy efficient: the energy consumed by the system has to be as low as possible.
- The sensor has to be correlated with some aspects of train operation (if the speed of train is limited up to 10–20 km/h, the case of marshalling yards, the principles could be changed and it is possible to us use a track circuit without any fail safety approach).
- The assessment of risk for railway environment (frequent damage of railroad, weather conditions)

The functional architecture of a track circuit or a sensor for the detection of trains is based, primarily, on two major activities: the collection of users' aspirations and risk assessment. Based on this primarily information the principles presented in this chapter will be applied to define the functional architecture of the system or parts of this system.

Based on the process described in Fig. 14, a track circuit can be design using new or advanced technologies. For example, a virtual track circuit could be designed using CCTV cameras and the safety impact could be assessed using comparative assessment methods [33].

3.9 The " Interlocking Systems" Principle

In railway systems, interlocking is an arrangement of signals and tracks (junctions and crossings) that prevents any conflict between the movements of different trains and assures a certain level of safety. An interlocking system displays a permissive light (green, yellow or a combination of them) if the route or path is safe and all subsystems and components have checked the safety conditions in the coverage area. Railway interlocking systems implement constraints specified in the control tables, in accord with safety principles and regulations [26].

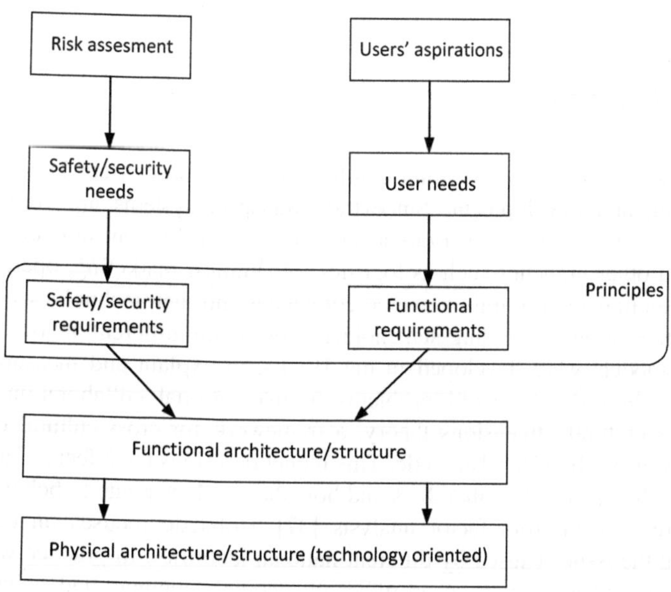

Fig. 14 The role of principles in the design of the system

The main principles of an interlocking system are:

- Fail-safe design for every components of the system
- No permission without the validation of applicable safety rules
- Defining a list of compatible routes and selection of the route based on this list
- Locking of all components and subsystems which are part of a selected route
- Permissive information displayed on the signal if the safety conditions are checked and all components are functioning well

The interlocking system and the signaling system are the most important tools to keep the level of safety and the level of efficiency as high as possible. This objective has to be achieved in all technical and natural conditions and using a given amount of resources [34].

4 External Factors

Besides technology, external factors must be taken into account when transforming organizations in order to provide new services. Most of these external factors are common in day-to-day business, but nevertheless their impact on developing and providing services is essential. This section gives examples of the most common and important external factors. This list may vary for different companies depending on their size, type of business, location and market position.

4.1 Cultural Diversity

As already shown in [35, 36] cultural differences may cause issues in collaborative engineering and development. Since train transport systems are very complex systems, covering not only various technologies from different domains, but also tasks from other branches such as logistics, economics, marketing, operations and many more, transition of train transport companies and their technology is always a collaborative project including stakeholders from many different areas.

Some models were developed in the 1980 s, to explain and measure cultural differences in order to avoid problems in international collaboration, such as Hofstede's cultural dimensions theory, a framework for cross-cultural communication developed by Geert Hofstede. This model describes the effects of a society's culture on the values of its members and how these values relate to behavior, using a structure derived from factor analysis [37]. Hofstede focuses on dimensions describing the issues caused by different national identities. In practice we see that many other aspects may have an effect on people's feelings and behaviors. As shown in [38, 39] there are several dimensions of cultural background that are not strictly bound to a nationality but may cause problems. The most important in context of transition projects in train transportation companies are:

(i) **Size of company**. Different sizes of companies adopt different types of organizations, which lead to different ways processes are organized. Thus, different ways of thinking are preferred in small, medium sized and large enterprises. In train transport infrastructure projects we deal with many different types of companies. Usually the infrastructure manager or provider and most of the train running companies are big enterprises, but in case of transformation projects and during operating a rail related service, many small and medium sized companies are involved as suppliers, consultants, technology providers and service providers, for such activities as teaching, ticketing and local construction. The different ways in organizing and thinking may cause misunderstandings, leading to additional failure, increased workload and disturbing the working atmosphere. This leads to a strong recommendation: that project managers and other employees should be trained in adoption of other processes and business practices.

(ii) **Level of education**. Different educational institutions and different education, such as vocational training or studies, shape different attitudes toward problem solving and different ways of finding solutions. This also leads to different understanding in the meaning of concepts while developing a solution. In train transport infrastructure projects and when running rail-related services, we deal with a great range of educational levels, from semi-skilled workers in construction or ticket sales, to employees in engineering or logistics who undertake studies, to doctorate employees developing methods and algorithms for such things as creating schedules or forecasting demand. This wide range of different ways of thinking about tasks and solutions can cause ambiguities and a lack of coordination, leading to missed project targets and processes that may interfere with the customers or cause higher cost. This leads to a strong recommendation: that project managers and line managers shall have the capability to identify and compensate for these kind of problems.

(iii) **Domain of education**. In addition to the level of education, the different domains in which people work and receive their educations results in different schools of thought, for example, preferring inductive or deductive approaches on problem solving. Running or transforming a rail transport organization entails a wide range of technologies and a wide variety of tasks; there are a lot of different domains involved in transport infrastructure projects and in running rail-related services. For example, the technical aspects are part mechanical, civil and electrical engineering as well as computer science and ICT-related. On the administrative side, we find economists, logisticians, product managers, key account managers, sales and marketing people. This illustrates a wide range of tasks and challenges, where each domain has developed a system of approaches and related terms, methods and ways of understanding its surroundings. These different understandings and the fact the companies usually are organized in departments reflecting these different domains, can lead to a number of problems in coordinating activities of different stakeholders from different domains and departments.

Besides general approaches in agile management (as proposed in [40]) and new ways in organizing companies (e.g. The Fifth Discipline/ Learning Organization [41]) a good approach in solving this problem is accepting the cultural variety and minimizing the problems caused thereby by training the members of multidisciplinary project teams and their managers to give them intercultural competencies.

(iv) **Nationality**. Different nations have different value systems causing different ways in making and judging decisions or perceptions of project partner behavior. Today, both the market for rail technology and the market for transport services are international markets. This causes a high rate of interaction of people from different nations. During technical transformation of projects, railway companies deal with technology providers and engineering service providers from different countries, employing people from all over the world. On the customer side we find people from all areas. That makes intercultural competence a key competence for all stakeholders and their employees.

As seen in the preceding enumeration, there are many requirements on employees, project and line managers caused by the cultural diversity of people. Therefore, dedicated skills management related to this demand is essential. Besides registering existing skills and capabilities, a structured training and education program filling the gaps by internal or external training is needed. In order to minimize cost and effort this program should be a part of the overall *Competence Management,* also including training and education programs as described in the next section.

4.2 Tendering Policies

The most significant way of doing businesses in the railway domain is the tender process, especially with regard to infrastructure. This process has to be correlated with the stages of the railway infrastructure implementation project:

- The elaboration of ToR (terms of reference) for every system, subsystem and component
- The design of the infrastructure or parts of it
- The installation of the infrastructure or parts of it
- The maintenance activities (annual and multi-annual)
- The mitigation of new technologies and new services (i.e. high-speed train)
- The upgrading of infrastructure

All these stages must be under the assessment and monitoring of a national safety authority which is represented at the European level by the European Railway Agency. Every stage has a special procedure and a specific approach in accord with the European, regional and national standards and regulations.

The European Commission is focused on three major areas, which are all crucial for developing a strong and competitive rail transport industry:

- Opening of the rail transport market to competition
- Improving the interoperability and safety of national networks
- Developing rail transport infrastructure

The second point, the improvement of the interoperability and safety of national networks, has become the task of the European Railway Agency, together with all national authorities, with strong cooperation of the European Commission.

The main issue of the tender process is not only to provide a service or to deliver a product, but also to do the work under maximum safety conditions based on European and national regulations.

In every tender procedure, access to information is crucial for the success of the process. Computer based tools could be a real option in communicating and selecting the best options for new works in a railway network. As an example, TfL has launched this type of e-Tendering tool [42].

In measuring the success of a tendering system, it is important to define three categories of key performance indicators: the cost of the process, the effect on transport and additional services, and the revenue.

5 Competence Management

Today's trend in education in general, and within the so-called Bologna process in higher education in particular, is planning education and courses to be focused on training competencies, not just imparting knowledge [43, 44]. In terms of the well-known distinction between formal and informal knowledge, the focus shifts to informal knowledge, and application know-how. In contrast to the general trend in the railway industry, there is a high demand on formal qualifications, such as the certification of competence. Although a well-managed rail system is very safe, the impact of an incident may harm or kill people. Therefore, in most countries, for most tasks related to running the system, we need employees holding a formal qualification. Those qualifications are usually bound to an existing technology and/or its underlying methodology.

So, in fact, railway companies in transition face two challenges: (a) empowering staff to run the new processes and technology, and (b) proving the formal qualification against authorities. As an additional challenge, they have to face the fact most qualifications are valid in only one country, making offering services on an international market more complicated, even if the requirements on interoperability are fulfilled on a technical level. This means companies in the rail sector need to manage qualifications and competencies of their employees and implement a reliable forecast on future demand in this area. This also includes establishing and maintaining a network of providers of teaching and training.

In addition, a company's competence managers must deal with many different levels of education and the impact of transition on salary schemes. Therefore, a universal and country-independent structure to sort, appraise and (at the end of the day) price qualifications and their holders is needed. Despite all the criticism of its implementation in Europe, the so-called European Qualifications Framework (EQF) can be a good starting point for such a structure. The European Qualifications Framework is a translation tool that helps communication and comparison between qualifications systems in Europe. Its eight common European reference levels are described in terms of learning outcomes: knowledge, skills and competencies. This allows any national qualifications systems, national qualifications frameworks (NQFs) and qualifications in Europe to relate to the EQF levels. Learners, graduates, providers and employers can use these levels to understand and compare qualifications awarded in different countries and by different education and training systems [45, 46].

A competence management system should be part of, or strongly linked with, human resource management and should define current demands and forecast future demands for each level. To allow a valid forecast, competence management should be involved in the very early steps of service development and transition planning. In this way, future gaps in competencies can be identified and programs to fill this gaps can be developed and implemented.

In general there are two sources for competencies: (a) educating current staff and (b) hiring people with the needed competencies from the market or universities.

5.1 Role of Educational Institutions and Universities

As shown in [47, 48] due to the complexity of growing railway systems, the demand for higher educated employees is rising, making universities and colleges a provider for future employees. Rail Education and training falls within the context of Applied Science. For students and trainees, rail operations and technology laboratories equipped with modern software are needed. These are currently missing in many universities who claim to have specialized in rail education and training. For the efficiency of rail education and training it is important to secure contacts within the world of rail operation and development. Therefore it is crucial to organize technical visits to rail-focused organizations and train operating companies. Access to real world data and case studies is crucial; this is an area which needs to improve constantly.

Regardless of the characteristics of the rail sector, rail education and training is driven by an intended learning outcome, which needs to be achieved through a learning process. It is now known that learning happens when, as a result of brain processes, knowledge transitions from short- to long-term memory. The rail educator acts as a connecter, making sure that students/trainees have stored some vital knowledge for later recall after successful completion of a rail learning activity.

Therefore, leading edge methodology should be used in teaching (e.g. [49–51]) as well as in planning the courses (e.g. [43, 44, 52]). Rail is one of the fastest growing industries in the world, engaging with modernization and implementation of new technologies. It is of prime importance to lay a foundation in the student's/trainee's mind so that new practical rail-orientated techniques, skills and information can be linked to prior knowledge of rail systems at any time in the future. The ultimate goal of education and training is to pave the way towards lifelong learning, and the rail sector is no different. Therefore, study courses have to attract potential students by following current trends, such as providing international joint degrees as described in [52–54]. Of course the rail sector as a whole has to offer attractive job opportunities and career chances to a wider range of potential employees by means of specific campaigns and joint actions, such as the international rail talents network (http://www.railtalent.org/).

At present, rail education and training at universities include mainly Professor-led/Senior trainer-led methods, which seem not to give the opportunity for young students/trainees to take part in the building of their own rail related program/course curricula [44]. This situation is not favorable to the innovation and modernization of rail education and training, as it seem not to envisage comprehensive learning through interchangeable roles and activities for the building and implementation of new methods and knowledge. A more dynamic, student/trainee-led method for rail education and training is necessary to offer innovation and modernization for new rail-specific skills. As shown in [55] those activities can boost success in earning competencies during studies. This will lead to the development and implementation of future educational concepts, innovative approaches, new policies and practices for a rail sector which is vibrant and ever-growing. A steady career path for every rail professional will then be secured, which will have a positive impact on the performance of the sector as a whole [44].

5.2 *Educating Current Staff*

Besides recruiting graduates, a good source of competencies is for a company to educate its own staff. This type of education has many advantages, for example, the curriculum can be shaped to the actual needs of the company. When identifying a lack in competencies training, the current staff is the best solution to keep know-how in the company while facilitating the ability to provide state-of-the-art services. In today's war for talents, a company must build up a reputation of being a good and attractive employer. By giving people the chance to develop professionally by enhancing their personal skills, this reputation gets an additional boost.

Today there are many different types of internal training in companies, for example:

- Vocational training targeting a predefined vocational qualification
- Training-on-the-job programs, such as tutoring targeting an additional qualification in order improve quality in the current position
- In-company training for a new technology or process
- Study courses in cooperation with universities in order to train company-fitting graduates

During those programs different types of teaching and learning can be used, such as:

- Distance learning and e-learning online units
- In-house face-to-face units provided by internal departments
- In-house face-to-face units provided by external education or technology providers
- Learning-on-the-job units
- Lectures
- Exercise units
- Practice units

After identifying current or future gaps in competencies, the competence management group needs to characterize the demand and find a suitable format to educate people.

Figure 15 gives a hint as to which formats may fit as a function of size of competence gap and time to fill the gap.

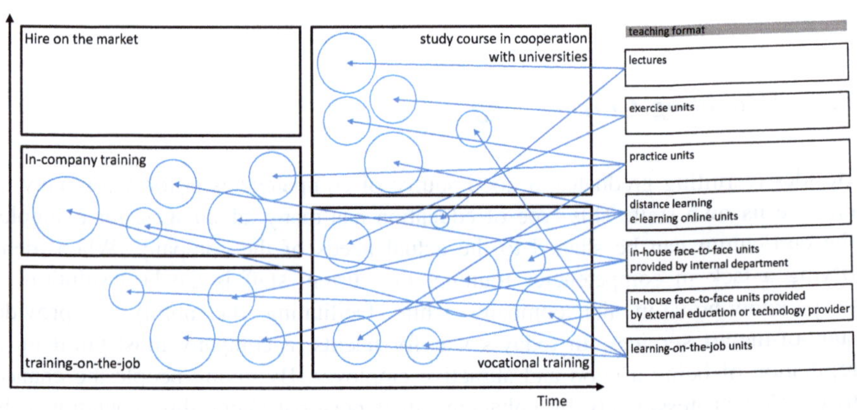

Fig. 15 Course and Teaching format in relation to size of gap in competencies and time to fill it

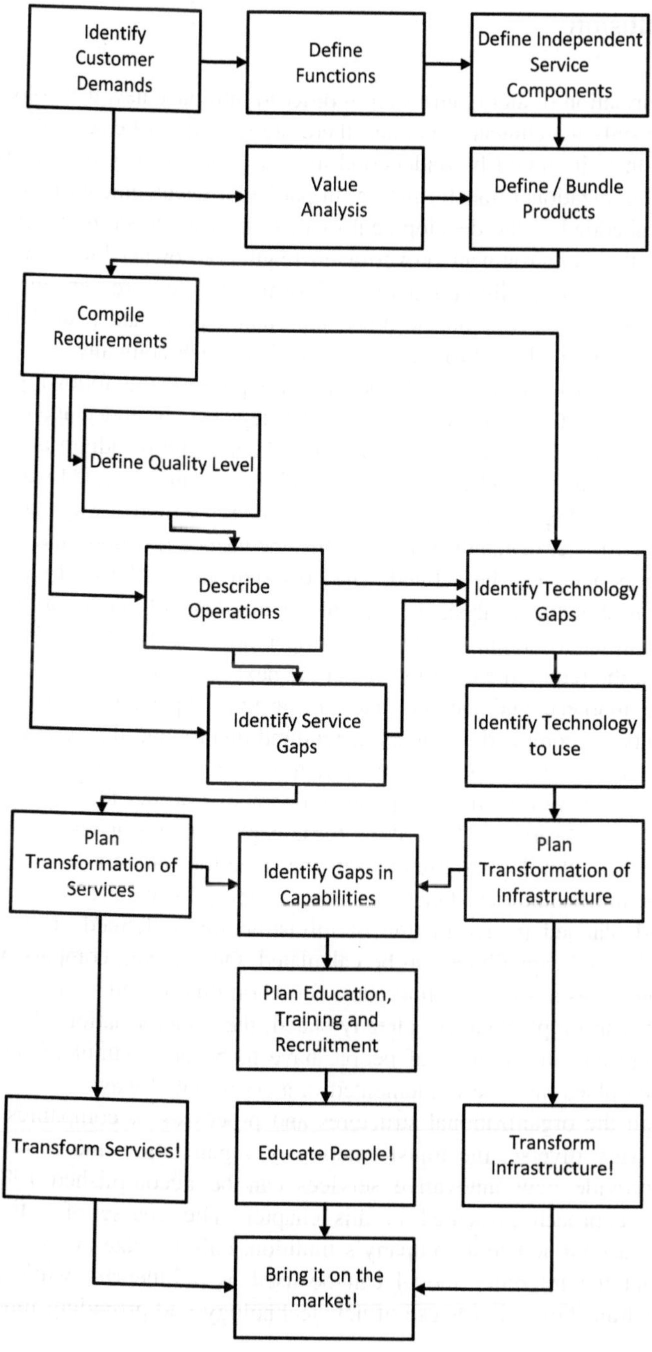

Fig. 16 General procedure for developing and introducing new services and related technology and processes

6 Conclusion

The transformation of an organization in order to introduce new technologies is far from being only a technical issue, and therefore should not be seen as a technical project. Rather, it should be understood as a step on the way to new and better services. Organizational transformation, in particular development of its own staff, must be embedded in the development of new services. Since train service companies and their environment (government regulation, ownership structure, company size, company tradition, existing infrastructure, etc.) are very different, this chapter gives an overview on the most important aspects and related steps. This approach has to be adapted to the special needs of each company.

In order to provide a general checklist we propose the following reference process (see Fig. 16), where company-specific steps may be added and unnecessary steps may be skipped. The main aspect to be kept is the paradigm of developing services, infrastructure and employees' capabilities in an integrated way.

As shown in Fig. 16 the transformation process should always start with an analysis of customer demands. Based on this, the needed functions and independent service components can be defined. The next step, in parallel to the definition of functions, is a value analysis based on the outcome of the customer demand analysis in order to define suitable and proper prices of products. This approach ensures that the requirements compiled in the next steps are linked to the customer needs (the market), and can be rated respecting expected benefits. Once the requirements are compiled, the level of required quality and the needed operations can be described. A comparison of current state and expectations described by requirements, level of quality and planned operations allows identification of gaps in technology and services. Describing these gaps explicitly gives a good base for identifying the technology to use and sets the stage for planning the transformation of services and technology. Once the target in terms of planned service transformation and planned transformation of infrastructure is defined, a needed set of competencies and capabilities can be calculated. Once again, comparison with the current state gives a list of existing gaps. Based on this list, education, training and recruitment can be planned. As already stated, the transformation of services and infrastructure and the training of people have to be done simultaneously, so the three related plans must be implemented in a coordinated way.

Although the organizational structures and processes of companies in the rail sector are very diverse, the transformation of organization and infrastructure in order to provide new innovative services can be accomplished following the methodical approach presented in this chapter. The course of action may be adjusted to accommodate a company's tradition, culture, size and existing infrastructure, but the reference model can be used as a blueprint while planning a company's transition towards use of new technology and providing innovative rail services.

References

1. ERTMS Unit-European Railway Agency. ERTMS operational principles and rules draft version 2. European Railway Agency; 2029 Mar. Report No. ETCS 2.3.0d AND GSM-R 7.0
2. Vinck K (2015, May) ERTMS—Work plan of the European Coordinator Karel Vinck. European Commission—Directorate General for Mobility and Transport Directorate B—European Mobility Network Unit B1—Trans European Network
3. Administration USFR (2009) Vision for high-speed rail in America: high-speed rail strategic plan. Federal Railroad Administration
4. High-speed Europe: A sustainable link between citizens. Publications Office of the European Union (2010)
5. Schneider J (2013, Aug) Aus dem Verkehr gezogen. Süddeutsche Zeitung. 8
6. NN. Bahn will Chaos in Mainz bis Ende August beheben. 13. August 2013
7. Stratmann K (2013, Aug) Eisenbahn-Bundesamt rüffelt die Deutsche Bahn. Handelsblatt
8. Kuhr D (2013, Sept) Probleme in Mainz waren lange bekannt. Süddeutsche Zeitung. 2
9. Schlingensiepen J, Nemtanu F, Mehmood R, McCluskey L (2016) Autonomic transport management systems—enabler for smart cities, personalized medicine, participation and industry grid/industry 4.0. In: Intelligent transportation systems—problems and perspectives. Springer International Publishing, pp 3–35
10. Schlingensiepen J (2015, Oct) Utilise public transport for disabled people. In: Proceedings of 22nd World Congress and exhibition on intelligent transport systems and services. ITS—Europe
11. Schlingensiepen J, Stockmanns G, Naroska E, Christen O, Bolten T (2015) Personal smart travel agent for empowering persons with disabilities using public transport. In: Transport Problems, vol. 10 (Special Edition), pp 5–14
12. Posselt T, Förstl K (2011) Success factors in new service development: a literature review. Productivity of services next gen-beyond output/input Fraunhofer Center for Applied Research and Supply Chain Service, Germany. Available: http://reser.net/materiali/priloge/slo/posselt_et_al.pdf
13. Fachbericht D (1998) 75 "Service Engineering", Entwicklungsbegleitende Normung (EBN) für Dienstleistungen, DIN Deutsches Institut für Normung eV. Beuth Verlag, Berlin
14. Schneider K, Daun C, Behrens H, Wagner D (2006) Vorgehensmodelle und Standards zur systematischen Entwicklung von Dienstleistungen. Service Engineering. Springer, Berlin, pp 113–138
15. Kunau G, Junginger M, Herrmann T, Krcmar H (2005) Ein Referenzmodell für das Service Engineering mit multiperspektivischem Ansatz. Konzepte für das Service Engineering. Physica-Verlag HD, pp 187–216
16. Ramaswamy R (1996) Design and management of service processes: keeping customers for life. Addison-Wesley Publishing Company
17. Cockburn A (2006) Agile software development: the cooperative game. Pearson Education
18. Beck K, Beedle M, Van Bennekum A, Cockburn A, Cunningham W, Fowler M et al. (2001) Manifesto for agile software development. Available: http://academic.brooklyn.cuny.edu/cis/sfleisher/Chapter_03_sim.pdf
19. Jaschinski C (1998) Qualitätsorientiertes Redesign von Dienstleistungen. Shaker
20. Akao Y (1992) QFD-quality function deployment. Landsberg/Lech
21. Chakraborty A (2009) Fault tolerant fail safe system for railway signalling. In: Proceedings of the World Congress on engineering and computer science, pp 20–22
22. Xue W, Zhao Y, Xiao J, Zhang M (2014) The research and application of fail-safe technologies in rail transit train operation control system. In: 10th international conference on reliability, maintainability and safety (ICRMS). IEEE, New York, pp 1100–1104
23. Kim H, Jeon H-J, Lee K, Lee H (2002) The design and evaluation of all voting triple modular redundancy system. In: Reliability and maintainability symposium, proceedings annual, pp 439–444

24. Winter P (2009) International Union of Railways, compendium on ERTMS. EurailPress, Hamburg
25. ERTMS Online (2015, Mar) In: ERTMS Online [Internet]. Available: http://ertmsonline.com/
26. Kanso K, Moller F, Setzer A (2009) Automated verification of signalling principles in railway interlocking systems. Electron Notes Theor Comput Sci 250:19–31
27. McLeod RW, Walker GH, Moray N (2005) Analysing and modelling train driver performance. Appl Ergon 36:671–680
28. Glossary of Terms and Abbreviations (2012, Mar) ERA-UNISIG; Report No. 3.0.0
29. CENELEC. Railway applications—communication, signalling and processing systems—safety-related communication in transmission systems. CENELEC; 2001. Report No. EN 50159
30. Evaluating Transportation Resilience (2014, Apr) In: TDM Encyclopedia—Victoria Transport Policy Institute. Available: http://www.vtpi.org/tdm/tdm88.htm
31. Herrera I, Schraagen JMC, Vorm J van der, Woods DD (2014) Developing resilience signals for the Dutch railway system. Faculty of Behavioural, Management and Social sciences (BMS). Available: http://doc.utwente.nl/91398/
32. Thurston D, Kozol B (2010) Axle counters vs. track circuits–safety in track vacancy detection and broken rail detection. American Railway and Maintenance-of-Way Association. In: Annual conference and exposition, Orlando, Florida
33. Nemţanu FC, Bureţea DL, Obreja LG (2015) Comparative assessment of virtual track circuit based on image processing. In: Urban Rail Transit, vol 1. Springer, Berlin, pp 131–137
34. Theeg G, Vlasenko S (2009) Railway signalling and interlocking: international compendium. Eurailpress
35. Schlingensiepen J, Phillipson J (2009) Integration of cultural advisory systems in product data management for supporting international collaborations in product engineering. In: 2nd IPROMS international researchers symposium (ISBN 978-88-95028-38-5)
36. Phillipson J, Schlingensiepen J (2009) Cultural support during product engineering in international collaborations. In: Proceedings of the international conference on collaborative mechatronic engineering, pp 151–156
37. Hofstede G (1984) Culture's consequences: international differences in work-related values. Sage
38. Jamieson R, Pearce KF, Harvey J (2009) Risks and engineering design across cultures. In: Reliability, risk, and safety, three volume set
39. Pearce K, Jamieson R, Harvey J (2009) Cultural differences in design perceptions of consumer products. In: Reliability, risk, and safety, three volume set
40. Moran A (2015) Managing agile: strategy, implementation, organisation and people. Springer, Berlin
41. Senge PM (2006) The fifth discipline: the art and practice of the learning organization. Doubleday/Currency, New York
42. TfL. Transport for London—e-Tendering (2015) In: TfL—Electronic tendering. Available: https://eprocurement.tfl.gov.uk/epps/home.do
43. Schlingensiepen J (2014) Competence driven methodology for curriculum development based on requirement engineering. Proc Soc Behav J. Elsevier
44. Marinov M, Fraszczyk A (2014) Curriculum development and design for university programmes in rail freight and logistics. Proc Soc Behav Sci 141:1166–1170
45. Descriptors defining levels in the European Qualifications Framework (EQF). In: European Commission: Learning opportunities and qualifications in Europe [Internet]. Available: https://ec.europa.eu/ploteus/content/descriptors-page
46. Information on the EQF, NQF's. In: European Commission: Learning opportunities and qualifications in Europe [Internet]. Available: https://ec.europa.eu/ploteus/search/site?f[0]=im_field_entity_type%3A97
47. Rizzetto L, Ricci S, Marinov M, Zunder T, Schlingensiepen J, Karagyozov K et al. (2011) A structured survey on MScs in transport and logistics for designing a new programme. In:

Proceedings of the 20th international scientific conference transport. Todor Kableshkov University of Transport (VTU)
48. Marinov M, Fraszczyk A, Zunder T, Rizzetto L, Ricci S, Todorova M et al. (2013, Apr) A supply-demand study of practice in rail logistics higher education. J Transp Lit Sociedade Brasileira de Planejamento dos Transportes 7(2):338–351
49. Schlingensiepen J (2014) Innovation in distance, e- and blended learning in educational mass production using inverted class room model (icm). Proc Soc Behav J
50. Schlingensiepen J (2014) Video killed the radio star—lecturing using new media—a case study. In: Hagel G, Mottok J (eds) ECSEE. See on: Shaker; 2014. doi:10.2370/9783844030679
51. Marinov M, Zunder T, Schlingensiepen J, Ricci S, Karagyozov K, Razmov T et al. (2010) Innovative concepts for knowledge exchange, mobility and expertise in rail freight and logistics. Highlight
52. Dzhaleva-Chonkova A, Todorova M, Marinov M, Razmov T (2014) Internationalisation of MSC programmes in rail transport and logistics. In: EURO-ZEL 2nd international symposium, Žilina
53. Marinov M, Ricci S (2012) Organization and management of an innovative intensive programme in rail logistics. Proc Soc Behav Sci 46:4813–4816
54. Fraszczyk A, Dungworth J, Marinov M (2015) An evaluation of a successful structure and organisation of an intensive programme in rail and logistics. In: Proceedings of the 3rd UIC World Congress on rail training
55. Fraszczyk A, Dungworth J, Marinov M (2015) Analysis of benefits to young rail enthusiasts of participating in extracurricular academic activities. Soc Sci. mdpi.com. Available: http://www.mdpi.com/2076-0760/4/4/967/htm

[illegible] Management [illegible] [illegible]
[illegible] [illegible] [illegible] (2010) [illegible] [illegible] [illegible] [illegible] [illegible] [illegible] [illegible] [illegible] [illegible] [illegible] [illegible] Rev [illegible]
[illegible] [illegible] [illegible] [illegible] [illegible] (2013) [illegible] [illegible] [illegible] [illegible] [illegible] [illegible] [illegible] [illegible] [illegible] [illegible] [illegible] [illegible] [illegible] [illegible] [illegible] [illegible]
[illegible] [illegible] [illegible] [illegible] [illegible] [illegible] (2011) [illegible] [illegible] [illegible] [illegible] [illegible] [illegible] [illegible] [illegible] [illegible] [illegible] [illegible] [illegible] [illegible]

Train Protection Systems in Different Railway Gauges

Lionginas Liudvinavičius and Aleksander Sładkowski

Abstract This chapter analyzes the train traffic control systems for 1435- and 1520-mm railway gauges, as well as their compatibility issues. The British Rail Traffic control system is analyzed. European train control systems (ETCS) and ETCS levels are described. Differences between European train control systems in 1435- and 1520-mm railway gauges related technical problems and proposed solutions are presented with regard to ETCS implementation in the Baltic states. The existing train control systems do not meet requirements of traffic safety in light of increased train speeds.

Keywords Train protection system · Interlocking · Track circuit · Moving block · Balises · Train speed · Axles counters · Global positioning system (GPS)

1 Introduction

Today, many countries apply interval train traffic control, where trains in a section are separated using fixed-block principle. Increased train speeds dictate the necessity to improve train control in order to reach ETCS level 3. This may be achieved by using the principles of moving block and radio transmission instead of interval train control. At the same time, reliable automated braking systems for high-speed trains have to be developed. Examples of such systems used for stopping high-speed trains of Japan, France and Italy are analyzed in a dedicated chapter. This chapter also presents new opportunities arising from the modernization of Lithuanian railways

L. Liudvinavičius (✉)
Department of Railway Transport, Vilnius Gediminas Technical University, J. Basanavičiaus g. 28, Vilnius, Lithuania
e-mail: lionginas.liudvinavicius@vgtu.lt

A. Sładkowski
Faculty of Transport, Department of Logistics and Industrial Transportation, Silesian University of Technology, 40-019 Krasinskiego 8, Katowice, Poland
e-mail: aleksander.sladkowski@polsl.pl

A. Sładkowski (ed.), *Rail Transport—Systems Approach*, Studies in Systems, Decision and Control 87, DOI 10.1007/978-3-319-51502-1_7

with microprocessor traffic control systems, new-generation traffic control centers and Global System for Mobile Communications-railway (GSM-R) radio communication, among others.

2 Global Navigation Systems

Global navigation systems. The Global Positioning System (GPS) is a universal positioning system comprising Earth-orbiting satellites [1]. GPS is the US satellite radio navigation system, and GLONASS (Russian *Globalnaja navigacionaja sputnikovaja sistema—Global Navigation Satellite System*) is the Russian system. In the GPS system, the exact location of a train is established by a GPS receiver mounted in the train. Using the signals received from satellites, the GPS receiver establishes geographic coordinates and speed, and transmits this information to the control and memory block [2–5]. In the GALILEO system, this service will be implemented on higher-quality precision, universality and reliability levels compared to the GPS or GLONASS systems [6–8].

The GALILEO system is part of an interconnection program of the Trans-European Transport Networks. The main purpose of the GALILEO system is the improvement and development of transport systems of all types. The GALILEO service will be used, for example, by geodesists, operators of communication networks, dispatchers of transport systems (including train traffic control). A diagram of the operation and control of navigation systems is shown in Fig. 1. Implementation of global navigation-based control systems suited for modern

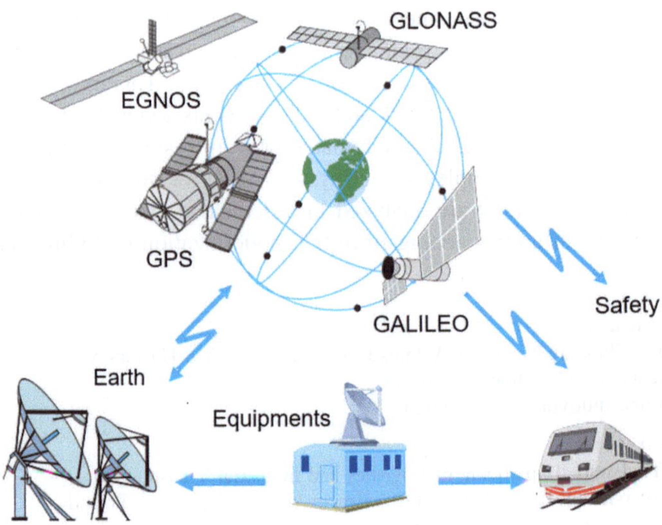

Fig. 1 Diagram of operation and control of navigation systems

transport will provide essentially new traffic control capabilities of the first, second and third levels of ERTMS.

3 GSM-R Radio Communication System

The GSM-R radio communication system is based on a standard GSM, but with different specific frequencies used in the railway system and some programmed functions. It is a radio communication system, which is meant for information to be transmitted and received between the rail tracks and the train (voice and data) [9, 10]. A general diagram of GSM-R network architecture is shown in Fig. 2. Information to a train is transmitted by radio communication GSM-R. The GSM-R radio communication system is intended for voice communication and fast transmission of people data and train traffic safety non-critical data. The GSM-R system includes the following subsystems: base station subsystem (BSS), network and switching subsystem (NSS), network management system (NMS), package data transmission subsystem and fixed dispatcher network (FDN) subsystem [11–13] (Fig. 3).

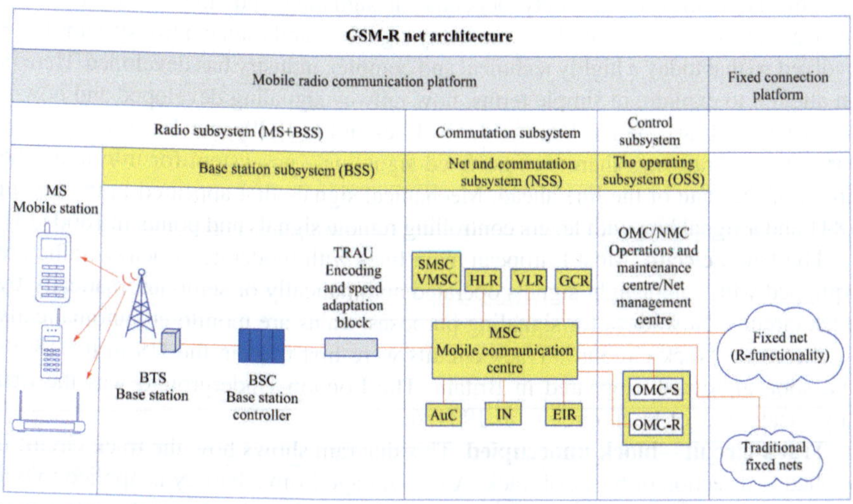

Fig. 2 General diagram of GSM-R network architecture

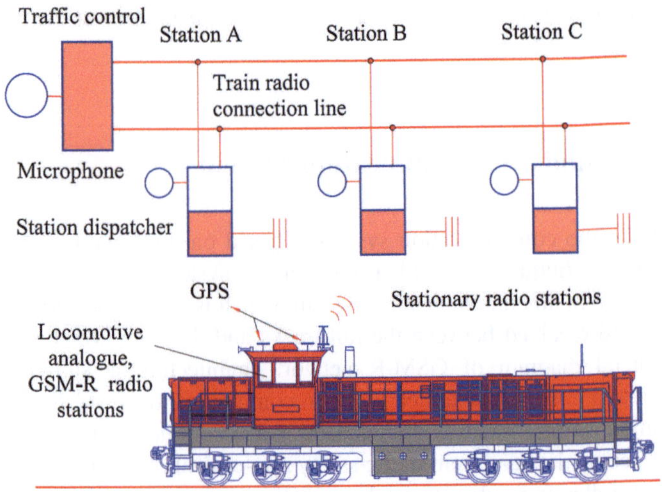

Fig. 3 Block diagram of analogue and digital GSM-R communication systems

4 British Rail Traffic Control System

Signaling is one of the most important of the many parts that make up a railway system. Train movement safety depends on signaling and thus the control and management of trains. Over the years many signaling and train control systems have evolved so that today a highly technical and complex industry has developed. Here is an attempt to explain, in simple terms, how railway signaling developed and how it really works, based on the UK standards. Back in the 1830s and 1840s in the very early days of railways there was no fixed signaling—no system for informing the driver of the state of the line ahead. Mechanical signals first appeared in the UK in 1841 and a signal box with levers controlling remote signals and points in 1860 [14].

The track circuit. Most European main lines with moderate or heavy traffic are equipped with colour light signals operated automatically or semi-automatically by track circuits. Nowadays for signaling purposes, trains are monitored automatically by means of "track circuits". Track circuits were first tried in the US in the 1890s and soon afterwards appeared in Britain. The London Underground was the first large-scale user of track circuits in 1904–6.

Track circuit—block unoccupied. This diagram shows how the track circuit is applied to a section or block of track. A low voltage from a battery is applied to one of the running rails in the block and returned via the other. A relay at the entrance to the section detects the voltage and energizes to connect a separate supply to the green lamp of the signal [14]. A track circuit diagram for unoccupied block is provided in Fig. 4.

Track circuit—bock occupied. When a train enters the block (above), the leading wheelset short circuits the current, which causes the relay to de-energize

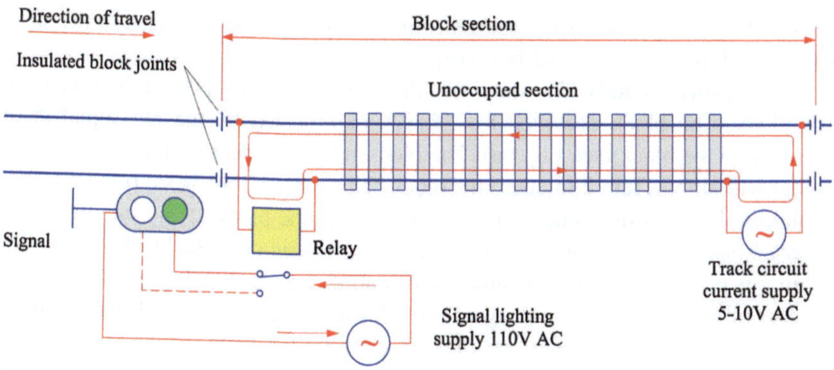

Fig. 4 Track circuit diagram for unoccupied block

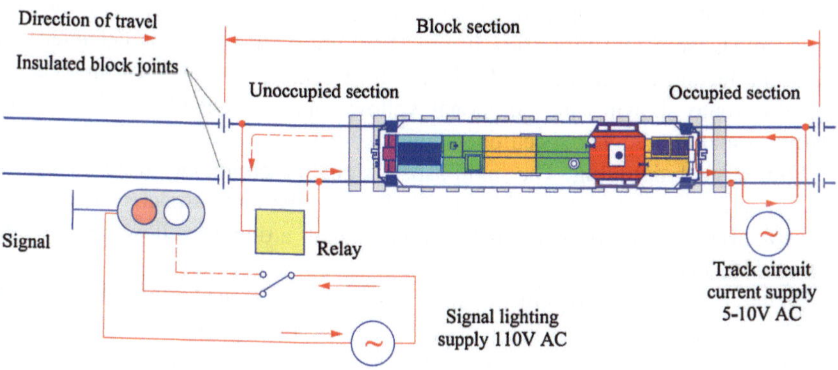

Fig. 5 Track circuit diagram for occupied block

and drop the contact so that the signal lamp supply circuit now activates the red signal lamp. The system is a "fail-safe", or "vital" as it is sometimes called, because any break in the circuit will cause a danger signal to be displayed [14].

The above is a simplified description of the track circuit. The reality is somewhat more complex. A block section is normally separated electrically from its neighboring sections by insulated joints in the rails. However, more recent installations use electronics to allow jointless track circuits. Also, some areas have additional circuits that allow the signals to be manually held at red from a signal box or control center, even if the section is clear. These are known as semi-automatic signals. Even more complexity is required at junctions. Track circuit diagram for occupied block is provided in Fig. 5.

Interlocking. Another safety feature introduced in the mid-nineteenth century was mechanical interlocking of points and signals. The purpose was to prevent setting up a train route and clearing its protecting signal if there was already another conflicting route set up and the protecting signal for that route cleared. Interlocking was performed by a series of mechanically interacting rods connected to the signal operating levers in the signal box [15].

Multi-aspect signals. The simplified diagram above shows the basic principle of the block. The block occupied by Train 1 is protected by the red signal at the entrance to the block (see Fig. 6a). The block behind is clear of trains, and a green signal will allow Train 2 to enter this block. This enforces the basic rule or railway signaling that says only one train is allowed onto one block at any one time. This diagram (see Fig. 6b) shows a line with three-aspect signals. The block occupied by Train 1 is protected by the red signal at the entrance to the block. The block behind is clear of trains, but a yellow signal provides advanced warning of the red aspect ahead. This block provides a safe braking distance for Train 2. The next block in rear is also clear of trains and shows a green signal. The driver of Train 2 sees the green signal, and knows that he has at least two clear blocks ahead and can maintain the maximum allowed speed over this line until he sees the yellow.

A schematic of the signal block section is shown in Fig. 6a. When a block is unoccupied, the signal protecting it will show green. If a block is occupied, the signal protecting it will show red. Figure 6b presents a schematic of three-aspect signaled route showing the additional yellow aspect provided to allow earlier warnings and thus higher speed operation.

Four-aspect signaling. The multi-aspect signaling commonly used in the UK today is a four-aspect system. It works similarly to the three-aspect system except that two warnings are provided before a red signal, a double yellow and a single yellow [16, 17].

Figure 7 provides a schematic for a four-aspect signaled route showing the double-yellow aspect. This diagram shows four-aspect signals with (in the upper diagram) a high-speed train with three clear blocks ahead of it and (lower diagram) a slower train with two clear blocks ahead of it. The lower speed trains can run closer together, so more trains can be operated over a given section of line.

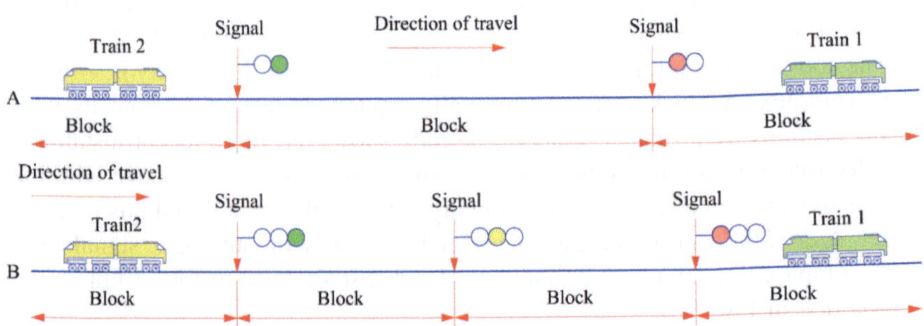

Fig. 6 Schematic diagram of two- and three-aspect train protection system

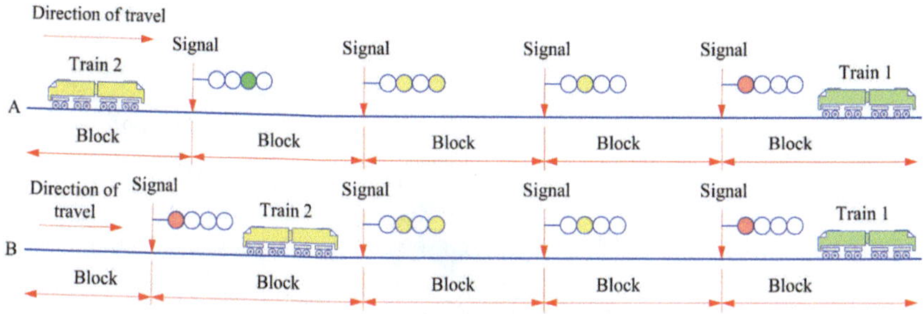

Fig. 7 Schematic diagram of four-aspect train protection system

5 The 1520-mm Rail Gauge Traffic Control System

Automatic track interlocking. In the case of automatic track interlocking, a section is divided into blocks (interlocked sections). Blocks are separated by traffic lights. Blocks are equipped with electric track circuit, which detects the presence or absence of a train in the block. The track acts as a conductor in a track circuit. Traffic lights automatically react to the train location. In the case of automatic track interlocking, three-aspect or four-aspect traffic signaling is used. Automatic track interlocking systems are divided into *direct current* and *alternating current* (*coded*), depending on which type of current is used to supply track circuits.

The track circuit. Two parallel track circuits are separated by insulated joints. A power source [battery, Fig. 8(3)] is connected to one end of the track circuit, and a receiver [track relay, Fig. 8(1)] is connected to the other. When the interlocked section is unoccupied, current flows from the power source by one track line, i.e. through track relay coil windings [Fig. 8(7)], then by the other track line and returns to the power source [Fig. 8(3)]. The track relay is activated by the current flowing through the coil windings and its contacts close the supply circuit of sectional traffic light, thus the permission signal is active [19]. When a train enters the block, its wheelset links two opposite track lines. Wheelset resistance is many times lower than the resistance of the track relay coil, so the armature of the track relay moves and with one set of contacts opens the circuit of the permission signal of the sectional traffic light, and with other contacts it closes the supply circuit of the red traffic light, so the traffic light prohibitive signal is activated. A block diagram of a DC current automatic track interlocking is shown in Fig. 9.

Coded automatic interlock. A coded automatic interlock is usually used in electrified rail sections of a 1520-mm gauge. In this case, AC current coded track circuits are applied. Coded automatic interlocks between adjacent traffic lights are linked by the track circuits to which coded signals with certain information are sent. Each traffic light (green, yellow, red) corresponds to a specific pulse combination (**code**) in the coded automatic interlock. The general layout of coded automatic interlock equipment is shown in Fig. 10 [19]. There are three codes displayed in the cab signal corresponding to the aspect of the trackside signal ahead. In the case of

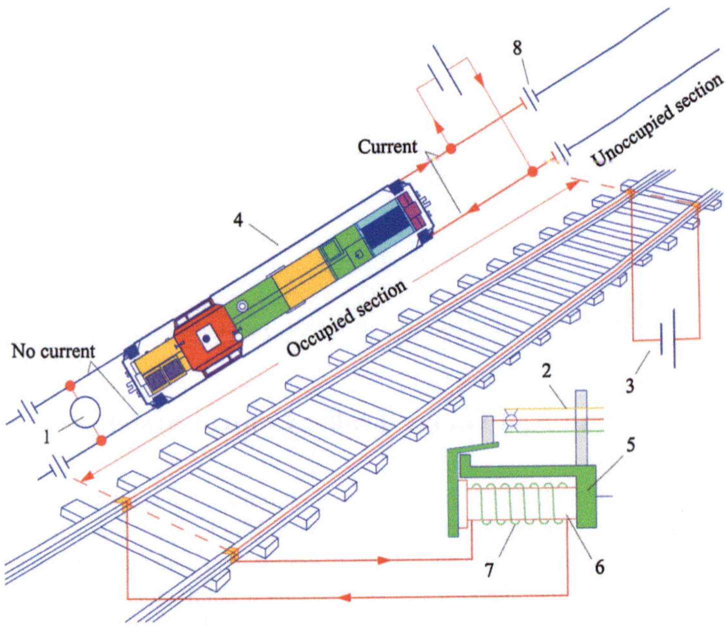

Fig. 8 Track circuit: *1*—track interlock relay; *2*—relay contacts; *3*—accumulator battery; *4*—wheelset; *5*—relay armature; *6*—solenoid core (magnetic conductor); *7*—track interlock relay coil windings; *8*—insulated joints

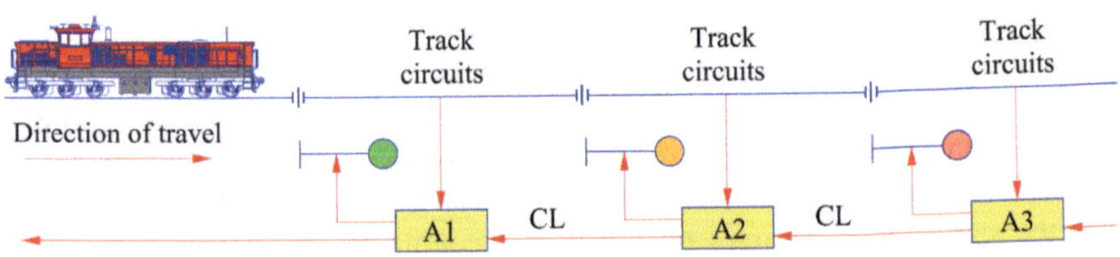

Fig. 9 Block diagram of DC current automatic track interlocking: *A1–A3*—automatic signal processing unit; *CL*—interconnection line of adjacent traffic lights of a section

three-aspect signaling these codes are red signal ahead (results in cab signal red-yellow), yellow signal ahead (results in cab signal yellow), green signal ahead (results in cab signal green). A green traffic light corresponds to a three-pulse long interval sequence; yellow traffic light to a two-pulse sequence; red to a one-pulse sequence. The special device (*code generator*) connected to one end of the interlocked section sends coded current pulse sequences to the track circuit. The ***decoder*** decodes the code sequences received by the track relay, and as a result, the corresponding signal light is activated in the interstation traffic light. The same

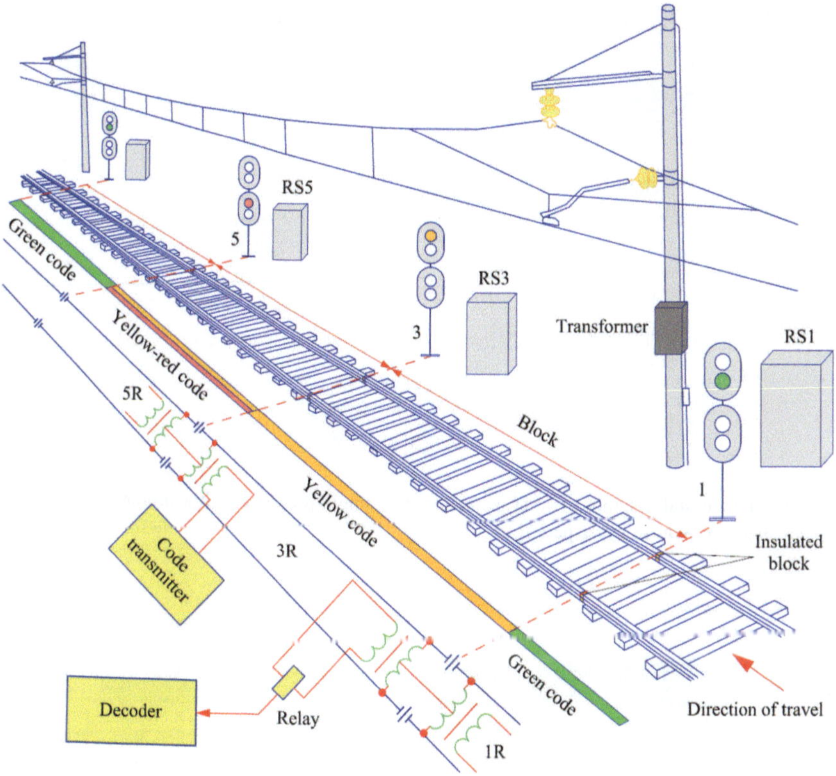

Fig. 10 General layout of coded automatic interlock equipment

automatic interlocking coded AC signals are used to transmit signals of track traffic lights to the cab. Code generators installed in individual blocks generate coded pulse cycles of different durations.

Code generators. Code generators are used in systems of automatic code interlocking, electric interlocking and automatic locomotive signaling to convert continuous AC current into a pulse code sent to a track circuit. Generators KPTŠ-5, KPTŠ-7, KPTŠ-515, KPTŠ-715 and KPTŠ-8, KPTŠ-9, KPTŠ8-15, KPTŠ-915 are used in automatic interlocking AB equipment to coded track circuits of stations. A block diagram of a code generator is shown in Fig. 11.

From Fig. 11, the code generator includes single-phase asynchronous motor M and gear-box 2, with a shaft connected to camshaft 3, whose rotation connects or disconnects circuit contacts green G, yellow Y, yellow-red YR. Contact time of connection/disconnection depends on motor speed, gear-box ratio, cam profile and the gap between moving and stationary contacts.

A time diagram of code signals using code generator KPTŠ-515 is presented in Fig. 12. The structure of code signals does not change when using code generator

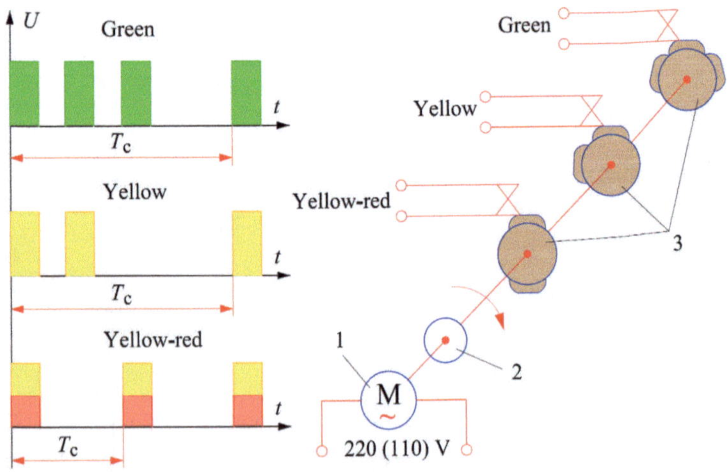

Fig. 11 Functional diagram of code generator: *M*—single-phase capacitor motor; *2*—gearbox; *3*—camshaft; *Tc*—pulse cycle period, *s*

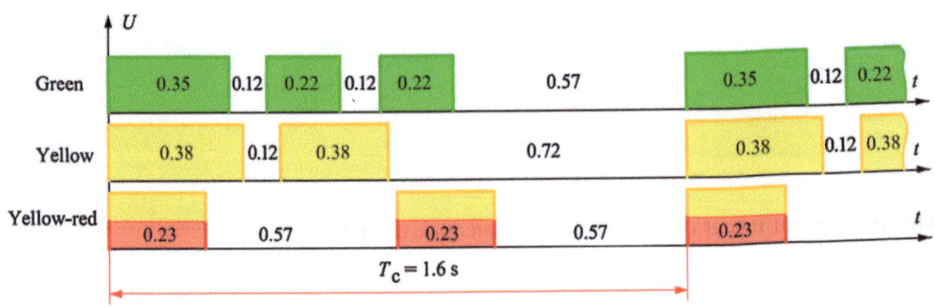

Fig. 12 Signals of code generator KPTŠ-515

KPTŠ-515. Only the duration of pulses, pauses and code cycle Tc (1.6 s) are changing.

The period of the code transmitted by the track circuit is 1.60 or 1.86 s.

6 High-Speed Train Braking Control

Considering the increasing speeds of trains, implementation of a moving block and radio transmission system is planned. When the current interval train control principle is replaced with the new system, trains will run separated by minimal distances one from another. Therefore, it is very important to develop computer software and algorithms that will use train identification data, train location point, direction and speed data received by radio transmission to calculate safe train separation distances and braking. It is necessary to assess the reliability of systems used by Japan, France, Italy for braking of high-speed trains. Then, considering the results of this assessment, develop modern automatic braking systems for high-speed trains, adapted to a moving block and radio transmission system.

Italian Coded Current Automatic Block (BACC) for conventional and high-speed traffic. On Italian conventional lines and lines for increased speed up to 200 km/h, a cab signaling system with four different aspects is applied. As track circuits have almost equal length (1350 m), it is possible to calculate and supervise the braking curve. A stop is, therefore, announced 2700 m behind the stop position [17].

The braking distance is more than two track circuits if the train speed is higher than 200 km/h; therefore, the system had been upgraded for the high-speed line (Fig. 13a). The system is downwards compatible, which means that high-speed trains can run on conventional lines and conventional trains on high-speed lines using only the 50 Hz code and at speeds not higher than 200 km/h. This compatibility is necessary, as only one line (Rome-Florence) is equipped with a BACC high-speed system and trains continue onto the conventional network. More information on BACC can be found in 1985 [20].

Japanese ATC for high-speed rail. The first Japanese high-speed line opened in 1964; therefore, the ATC system used there is the oldest cab signal and train protection system for high-speed rail in the world. The system is adapted to the requirements of a high-speed network, which is technically and operationally separated from the conventional network. The capacity and availability requirements are high and the required flexibility regarding different types of traffic and vehicles is low, which makes a limited number of signal aspects sufficient. The trains are driven semi-automatically: the safety-related braking for a signaled stop is controlled automatically, whereas non-safety-related processes such as acceleration or platform stops are controlled under the responsibility of the driver. Information transmission from track to train is achieved by coded track circuits. Each block section is covered by two track circuits. All track circuits belonging to the same line have the same length; therefore, the ends of the track circuits are used as fixed

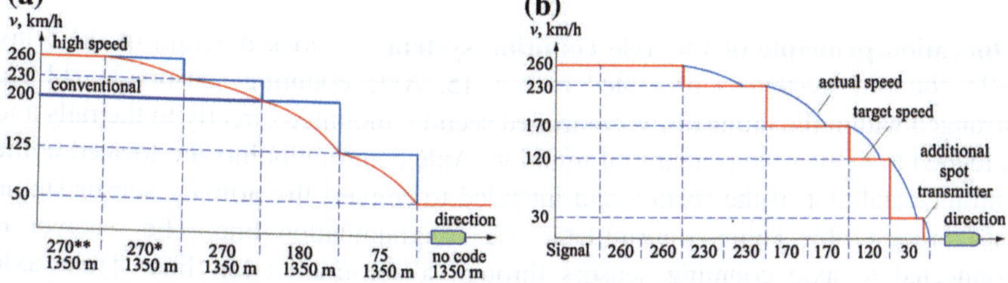

Fig. 13 Braking supervision control in BACC for conventional and high-speed traffic (**a**) and brake control in Japanese automatic train control (ATC), example (**b**) [18]

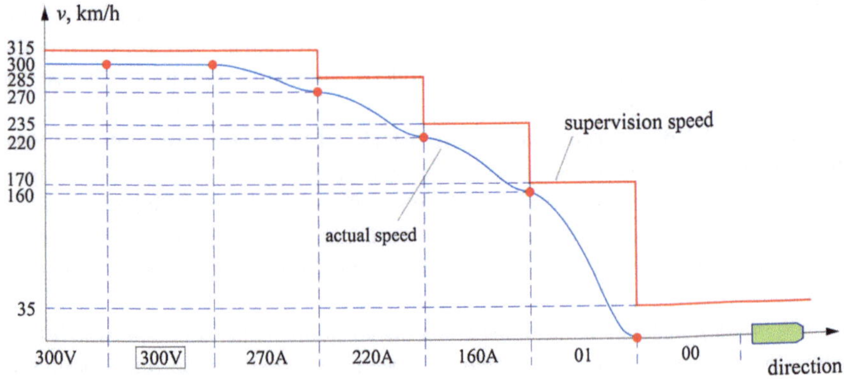

Fig. 14 Brake control in French TVM 300

intermediate points for speed reduction. The result is a multi-step cascade braking curve (see Fig. 13b) behind a signaled stop [21].

French Transmission Voie-Machine (TVM) 300. TVM 300 (see Fig. 14) is installed on the older French high-speed lines, the first of which opened in 1981. Like in Japan, the high-speed lines are regularly used by high-speed trains only, but these also pass into the conventional rail network [17].

Regarding the data transmission and control, the TVM 300 has large similarities with the Japanese ATC. One main difference is that the driver regulates the braking process and is supervised by the technical system. The staircase speed information is a limit speed for supervision, and each block section is covered by exactly one track circuit. The length of the block sections is adapted to the gradient. Resulting from the staircase supervision pattern, the required overlap has the length of a complete block section.

7 Axle Counting System

Operation principle of the axle counting system. A block diagram of the relay axle counting system is presented in Fig. 15. Axle counting sensors J1, J2 are arranged within the limits of the controlled section, mounted directly to the rails and intended to deliver the primary information. Axle counting points are located at the entrance and exit of the section and intended to convert the primary sensor signal into process able signal transmittable via communication line. The receiver is connected to axle counting sensors through a communication line. Train axle number at entrance and exit of the controlled section is used to generate control signal for relays of occupied and unoccupied track section.

The Frauscher axle counting system is usually used as a component of a train positioning system. It consists of track equipment: sensor, cables, data processing

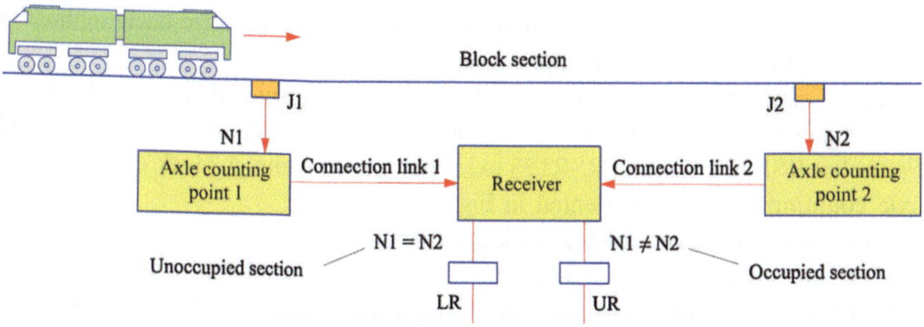

Fig. 15 Block diagram of a relay axle counting system. *J1*, *J2*—axle counting sensors; N_1—axle number at the entrance of the controlled section; N_2—axle number at the entrance of the controlled section; *LR*—unoccupied section control relay; *UR*—occupied section control relay

Fig. 16 Frauscher axle counting sensor

and transmitting devices. The axle counting sensor is mounted to a rail. The Frauscher axle counting sensor is shown in Fig. 16.

For many years, Frauscher axle counting sensors have been reliably operated in over 70 countries. They impress with their extremely high availability, even when subjected to extreme temperatures, high vibration levels and various types of electromagnetic interference. The Frauscher axle counting sensor provides a wide range of different functions for numerous applications.

Axle counting system ACS2000. They are supplied without an evaluation board, and evaluation of the signal falls to the system integrator, who is thereby able to adjust the interpretation in accordance with its individual requirements. The sensors consist of two independent sensor systems built into a single housing, which generate an analogue signal based on inductive processes. This is proportionate to the damping by the wheel flange and is supplied in the form of a direct current signal. From this signal, different data and information for each application can be acquired for non-vital applications.

Track equipment consists of a contactless sensor mounted on one of running rails, a junction box and connection cables. This equipment is compatible with Frauscher RSR180/181 and RSR123 axle counting systems. A block diagram of the axle counting system is presented in Fig. 17.

Track occupation and rolling stock passage control equipment includes A station axle counters and B station axle counters, a comparator (signal receiver) and sensors controlling rolling stock passage through track sections. Data from A station metering equipment and B station metering equipment are continuously supplied to the axle number comparator, which generates a control signal if the axle number in A and B stations does not coincide. A Thales axle counting sensor and signal processing box (a) and an analogue-digital signal converter of a Siemens axle counting system ZP43 (b) are shown in Fig. 18 [17]. The analogue-digital converter converts analogue signals of axle counting sensors into digital signal for convenience of data processing.

Analogue signals of axle counting sensors are processed by the analogue-digital signal converter of Siemens axle counting system ZP43 (see Fig. 18). The train axle number at entrance and exit of the controlled section is used by converter ZP43 to generate a digital control signal of occupied/unoccupied track sections.

Control algorithm of an axle counting system: If at the beginning the controlled section was unoccupied, and the number of entered and exited wheelsets has coincided, the section is registered as unoccupied. Otherwise, this section is considered occupied.

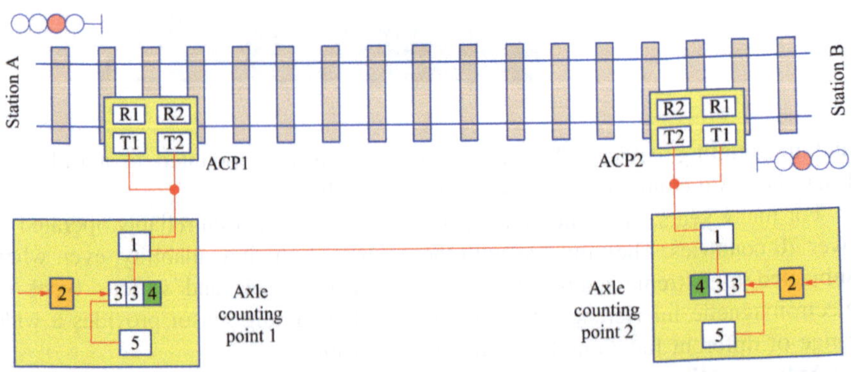

Fig. 17 Block diagram of axle counting system: *ASP1*, *ASP2*—axle counting points; *1*—signal comparator; *2*—uninterruptible supply unit; *3*—electronic unit board of axle counting point; *4*—power supply unit; *5*—error elimination unit; *R1*, *T1*, *R2*, *T2*—redundant axle counting sensors

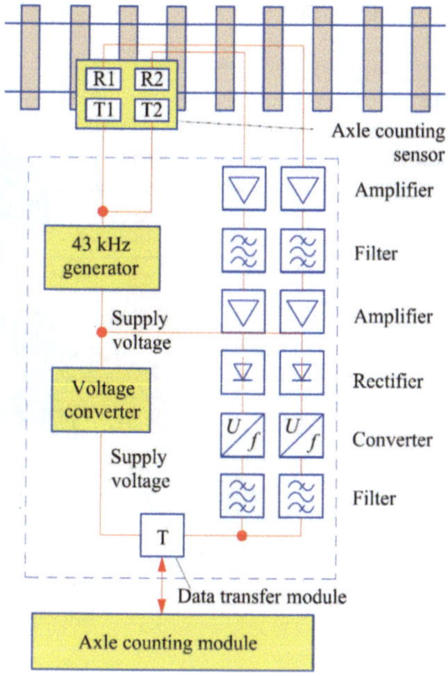

Fig. 18 Thales axle counting sensor and signal processing box (**a**) and analogue-digital signal converter of Siemens axle counting system ZP43 (**b**): *R1*, *T1*, *R2*, *T2*—redundant axle counting sensors; *T*—data transmission module

8 The European Train Control System

In order to reduce train interaction technical obstacles, representatives of the rail sector and the European Commission signed memoranda of understanding in 2005, 2008 and 2012 on implementation and development of the European Rail Traffic Management System (ERTMS/ETCS) [22–24]. The objective of the system is to harmonize about 20 different European signaling systems and to implement a uniform automatic speed control system developed in consideration of state-of-art telecommunication technologies. All types of safety—***control, management and signaling***—are based on the principle of mitigation or elimination of driver's fault and accident risk during train movement. Signaling, control and train protection systems used in different European countries are presented in Fig. 19 [25]. At this time, more than 20 different train protection systems (TPS) and five catenary systems are used in EU railways (see Fig. 19); for example, FAB/KHP UZ in Poland, INDUSI/LZB in Germany, TVM/KVB Krokodil in France, ASFA/LZD in Spain, ATC in Sweden, Denmark and Norway, and the ALSN system is used in countries including Russia, Belarus, Ukraine, Lithuania, Latvia, Estonia and Norway. They systems are developed within state boundaries and use different technical means.

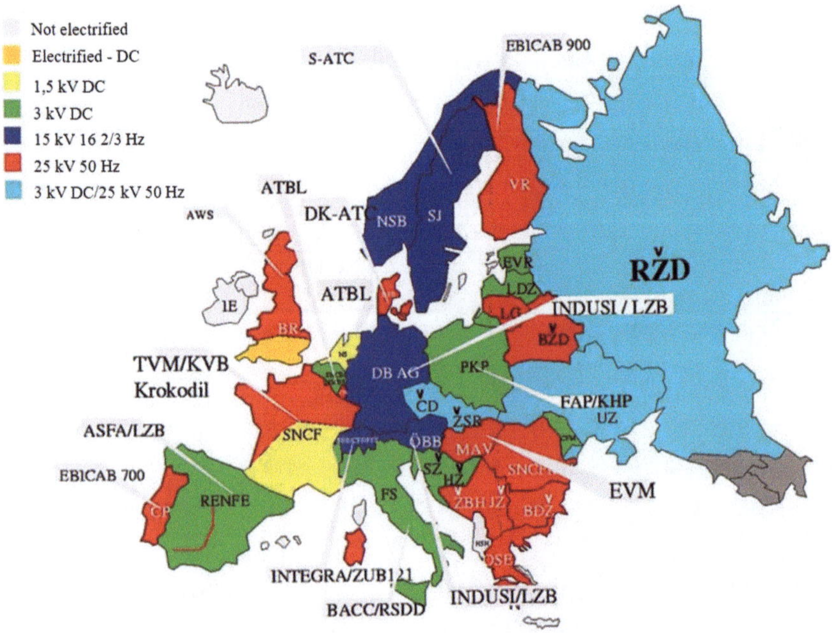

Fig. 19 Signaling, control and train protection systems used in different European countries

Train protection systems: ALSN (Russian Federation, Belarus, Estonia, Latvia, Lithuania, Ukraine), Ansaldo L10000 (Sweden), ASFA (Spain) ATB (Netherlands), ATC (Sweden, Denmark, Norway, Brazil, South Korea, Japan, Australia [Queensland], Indonesia), ATP (UK, USA, Brazil, Australia [Queensland], Indonesia, Ireland), AWS (UK, Australia [Queensland, South Australia], Indonesia), BACC (Italy), CAWS (Ireland), CBTC (Brazil, USA, Canada, Singapore, Spain, Gabon), CONVEL (Portugal), Crocodile/Memor (Belgium, France, Luxembourg), EBICAB 700 (Bulgaria, Norway, Portugal [called CONVEL by the Portuguese railways], Sweden), EBICAB 900 (Finland, Spain, Sweden, Australia [Queensland, Western Australia]), ETCS (Austria, Belgium, Bulgaria, Italy, Switzerland, EVM 120 (Hungary), HKT (Denmark), Integra-Signum (Switzerland), KVB (France, Channel Tunnel Rail Link), LZB (Germany, Austria, Spain), LS (Czech Republic, Slovakia), MEMOR II + (Luxembourg), PZB Indusi (Germany, Austria, Romania, Slovenia, Croatia, Bosnia-Herzegovina, Serbia, Montenegro, Macedonia, Israel), FAB/KHP UZ (Poland), SCMT (Italy), SELCAB (Spain), SHP (Poland), TASC (Japan), TBL (Belgium, Hong Kong), TPWS (UK, Australia [Victoria]), Tren Denetim Sistemi (TDS; Turkey), TVM (France, Belgium, UK, Channel Tunnel Rail Link, South Korea), ZUB 121 (Switzerland), ZUB 123 (Denmark), ZUB 262 (Switzerland).

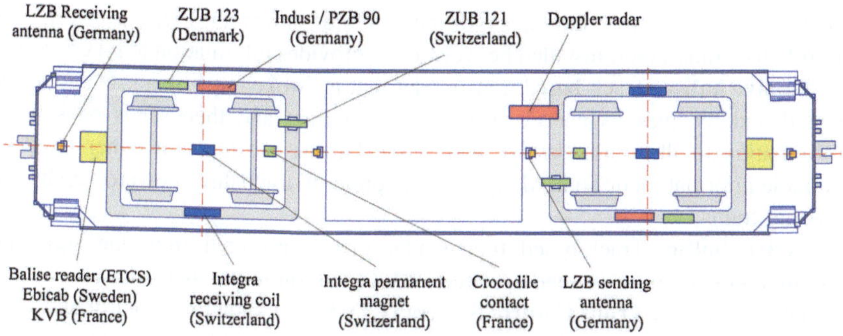

Fig. 20 Positions of vehicle communication units in different train protection systems [17]

Fig. 21 Indusi prototype on a steam locomotive in May 1930 [26]

Figure 20 shows an example of positions of train-side antennas for various systems on the bottom of the locomotive. A prototype of the PZB (induction) train traffic safety and signaling system was first used in locomotives in 1930 (see Fig. 21).

A diagram of track signal communication for locomotives is presented in Fig. 20. We can see from Fig. 20 that track signal transceivers in locomotives are arranged in different points. Transceivers of track signals in a vehicle are arranged perpendicularly to trackside stationary equipment (balises, etc.), where they transmit or receive communication signals, because electromagnetic signals are of maximum strength in this arrangement. In sections equipped with track signaling systems (traffic lights, information stands and LED boards informing drivers on permitted speed), data are transmitted to a train from standard electromagnetic inductors (called "balises") installed on a track. PZB, or Indusi, is an interval cab signaling system and train protection system used in Germany, Austria, Slovenia, Croatia, Romania and Israel, and on one line in Canada. A balise is a device in a track designed to transmit data (ALS signals, speed limitation, train location, etc.) to the passing train locomotive (passive) or to transmit and receive data (active). A balise used for ETCS is called a "Eurobalise". Examples of balises are shown in Fig. 22. An ETCS Eurobalise transceiver, installed between rails, provides information to ETCS trains.

Passive balise: Track-based transponder that is "woken up" by a low frequency signal and receives its energy from a passing train and then sends packets of information to the train. Passive and active balises can transmit either fixed or variable information or both. Many railways prefer to use balises powered-up by the passing trains.

Active balise: Track-based transponder that is powered from the signaling supply and continuously sends packets of information to passing trains.

The European Train Control System (ETCS) is a signaling, control and train protection system designed to replace the many incompatible safety systems currently used by European railways, especially on high-speed lines. ETCS requires standard trackside equipment and a standard controller within the train cab. In its final form, all lineside information is passed to the driver electronically, removing

Fig. 22 Examples of Eurobalises **a** ETCS "Eurobalise transceiver", installed between rails, provides information to ETCS trains [27], **b** Croco + TBL + ETCS balises at the same signal [28], **c** Siemens Eurobalise EBICAB [29], **d** Siemens balise [30]

the need for lineside signals, which at high speed could be almost impossible to see or assimilate. By using the European Train Control System from the equipment in the rail track to the train, the information is transferred, according to which maximum permissible speed is constantly calculated. In the sections where a signalization system is installed (traffic control light signals and information stands and displays, which are used to inform the locomotive driver of the permissible speed). Information is transmitted to the train from the standard electromagnetic inducers (the so-called Eurobalises) that are installed in the track. In the cases of all three levels, by using Eurocab in the train computer, the speed of the train is compared to the maximum permissible speed and automatically by exceeding it [31–35].

ETCS is specified at four different levels [36]. Level 0: ETCS-compliant locomotives or rolling stock interact with lineside equipment that is non-ETCS compliant. Level 1: ETCS is installed on lineside (possibly superimposed with legacy systems) and on board; spot transmission of data from track to train via ETCS balises. Level 2: Same as level 1, but ETCS data transmission is continuous; the currently used data carrier is GSM-R. Level 3: Same as level 2, but train location and train integrity supervision no longer rely on trackside equipment such as track circuits or axle counters.

ETCS level 1 schematic. Information to the passing train locomotive is transmitted by radio communication GSM-R. The ETCS level 1 signaling system can be implemented for the national signaling system by leaving the existing system. In the ETCS level 1 signaling system, the Eurobalise transmits route data by GSM-R radio communication from the rail track via signal adapters to the approaching train and to the traffic control authority. Block diagrams of ETCS level 1 are presented in Figs. 23 and 24.

Eurobalise radio beacons pick up signal aspects from the trackside signals via signal adapters and telegram coders (lineside electronics unit—LEU) and transmit them to the vehicle as a movement authority together with route data at fixed points.

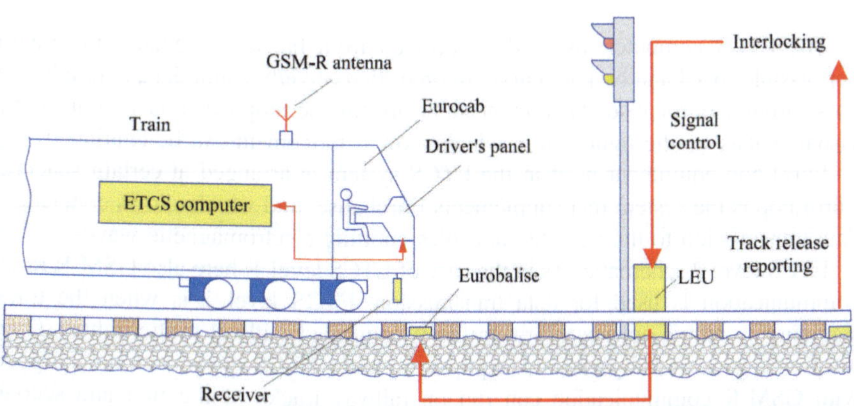

Fig. 23 Block diagram of European train control system ETCS level 1: *LEU*—lineside electronics unit [37]

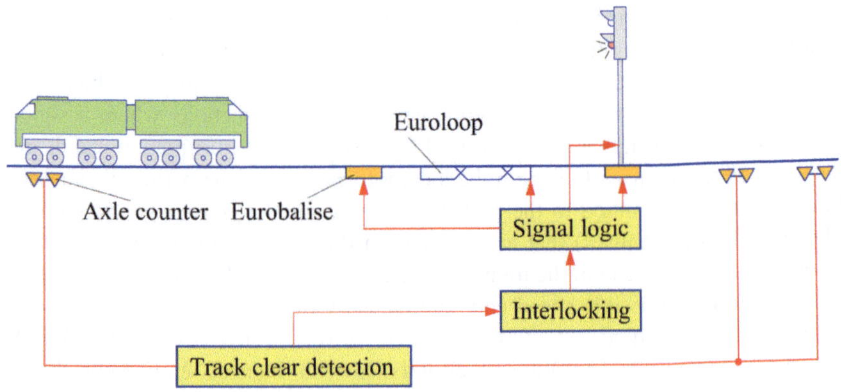

Fig. 24 Block diagram of the ETCS level 1 using axle counters, Eurobalises and EuroLoop equipment [17]

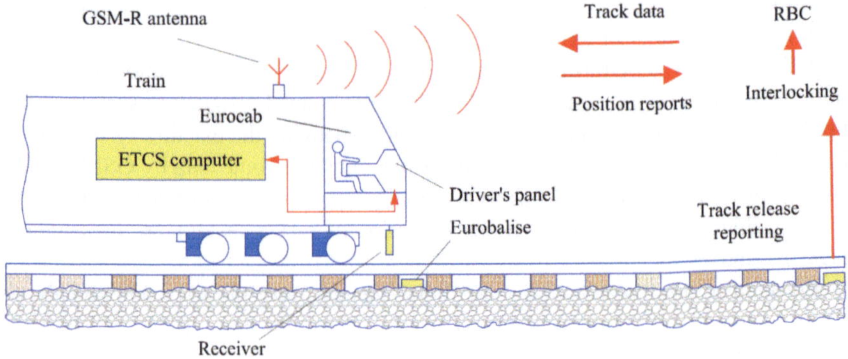

Fig. 25 Block diagram of European Train Control System (ETCS) level 2 [38]

The ETCS computer, using data received from balises, calculates maximum permissible speed and braking curve. In order to receive/transmit data, a train has to pass through Eurobalise. Movement authority for the stopped train is granted by visual signals (traffic lights, etc.), which show if movement can be continued.

EuroLoop equipment used in the ETCS system is arranged at certain spacing. EuroLoop is the system that supplements Eurobalise, and which allows contactless data transmission to the vehicle via cables emitting electromagnetic waves.

ETCS level 2 schematic. As in the case of ETCS level 1, here also GSM-R radio communication is used for data transmission. ETCS level 2 is when the track signaling system is not necessary (traffic is controlled without traffic lights). Train location is positioned by trackside equipment. In the ETCS system, a train equipped with GSM-R communication can run on railway tracks of the first and second levels. Block diagrams of ETCS level 2 are presented in Figs. 25 and 26.

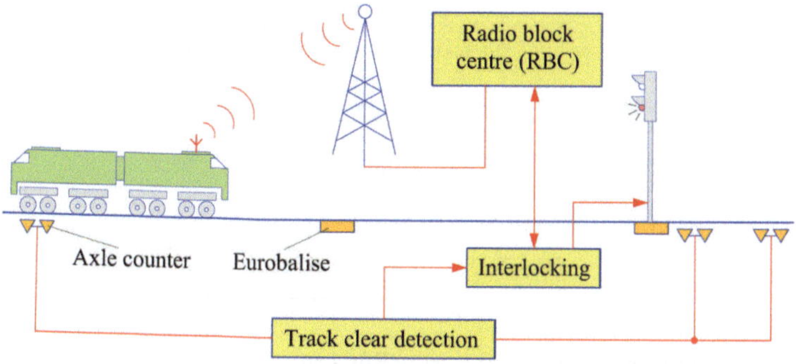

Fig. 26 Block diagram of ETCS level 2, when radio block centers (RBCs) are used for train control, and information from trackside equipment is transmitted by signals of axle counters and Eurobalises [17]

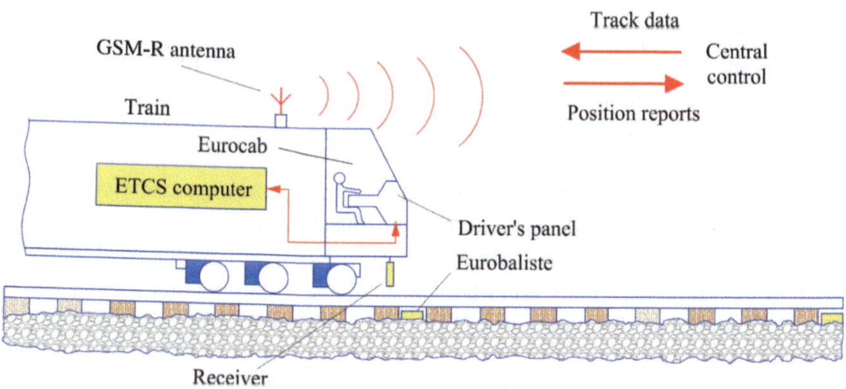

Fig. 27 Block diagram of ETCS level 3

However, train detection and train integrity supervision still remain in place at the trackside. Train movements are monitored continually by the radio block center (RBC) using this trackside-derived information. The movement authority is transmitted to the vehicle continuously via GSM-R together with speed information and route data. The Eurobalises are used at this level as passive positioning beacons or "electronic milestones". The train determines its position between two positioning beacons via sensors (axle transducers, accelerometer and radar). The positioning beacons are used in this case as reference points for correcting distance measurement errors. The onboard computer continuously monitors the transferred data and the maximum permissible speed.

ETCS level 3 schematic. ETCS level 3 is the same as level 2, but train positioning and train integrity control cannot rely only on data of trackside equipment such as track circuits or axle counters. ETCS level 3 is currently under development. Block diagrams of ETCS level 3 are presented in Figs. 27 and 28. In certain

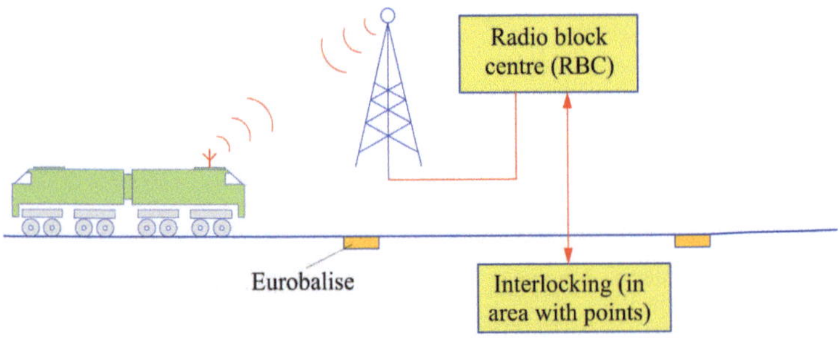

Fig. 28 Block diagram of ETCS level 3, when RBCs are used for train control and information from trackside equipment is transmitted by signals of Eurobalises [17]

countries (Sweden and others), the first ETCS level 3 systems are under implementation, and experience with these systems will be used for system improvement. By transmitting the positioning signal to the RBC, it is always possible to determine which point on the route the train has safely cleared. The following train can already be granted another movement authority up to this point. The route is thus no longer cleared in fixed track sections. In this respect level 3 departs from classic operation with fixed intervals: given sufficiently short positioning intervals, continuous line-clear authorization is achieved, and train headways come close to the principle of operation with absolute braking distance spacing ("moving block"). Within the framework of ETCS implementation, RBCs will be installed and CPU units will be developed to control traffic of locomotives in a section by exchanging data with them. In the case of ETCS level 3, accurate train positioning data transmittance from the train must be ensured in order to optimize sections throughput and reduce the amount of trackside equipment.

Fixed train detection devices (GFM) are no longer necessary. As in the case of the ETCS level 2 signaling system, Eurobalises and sensors (axle transducers, accelerometer and radar) are used for train positioning. ETCS level 3 must ensure train integrity determination with the highest possible degree of reliability. The transmitted train positioning signal in the RBC always allows determining which point in the route ensures safe forward movement of the train. RBC will receive data on prepared routes, traffic light signals, shunt positions, train position, etc. It then analyzes the occupation of forward sections and generates train movement authority up to a certain destination. A block diagram of ETCS level 3 when RBCs are used for train control, and information from trackside equipment is transmitted by signals of Eurobalises, is presented in Fig. 28.

For all three levels, the ETCS computer calculates train speed and compares it with maximum permissible speed, and brakes the train if the speed is exceeded.

9 Automatic Train Protection System

Train protection systems are continuously improved. It is necessary to find methods and technical means in order to reach ETCS level 3. The interval train control principle should be abandoned and replaced with a moving block system. This chapter presents block diagrams of train protection systems using beacons, train positioning signal transmission and processing principles. A block diagram of a train protection system using coded track circuit is presented in Fig. 29. Generated codes are received by an antenna. Train speed variation signals are generated by speed sensors installed on a wheelset.

We have seen in the previous articles that the ATP signaling codes contained in the track circuits are transmitted to the train. They are detected by pick-up antennae (usually two) mounted on the leading end of the train under the driving cab. These data are passed to an onboard decoding and safety processor. The permitted speed is checked against the actual speed, and if the permitted speed is exceeded, a brake application is initiated.

Operation with beacons. A block diagram of a train protection system using beacons is presented in Fig. 30. The figure shows links between electronic beacons placed at intervals along the track, train and traffic light.

Electronic beacons are placed at intervals along the track. The beacons are sometimes referred to as "balises" after the French word for beacon. Data processing and the other ATP functions are similar to the continuous transmission system. The beacon system operates as shown in the simplified diagrams below. In the diagram (Fig. 31 A), the beacon for red Signal A2 is located before Signal A1 to give the approaching train (2) room to stop. Train 2 will get its stopping command here so that it stops before it reaches the beacon for signal A3. In the diagram in

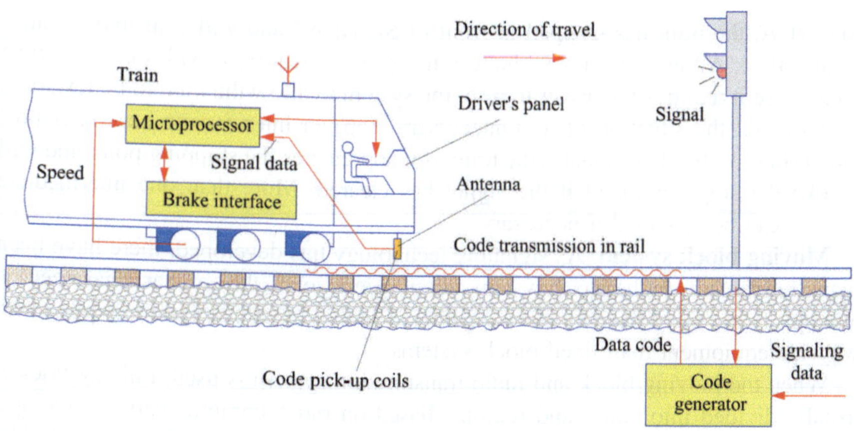

Fig. 29 Block diagram of a train protection system using coded track circuit [40]

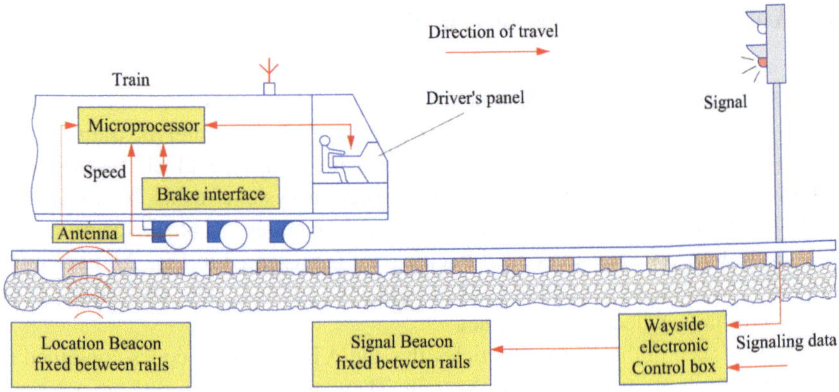

Fig. 30 Block diagram of a train protection system using beacons [41]

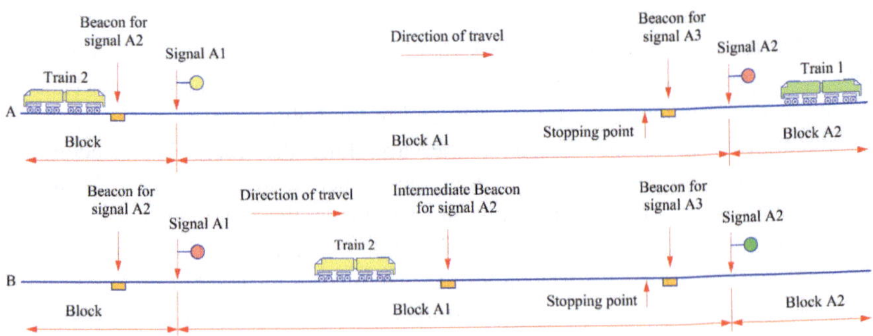

Fig. 31 Train traffic control schemes: *A*, *B* [39]

Fig. 31 A, the train has stopped in front of Signal A2 and will wait until Train 2 clears Block A2 and the signal changes to green. In reality, it will not move even then, since it requires the driver to reset the system to allow the train to be restarted.

To avoid the situation of an unnecessary stop, an intermediate beacon is provided (Fig. 31 B). This updates the train as it approaches the stopping point and will revoke the stop command if the signal has cleared. More than one intermediate beacon can be provided if necessary.

Moving block system. As signaling technology has developed, there have been many refinements to the block system, but in recent years, the emphasis has been on attempts to get rid of fixed blocks altogether. The moving block system requires less wayside equipment than fixed block systems.

When the moving block and radio transmission system is used, railway track is usually divided into zones and regions, based on radio communication coverage. RBCs are arranged so that controlled blocked sections would be reliably covered by radio communication. Train identification, location, direction and speed data from

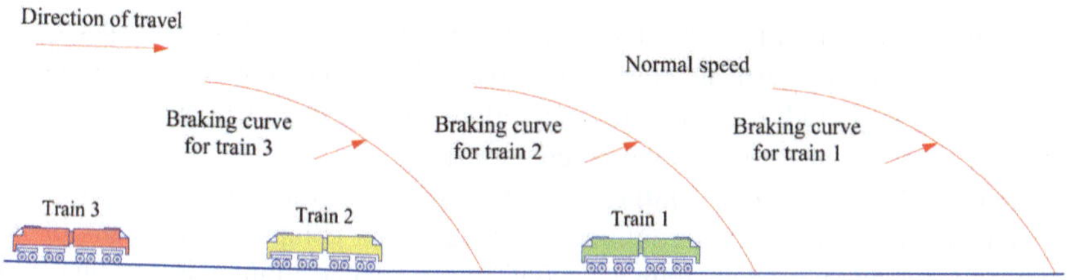

Fig. 32 Train movement theoretical diagram using the moving block system [39]

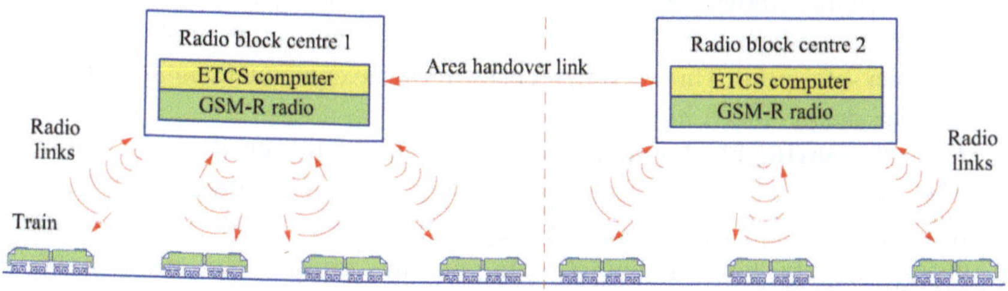

Fig. 33 Train protection system using moving block and radio transmission [39]

RBCs are processed by a computer. Each blocked section has its own radio transmission system. Each train transmits its identification, location, direction and speed data to the computer, which computes safe train separation distance and transmits data to a train. A train movement theoretical diagram using the moving block system is presented in Fig. 32. The diagram shows trains moving in one direction separated just by braking distance. The question is what technical measures could be used to ensure train traffic safety.

This flexibility requires radio transmission, sometimes called "communications-based train control" (CBTC) or "transmission-based signaling" (TBS), rather than track circuit transmission, to detect the location, speed and direction of trains and to tell trains their permitted operating speed.

Moving block and radio transmission. A train protection system using moving block and radio transmission is presented in Fig. 31. On a moving block-equipped railway, the line is usually divided into areas or regions, each under the control of a computer and each with its own radio transmission system. Each train transmits its identity, location, direction and speed to the area computer, which makes the necessary calculations for safe train separation and transmits this to the following train, as shown in Fig. 33. Train spacing data are transmitted by radio channel to a computer. Therefore, the computer always knows the position of all trains in the controlled section. It transmits to each train the position of the preceding train and computes the braking curve for the train enabling it to stop, while it does not reach the preceding train. When the moving block and radio transmission system is used, train separation distance system is of a dynamic nature. In this case, communication between individual trains executes CBTC. To ensure reliability of speed and braking distance between all trains, one fixed block separating the trains can be used. In case of radio communication failure, the newest data will force train braking earlier than it reaches the preceding train.

Moving block location updates. As we have seen, trains in a moving block system report their position continuously to the area computer by means of the train-to-wayside radio. Each train also confirms its own position on the ground from beacons, located at intervals along the track, which recalibrate the train's position compared with the onboard computerized line map. A train protection system using moving block location updates is presented in Fig. 34.

This is another moving block and radio transmission version, when positioning computers are on board. Each train knows where it is in relation to other trains and sets its own safe speed. This version reduces trackside equipment, but the amount of transmitted data is much higher.

An early moving block system. One system that claims the distinction of being the first moving block system is marketed under the name Seltrac by Alcatel. It is used in Canada and on the Docklands Light Railway in London. It has the ingredients of moving transmission of data, but the transmission medium is the track-mounted induction loops, which are laid between the rails and which cross every 25 m to allow trains to verify their position. Data are passed between the vehicle onboard computer (VOBC) and the vehicle control center (VCC) through the loops. The VCC controls the speed of Train 2 by checking the position of Train 1 and calculating its safe braking curve.

The Seltrac Alcatel moving block system is shown in Fig. 35. In the Seltrac Alcatel system, a train is controlled automatically. A driver does not participate

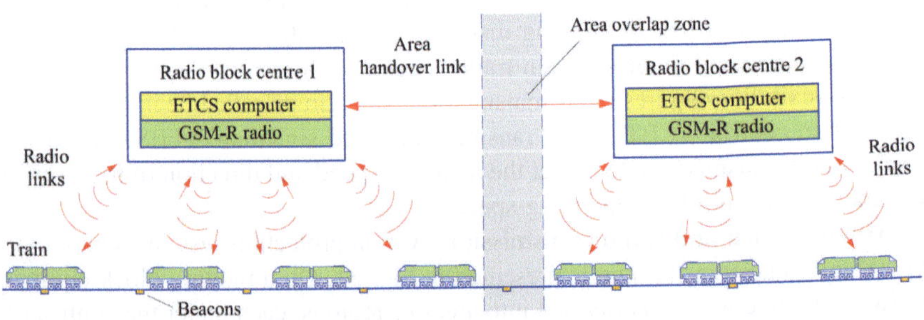

Fig. 34 Train protection system using moving block location updates [39]

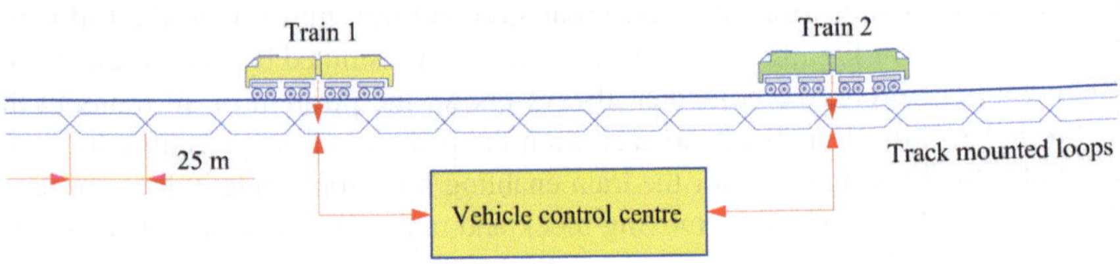

Fig. 35 Seltrac Alcatel moving block system [39]

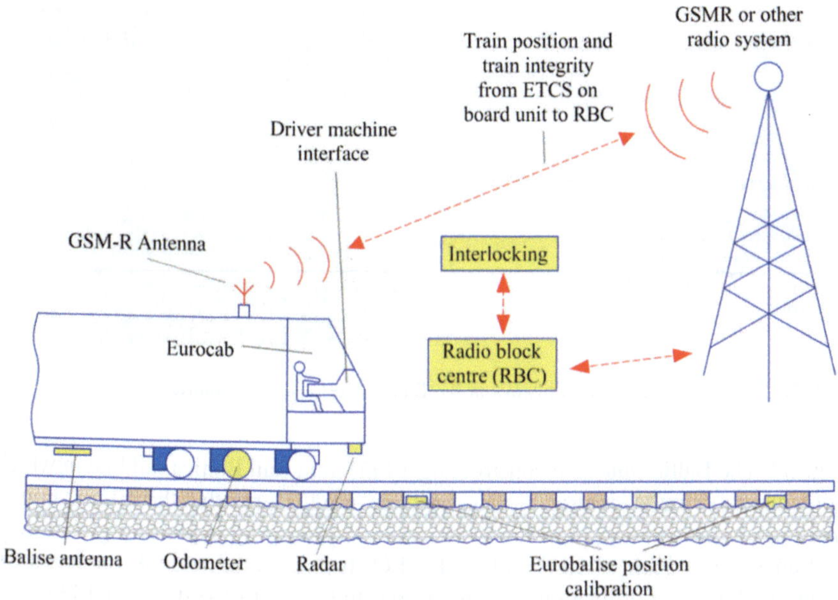

Fig. 36 The simplified block diagram Siemens Trainguard 300 project ETCS level 3

directly in train control. In case of system failures, the train is controlled manually. In this case, train position is determined by axle counters. The biggest disadvantage of the Seltrac system is that cables are laid between rails and can be damaged during track maintenance works or otherwise. The principal difference between this system and other modern systems is that Seltrac uses electromagnetic cables for data transmission, while radio systems use antennas.

Trainguard 300. The simplified block diagram Siemens Trainguard 300 project ETCS level 3 is presented in Fig. 36. Siemens in its project, Trainguard 300 ETCS Level 3, proposes to use all complex of equipment necessary for ETCS level 3, which is not fully standardized at the moment.

The Trainguard 300 ETCS Level 3 [40] is a uniform complex (produced by a single company) including locomotive computer control and track equipment, which would enable operation of such trains in *fixed-block or moving-block operation* railways. An *odometer* is a device for measuring the distance traveled by a vehicle (see Fig. 36).

10 Ten Baltic State Rail Traffic Control Aspects

The railways of the Baltic states have inherited a railway infrastructure of 1520-mm railway gauge and a typical Russian standard traffic control system. The railways of different railway gauge width in the Baltic states and the EU are presented in

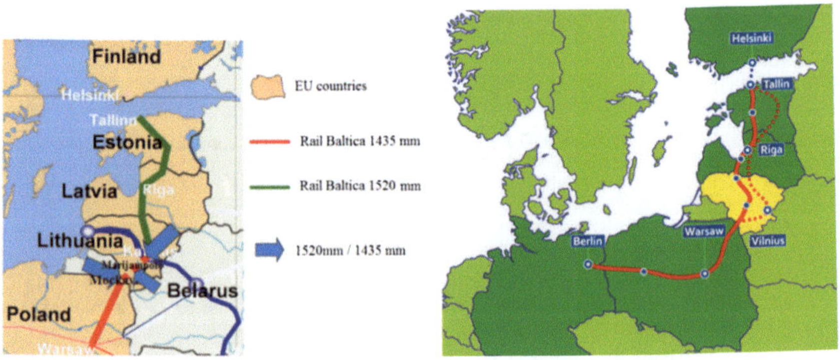

Fig. 37 Different gauge-width railways in the Baltic states and EU countries

Fig. 37. The Baltic states are constructing a 1435-mm Rail Baltica rail line in which contemporary traffic control systems will be used. Currently, the EU is implementing the installation of smart traffic control systems by exploiting global navigation systems such as GPS and GALILEO. It is essential to look for technical solutions for the coordination of different rail traffic control systems of 1435- and 1520-mm width. To master passenger train routes between the railways of 1435- and 1520-mm width, it will be essential to evaluate not only the differences of the gauge and contact network standards, but also those of traffic control systems. The kinetic energy of trains is not fully used to reduce the locomotive fuel or electric energy consumption for traction. Production of hybrid locomotives with energy saving and accumulation systems is only just being developed. At the detachment of JSC Lietuvos geležinkeliai subsidiary Vilniaus lokomotyvų remonto depas UAB, along with Russian and CZ LOKO companies, has produced the first diesel-electric hybrid powered locomotive with electric power TEM-35 for running on 1520-mm gauge, in which energy collection and saving systems are installed.

10.1 Rail Traffic Control System Analysis

Railway automation remote mechanics equipment increases the throughput of trains, simplifies train traffic control, improves the overall quality of the entire railway services and increases their competitiveness. Currently, LG has installed microprocessor traffic control systems of the companies Bombardier and Siemens. The main technical measure of interval train traffic control is an automatic track control. Where an automatic interlocking system is used for train track traffic control, the waystations are divided into block zones, the length of which may be up to 1000–2600 m. Block zones are fenced off by light signals of waystations. An electric track circuit, which is an information sensor certifying the presence or absence of a train, is installed in block zones. The role of conductors in the electric track circuit is taken by rails (Fig. 38a). A relay-based traffic control system, where a chain of rails is used and microprocessor control on the centralized information screen are presented in Fig. 38b.

A track circuit is the body of the railway section rail lines and the equipment attached to it, which is meant for transmitting the signals. Should the train enter the block zone, it connects two opposite rail lines by its wheelsets [41]. Therefore, the track relay breaks the circuit of the permissible signal of the side track light, and by others it creates the power circuit of the red-light signal. This way a restrictive light signal is switched on. The main interval train movement instruments at waystations and stations are an automatic, semi-automatic track block, and automatic locomotive signalization, automatic level crossing signalization, electric switch and signal centralization EC, traffic control centralization and train traffic control in EC.

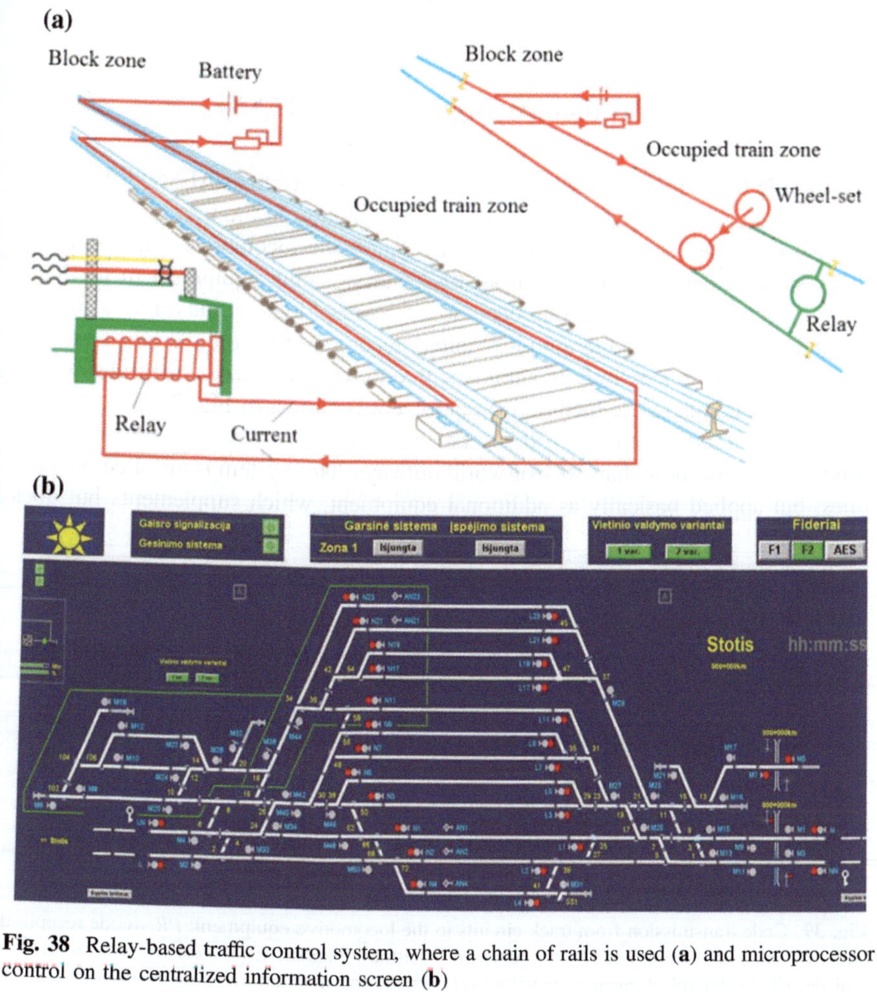

Fig. 38 Relay-based traffic control system, where a chain of rails is used (**a**) and microprocessor control on the centralized information screen (**b**)

To control the technical condition of the running train rolling stock, automatic control measures RAKP are used. To record the intensity of the track section by a train at intervals, traffic regulation system track circuits are used. The automatic track interlocking divides the waystation into interlocking zones by light signals of the waystations. The light signals change automatically depending on where the train is located.

Microprocessor centralized traffic control system. The microprocessor electronic centralized traffic control (CTC) system has replaced the relay system (RCTC). Relays were replaced by contactless systems created in the microprocessor base [41]. The CTC system consists of a centralized train dispatcher office that controls railroad interlocking and traffic flows in portions of the rail system designated as CTC territory. One hallmark of CTC is a control panel with a graphical depiction of the railroad. On this panel the dispatcher can keep track of train locations across the territory that the dispatcher controls.

10.2 *Automatic Locomotive Signalization (ALSN)*

Automatic locomotive signalization (hereinafter referred to as ALSN) is trackside locomotive equipment intended to transmit information to the train's driver about permissible traffic speed in block sections. Automatic locomotive signalization is the system of data transmission from trackside signaling equipment to the locomotive intended to facilitate train operation, to control logic of the driver's actions, to limit speed and brake the train if required, record speed and the driver's actions. The system is operating only in automatic interlocking sections. Code transmission from track circuits to the locomotive equipment is shown in Fig. 39.

ALSN is used on almost 100,000 km of track in the countries of the former Soviet Union, or more than 10% of world railways. This system is installed on main lines, but applied basically as additional equipment, which supplements but does

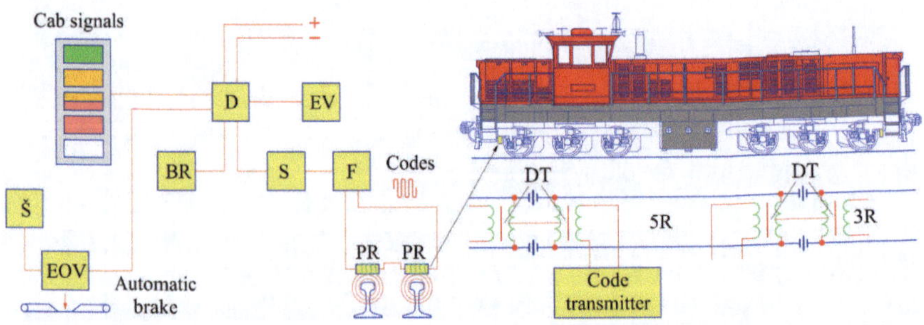

Fig. 39 Code transmission from track circuits to the locomotive equipment: *PR*—code reception coils; *F*—filter; *S*—amplifier; *D*—decoder; *BR*—alertness handle; *EOV*—electric air valve; *Š*—whistle; *EV*—solenoid of electric air valve; *DT*—choke transformer; *3R*, *5R*—block sections

not replace trackside signals in most cases. If there is a disagreement between trackside and cab signals, the driver must obey the trackside signal.

The system was developed in the 1930s in the Soviet Union using the experience of the first coded track circuits in the USA.

ALSN operation principle. Coded track AC circuits of automatic interlock are used in the ALS system to transmit track traffic light signals to a moving train (driver's cab). ALS track equipment in front of a running train sends circuit-coded AC current signals to a track, used in coded automatic interlocking AB. Equivalents of locomotive traffic light and track traffic light are shown in Fig. 40.

Codes of ALSN. According to new technical requirements, stations and open lines will have high-frequency track circuits. This implies new functions for ALSN. When the track section is clear, the track circuit carries no code and serves for track clear detection only. Only in these sections where a train is detected or expected soon, is the code applied. As soon as the train occupies a new section, the coding in the previous section is switched off. Likewise, generally in ALSN, only these station tracks which are provided for nonstop train passage are coded. The section beyond a signal at Stop is not coded; therefore, the train will be emergency stopped (cab signal red). This is in accordance with the failsafe principle. Passage of a Stop signal can be authorized with the driver's special action at a maximum speed 20 km/h [17].

Unified complex system of locomotive protection KLUB-U. KLUB-U, in addition to the main ALS functions, by using GPS allows the following: set and correct the actual time according to astronomic time; determine train running

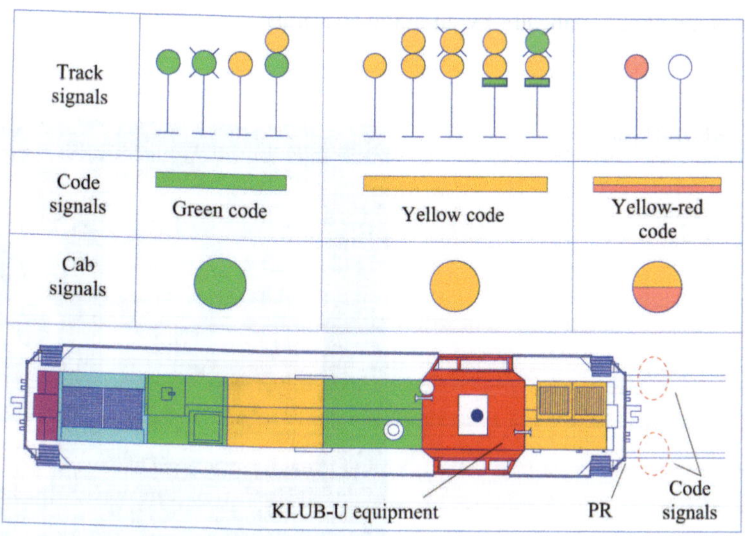

Fig. 40 Equivalents of locomotive traffic light and track traffic light [42]

parameters, coordinates, speed; control maximum permissible speed and generate the corresponding automatic braking signal; control train self-rolling; receive and record to the internal memory the data of track electronic map, etc. KLUB-U equipment is shown in Figs. 41, 42, 43 and 44. The driver's physical condition is monitored by measuring resistance with a special bracelet (see Fig. 43), which is worn by the driver during train operation. Signals of driver's physical condition are sent to the driver's alertness module TSKBM. In case the driver's condition is poor, forced braking of the train is activated.

KLUB-U equipment ensures the following functions: Reception and decoding of ALSN signals; indication of traffic light signals to a driver; telemechanic monitoring of driver's vigilance (TSKBM), indication of actual running speed, regular control of driver's alertness using indication and signalization, indication of permissible speed in the section to a driver, protection against vehicle self-rolling, indication of braking curve to a driver, continuous monitoring of brake system condition, automatic activation of emergency braking, recording of running parameters in memory storage, data (actual train running speed, train position GPS coordinate, locomotive traffic light indication) transmittance via GSM-R channel, command (train forced braking, speed limitation, permission to pass the prohibitive traffic light signal) receiving via GSM-R radio channel.

Control of locomotive driver's alertness. Alertness of the locomotive driver is controlled as the train approaches a restrictive light signal, when the green light signal of the locomotive changes to yellow. In this case, the locomotive driver must confirm his alertness once by pressing the alertness handle. Later on, KLUB-U carries out period checks (every 30–40 s) of the alertness of the locomotive driver. In all cases, if the locomotive driver does not press the alertness handle in time, the train is stopped automatically by an automatic brake.

Fig. 41 KLUB-U panel

Fig. 42 KLUB-U information panel BIL-U

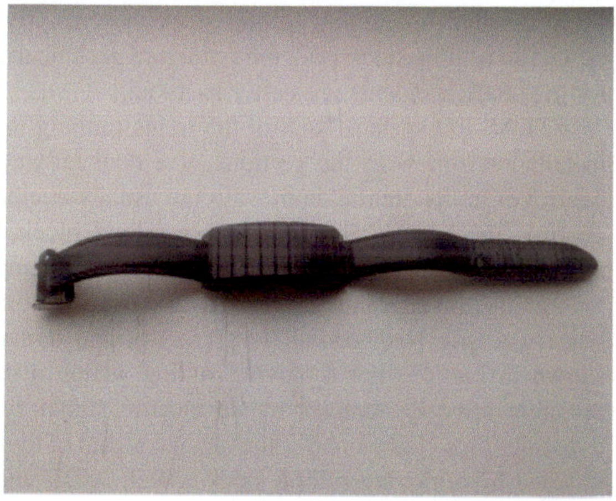

Fig. 43 Driver's alertness checking bracelet

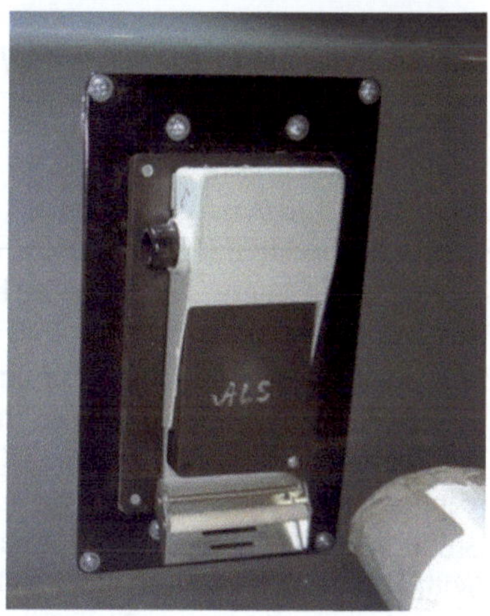

Fig. 44 KLUB-U memory storage is intended to record train operation parameters

10.3 Automatization Systems of Locomotive Control

Signalization equipment is used for transferring information one way and both ways from the track to the locomotive and from the locomotive to the track equipment. To this end, along with the induced channels, radio and satellite communication channels are used. The information transferred from track equipment to the locomotive is used in the signalization systems of the locomotive, in train automatic braking systems, in the safety complexes of the locomotive equipment, and also in various optimum control unified systems with different levels of locomotive control automatization—in this case, the locomotive is controlled by an onboard computer, which receives relevant information.

Automatic rolling stock control system. Modern rolling stock axle box heating (temperature) control, blocked wheel and wheel geometry defect detection have been installed in running rolling stock systems. By installing these systems, the costs for rolling stock and infrastructure technical maintenance is reduced, thus rolling stock accidents are easier to avoid.

ATLAS-LG systems control the trains running in both directions in the system installation zones. In the sections of a dual railway system, the systems for the control of trains running both ways are usually equipped in one control post, which controls the heating of the axle boxes and blocked wheel temperature and the surface defects of wheels rolling on the trains passing by, dynamic wheel forces to the rail and the alternating forces (load). The structural scheme of the control of the wheelsets axle box heating (temperature) and dynamic wheel force to the rail is shown in Fig. 45. For a moving rolling wheel in the ATLAS-LG system, acting dynamic force Q changes by an electric signal system and a diagram of total dynamic force change: ΔQ—the change signal of dynamic force Q; S—the distance traveled by a moving rolling stock wheel; A/D—analog/digital signal converter.

10.4 New JSC Lietuvos Geležinkeliai Traffic Control System Opportunities

Once JSC Lietuvos geležinkeliai (LG) started planning the procurement of new generation western locomotives by completing locomotive modernization, it had to decide what safety systems will be used. On the one hand, locomotive safety systems (LSS) have to operate with the technically outdated locomotives, the cabins of which have traffic control equipment for the 1520-mm gauge width railways installed, which is different from the traffic control equipment used on tracks with 1435-mm gauge width used in EU countries. On the other hand, the LSS conception must allow moving on to the second and third ECTS level technical solutions. By creating the LSS structure, not by reducing the system interoperability with ETCS equipment, the national peculiarities of the railway infrastructure were taken into consideration. Therefore, the LSS of the Lithuanian railways has certain

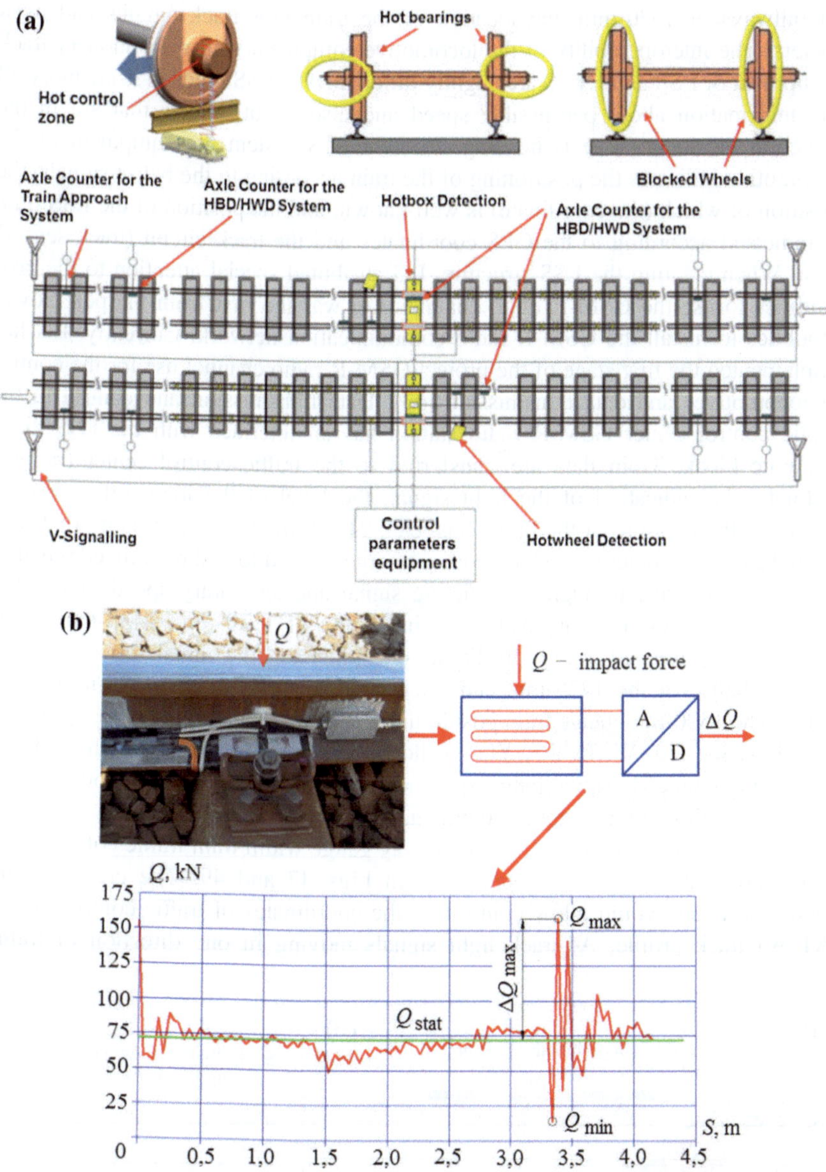

Fig. 45 The structural scheme of wheel set axle box heating (temperature) control and dynamic force to rail measurement

differences as compared to those of the second level ETCS requirements. First of all, the LSS is meant for work in railways with 1520-mm railway gauges and for use in locomotives operating on a 1520-mm railway gauge, in which the train location is indicated not by means of balises, which are widespread in EU railways.

In railways of 1520-mm, the location of the train uses track circuits and axial meters. The interoperability of the locomotive equipment with the stationary track equipment of LSS and ETCS are slightly different. The LSS holds in its memory all the information about permissible speed and also about objects that are in the direction the locomotive is heading. In the ETCS system the equipment of the locomotive indicates the positioning of the train according to the balise signals, the position of which (the coordinate) is well known, and the position of the LSS train is indicated according to the GPS coordinates and the track circuit (track sensor) data. When creating the LSS structure, LG attributed special attention to the possibilities to use the GSM-R radio channel. LG was the first from the post-Soviet republics to install the GSM-R radio communication network. Currently, LG has implemented the first stage of the project GSM-R radio channel use for the control of locomotives and data transmission [43]. Data transmission and control to the traffic control center GSM-R radio station has an interface with the LSS via a systemic block. Train data are transferred to the traffic control center (railway ordinate, the indication of the light signal, the level of the locomotive driver's alertness, the pressure of the brake bus, etc.; 21 information channels in total) and from the control center to the locomotive emergency and forced braking commands, permission to drive through a restricting signal and temporary speed limits. The structural schemes of train positioning indication of 1520- and 1435-mm gauge railways are presented in Fig. 46. Figure 46b illustrates the signal structures using the Eurobalise of the 1435-mm and track circuit signals of the 1520-mm gauge trains. The 1520-mm gauge train positioning control system structure: G—green, Y—yellow, R—red, W—white, YG—yellow-green transmitter code signals; PR—track equipment code signal input equipment; TCS—track equipment code signals; EIE—Eurobalise signals input equipment.

A structural scheme for 1520-mm railway gauge width train traffic control using global positioning systems is presented in Figs. 47 and 49. The created traffic control structure would allow controlling the coordinates of traffic control objects (A1–A3 track profile, A4-track light signals moving in one direction of trains,

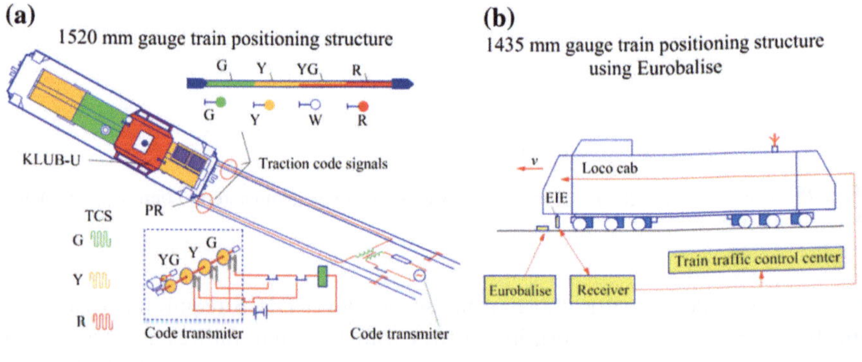

Fig. 46 1520- and 1435-mm gauge train location indication structural schemes using Eurobalise and track circuit signals

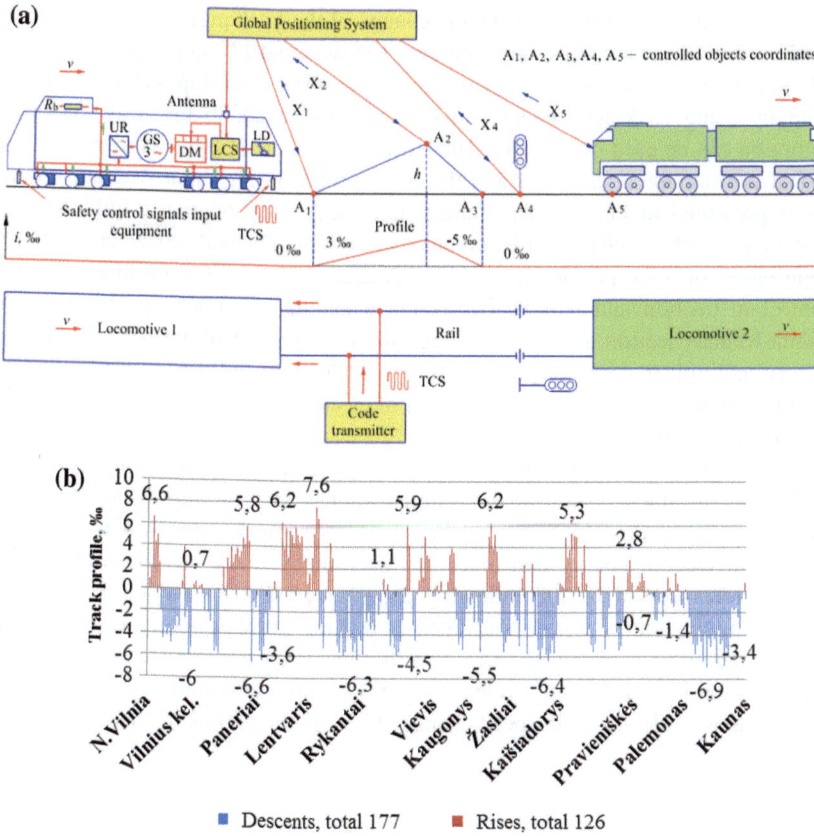

Fig. 47 The structural scheme of traffic control by using global positioning systems (**a**) and the track profile in the Naujoji Vilnia–Kaunas section (**b**): R_b—braking resistor; *L*—distance; *TCS*—traction code signals

individual locomotives, etc.). By controlling the change of coordinates of track profile, one can create locomotive control attachments (modules), which will provide information directly to the locomotive control system LCS [44–47]. In this way, by evaluating the change of the track profile coordinates, independently from the position of the locomotive controller LD, the speed of the locomotive will be controlled and fuel consumption will be significantly reduced. Currently, locomotive drivers control the train manually and cannot fully evaluate the change of the track profile. For example, in the Kaunas-Vilnius section there are over 160 rises and 140 slopes bigger than two parts per thousand. However, a locomotive driver can only memorize the greatest rises and slopes. In the structure of a typical AC/DC current system locomotive there is a diesel engine (DM), the crankshaft of which is connected to the traction generation GS of alternating current (AC). The speed of locomotive DC traction motor TM, which is changed by changing the output tension of uncontrolled rectifier. Currently, locomotive drivers control trains manually and cannot fully evaluate the change of the track profile. For example, in the N. Vilnia-Kaunas section there are over 126 rises and 177 slopes [48]. However, a locomotive driver can only memorize the greatest rises and slopes and controls the train by using its kinetic energy, and fuel is not saved. In this way, fuel consumption can be used only partly. The structural scheme of traffic control by using global positioning systems (a) and the track profile in N. Vilnia–Kaunas section (b) is presented in Fig. 47. The authors have suggested creating a traffic control system in which the change of locomotive speed would be realized according to the coordinates of track profile change. The diagrams of electric train kinetic energy control in traction and braking cycles by evaluating the track profile in the N. Vilnia-Kaunas section are shown in Fig. 48. The diagram of electric train kinetic energy control traction and braking by evaluating the track profile in the Vilnius-Kaunas section shows the amount of electric energy that we can save at the expense of kinetic energy.

The diagram of electric train kinetic energy control traction and braking by evaluating the track profile in the Vilnius-Kaunas section shows the amount of electric energy that we can save at the expense of kinetic energy. The aggregate amount of electric energy consumed for traction by using energy accumulation systems is expressed by the following dependence: $P_{\Sigma}(t) = P_{T}(t) - P_{B}(t)$. Here, $P_{T}(t)$ is the amount of power consumed in traction cycles t_1–t_2, t_3–t_4, t_4–t_5, t_7–t_8; $P_{B}(t)$ is the amount of power consumed that has been accumulated in the power accumulation batteries in the cycles of electrodynamic braking t_4–t_5, t_6–t_7. $v(t)$ are the diagrams of electric train speed change: 1, 2—in the parking cycle; 3, 4—in the

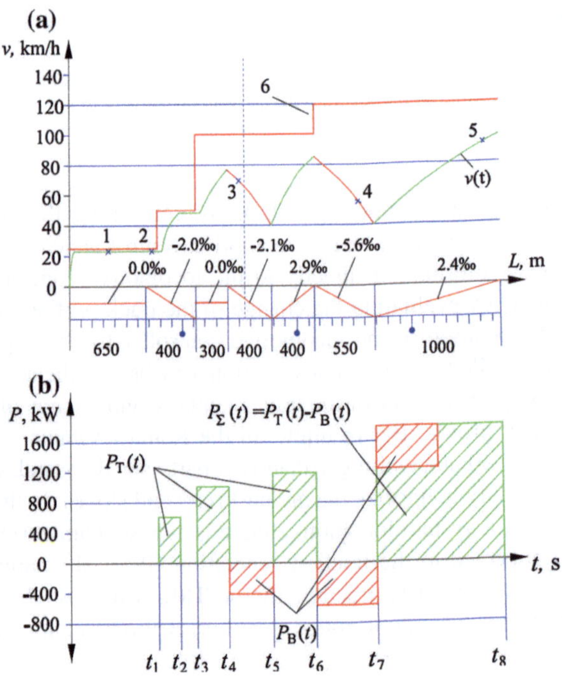

Fig. 48 The diagrams of electric train kinetic energy control traction in electrodynamic braking cycles by evaluating the track profile (**a**, **b**)

electrodynamic braking cycle; 5—in the traction cycle; 6—speed limit diagram in the section. The amount of power saved by an electric train is marked in Fig. 48. The diagrams of electric train kinetic energy control traction in electrodynamic braking cycles by evaluating the track profile show that the aggregate amount of the power for traction consumed by using energy accumulation systems will be reduced, since it is expressed by the following dependence: $P_{\Sigma}(t) = P_{T}(t) - P_{B}(t)$.

From the amount of power consumed in the traction cycles, the accumulated amount of energy consumed in the accumulation batteries will be deducted. The structural schemes of the LG traffic control center by using global positioning (a) and transmission data via a GSM-R radio channel (b) for the locomotive systems is presented in Fig. 49 [49–51].

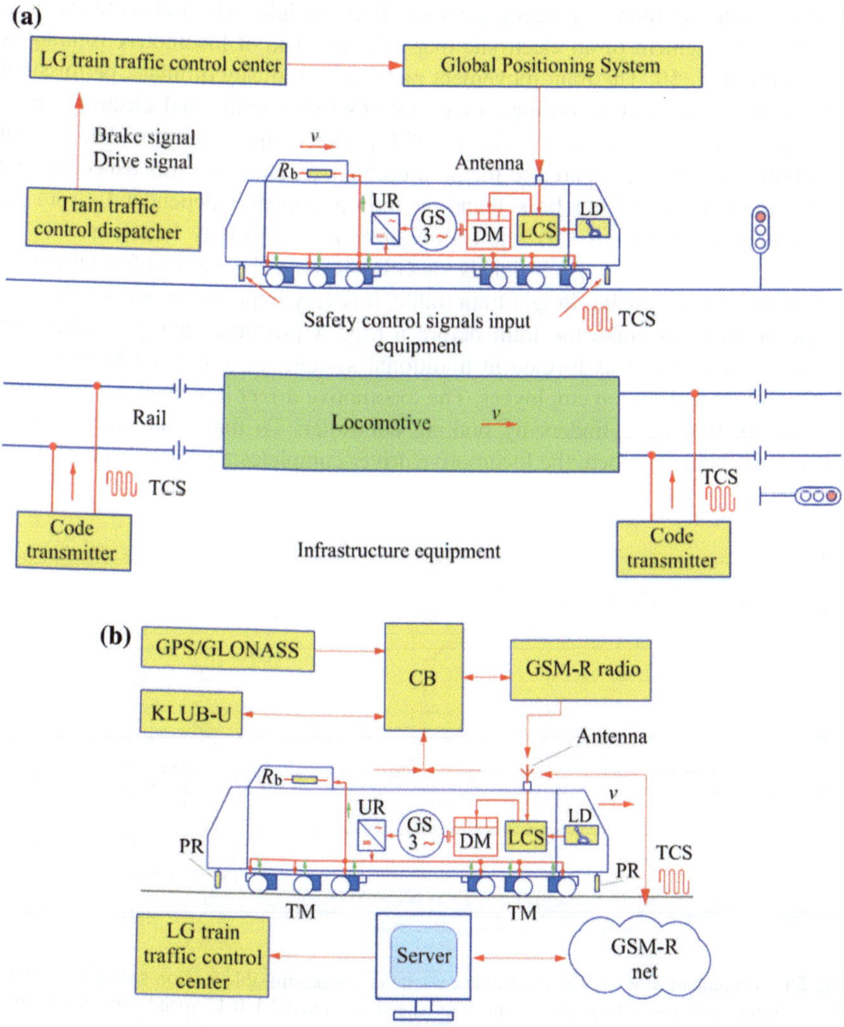

Fig. 49 The structural schemes of LG traffic control center by using global positioning (**a**) and transmission data via GSM-R radio channel (**b**) for the locomotive systems: *CB*—commutation block

In 1520-mm gauge railways, the train traffic control systems are made of stationary equipment in infrastructure and mobile equipment in locomotives (the locomotive signalization ALS). The LSS completes information exchange with the track equipment, receives automatic locomotive signalization (ALS) signals, receives information about speed limits, about whether employees are working on the track, etc. The equipment of the locomotive transmits information to the LSS about the traction on/off switch, control re-switching from one cabin to another, the position of the electric air valve (EAV), the pressure of the supply and brake bus. Current time is corrected according to GPS signals [52].

Data for the track electronic map and the train running schedule are recorded into the LSS interior energy-wise independent memory. The permissible speed in the section, independent from the engine driver braking is completed according to the data of electronic map and train parameters (train weight, wheelset number, etc.). The typical fragment of an electronic map with the data of locomotive running is depicted in Fig. 50. The train movement parameters (railway ordinate, permissible speed) are established according to GPS/GLONASS satellite and electronic map data and also according to the signals of locomotive track sensors. The factual locomotive speed is compared to the permissible one, and once it is exceeded, the supply of the EAV valve is broken and the train is stopped independently from the locomotive driver. In this way, the LSS realizes its main safety functions.

A locomotive receives ALS signals via GSM-R radio channel and from the track equipment. By organizing freight train traffic, it is very important to reduce the time of the attempts to brake the train before a trip. A practical attempt of the train braking system that was formed in traditional systems is completed by the locomotive driver and station employees. The locomotive driver indicated the necessary pressure in braking cylinders by braking controller. To make sure all the front wagons are being stopped, the locomotive driver completes a visual check of them.

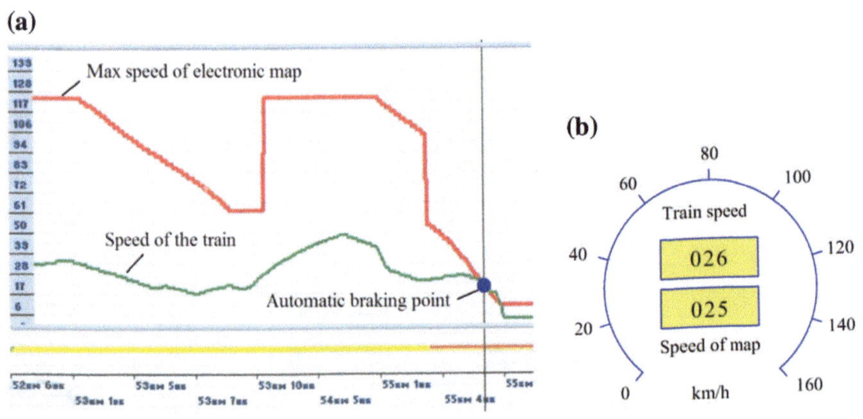

Fig. 50 A fragment of KULB-U electronic cassette of the permissible section speed according to an electronic map and a fragment of the train speed (**a**) and KLUB-U speed indications in the shield during the automatic braking (**b**)

Where the weight of the train is around 6000 t, the number of wagons is more than 100 units. Such a train is very long, and the time of their practical inspection of its braking system functioning is very long. The structural schemes of freight train set braking attempt automatization and the integrity control of the train by using GPS and GSM-R is provided in Fig. 51.

By using the system provided for freight train braking attempts, one can significantly reduce the time of the tests, the test results may be transferred by radio communication channel to the locomotive driver, station employees, who prepare the documentation of the train and LG train traffic control dispatcher. Figure 51: A —train front coordinate; B—train end coordinate; L—locomotive coordinate. Train set braking system test automatization allows reducing the costs by 25%, the time of the train set braking system test, archiving of the test parameter, sending the

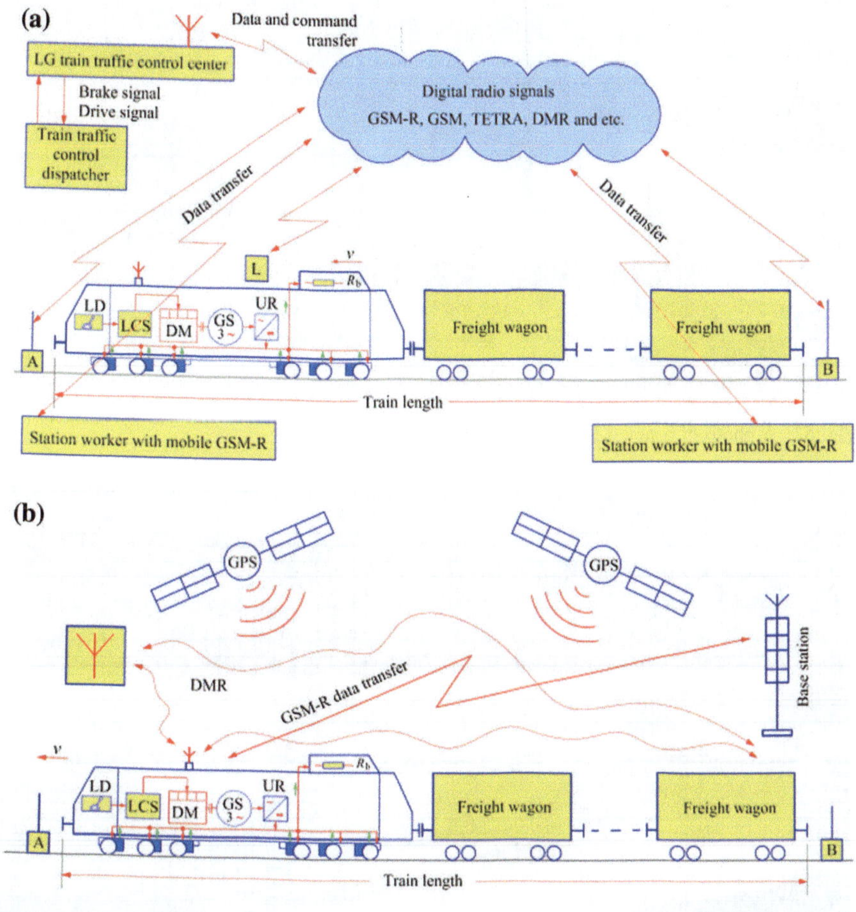

Fig. 51 The automatic control schemes of the LG train braking (**a**) and integrity (**b**) systems test using GPS and GSM-R

parameters in an electronic form (reduces documentation preparation time). A photo of a new LG traffic control center is presented in Fig. 52.

Train integrity control. The integrity of the freight train is checked by controlling the speed of the last wagon, the route of the last wagon, the pressure of the braking bus of the last wagon, the changes in pressure of the braking bus of the last wagon braking. A photo of LG contact network equipment remote control is presented in Fig. 53.

Fig. 52 A photo of a new LG traffic control center

Fig. 53 A photo of LG contact network equipment remote control

Before the launch of the Da Vinci system, there were 33 stations in LG, which could have been controlled from an old traffic control center EVC. Out of these, 21 stations were controlled continuously. In the new traffic control center, the control of 91 stations is completed. Of these, traffic control of seven border stations traffic is carried out; traffic control of 13 distribution, freight, passenger stations; 11 stations (five of which are included in the Rail Baltica project) have no traffic centralization EC.

11 Conclusions

1. Train control systems use 1435- and 1520-mm track gauges, and the main difference is that in the 1520-mm gauge railway, train position is determined using track circuits, in the 1435-mm gauge railway, train position is determined using Eurobalise signals.
2. To master passenger train routes between the railways of 1435- and 1520-mm width, it will be essential to evaluate not only the differences of the gauge and contact network standards, but also those of traffic control systems.
3. It is essential to look for technical solutions to the coordination of different rail traffic control systems of 1435- and 1520-mm width.
4. It is necessary to find technical solutions to obtain compatibility of train traffic control systems for 1435- and 1520-mm track gauges.
5. In light of improving train traffic control systems, it is necessary to abandon interval-based train control principle.
6. In light of improving train traffic control systems, it is necessary to abandon fixed block system principle.
7. We are working towards implementing moving block and radio transmission system in train traffic control systems.
8. Train traffic control systems specifically for track 1520-mm gauge has to be considered during integration of Baltic states railways into EU network.
9. Driver alertness monitoring specifics must be considered during integration of Baltic states railways into EU network.
10. Features of power units and electrical drives of locomotives operated in railways with 1520-mm gauge track must be taken into account in the development of ETCS computer TSS.
11. Train traffic control systems must be developed by implementing power saving and storage systems.
12. Train traffic control systems must be improved considering the development of GPS, GLONASS and, GALILEO.
13. Train identification, train position, direction and speed, and other data have to be transmitted using GSM-R radio communication.
14. About 20 different European signaling systems must be made uniform, and a uniform automatic speed control system must be implemented in the framework of train traffic control system improvement.

15. In is necessary to assess experience of systems used for braking of high-speed trains by Japan, France, Germany, Italy, etc.

References

1. Divis DA (2002) Military role for Galileo emerges. GPS World 13(5):10
2. Johnson C (2004) U.S., EU to Sign Landmark GPS-Galileo Agreement. Available at: http://iipdigital.usembassy.gov/st/english/article/2004/06/200406241609361acnosnhoj0.5679285.html#axzz4GfaneANy
3. Hein GW, Godet J, Issler J-L, Martin J-C, Erhard Ph, Lucas-Rodriguez R, Pratt T (2002) Status of Galileo frequency and signal design. In: Proceedings of ION GPS. Available at: http://ec.europa.eu/dgs/energy_transport/galileo/doc/galileo_stf_ion2002.pdf
4. Salmi P, Torkkeli M (2009) Inventions utilizing satellite navigation systems in the railway industry—an analysis of patenting activity. J Technol Manag Innov 4(3):46–58
5. Rajkumar RI, Sankaranarayanan PE, Sundari G (2013, Sept) GPS and Ethernet based real time train tracking system. In: Proceedings of 2013 international conference on advanced electronic systems (ICAES) pp 282–286. Available at: http://ieeexplore.ieee.org/stamp/stamp.jsp?tp=&arnumber=6659409
6. Psiaki ML (2001) Block acquisition of weak GPS signals in a software receiver. In: Proceedings of ION GPS 2001, the 14th international technical meeting of the satellite division of the institute of navigation, Salt Lake City, Utah, pp 2838–2850
7. The Galilei Project. Galileo Design Consolidation (2003) European Commission. Available at: http://ec.europa.eu/dgs/energy_transport/galileo/doc/galilei_brochure.pdf
8. Van Der Jagt CW (2002) Galileo: the declaration of European Independence: a dissertation. American Graduate School of International Relations and Diplomacy
9. The GSM-R Frequency Workshop (2009) UIC eNews. No. 143. Available at: http://www.uic.org/com/uic-e-news/143/article/the-gsm-r-frequency-workshop?page=thickbox_enews
10. Technologies—GSM-R (2012) Willtek. Available at: http://archive.is/OS7N
11. GSM-R Technology (2011) GSM-R Industry Group. Available at: http://web.archive.org/web/20110711132640, http://www.gsm-rail.com/node/10
12. Implementation planning and progress (2007) GSM-R. Available at: http://web.archive.org/web/20071014133845, http://gsm-r.uic.asso.fr/implement_map.html
13. Digital Train Radio System (2013) PTV Corporate. Government of Victoria. Available at: http://web.archive.org/web/20130629002926, http://corp.ptv.vic.gov.au/projects/rail-projects/digital-train-radio-system/
14. Connor P (2013) A job for life: changes seen in a 50-year career on London Underground 1916–1966. Available at: http://www.railway-technical.com/Paper%202%20-%2050%20Years%20on%20LU.pdf
15. Connor P, Harris NG, Schmid F (eds) (2015) Designing and managing urban railways London: A & N Harris
16. Connor P (2015) London underground electric train. Crowood Press, London
17. Theeg G, Vlasenko S (eds) (2009) Railway signalling & interlocking—international compendium. Eurailpress
18. Suwe KH (1988) Signaltechnik in Japan. Signal + Draht 80(12) pp 284–288
19. Lingaitis LP, Liudvinavičius L, Butkevicius J, Podagėlis I, Sakalauskas K, Vaičiūnas G, Bureika G, Gailienė I, Petrenko V, Subačius R (2009) Geležinkeliai. Bendrasis kursas: vadovėlis. Vilnius: Technika [In Lithuanian: Railways. General Course: text book]
20. Bianchi C (1985) Die Fuhrerraumsignalisierung auf der Direttissima Rom-Florenz. Signal + Draht 77(1/2) pp 9–16
21. Yamanouchi S (1979) Safety and ATC of Shinkansen. In: Japanese railway engineering

22. Directive 96/48/EC99 of 23 July 1996 amending Council Directive 96/48/EC on the interoperability of the trans-European high-speed rail system and Directive
23. EC sets out ERTMS deployment deadlines. Available at: http://www.railwaygazette.com/news/single-view/view//ec-sets-out-ertms-deployment-deadlines.html
24. 2001/16/EC of the European Parliament and of the Council on the interoperability of the trans-European conventional rail system
25. Liudvinavičius L, Lingaitis LP, Dailydka S (2010) Traukos riedmenų elektros pavaros ir jų valdymas: bendrasis aukštųjų mokyklų vadovėlis. Vilnius: Technika. [In Lithuanian: Traction rolling stock power drives and their control: common high school text book]
26. Indusi prototype on a steam locomotive in May 1930. Available at: https://upload.wikimedia.org/wikipedia/commons/9/95/Bundesarchiv_Bild_102-09857%2C_Induktive_Zugsicherung_%28Indusi%29_an_D-Zug-Lok.jpg
27. ETCS "Eurobalise" transceiver, installed between rails, provides information to ETCS trains. Available at: https://upload.wikimedia.org/wikipedia/commons/7/73/Balizok_az_%C5%90rs%C3%A9gi_vas%C3%BAton2.jpg
28. Croco + TBL + ETCS balises at the same signal. Available at: https://upload.wikimedia.org/wikipedia/commons/6/6d/Croco%2BTBL1%2BETCS.jpg
29. EBICAB balise in the Mediterranean Corridor. Available at: https://upload.wikimedia.org/wikipedia/commons/f/f3/EBICAB.jpg
30. A Siemens Eurobalise in Germany. Available at: https://upload.wikimedia.org/wikipedia/commons/0/0c/Siemens_Eurobalise.jpg
31. Rail Traffic Management and Signaling. Available at: http://www.navipedia.net/index.php/Rail_Traffic_Management_and_Signaling
32. Recommendation on ERTMS delivered to European Commission, European Railway Agency. Available at: http://www.era.europa.eu/Communication/News/Pages/New-ERTMS-Recommendation-delivered.aspx
33. Commission decision on the technical specification for interoperability relating to the control-command and signalling subsystems of the trans-European rail system. 2012-01-25. 2012/88/EU
34. Commission decision amending Decision 2012/88/EU on the technical specifications for interoperability relating to the control-command and signalling subsystems of the trans-European rail system. 11/6/2012. 2012/696/EU
35. System Requirements Specification (SUBSET-026). European Railway Agency. Available at: http://www.era.europa.eu/Document-Register/Pages/System%20Requirements%20Specification%20(Recommendation).aspx
36. ERTMS Levels. Available at: http://www.ertms.com/media/2428/fact-3.pdf
37. ETCS Level 1 schematic. Available at: https://upload.wikimedia.org/wikipedia/commons/thumb/8/83/ETCS_L1_en.svg/640px-ETCS_L1_en.svg.png
38. ETCS Level 2 schematic. Available at: https://upload.wikimedia.org/wikipedia/commons/thumb/9/9d/ETCS_L2_en.svg/640px-ETCS_L2_en.svg.png
39. Railway technical web pages. Railway systems, technologies and operations across the world. Available at: http://www.railway-technical.com/
40. Trainguard 300—the solution for ETCS Level 3 applications. Available at: https://www.mobility.siemens.com/mobility/global/SiteCollectionImages/rail-solutions/rail-automation-new/automatic-train-control-systems/etcs-level-3-large-en.jpg
41. Colburn R (2013). A history of railroad signals. Institute of Electrical and Electronics Engineers. Tech Focus
42. Sapozhnikov VV, Kokurin IM, Kononov VA, Lykov AA, Nikitin AB (2006) Operational bases of automation and telemechanics. Moscow: Marshrut
43. Voronin VS (2009) Intelligent transport management systems. Railway transport, No. 3 pp. 140–143
44. BNSF Railway Company (2005) General Code of operating rules, 5th edn
45. Calvert JB (1999) Centralized traffic control. Available at: http://mysite.du.edu/~jcalvert/railway/ctc.htm

46. Brian FW (2006). Railroad's traffic control systems. Trains. Available at: http://trn.trains.com/railroads/abcs-of-railroading/2006/05/railroads-traffic-control-systems
47. Lundsten CS (2000, Oct) North American Signaling: Absolute Permissive Block. Available at: http://www.lundsten.dk/us_signaling/abs_apb/
48. Liudvinavičius L (2012) Investigation of locomotives electrodynamic braking. Doctoral dissertation. Vilnius: Technika. ISBN 978-609-457-333-0
49. Bandemer B, Denks H, Hornbostel A, Konovaltsev A, Ribeiro Coutinho P (2005) Performance of acquisition methods for Galileo SW receivers. Eur J Navig 4(3):17–29
50. Mendizabal Samper J, Berenguer Pérez R, Meléndez Lagunilla J (2009) GPS & Galileo: dual RF front-end receiver and design, fabrication, and test. McGraw Hill. ISBN 978-0-07-159870-5
51. Mendizabal J, Berenguer R, Melendez J (2009) GPS and Galileo. McGraw Hill. ISBN 978-0-07-159869-9
52. Ishchenko VA (2009) Organization of communication on low-density lines. Rail transport

Modeling of Traffic Smoothness for Railway Track Closures in the Rail Network

Grzegorz Karoń

Abstract This article presents a discussion of railway track section closures and loss of traffic. A method for rational planning and optimal coordination of track closures is introduced, along with estimations of traffic smoothness, including a few example calculations.

Keywords Railway track closures · Traffic smoothness · Railway capacity · System for assessment of track configuration (SOUT)

1 Introduction

Maintaining railways requires periodic closures [2, 5, 24]. The proposed rationale for planning of railway track closures requires estimation of railway capacity limitations. According to the theory of traffic flow, traffic smoothness capacity of an element of a railway network (railway junction, rail station or railway line section) depends on the number of trains in a given period of time (usually during the day) in which traffic is moving at maximum smoothness. The definition of optimal capacity can be expressed as follows: the optimal capacity (traffic smoothness capacity) of the railway junction (rail station) is the a number of trains—referred to as the optimal traffic intensity in terms of traffic smoothness—for which the average traffic smoothness flow is the greatest [27–29]. Traffic smoothness capacity treated as optimal capacity requires an assessment of potential railway traffic disruption resulting from the intersection of train routes at rail stations. The probability of disruption and the average length of disruption are obtained by computer simulation of train traffic.

G. Karoń (✉)
Faculty of Transport, Silesian University of Technology,
Krasinskiego 8, 40-019 Katowice, Poland
e-mail: grzegorz.karon@polsl.pl

A. Sładkowski (ed.), *Rail Transport—Systems Approach*, Studies in Systems, Decision and Control 87, DOI 10.1007/978-3-319-51502-1_8

Closure of railway sections causes losses of traffic, expressed as the number of canceled trains at the time of closure. This number is dependent on the daily train intensity, capacity reserve and duration of the closure during the day. Because trains run on more than one railway section, the removal of a train from a closed railway section results in its removal from other sections of the railroad. In light of this fact, scheduled closures of railway sections can be coordinated so that they occur at the same time. For this purpose, it is necessary to designate, for each pair of railway sections on the network, traffic compounds. The essence of optimal distribution of closures is to minimize the expected total traffic losses—number of canceled trains.

2 The Issues of Planning and Coordination of Railway Track Closures

2.1 Repairs of Railway Track Sections and Repair Cycles

Maintaining railways requires periodic closure of track sections for maintenance and repair, which is widely associated with rail traffic disruption.

The average duration of repair cycles for railway track sections is 8–10 years. Major repairs (MR) are carried out at the beginning of the cycle. Between major repairs, medium-level repair (MeR) is conducted, with current repair (CR) of railway track sections carried out once a year. For each type of maintenance and repair, the number and duration of railway track closures can be specified. Typically, major repairs and average repairs have a predetermined duration of one railway track closure, while current repairs have limited advance notice of track closure.

Problems of monitoring and statistical analysis of railway track closures have been studied in order to facilitate their planning in terms of reduced rail traffic disruptions. In analyzing the functioning of the railway track, it can be assumed that at a certain time, t, the track will be in one of the following states [29]:

- full operational suitability lasting together $T_1(t)$;
- limited operational suitability, with speed limit of traffic, lasting together $T_2(t)$;
- renewal during planned technological closures, lasting together $T_{op}(t)$;
- renewal during unplanned technological closures, lasting together $T_{on}(t)$;
- renewal during emergency closures, lasting together $T_{oa}(t)$.

It is assumed that the smallest section of the railway track is the closed railway section between each railway junction, i.e. the track railway section or the station track. Due to maintenance and repair technology, it is possible to close two, and sometimes three, railway sections, resulting in three cases of closures:

1. Closure of only the railway section track or the station track between consecutive railway junctions. This is the case occurring most often.
2. Closure of two adjacent railway sections at both sides of a railway junction, which means closure of railway section track and station track, or closure of two railway section tracks contiguous to the railway junction. This case is rare and occurs in major and medium repairs. If one of the closed sections is the station track, it is comparable to movement disruption from the closure of only the railway section, because the organization of train traffic can be adjusted by the station track system. If, however, two consecutive tracks of railway section are closed adjacent to a railway junction, it causes major disruption to train traffic movement.
3. Closure of three consecutive railway sections, which means: closure of railway section track, station track and the next railway section track. This case is the most troublesome from the standpoint of train traffic but it is very rare, occurring only during major repairs due to technology upgrades/repairs, and it does not occur in lines with extensive use.

2.2 *Railway Capacity and Optimal Intensity*

The proposed rationale for planning of railway track closures requires estimation of railway capacity limitations. The general definition of railway capacity is: *a measure of the ability to move a specific amount of traffic over a defined rail line with a given set of resources under a specific service plan* (e.g. number of tons moved, average train speed, on-time-performance, maximum number of trains per day, etc.). The term railway capacity is widely presented in the literature, e.g. [1, 3, 6, 14, 19, 22], as well as the terms train schedule/timetable [4, 5, 21, 23, 26] and capacity planning [15–17, 30].

According to the theory of traffic flow, the traffic smoothness capacity of an element of a railway network (railway junction, rail station or railway line section) is the number of trains in a given period of time (usually during the day) during which traffic is moving at maximum smoothness [29]. This number of trains is calculated using an iterative simulation technique for different variants of train operation plans. During calculations, the average number of trains disturbed is determined, i.e. those passing through the railway network test component which had to be redirected during the elaboration of the train operation plans due to the passage of another train (e.g. with higher priority)—this event is referred to as "*meet delay or pass delay*". After completion of calculations simulating the process of constructing traffic train operation plans, an expected railway traffic smoothness function is obtained, which assumes the maximum value for the traffic smoothness capacity of a railway network element (Fig. 1) [12, 28, 29].

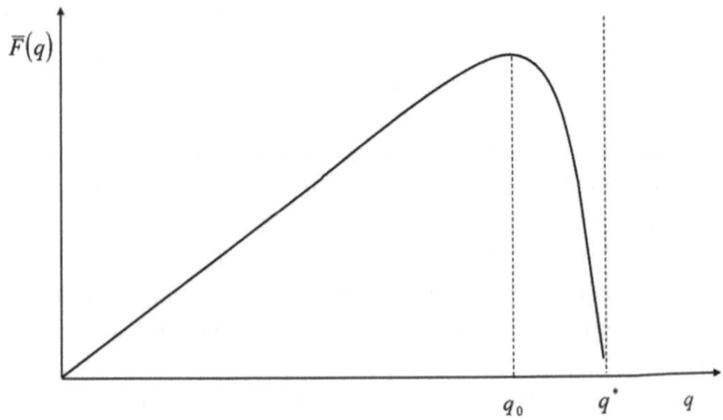

Fig. 1 The function of the expected traffic smoothness flow; based on [29]

To understand the traffic smoothness capacity requires an assessment of potential railway traffic disruption resulting from the collision of train routes at a rail station. The probability of disruption and the average duration of disruption are called regulation characteristics and are obtained by computer simulation of the train traffic.

The simulation program [28, 29] developed for traffic smoothness capacity estimation enables computer simulation of train traffic for a given railway junction (rail station), which produces a regulation characteristic, among others, the average frequency of delay (regulation) $\overline{p}(q)$ as a function of traffic intensity q. The definition of optimal capacity can thus be expressed as follows: optimal capacity (traffic smoothness capacity) of the railway junction (rail station) is the number of trains q_O, referred to as the optimal traffic intensity in terms of traffic smoothness flow, for which the average traffic smoothness flow:

$$\overline{F}(q) = (1 - \overline{p}(q)) \cdot q \quad (1)$$

is the greatest. Since the explicit form of the function $\overline{p}(q)$ is not known, an explicit formula for capacity cannot be given. Using computer simulation, one obtains only estimates $\overline{p}(q)$ for a predefined value of q. Then, by receiving different estimates for different application intensity values of q, it is possible in a numerical and statistical way to appoint the searched value of the optimal traffic intensity value q_O.

During computer simulation of train traffic for a given junction of the railway network, $\overline{p}(q)$, among others, is obtained—$\overline{p}(q)$ is the average frequency of the delay, which is an estimate of the delay probability. This value is calculated as the quotient of the number of delayed trains to the total number of trains q. Knowing the average frequency of delays allows calculation of the average traffic smoothness flow, which is an estimate of the expected traffic smoothness flow $\overline{F}(q)$, as the subtraction of the average number of all trains q, and the average number of delayed trains $q \cdot \overline{p}(q)$. Figure 1 shows the function of the expected traffic smoothness flow

$\overline{F}(q)$ with characteristic values marked: the optimal intensity q_O for which $\overline{F}(q)$ achieves the greatest value and intensity q^*, which is the capacity value calculated without taking into account the traffic smoothness flow. Scheduled railway traffic trail capacity q_i may exceed the optimal intensity of railway traffic until reaching values q^*, which causes a decrease in traffic smoothness flow on the rail line section.

2.3 An Example of the Optimal Intensity Estimation

The example of optimal intensity estimation with use of computer software (System for Assessment of Track Configuration [SOUT]) was prepared for railway line 139 Bielsko-Biała – Zwardoń (in Poland), section Łodygowice Station–Żywiec Station–Węgierska Górka [11–13]. The base characteristics of the railway line are presented in Fig. 2, and modeled configurations of station tracks is presented in Fig. 3. Computations were done for two options of the train control system. The first option is for a mechanical and relay train control system and second option is for a computer train control system [18, 20, 25].

The calculations provide the function of the expected traffic smoothness flow and optimal intensity for both options, as presented in Fig. 4.

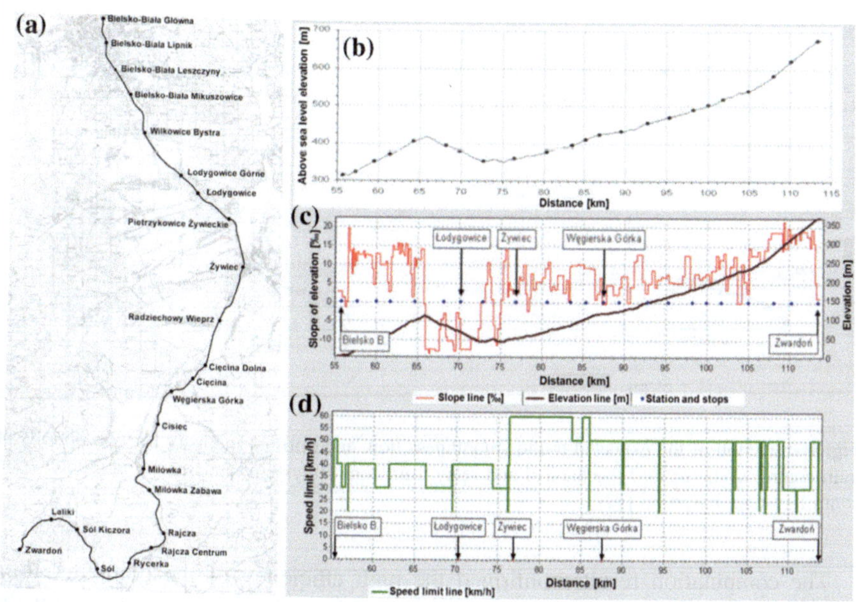

Fig. 2 Railway line 139 Bielsko-Biała–Zwardoń (in Poland), section Łodygowice Station–Żywiec Station–Węgierska Górka: **a** route of line, **b** elevation of line, **c** slope of elevation, **d** speed limit; based on [11]

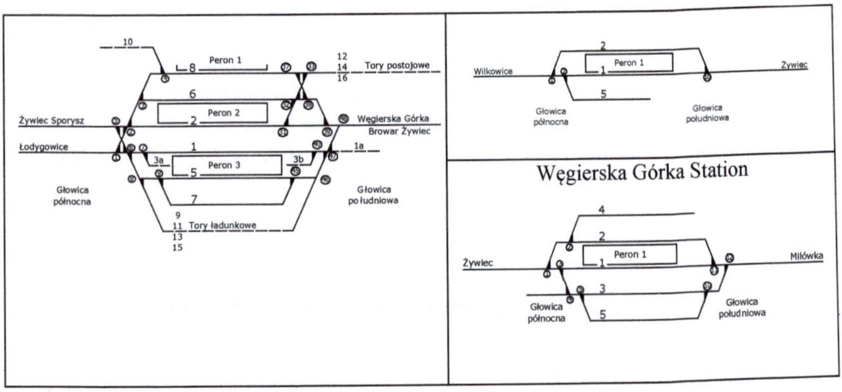

Fig. 3 Modeled configurations of station tracks; based on [12]

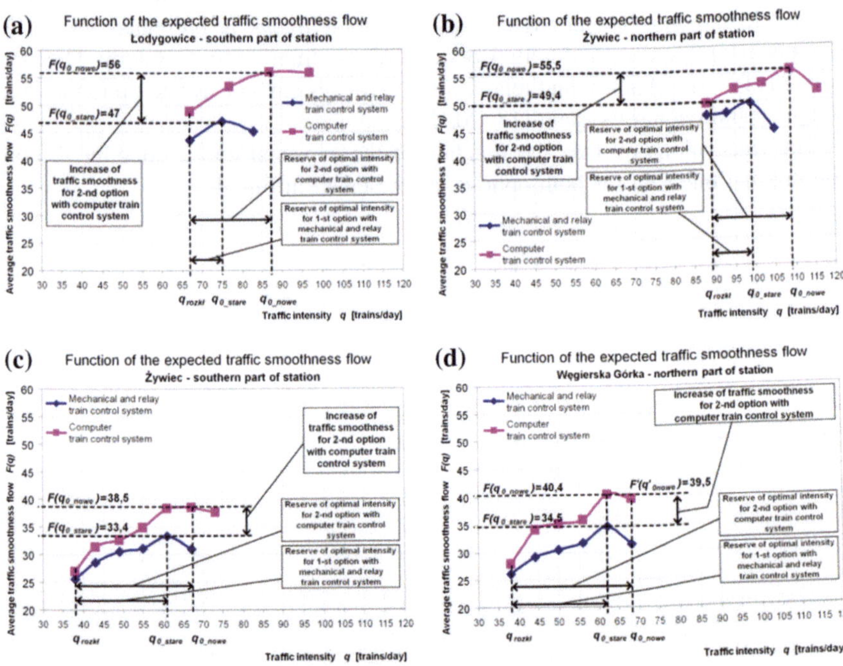

Fig. 4 Function of the expected traffic smoothness flow and optimal intensity for two options of train control systems: *first* mechanical and relay train control system; and *second* computer train control system; based on [12]

The computation results confirmed the high efficiency of the computer train control system—see values of reserve of optimal intensity $q_{\text{o\_nowe}} - q_{\text{\_rozkł}}$ for the 2nd option (Table 1).

Table 1 Comparison of results of optimal intensity for two types of train control systems

Station part	Direction (relation)	Adjacent station	Scheduled intensity $q_{\_rozkł}$ [trains/day]	1st option mechanical and relay train control system			2nd option computer train control system		
				Optimal intensity q_{o_stare} [trains/day]	Reserve of optimal intensity $q_{o\_stare} - q_{\_rozkł}$		Optimal intensity q_{o_nowe} [trains/day]	Reserve of optimal intensity $q_{o\_nowe} - q_{\_rozkł}$	
					[trains/day]	[%]		[trains/day]	[%]
Łodygowice—southern part of station	From	Żywiec	33	36	3	9.1	41	8	24.2
	To		34	39	5	14.7	46	12	35.3
Żywiec—northern part of station	From	Sporysz	18	18	0	0.0	18	0	0.0
	To	Sporysz	4	4	0	0.0	4	0	0.0
	From	Łodygowice	34	39	5	14.7	46	12	35.3
	To		33	38	5	15.2	41	8	24.2
Żywiec—southern part of station	From	Węgierska Górka	19	31	12	63.2	34	15	78.9
	To		18	30	12	67.7	33	15	83.3
Węgierska Górka—northern part of station	From	Żywiec	19	31	12	63.2	31	15	78.9
	To		19	31	12	63.2	31	15	78.9

2.4 Assessment of Capacity Loss

Based on the optimal traffic intensity of trains, capacity loss is assessed according to the following data [2]:

- q scheduled traffic intensity of trains (on the trail);
- q_O optimal traffic intensity (traffic smoothness capacity) of trains (on the trail) during scheduled railway traffic organization;
- q_I optimal traffic intensity (traffic smoothness capacity) of trains when one railway section track is being closed;
- q_{II} optimal traffic intensity (traffic smoothness capacity) of trains when two tracks (railway section track and station track or two tracks of railway section adjacent to the railway junction) are being closed.

These values are expressed as the number of train paths per day. For each, the sections of rail line for each repair *MX* (MR—major repair, MeR—medium repair, CR—current repair) is determined [2]:

- the number of train paths (in the train operation plans) lost per unit of closure:

$$l_i^{NX} = \frac{(q - q_i) \cdot C_{NX}}{24}, \tag{2}$$

  where:
  i—the type of railway traffic organization; I—railway traffic organization when one railway section track is being closed; II—railway traffic organization when two tracks (railway section track and station track or two tracks of railway section adjacent to the railway junction) are being closed; C_{NX}—the number of hours of one closure for the repair of type: *NX* (MR—major repair, MeR—medium repair, CR—current repair),
- the reserve of train paths for the rest of the day:

$$l_q^{NX} = \frac{(q_O - q) \cdot (22 - C_{NX})}{24}, \tag{3}$$

  where $(22 - C_{NX})$ is the number of hours between two consecutive closures for the repair of a kind *NX* taking into account 2 h of losses due to so-called fortuitous events.

The proposed requirement for permissible capacity of railway traffic for railway lines due to repair cycles for pavement repair is of the form:

$$l_i^{NX} \leq l_q^{NX}, \tag{4}$$

indicating that the number of train paths lost for one closure for a repair of type *NX* should not exceed the reserve of paths between consecutive closures. Sections of rail line for which the condition is not met, even for one type of repair, are called "critical sections" (bottlenecks). Values l_i^{NX} and l_q^{NX} are fundamental decision-making values of loss assessments, enabling assessment of whether a given section repair can be conducted according to the scheduled repair cycle without worrying about failure to comply with condition (4).

The methodology assumes that the duration of the closure is 24 h, which means that all trains from the closed track should be directed to an alternate path or be replaced by road transport. Synchronization of rail closures, which determines which sections should close at the same time and which are releasable, is divided into four classes:

- Class I—recommended synchronization: the common closure of two sections is more favorable that disjointed closures;
- Class II—closing independent of each other: the synchronization does not improve organization of railway traffic;
- Class III—unfavorable synchronization, but optimal intensity is not exceeded;
- Class IV—incorrect synchronization: causes an increase in global travel time and deterioration of traffic smoothness.

Analysis of closure synchronization showed that in the area of low train route network density, there is a pair of track sections that are both worth closing and a pair of sections whose simultaneous closure would cause a significant deterioration in traffic flow smoothness. In relation to the distribution of several closures the following should be considered:

- Closure of the two pathways should not be made simultaneously; one of the pathways can serve as an alternate line segment to the traffic carried over from the second pathway,
- Two closures located on one line should be separated by at least one railway station.

The problem of closure planning is to determine the time standards for a single closure. This ensures rationality of repair work progress and the reasonableness of capacity loss. However, the same closing time for two technically and similarly traversed sections of the railway network, but localized in different areas, may cause different exploitation effects. It cannot, therefore, be expected that the same closing time is equally rational in both cases.

Two classes of problems are related to the planning of track closures:

- Railway traffic organization for the assumed locations of closures, involving determining the daily intensity of trains for each main track of the railway section (without any detailed consideration of the order of traffic operations);
- Regulation of train traffic at the respective closure sites for specified traffic organization, which consists of developing a train operation plan.

Allocation of rail closures in the rail network for certain work schedules requires coordination of works and determining the number and the time of a single closure for each of the coordinated works. Determination of a single closure time requires an assessment of losses in train traffic (reduction of train traffic), and, then, the transfer excess of train traffic on the alternate paths with reserve capacity.

Based on the structure of the train operation plans, the time of day for the closure should be determined as a method of studying changes in traffic organization. Finding a reasonable closing time by this method for many different values of the closure time located at different periods per day is very difficult. The issue is complicated further after taking into account various possible changes in the train traffic organization for a predefined value of closure time. Therefore, in practice, the aim is to determine the maximum permissible train traffic organization, to a predetermined location of closure time per day.

Assuming that an efficient manner of determining necessary changes in train traffic organization during the closures is known, the determination of the rational time of a single closure requires consideration of the technology of the conducted works. Different technologies of track works, various railway traffic organizations and different train traffic regulations provide a picture of the complexity of the closure planning problem, for which the reasonable solution is the proper decomposition of the problem and use the successive approximations method [2, 7, 10].

Decomposition of the closure planning issues gives the following components:

- Schedule of works;
- Schedule of closures;
- Railway traffic organization during the closures;
- Train operation plans during the closures.

The first problem (task 1) of railway traffic organization is coordination of track works sharing the same closure. Allocation of closures in the network on individual days (task 2) requires an assessment of losses of train traffic for the specified closure time of day (task 3), and this in turn requires determining the volume of train traffic transferred and lost, and a search for alternate paths (task 4). During multiple simultaneous closures, this leads to the need for coordination of closures in terms of traffic volume loss (task 5). The result of tasks 3, 4 and 5 for all scheduled track works is organization of railway traffic during periods of closure, while minimizing the size of expected loss train traffic.

In the organizational problem, there is still the problem of an acceptable level of train traffic loss. With the level of allowable traffic loss set high, it may be that changing traffic organization is not required and the allocation of closures can be arbitrary. When only a low level of traffic loss is acceptable, it can become necessary to search for alternate lines and the distribution of simultaneous closures so as to minimize expected train traffic loss.

Another problem is the daily allocation of closure time (task 6), demanding a full analysis and regulation of train operation plans when assessing the expected losses of railway traffic in a specified range of a train operation plan (task 7). Also required is the designation of trains on alternate paths, the designation of canceled trains (task 8) and the coordination of closures per day in terms of train traffic disruption (task 9). If organizational problems have not been solved in advance, e.g. closures onto the network on the individual days were placed arbitrarily, problems will appear when determining alternate train paths with alternate paths indicated on the basis of relevant train operation plan analysis. Furthermore, during closure coordination, the distribution of different railway track closures on the network on individual days should be determined, in addition to the coordination of time intervals.

2.5 *Evaluation of Railway Network Capacity Reserve*

To determine the train network capacity reserve, the optimal intensity values should be calculated:

qo_i when both tracks of the rail line section are active;
qz_{ik} when the *k*th track of the *i*th rail line section is closed.

In the remainder of the text, by the marking and patterns, the following indexes are used:

i, j indexes relating to the railway sections; because of its use in the issue of traffic compound coordination between railway sections (expressed by a number of common train routes), two indexes, i, j, were used, designating a pair of sections;
k railway section track index;
t the subsequent day of repair period T.

The average values of indicators have a top dash. Description of *i*th railway sections necessary to carry out the process of coordinating thc closure includes the following values [2, 9]:

- q_i—intensity of train traffic according to the timetable for the *i*th railway sections [trains/day];
- qo_i—the optimal intensity of train traffic under normal conditions [trains/day];
- qz_{ik}—the optimal intensity of train traffic during the closure of the *k*th track of the *i*th railway section (for one track $q_{Z_{ik}} = 0$) [trains/day].
- The average number of scheduled trains per hour:

$$\overline{q_i} = \frac{q_i}{24}\ \text{[trains/h]} \tag{5}$$

- Average hourly capacity reserve of the *i*th railway section:

$$\overline{r_i} = \frac{qo_i - q_i}{24}\ \text{[trains/h]} \tag{6}$$

- Average hourly intensity of traffic on the railway section ensuring traffic flow smoothness during closure of the *k*th track of the *i*th railway section:

$$\overline{q}_{Z_{ik}} = \frac{q_{Z_{ik}}}{24}\,[\text{trains/h}] \tag{7}$$

- Average hourly number of trains canceled due to the closure of the *k*th track of the *i*th railway section:

$$\overline{q}_{S_{ik}} = \frac{q_i - q_{Z_{ik}}}{24}\,[\text{trains/h}] \tag{8}$$

- Allowable closure time $t_{Z_{ik}}$ of the *k*th track of the *i*th railway section depending on the intensity and capacity reserve on the railway section:

$$t_{Z_{ik}} = \begin{cases} 24 \cdot \frac{q_{O_i} - q_i}{q_{O_i} - q_{Z_{ik}}} & for \quad \overline{r}_i > 0 \quad and \quad \overline{q}_{S_{ik}} > 0 \\ 24 & for \quad \overline{r}_i > 0 \quad and \quad \overline{q}_{S_{ik}} \leq 0 \end{cases} \tag{9}$$

Closure of the track of the railway section causes losses of traffic expressed as the number of canceled trains at the closure time. This number is dependent on the daily train intensity, capacity reserve and duration of the closure during the day. In the case of railway sections which have capacity reserve, duration of the closure during the day determines the amount of losses of traffic. For railway sections with capacity reserve ($\overline{r}_i = \frac{q_{O_i} - q_i}{24} > 0$) and losses of traffic ($\overline{q}_{S_{ik}} > 0$), an acceptable value of this time can be defined above which the losses occur. Assuming that the time allowed for closure on the trail should be such that the amount of losses is equal to the capacity reserve, trains canceled at the closure time could be handled during the rest of the day:

$$t_{Z_{ik}} \cdot \overline{q}_{S_{ik}} = (24 - t_{Z_{ik}}) \cdot \overline{r}_i \tag{10}$$

obtaining the value of admissible closure time of *k*th track of the *i*th railway section $t_{Z_{ik}}$ and depending on the train intensity and capacity reserve on the railway section:

$$t_{Z_{ik}} = 24\frac{\overline{r}_i}{\overline{q}_{S_{ik}} + \overline{r}_i}\,[\text{h}] \tag{11}$$

Substituting the patterns (6) and (8) into (11) gives (9).

For non-positive losses of traffic during closures ($\overline{q}_{S_{ik}} \leq 0$), the capacity of the railway section is not less than the scheduled intensity, i.e. allowable closure time $t_{Z_{ik}}$ per day equals 24 h.

3 Problems of Coordinating Track Closures in the Railway Network

3.1 Closure of a Pair of Railway Sections

To simplify the discussion of closure coordination the simple case of one pair of railway sections (*i*, *j*) is shown. In addition, it is assumed that the closures do not last longer than one day. The closure of railway section track means that trains should be canceled or moved to alternate paths in cases where capacity reserve is lacking. The number of trains canceled because of the closure constitutes the loss in traffic. Denoting daily losses of traffic on the *i*th railway section by s_i, and daily losses on the *j*th railway section by s_j, the total loss of traffic resulting from the closure of these trails can be written as [7]:

$$s_0 = s_i + s_j \,[\text{trains/day}] \tag{12}$$

Equation (12) is right when railway section closures are not held on the same day. Because some trains from the *i*th railway section also operate on the *j*th railway section—these trains share a pair of trails (*i*, *j*), the redirection of them from the *i*th railway section will also result in cancellation of them from the *j*th track. The advantage of this fact can be taken and the planned closure at the *i*th and *j*th railway sections can be coordinated to be conducted on the same day. Rail traffic is planned and organized. Hence, for a pair of railway sections (*i*, *j*), traffic compounds z_{ij} can be determined; the number of trains shared for railway section *i* and *j* can also be determined. Denoting the daily intensity of the *i*th railway section by z_{ii}, and the daily intensity of the *j*th railway section by z_{jj}, the probability that a randomly selected train from one railway section will be a train from the second railway section can be calculated, namely [7]:

- The probability that a randomly selected train from the *i*th railway section is a train from *j*th railway section is:

$$p_{ij} = \frac{z_{ij}}{z_{ii}} \tag{13}$$

- The probability that a randomly selected train from the *j*th railway section is a train from the *i*th railway section is:

$$p_{ji} = \frac{z_{ji}}{z_{jj}} \tag{14}$$

Closures at the *i*th and *j*th railway sections cause the occurrence of traffic losses, respectively, s_i and s_j, expressed as a number of canceled trains during the day. For railway sections with capacity reserve $\bar{r}_i \geq 0$, daily traffic losses depend on the relationship between the duration b_i of a single closure for the ith repair, and the allowable $t_{Z_{ik}}$ closure time of the kth track of the ith railway section. It should be noted that the i index refers to railway section as well as for the repair (or more precisely, for closure due to repair), because the ith repair is always at the ith railway section and causes its closure. If the duration of a single b_i closure does not exceed the admissible duration $t_{Z_{ik}}$, the losses do not occur. Exceeding the admissible duration of the closure, there are losses in the number of $(b_i - t_{Z_{ik}}) \cdot \overline{q}_{S_{ik}}$ canceled trains. On the railway section without a capacity reserve, each closure causes losses of traffic in the number of $b_i \cdot \overline{q}_{S_{ik}}$ canceled trains. Therefore, the value of the daily losses is determined as follows [2, 9]:

$$s_i = \begin{cases} 0, & for \quad b_i \leq t_{Z_{ik}} \\ (b_i - t_{Z_{ik}}) \cdot \overline{q}_{S_{ik}}, & for \quad b_i > t_{Z_{ik}} \\ b_i \cdot \overline{q}_{S_{ik}}, & for \quad \overline{r}_i \leq 0 \end{cases} \tag{15}$$

where:

$\overline{r}_i$ the average capacity reserve of the ith railway section (6),
$\overline{q}_{S_{ik}}$. the average number of canceled trains per hour of closure of the kth track of ith railway section (8)

Knowing these losses and probabilities (13) and (14) allows the calculation of expected losses associated with the closure of one railway section of the pair (i, j), for which the traffic compounds are non-zero, i.e. calculation:

- the expected number of canceled trains of the jth railway section due to the closure of ith railway section:

$$s_{ij} = p_{ij} \cdot s_i \, [\text{trains/day}] \tag{16}$$

- the expected number of canceled trains of the ith railway section due to the closure of jth railway section:

$$s_{ji} = p_{ji} \cdot s_j \, [\text{trains/day}] \tag{17}$$

Closing the ith and jth railway sections on the same day can reduce the total loss of traffic (12), considering the fact that at the load of both railway sections also comprises shared trains in the number of z_{ii} trains. The existence of z_{ij} traffic compounds also means that among the canceled s_i and s_j trains are shared railway sections (16) and (17); therefore, formula (12) will determine the overvaluation of losses of shared traffic losses, which are:

$$ws'_{ij} = \max\{s_{ij}, s_{ji}\} \, [\text{trains/day}] \tag{18}$$

Then, for a couple of trails closed on the same day, formula (12) is modified to form:

$$s' = s_i + s_j - ws'_{ij} \, [\text{trains/day}] \tag{19}$$

3.2 Example of Coordination of a Pair of Railway Sections

Track closures have been planned on two railway sections of the railway network on the basis of the adopted traffic for railway sections z_{11} and z_{22}, the train timetables set z_{12} traffic compounds and the losses caused by the lack of sufficient capacity reserve on each of the railway sections. Table 2 summarizes other data necessary to evaluate the closure coordination. The expected loss of traffic due to the two closures should be calculated t:

- in case of lack of coordination, i.e. scheduling closures on two different days;
- coordinating closures, i.e. planning them on the same day.

If the closures described in Table 2 will be scheduled on two different days, then losses in the traffic amount on the basis (12) are:

$$s_0 = s_1 + s_2 = 35 + 50 = 85 \, [\text{trains/day}]$$

Planned closures on the same day will let us use shared traffic losses (18):

$$ws'_{12} = \max\{s_{12}, s_{21}\} = \max\{21, 24\} = 24 \, [\text{trains/day}]$$

and to reduce the traffic losses to the content of (19):

$$s' = s_1 + s_2 - ws'_{12} = 35 + 50 - 24 = 61 \, [\text{trains/day}]$$

The expected shared traffic losses (18) in the amount $ws'_{12} = 24$ trains means that for trains canceled from railway section 1, $s_1 = 35$ trains, and for trains canceled from railway section 2, $s_2 = 50$ trains. Twenty-four trains will be common to both railway section, so closing railway sections on the same day is more beneficial than implementing closures on two different days.

Because the losses s_{ij} and s_{ji} are expected values calculated taking into account probabilities p_{ij} and p_{ji}, a situation may occur such that the expected shared traffic losses calculated from formula (18) will exceed traffic losses on one of the railway sections. This would mean that the number of canceled shared trains is greater than the total number of trains canceled in one of the railway sections. In this example, this situation would occur when the railway section 2 capacity reserve is smaller, and hence losses caused by its closure amount to $s_2 = 100$ trains/day. The expected traffic losses s_{21} would increase from 24 to 48 trains/day, and shared traffic losses calculated from the formula (18) amount to:

Table 2 Illustrative data describing two trails on the network

No.	Data	Description	Value	Unit
1	Intensity of railway section 1	z_{11}	200	[trains/day]
2	Intensity of railway section 2	z_{22}	250	[trains/day]
3	Number of shared trains—traffic compounds	z_{12}	120	[trains/day]
4	The probability that a randomly selected train from railway section 1 will also be a train from railway section 2 (Eq. 13)	p_{12}	0.60	[–]
5	The probability that a randomly selected train from railway section 2 will also be a train from railway section 1 (Eq. 14)	p_{21}	0.48	[–]
6	The losses caused by the closure of railway section 1—the number of trains canceled due to lack of capacity reserve	s_1	35	[trains/day]
7	The losses caused by the closure of railway section 2—the number of trains canceled due to lack of capacity reserve	s_2	50	[trains/day]
8	The expected number of trains canceled on railway section 2 due to the closure of railway section 1 (Eq. 16)	s_{12}	21	[trains/day]
9	The expected number of trains canceled on railway section 1 due to the closure of railway section 2 (Eq. 17)	s_{21}	24	[trains/day]

$$ws'_{12} = \max\{s_{12}, s_{21}\} = \max\{21, 48\} = 48 \text{ [trains/day]}$$

and will exceed $s_1 = 35$ trains/day.

To ensure that the expected shared traffic losses do not exceed the traffic losses on each of the railway sections separately, the upper limit to the formula (18) should be entered, which gives a new formula for the shared traffic losses in the form:

$$ws_{ij} = \min\{s_i, s_j, \max\{s_{ij}, s_{ji}\}\} \text{ [trains/day]} \tag{20}$$

In this way, the calculated shared traffic losses for the set example will be:

$$ws_{12} = \min\{s_1, s_2, \max\{s_{12}, s_{21}\}\} = \min\{35, 100, \max\{21, 48\}\} = 35 \text{ [trains/day]}$$

After taking into account (20), a new formula for total traffic losses for a pair of closures (i, j) can be written:

$$s = s_i + s_j - ws_{ij}\ [\text{trains/day}] \tag{21}$$

In conclusion, formulas (12)–(21) presented above concern a pair of closures (i, j) lasting no longer than one day. Coordination of closures in the network will be discussed in the following.

3.3 Closures in the Railway Network

Closures are planned during overhaul periods and last several days for a fixed number of hours per day. Therefore, additional variables describing the issue of closure coordination in the network must be entered. The aforementioned considerations and formulas describe the coordination of closures for a pair of trails (i, j). The remainder of this paper will present closure coordination in a network consisting of N railway sections.

Because trains run on more than one railway section, removing a train from a closed railway section results in its removal from other railway sections in the railroad. Using this fact, the scheduled closures can be coordinated so that they can take place simultaneously. For this purpose, it is necessary to designate traffic compounds for each pair of railway sections on the network.

Traffic compounds of railway sections have been written in the form of a square, symmetrical ($z_{ij} = z_{ji}$) matrix **Z** called the matrix of traffic compounds or, for short, the compound matrix [7, 32]:

$$\mathbf{Z} = \begin{bmatrix} z_{11} & z_{12} & \cdots & z_{1N} \\ z_{21} & z_{22} & \cdots & z_{2N} \\ \cdots & \cdots & \cdots & \cdots \\ z_{N1} & z_{N2} & \cdots & z_{NN} \end{bmatrix}, \tag{22}$$

where:

N is the number of railway sections in the railway network;
z_{ij} is the number of shared trains on i and j railway sections;
z_{ii} is the number of trains on the railway section i—daily train intensity.

For each pair of railway sections (i, j) in the railway network the following can be designated:

- probabilities p_{ij} and p_{ji} using formulas (13) and (14);
- expected losses s_{ij} and s_{ji} from formulas (16) and (17);
- shared traffic losses ws_{ij} from formula (18).

Using formula (21), it can be written that for n railway sections closed simultaneously, total daily traffic losses are as follows:

$$S' = \sum_{i=1}^{n} s_i - \sum_{(i,j)\in P} ws_{ij}, \tag{23}$$

where:

$$P = \{(i,j) : i = 1, 2, \ldots, n-1,\ \ j = 2, \ldots, n; \wedge i < j \wedge s_i > 0 \wedge s_j > 0\}$$

is a collection of ordered index pairs for congested railway sections.

Allocation of closures on the network during the repair period describes the matrix of decision variables [7, 32]:

$$\mathbf{X} = \begin{bmatrix} x_{11} & x_{12} & \ldots & x_{1T} \\ x_{21} & x_{22} & \ldots & x_{2T} \\ \ldots & \ldots & \ldots & \ldots \\ x_{n1} & x_{n2} & \ldots & x_{nT} \end{bmatrix}, \tag{24}$$

whose x_{it} element refers to the ith railway section in the tth day of the repair period T. The matrix elements assume two values: $x_{it} = 0$ when the ith railway section is opened on the tth day of the repair period or $x_{it} = 1$ if the railway section is closed. Rows of the matrix describe the individual railway section on which there are closures and the columns are the subsequent days of the repair period T.

The deployment of closures was subjected to the following restrictions:

(a) Limitation in the continuity of closures to ensure that closure of the ith railway section, for which d_i days were provided, will be implemented in the coming days from the date $t = t_i$:

$$\sum_{t=1}^{t_i-1} x_{it} = 0, \quad \sum_{t=t_i}^{t_i-1+d_i} x_{it} = d_i, \quad \sum_{t=_i+d_i}^{T} x_{it} = 0, \tag{25}$$

This limitation must always be formulated.

(b) The closure order to ensure that at the end of the closure, i starts the closure j:

$$x_{it} = 1 \wedge x_{it+1} = 0 \Rightarrow x_{jt} = 0 \wedge x_{jt+1} = 1 \tag{26}$$

This limitation results from the needs of the rational use of the repair potential.

(c) Limitation of closure exclusion:

$$x_{kt} = 1 \Rightarrow \sum_{i \in M_k} x_{it} = 0 \tag{27}$$

is specified for a repairs system with the indexes $M_k \subset \{1, \ldots, n\}$, and means that for repair of this set on any given day, only one repair can be carried out. This is due to the repairs, for which alternate paths are being determined, and this means that the railway section and railway sections of alternate paths should not be closed.

Taking into account the entire period of repair, it can be written that the allocation of closure **X** is permitted under the conditions a, b and c. The allocation of **X** is optimal if the expected traffic losses determined by the formula

$$S = \sum_{t=1}^{T} \left(\sum_{i=1}^{n} s_i \cdot x_{it} - \sum_{(i,j) \in P} x_{it} \cdot x_{jt} \cdot ws_{ij} \right) \tag{28}$$

are minimal. Formula (28) can be transformed to [7]:

$$S = \sum_{t=1}^{T} \sum_{i=1}^{n} s_i \cdot x_{it} - \sum_{t=1}^{T} \sum_{(i,j) \in P} x_{it} \cdot x_{jt} \cdot ws_{ij} \tag{29}$$

For a given repair plan, i.e. a specific list of n repairs (closures), a fixed number of closure days d_i for every repair and established traffic losses s_i a first component of the formula (29) is constant:

$$\sum_{t=1}^{T} \sum_{i=1}^{n} s_i \cdot x_{it} = \text{const} \tag{30}$$

Accordingly, the loss minimization S is equivalent to maximizing the value of the second component of formula (29)—shared traffic losses for simultaneous closures called by the indicator of the quality of track closure coordination K [7, 10]:

$$K = \sum_{t=1}^{T} \sum_{(i,j) \in P} x_{it} \cdot x_{jt} \cdot ws_{ij} \tag{31}$$

3.4 The Formulation of the Closure Coordination Problem

In summary, the problem of closure coordination [7] can be written as follows. At the repair planning stage, the number and duration of each closure was determined, i.e. the *i*th closure is described as:

b_i duration of *i*th closure during the day [h];
d the numbers of days for *i*th closure of railway section;
tdp_i allowable day of the beginning of the *i*th closure;
tdk_i allowable day of the end of the *i*th closure.

The set of *n* closures forms the input data.
Similarly, the *i*th railway section is described as:

q_i scheduled traffic intensity [trains/day];
q_{O_i} optimal traffic intensity under normal conditions [trains/day];
$q_{Z_{ik}}$ optimal traffic intensity during the *k*th track of the *i*th railway section closure (for one track railway section, $q_{Z_{ik}} = 0$) [trains/day].

There is a given compound matrix **Z**:

$$\mathbf{Z} = \begin{bmatrix} z_{11} & z_{12} & \cdots & z_{1N} \\ z_{21} & z_{22} & \cdots & z_{2N} \\ \cdots & \cdots & \cdots & \cdots \\ z_{N1} & z_{N2} & \cdots & z_{NN} \end{bmatrix}, \tag{32}$$

where *N* is a number of all railway sections in the railway network.
There is a demand for closure designation **X**:

$$\mathbf{X} = \begin{bmatrix} x_{11} & x_{12} & \cdots & x_{1T} \\ x_{21} & x_{22} & \cdots & x_{2T} \\ \cdots & \cdots & \cdots & \cdots \\ x_{n1} & x_{n2} & \cdots & x_{nT} \end{bmatrix}, \tag{33}$$

which complies with the limits:

$$\sum_{t=1}^{t_i-1} x_{it} = 0, \quad \sum_{t=t_i}^{t_i-1+d_i} x_{it} = d_i, \quad \sum_{t=_i+d_i}^{T} x_{it} = 0 \tag{34}$$

$$x_{it} = 1 \wedge x_{it+1} = 0 \Rightarrow x_{jt} = 0 \wedge x_{jt+1} = 1 \tag{35}$$

$$\sum_{i \in M_k} x_{it} \leq 1 \tag{36}$$

and maximizes the function criterion **K**:

$$K = \sum_{t=1}^{T} \sum_{(i,j)\in P} x_{it} \cdot x_{jt} \cdot ws_{ij} \tag{37}$$

The size and complexity of the issue is such that not all available optimization methods may be used. The product of design variables in the K function makes the issue non-linear and prevents the use of well-known and well-designed algorithms for linear optimization. Heuristic methods are thus advised, which do not guarantee finding the optimal solution, but a solution close to optimal.

4 Example of Track Closure Coordination on the Railway Network

4.1 The Input Data

Figure 5 illustrates an example of a railway network consisting of six double-track trails: 1 (A–B), 2 (B–C), 3 (C–D), 4 (E–B), 5 (B–G) and 6 (C–F).

Table 3 summarizes traffic flows in a rail network described by the relations, trails, scheduled intensity and scheduled load of individual trails [trains/day]. Moreover, Table 4 summarizes the compound matrix (22) between trails. Its elements are the number of common railroads for each trails pair. For example, $z_{12} =$ 80 trains/day represents the number of trains the railroad runs through rails 1 and 2. It consists of 20 trains of relation ⟨A,B,C,D⟩, 20 trains of relation ⟨D,C,B,A⟩, 20 trains of relation ⟨A,B,C,F⟩ and 20 trains of relation ⟨F,C,B,A⟩. Elements on the main diagonal are a daily burden on individual trails. Using the compound matrix, p_{ij} (13) can be determined—probability that a randomly selected train will be from the trail load and will also be a train moving on the trail j.

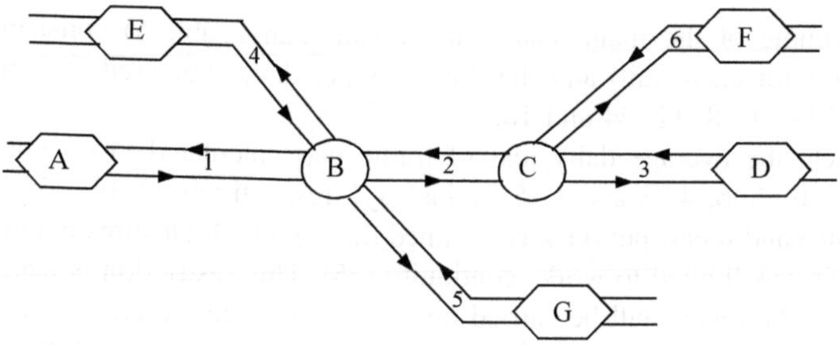

Fig. 5 A sample portion of a railway network consisting of five stations: A, D, E, F and G, and two railway junctions B and C

Table 3 Traffic flows in the railway network and load of each trail

No.	Relation	Trails	Schedule intensity of the relation [trains/day]	Trail	Intensity of trains on trails [trains/day]
1	⟨A,B,C,D⟩	1, 2, 3	20	1 (A–B)	40 + 40 + 20 = 100
2	⟨D,C,B,A⟩	3, 2, 1	20	2 (B–C)	40 + 40 + 50 + 60 = 190
3	⟨A,B,C,F⟩	1, 2, 6	20	3 (C–D)	40 + 50 = 90
4	⟨F,C,B,A⟩	6, 2, 1	20	4 (E–B)	70 + 50 + 60 = 180
5	⟨A,B,G⟩	1, 5	10	5 (B–G)	20 + 70 = 90
6	⟨G,B,A⟩	5, 1	10	6 (C–F)	40 + 60 = 100
7	⟨E,B,G⟩	4, 5	35	–	
8	⟨G,B,E⟩	5, 4	35	–	
9	⟨E,B,C,D⟩	4, 2, 3	25	–	
10	⟨D,C,B,E⟩	3, 2, 4	25	–	
11	⟨E,B,C,F⟩	4, 2, 6	30	–	
12	⟨F,C,B,E⟩	6, 2, 4	30	–	

Source Based on [7]

For example, the probabilities that a randomly chosen train from the third trail load will be a train moving on trails 1, 2 and 4 are as follows:

$$p_{31} = \frac{z_{31}}{z_{33}} = 0.44, \quad p_{32} = \frac{z_{32}}{z_{33}} = 1, \quad p_{34} = \frac{z_{34}}{z_{33}} = 0.55.$$

The certain event, that the train running on the third trail is also the train which runs on the second trail results from the timetable selected in this example in which all trains to and from the D station run by the second trail, i.e. there are no trains running only between stations C and D (Table 3, there are no relations ⟨C,D⟩ or ⟨D,C⟩). If such a relation was provided in the timetable, the load of the third trail would be higher than the current $z_{33} = 90$ trains/day by the number of additional trains, with the number of shared trails for the second and third trails remaining at the same level of $z_{32} = 90$ trains/day. The probability $p_{32} = \frac{z_{32}}{z_{33}} < 1$, and the value $1 - \frac{z_{32}}{z_{33}}$ would then be only trains running between C and D stations at the total third load (CD).

An example of the maintenance and repair plan (Table 5) consisting of six repairs (one for each trail) and the durations per day, expressed in hours, are in sequence: 24, 10, 8, 12, 24 and 16.

For each, the average daily loss of traffic was calculated (15), amounting to 11.25, 1.5, 0, 2.66, 4.98 and 1.25 [trains/day], respectively (Table 6). A 200-day maintenance and repair period was assumed during which closures could be made. Each repair was limited to works continuity (25). This restriction is necessary and ensures that the repair will be carried out during consecutive days of closure.

The limitation of work order (26), ensuring that after the completion of the *i*th repair, the *j*th repair begins, is not necessary in this repair plan because all repairs

Table 4 Compound matrix Z: the number of shared trains on trails [trains/day]

z_{ij} [trains/day]	1 (A–B)	2 (B–C)	3 (C–D)	4 (E–B)	5 (B–G)	6 (C–F)
1 (A–B)	100	40 + 40 = 80	40	0	20	40
2 (B–C)	40 + 40 = 80	190	40 + 50 = 90	50 + 30 = 80	0	40 + 60 = 100
3 (C–D)	40	40 + 50 = 90	90	50	0	0
4 (E–B)	0	50 + 30 = 80	50	180	70	60
5 (B–G)	20	0	0	70	90	0
6 (C–F)	40	40 + 60 = 100	0	60	0	100

Source Based on [7]

Table 5 The maintenance and repair plan

Number of repairs	Acceptable date of the repair start t_{dp}	Date of the repair start t_p	Number of repair days d	Date of the repair end t_k	Acceptable date of the repair end t_{dk}	Repair duration during a day [h] b
1	1	1	21	21	200	24
2	1	1	15	15	200	10
3	1	1	7	7	200	8
4	1	1	15	15	200	12
5	1	1	10	10	200	24
6	1	1	8	8	200	16

Source Based on [7]

Table 6 Assessment of traffic losses for the example network

No.	Parameter	Symbol or formula	Trails					
			1 (A–B)	2 (B–C)	3 (C–D)	4 (E–B)	5 (B–G)	6 (C–F)
Intensity of trains								
1	Scheduled intensity of trains	q_i [trains/day]	100	190	90	180	90	100
2	Optimal intensity without closures, both tracks active	q_{O_i} [trains/day]	150	250	150	250	150	150
3	Optimal intensity at a closed track 1	$q_{Z_{i1}}$ [trains/day]	70	100	70	100	70	70
4	Optimal intensity at a closed track 2	$q_{Z_{i2}}$ [trains/day]	70	100	70	100	70	70
Average values per hour								
5	The number of scheduled trains	$\overline{q_i} = \frac{q_i}{24}$ [trains/h]	4.17	7.92	3.75	7.50	3.75	4.17
6	Capacity reserve	$\overline{r_i} = \frac{q_{O_i} - q_i}{24}$ [trains/h]	2.08	2.50	2.50	2.92	2.50	2.08
7	Train intensity at a closed track 1	$\overline{q_{Z_{i1}}} = \frac{q_{Z_{i1}}}{24}$ [trains/h]	2.92	4.17	2.92	4.17	2.92	2.92

(continued)

Table 6 (continued)

No.	Parameter	Symbol or formula	Trails					
			1 (A–B)	2 (B–C)	3 (C–D)	4 (E–B)	5 (B–G)	6 (C–F)
8	Train intensity at a closed track 2	$\overline{q}_{Z_{i2}} = \frac{q_{Z_{i2}}}{24}$ [trains/h]	2.92	4.17	2.92	4.17	2.92	2.92
9	The number of trains canceled due to the closure of track 1	$\overline{q}_{S_{i1}} = \frac{q_i - q_{Z_{i1}}}{24}$ [trains/h]	1.25	3.75	0.83	3.33	0.83	1.25
10	The number of trains canceled due to the closure of track 2	$\overline{q}_{S_{i2}} = \frac{q_i - q_{Z_{i2}}}{24}$ [trains/h]	1.25	3.75	0.83	3.33	0.83	1.25
Acceptable times								
11	Acceptable time of track 1 closure	$t_{Z_{i1}}$ [h]	15.0	9.6	18.0	11.2	18.0	15.0
12	Acceptable time of track 2 closure	$t_{Z_{i2}}$ [h]	15.0	9.6	18.0	11.2	18.0	15.0
13	Duration of a single closure	b_i [h]	24	10	8	12	24	16
Daily traffic losses								
14	Traffic losses during the closure of track 1	s_i [trains/h]	11.25	1.50	0	2.66	4.98	1.25
15	Traffic losses during the closure of track 2	s_i [trains/h]	11.25	1.50	0	2.66	4.98	1.25

Source Based on [7]

can be carried out simultaneously. In the following section, an example of a maintenance and repair plan will be provided which, for the rational use of the repair potential, such a restriction was imposed. Because none of trails is being used as an alternate line for any of the closures, none of the repairs has an exclusion of work limitation (27).

Coordination of closures will be presented developed within the framework of the supporting program of track closure coordination KZT [8, 9], for which the main screen is shown in Fig. 6. At the top right of the screen, there is an array with

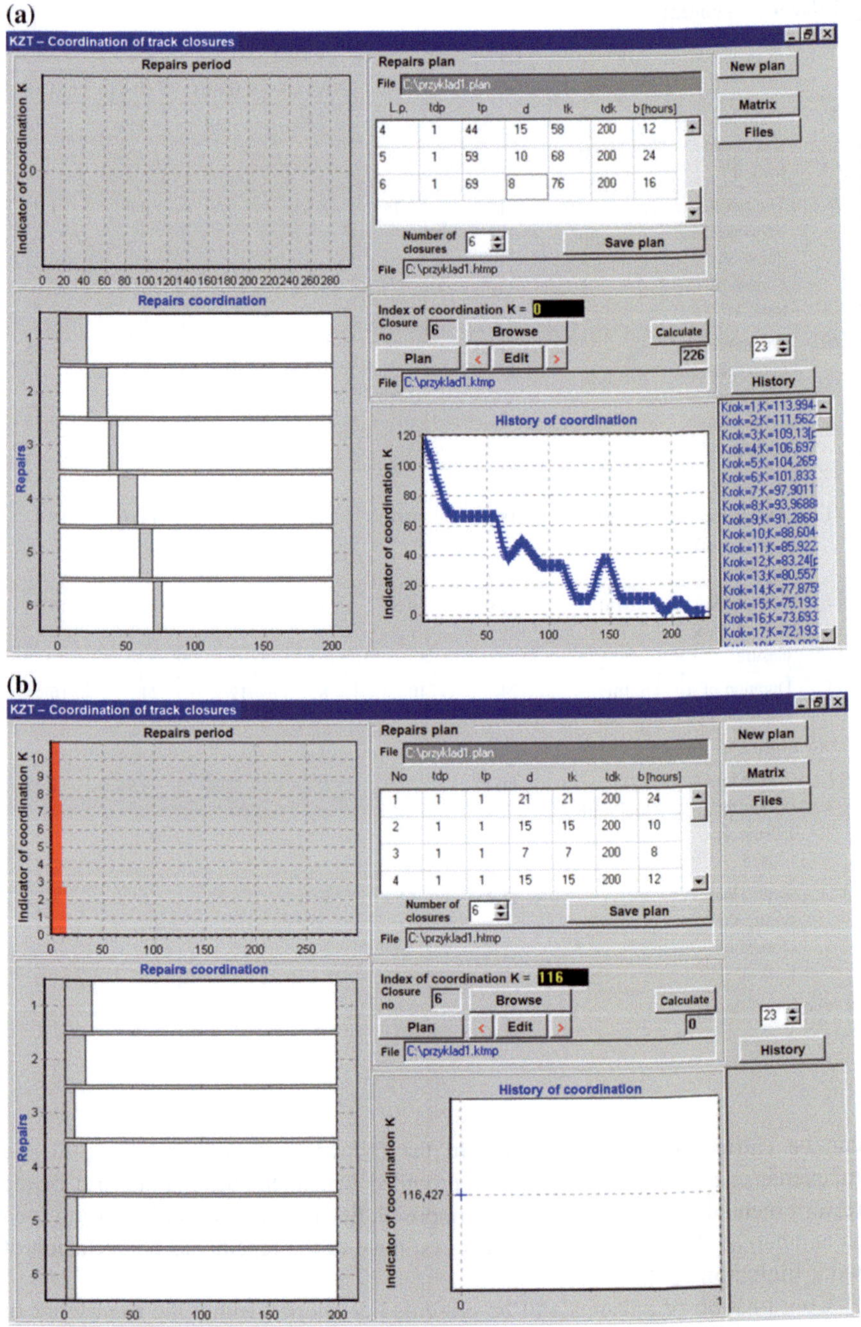

Fig. 6 Distribution of closures: **a** not simultaneously scheduled, lack of coordination, $K = 0$, **b** optimal, maximizes the common losses of traffic at $K = 116$

data describing the repairs; in the bottom left, is the deployment of closures in the repair period in the form of Gantt graphs.

The initial distribution of closures is presented in Fig. 3. Each starts in the first allowable date specified in the maintenance and repair plan. In this case, since the closures do not have order restrictions imposed, they have all been set to the first day of the repair period. For such a distribution rate of coordination (37), the criterion function is $K = 116$ trains. Its maximization corresponds to minimization of the expected traffic losses caused by closures (29).

4.2 Loss Analysis

On the basis of calculations listed in Table 6, it can be said that with both tracks of railway section opened, on every trail, there is capacity reserve amounting to:

$$q_{O_i} - q_i > 0 \tag{38}$$

The closure of one railway track section causes a decrease in the optimal intensity from the value of q_{O_i} [trains/day] to value $q_{Z_{i1}}$ or $q_{Z_{i2}}$ (depending on the number of tracks being closed) caused by the two-way traffic on one active track of the railway section, which limits the scheduled intensity to the value $q_{Z_{i1}}$ or $q_{Z_{i2}}$. The calculation of the average number of canceled trains per hour $\overline{q}_{S_{i1}}$ and $\overline{q}_{S_{i2}}$ and knowledge of the duration of the repairs during the day b_i allows one to set a daily traffic loss (15), the number of trains that should be canceled or moved to alternate lines. Time b_i has a significant impact on potential losses s_i, because when there is capacity reserve (37), then $t_{Z_{ik}}$ can be determined (9), the allowable duration of the closure during the day that avoids loss s_i. Time $t_{Z_{ik}}$ should be able to use the existing capacity reserve to handle canceled trains due to the one of the tracks being closed.

The closure that generates the greatest traffic losses is the closure related to repair 1. This follows from the assumed duration of the closure during the day equal to 24 h, while permissible closure time of the track on the trail 1 (A–B) $t_{Z_{1k}}$ specified in Table 6 is 15 h. Exceeding this time by 9 h, with average losses per hour of closure amounting to 1.25 trains/h, causes a loss of 11.25 trains/day. Closure of zero losses is the closure associated with repair 3. As seen in Table 6, the permissible time for track closure on the third trail (C–D) of 18 h has not been exceeded because the repair 3 takes 8 h during each of the intended 7 days. Of the remaining trails, s_i trains should be appealed, as in the example, alternate lines in the railway network do not exist for them. Relieving the specified trail for a s_i train will simultaneously relieve these trails with which the trail has nonzero traffic compounds (22). At the same time, trails going through the alternate line and acquiring additional load relieves other trails by their traffic compounds. It can be said that the closures which have shared traffic losses during closures attract (should be planned simultaneously), and closures with shared alternate lines (shared additional traffic) repel (should be scheduled on other days).

Closures 1 and 2 last one day. Losses of railway traffic which will then occur will depend on the method of closure planning:

(a) When the closures are planned for two different days, they will be implemented at different times.
(b) When the closures are planned for the same day, they will be coordinated and simultaneous.

In case (a), when closures at the first and second trails are held at different times, i.e. each on a different day, the total losses amount to:

$$s_0 = s_1 + s_2 = 11.25 + 1.5 = 12.75 \text{ trains}$$

Thus the number of trains expected to be canceled due to the closure is 11.25 on the first day, and 1.5 on the second day, giving a total of 12.75 trains.

Considering the case of (b) and using traffic compounds, the probability can be determined via (13) and (14):

$p_{12} = \frac{z_{12}}{z_{11}} = 0.8$ and $p_{21} = \frac{z_{12}}{z_{22}} = 0.42$; then, the expected number of trains will be reduced by the load on the second track because of the closure on the first trail (16):

$$s_{12} = p_{12} \cdot s_1 = 9 \text{ trains},$$

and the expected number of trains will be reduced by the load on first trail due to the closure on the second trail (17):

$$s_{21} = p_{21} \cdot s_2 = 0.63 \text{ train}$$

It can be said that on the first day, among an average of 11.25 trains canceled from the first trail, 9 trains also ply on the second trail. On the second day, among an average of 1.5 trains canceled from the second trail, an average of 0.63 train was also canceled from the first track. Thus, when the closure will be scheduled on the same day, it is possible to use the shared lost traffic (19):

$$ws_{12} = \min\{11.25, 1.5, \max\{9, 0.63\}\} = 1.5 \text{ trains}$$

and reduce the total traffic losses by canceled trains plying simultaneously on both trails (20):

$$s = s_1 + s_2 - ws_{12} = 11.25 + 1.5 - 1.5 = 11.25 \text{ trains}$$

In other words, the difference between the losses in the case of non-simultaneous and simultaneous closures is 12.75 − 11.25 = 1.5 trains; this should it be interpreted as the expected number of trains that need to be further appealed for non-simultaneous closures.

4.3 Indicator of Coordination

Mentioned losses relate to only one common day of first and second closures. For a full number of days of closures amounting to $d_1 = 21$ and $d_2 = 15$, respectively, we get:

(a) In the case of uncoordinated closures, because the coordination indicator (31):

$$K = \sum_{t=1}^{200} \sum_{(1,2) \in P} x_{1t} \cdot x_{2t} \cdot ws_{12} = 0 \text{ common trains,}$$

together, the traffic losses are (29):

$$S = \sum_{t=1}^{200} \sum_{i=1}^{2} s_i \cdot x_{it} - 0 = 258.75 \text{ trains,}$$

(b) In the case of coordinated closures (31):

$$K = \sum_{t=1}^{200} \sum_{(1,2) \in P} x_{1t} \cdot x_{2t} \cdot ws_{12} = 22.5 \text{ common trains;}$$

Together, the traffic losses are (29):

$$S = 258.75 - 22.5 = 236.25 \text{ trains.}$$

4.4 The Optimal Distribution

The essence of the optimal distribution of closures is to minimize the expected total traffic losses, which for the entire example plan consists of six repairs, calculated from formula (29):

$$S = \sum_{t=1}^{200} \sum_{i=1}^{6} s_i \cdot x_{it} - \sum_{t=1}^{200} \sum_{(i,j) \in P} x_{it} \cdot x_{jt} \cdot ws_{ij}$$

Part $\sum_{t=1}^{200} \sum_{i=1}^{6} s_i \cdot x_{it} = 358.45$ [of trains] has a constant value, independent of the closure distribution during repair periods (it is also the value of losses in the lack of coordination); thus, minimizing the expected traffic losses corresponds to maximizing the expected shared traffic losses, a coordination indicator (31):

$$K = \sum_{t=1}^{200} \sum_{(i,j) \in P} x_{it} \cdot x_{jt} \cdot ws_{ij}$$

Figure 6a shows the situation in which no closures have a common repair day. Hence, the coordination indicator K is zero. Closure distribution during the repairs such that work could began on the first day of the repair period (Fig. 6b) causes change of the coordination indicator value from zero to 116 trains, which is the expected shared lost traffic due to the closures. This is the maximum that can be achieved with that maintenance and repair plan.

In addition, due to the high degree of flexibility in the closure distribution (no order and simultaneity restrictions), there are several optimal solutions that maximize the K indicator. The total losses caused by such closure coordination amounts to:

$$S = \sum_{t=1}^{200} \sum_{i=1}^{6} s_i \cdot x_{it} - \sum_{t=1}^{200} \sum_{(i,j) \in P} x_{it} \cdot x_{jt} \cdot ws_{ij} = 358.45 - 116 = 242.45 \text{ trains}$$

Losses due to lack of coordination are reduced by 32.4% when optimal coordination is employed. When analyzing each pair of closures ws_{ij} the number of common losses (20) is used to determine which closures should be performed simultaneously, i.e. in the greatest possible number of common losses. For an example of the maintenance and repair plan, in descending order are: ws_{45} = 2.66 [trains/day], ws_{15} = 2.25 [trains/day], ws_{12} = 1.50 [trains/day], ws_{26} = 1.25 [trains/day], ws_{16} = 1.25 [trains/day], ws_{24} = 1.18 [trains/day] and ws_{46} = 0.89 [trains/day].

Additionally, it can be noted:

- the third closure has no effect on coordination, because it generates zero losses daily (s_3 = 0) and ws_{3j} = 0 for j = 1.6; therefore, it can be scheduled on any day of the repair period;
- pairs of closures (1, 4), (2, 5), (5, 6) have zero common losses (ws_{14} = 0, ws_{25} = 0 and ws_{56} = 0, respectively) because the traffic compounds of each pair are also zero.

Considering the above, for optimal distribution (with a maximum K indicator), it should be taken into account that the closure of non-zero shared traffic losses should overlap in as many days of the repair period as possible. In contrast, pairs of closures with zero shared losses can be set arbitrarily relative to each other.

4.5 Restrictions on the Works Order

Assuming the previously discussed maintenance and repair plan consists of only the first four repairs, and that there is a restriction applied to the order of repairs (26), the fourth repair can take place only after the end of the second, i.e.:

$$x_{2t} = 1 \wedge x_{2t+1} = 0 \Rightarrow x_{4t} = 0 \wedge x_{4t+1} = 1$$

Hence, for the fourth repair, a repair period of t_{dp} = 16 days is obtained.

Figure 7 shows the arrangement of the second and fourth closures, which cannot be covered, although to a preferred value ws_{24} = 1.18 trains/day, because a restricted order exists for them: 2–4. Setting the third closure has no effect on the coordination indicator, because s_3 = 0 trains/day. There remains only the question of the distribution of the first closure with respect to the second and fourth closures. Considering the fact that ws_{14} = 0 trains/day and ws_{12} = 1.50 trains/day, one of the optimal settings is shown in Fig. 8a; the number of days of first and second shared closures is 15, which after multiplying by ws_{12} = 1.5 rains/day, gives K = 22.5 trains of shared lost traffic. Thus, the optimal distribution for these four closures is reduced to mutual alignment of the first and second closures as to their mutual maximum coverage amounting to 15 days. In calculating the constant of losses caused by closures (30), we get:

$$\sum_{t=1}^{200} \sum_{i=1}^{4} s_i \cdot x_{it} = 298.65 \text{ trains.}$$

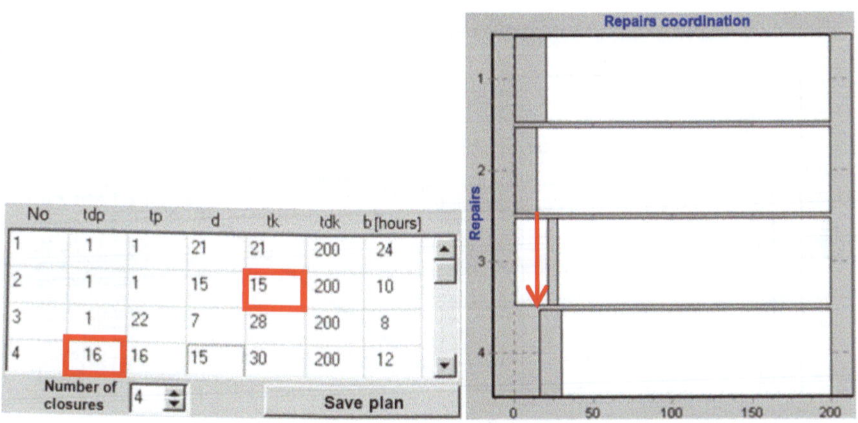

Fig. 7 Closures schedule with order restrictions imposed on the second and fourth repairs

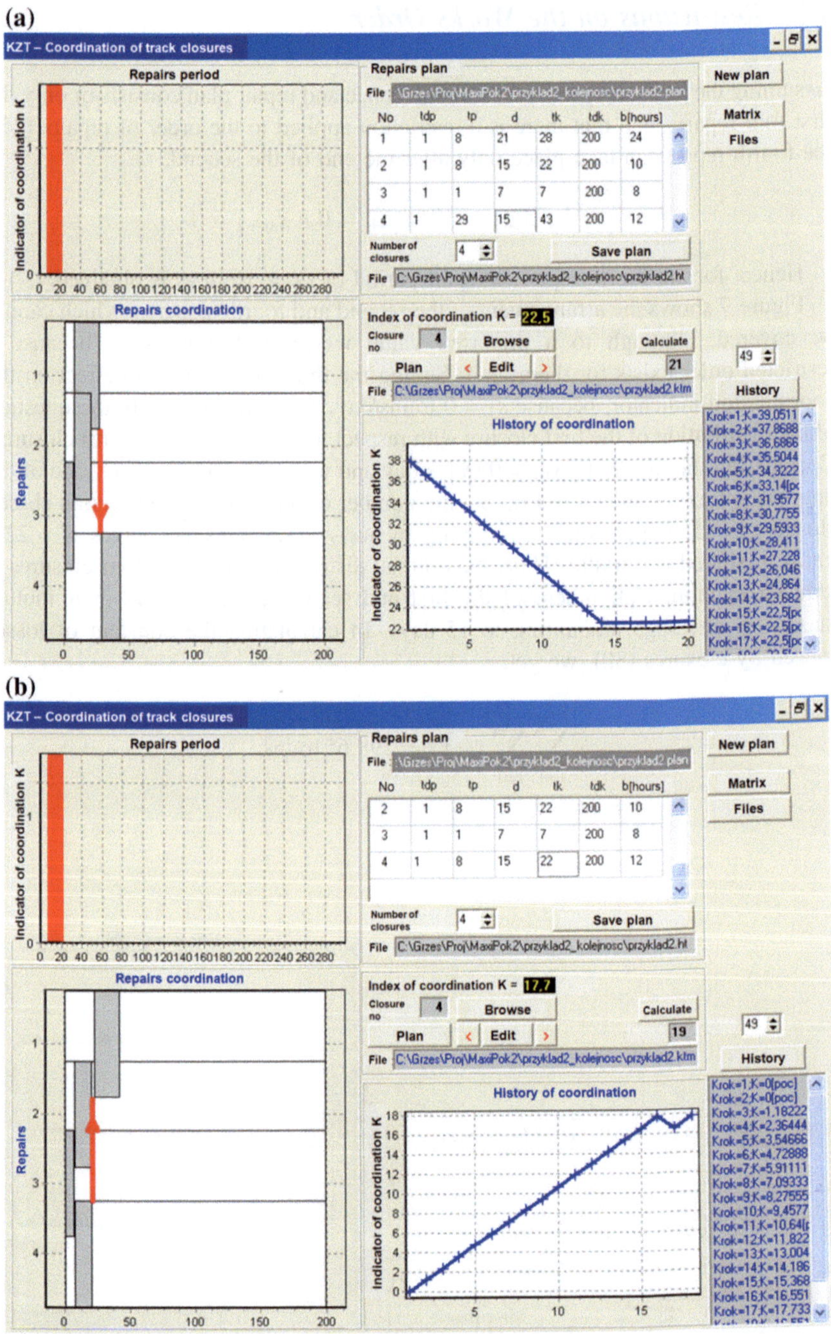

Fig. 8 Two variants of the optimal distribution closures: **a** the coordination of closures with regard to the order of 1–4, **b** the coordination of closures with regard to the order of 4–1

The total expected traffic losses (29) taking into account the coordination indicator are:

$$S = \sum_{t=1}^{200}\sum_{i=1}^{4} s_i \cdot x_{it} - \sum_{t=1}^{200}\sum_{(i,j)\in P} x_{it} \cdot x_{jt} \cdot ws_{ij} = 298.65 - 22.5 = 276.15\ \text{trains}$$

The K indicator also allows the study of the influence of different sequence variants of works at the expected traffic losses.

Example

Assumptions:

(1) due to the restrictions of renovation potential, the repair order 3–2 was imposed.
(2) due to the renovation potential, the first and fourth repair cannot take place simultaneously.

Calculated shared traffic losses (20) for pairs of closures are: $ws_{12} = 1.50$ trains/day, $ws_{13} = 0$ trains/day, $ws_{14} = 0$ trains/day, $ws_{24} = 1.18$ trains/day.

The preliminary analysis of shared traffic losses indicates that the closures (1, 3) have no common losses, because $s_1 = 0$ trains/day, and the pair of closures (1, 4) also have no common losses, because the traffic compounds are zero, i.e. $z_{14} = z_{41} = 0$ trains/day. Thus, coordination will take place between pairs (1, 2) and (2, 4), with a restriction of the order of 3–2.

Figure 8 shows two variants of the optimal distribution. The height of the Gantt charts was intentionally increased to this value for better visibility of the degree of coverage of coordinated closures. In both cases, due to the imposed order of 3–2, the first planned closure is the third closure, followed by the second closure. In variant (a), it was assumed that the order 1–4 received a coordinate indicator of $K = 22.5$ trains, and it was also assumed that the order of 4–1 received an inferior coordination variant, because in this case, $K = 17.7$ trains. When the second assumption is missed and the closures (1, 4) will be simultaneously coordinated, the indicator $K = 40.2$ trains (Fig. 9) will be received.

4.6 *Alternate Routes*

The discussed fragment of the railway network did not contain connections that could provide possible alternate routes for canceled trains during railway closures. Figure 10 shows a modified section of the network with added connections forming a potential alternate route.

Table 7 presents the traffic flows in the modified network. It is a modification of the content of Table 3 with additional relations:

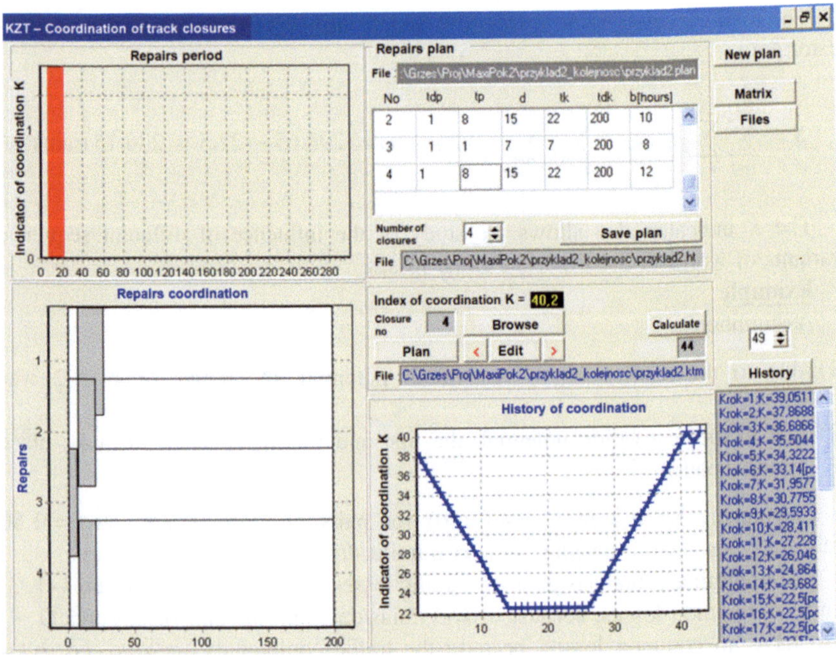

Fig. 9 Distribution of coordinated closures without the assumption about the closures exclusion (1–4)

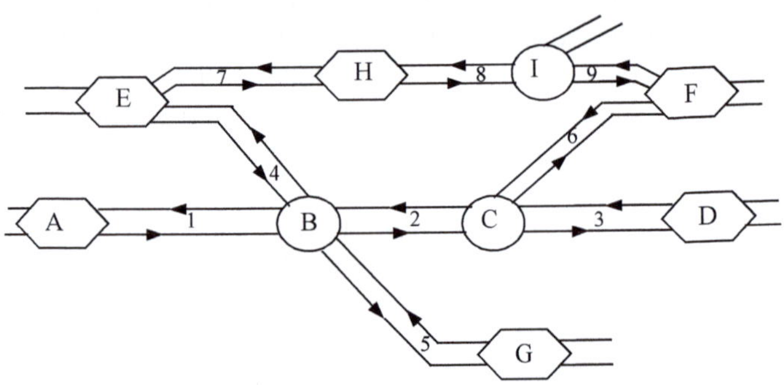

Fig. 10 The modified fragment of the railway network containing potential alternate routes

- from E to H,
- from H to E,
- from E to F,
- from F to E.

Three repairs were also introduced to the maintenance and repair plan, one for each added trail (Table 9). Conversely, Table 10 shows the characteristics of traffic

Table 7 Traffic flows in the railway network and the loads of individual trails

No.	Relation	Trail	Schedule intensity of the relation [trains/day]	Trail	Intensity on trails [trains/day]
1	⟨A,B,C,D⟩	1, 2, 3	20	1 (A–B)	40 + 40 + 20 = 100
2	⟨D,C,B,A⟩	3, 2, 1	20	2 (B–C)	40 + 40 + 50 + 60 = 190
3	⟨A,B,C,F⟩	1, 2, 6	20	3 (C–D)	40 + 50 = 90
4	⟨F, C, B, A⟩	6, 2, 1	20	4 (E–B)	70 + 50 + 60 = 180
5	⟨A, B, G⟩	1, 5	10	5 (B–G)	20 + 70 = 90
6	⟨G, A, B⟩	5, 1	10	6 (C–F)	40 + 60 = 100
7	⟨E, B, G⟩	4, 5	35	7 (E–H)	60 + 80 = 140
8	⟨G, B, E⟩	5, 4	35	8 (H–I)	80
9	⟨E, B, C, D⟩	4, 2, 3	25	9 (I–F)	80
10	⟨D, C, B, E⟩	3, 2, 4	25	–	
11	⟨E, B, C, F⟩	4, 2, 6	30	–	
12	⟨F, C, B, E⟩	6, 2, 4	30	–	
13	⟨E, H⟩	7	30	–	
14	⟨H, E⟩	7	30	–	
15	⟨E, H, I, F⟩	7, 8, 9	40	–	
16	⟨F, I, H, E⟩	9, 8, 7	40	–	

Source Based on [7]

of additional trails: 7 (E–H), 8 (H–I) and 9 (I–F). As the basis of Table 7, a new traffic compound matrix was determined (Table 8).

Trails 7, 8 and 9 may be alternate lines for the traffic transferred from trails 4, 2 and 6 and conversely, i.e. for traffic transferred from trails 7, 8 and 9 to alternate lines may be a string of trails 4, 2 and 6. There is, thus, the restriction of repair simultaneity (27) in the form:

$$x_{kt} = 1 \Rightarrow \sum_{i \in M_k} x_{it} = 0, \tag{39}$$

for the closures set: $M_2 = \{7, 8, 9\}$, $M_4 = \{7, 8, 9\}$, $M_6 = \{7, 8, 9\}$, $M_7 = \{4, 2, 6\}$, $M_8 = \{4, 2, 6\}$, $M_9 = \{4, 2, 6\}$, where k is the index of the repair, which was imposed by the simultaneity restriction.

For example, during closures on the second trail, on all trails from the set $M_2 = \{7, 8, 9\}$, closures should not be planned because it is an alternate line. Therefore, the t_{dp} and t_{dk} (Table 9) for items from the M_k sets will depend on t_p and t_k of the kth closure.

Table 8 Traffic compound matrix—the number of common railroads on trails

z_{ij} [trains/day]	1 (A–B)	2 (B–C)	3 (C–D)	4 (E–B)	5 (B–G)	6 (C–F)	7 (E–H)	8 (H–I)	9 (I–F)
1 (A–B)	100	40 + 40 = 80	40	0	20	40	0	0	0
2 (B–C)	40 + 40 = 80	190	40 + 50 = 90	50 + 30 = 80	0	40 + 60 = 100	0	0	0
3 (C–D)	40	40 + 50 = 90	90	50	0	0	0	0	0
4 (E–B)	0	50 + 30 = 80	50	180	70	60	0	0	0
5 (B–G)	20	0	0	70	90	0	0	0	0
6 (C–F)	40	40 + 60 = 100	0	60	0	100	0	0	0
7 (E–H)	0	0	0	0	0	0	140	80	80
8 (H–I)	0	0	0	0	0	0	80	80	80
9 (I–F)	0	0	0	0	0	0	80	80	80

Source Based on [7]

Table 9 The maintenance and repair plan

Repair No.	Acceptable date of the repair start t_{dp}	Date of the repair start t_p	The number of repair days d	Date of the repair end t_k	Acceptable date of the repair end t_{dk}	Repair duration during a day [h] b
1	1	1	21	21	200	24
2	1	1	15	15	200	10
3	1	1	7	7	200	8
4	1	1	15	15	200	12
5	1	1	10	10	200	24
6	1	1	8	8	200	16
7	1	1	7	7	200	24
8	1	1	10	10	200	24
9	1	1	5	5	200	24

Source Based on [7]

Transferring the traffic at alternate lines for closures, during which $s_i > 0$ losses occurs, requires the description of new traffic capacity conditions for alternate line trails with additional traffic:

$$q_{O_i} - q' \geq 0 \tag{40}$$

$$q_{Z_{ik}} - q' \geq 0, \tag{41}$$

where:

$$q' = q + s_i \tag{42}$$

Two cases of trails with capacity reserve can be distinguished:

1. The trail has capacity reserve only for both active tracks of railway sections, in this example trails 7, 8 and 9.
2. The trail has capacity reserve only for both active tracks of railway sections, and when one of the tracks is closed; this case does not occur in this example.

The condition for the transfer of additional traffic on the trail without sacrificing the flow of traffic is not to exceed the optimal intensity q_{O_i}. If condition (40) is fulfilled, then whether the intensity will be exceeded during the closure of one of the tracks $q_{Z_{1k}}$ or $q_{Z_{2k}}$ should be checked.

If the reserve capacity (41) is maintained, losses will be not generated; the trail then enters an alternate line, and there is also no imposed restriction of simultaneity of works with the loaded trail. Otherwise, losses will occur on the trail, caused by overloading and closure planning, and the restriction of simultaneity shall be imposed on the trail being loaded.

Table 10 Assessment of traffic losses of additional three trails of example railway network

No.	Parameter	Symbol or formula	Trails		
			7 (E–H)	8 (H–I)	9 (I–F)
Intensity of trains					
1	Scheduled intensity of trains	q_i [trains/day]	140	80	80
2	Optimal intensity without closures, both tracks active	qo_i [trains/day]	170	100	100
3	Optimal intensity at a closed track 1	qz_{i1} [trains/day]	130	70	70
4	Optimal intensity at a closed track 2	qz_{i2} [trains/day]	130	70	70
Average values per hour					
5	The number of scheduled trains	$\overline{q_i} = \frac{q_i}{24}$ [trains/h]	5.83	3.33	3.33
6	Reserve of the capacity	$\overline{r_i} = \frac{qo_i - q_i}{24}$ [trains/h]	1.25	0.83	0.83
7	Trains intensity at a closed track 1	$\overline{q}z_{i1} = \frac{qz_{i1}}{24}$ [trains/h]	5.42	2.92	2.92
8	Trains intensity at a closed track 2	$\overline{q}z_{i2} = \frac{qz_{i2}}{24}$ [trains/h]	5.42	2.92	2.92
9	The number of trains canceled due to the closure of track 1	$\overline{q}s_{i1} = \frac{q_i - qz_{i1}}{24}$ [trains/h]	0.42	0.42	0.42
10	The number of trains canceled due to the closure of track 2	$\overline{q}s_{i2} = \frac{q_i - qz_{i2}}{24}$ [trains/h]	0.42	0.42	0.42
Acceptable times					
11	Acceptable time of track 1 closure	$t_{Z_{i1}}$ [h]	18	16	16
12	Acceptable time of track 2 closure	$t_{Z_{i2}}$ [h]	18	16	16
13	Duration of a single closure	b_i [h]	24	24	24
Daily traffic losses					
14	Traffic losses during the closure of track 1	s_i [trains/h]	2.52	3.36	3.36
15	Traffic losses during the closure of track 2	s_i [trains/h]	2.52	3.36	3.36

Source Based on [7]

Closures on trails 4, 2 and 6 generate daily traffic losses of $s_4 = 2.66$, $s_2 = 1.5$ and $s_6 = 1.25$ [trains/day], respectively (Table 6). The transfer of these trains to a potential alternate line consisting of trails 7–8–9 is permitted when there is a capacity reserve for it. From Table 10, reserves for trails 7, 8 and 9 can be determined as $r_7 = 30$, $r_8 = 20$ and $r_9 = 20$ [trains/day], respectively, with a minimum reserve equal to 20 trains/day. Thus there is a possibility of transferring canceled trains from trails 4, 2 and 6 to the alternate lines 7–8–9, provided that closures of

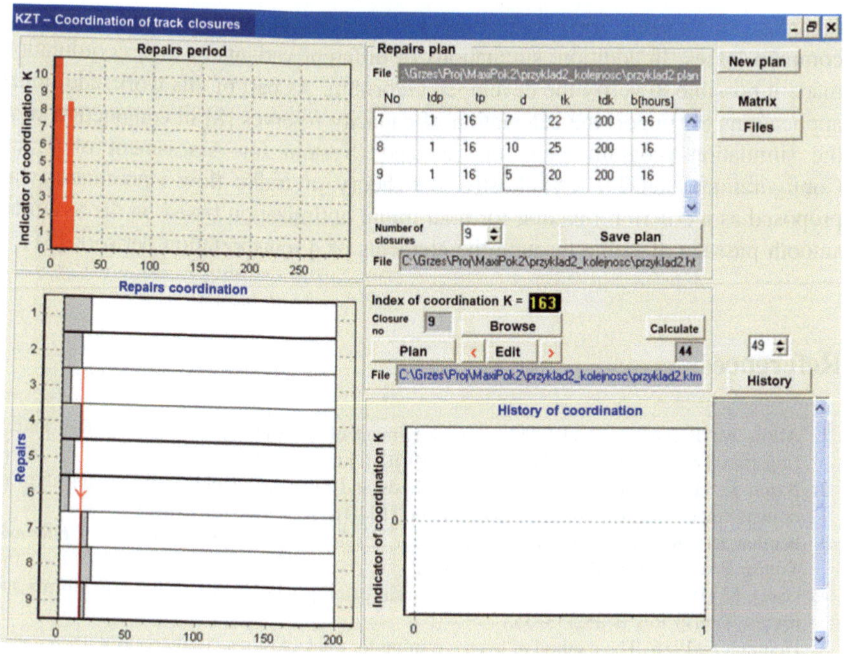

Fig. 11 Distribution of closures taking into account the alternate routes 7–8–9 for trains transferred from trails 2–4–6

these roads will be carried out separately. Figure 11 shows the schedule of coordinated maintenance and repairs, taking into account closure sets with the work simultaneity restrictions $M_2 = \{7, 8, 9\}$, $M_4 = \{7, 8, 9\}$ and $M_6 = \{7, 8, 9\}$.

5 Conclusions

This paper presents the problem of planning and coordination of closures in dense and complex transportation networks. In light of the fundamental differences in the organization and regulation of railway and road traffic, a method of coordination and a coordination algorithm were developed for the railway network. Planning of closures and their optimal distribution in such a network is a complex problem involving issues of traffic organization and traffic regulation. Effective decomposition of the problem helped to define the issue of closure coordination and to develop an algorithm to solve it [7]. The developed coordination method uses traffic compounds and the resulting common losses in traffic for connections being closed; hence, it is intended for planned traffic, as exemplified by railway traffic. In order to verify the methods and algorithm, the application was built to support the process of coordination, which made it possible to verify the validity of the assumptions regarding minimizing total traffic losses caused by closures by maximizing the common losses. In addition, simulations of different variants of their coordination made it possible to verify the developed algorithm. As part of this work, additional applications to support the calculation of capacity reserves [8] in conjunction with the simulation program were also created (System for Assessment of Track Configuration [SOUT]) [28]. Lastly, the theory of traffic flow smoothness was proposed as a criterion function for road traffic optimization based on the assumed smooth passage of traffic through the elements of a transportation network [31].

References

1. Abrila M, Barbera F et al (2008) An assessment of railway capacity. Transp Res Part E Logistics Transp Rev 44(5):774–806. doi:10.1016/j.tre.2007.04.001
2. Baron K, Heinrich L, Woch J (1984) Methods and tools for planning and organization of railway track closures. Project COBiRTK nr 3144/16. Katowice (in Polish)
3. Burdett RL, Kozan E (2006) Techniques for absolute capacity determination in railways. Transp Res Part B Methodol 40(8):616–632
4. Carey M (1994) A model and strategy for train pathing with choice of lines, platforms and routes. Transp Res B 28:333–353
5. Heinrich L (1986) The method of bypass routing for track closures. Mathematical modeling of transport—monograph. Cracow University of Technology, Cracow (in Polish)
6. Jin M, Zhang L (2011) Analyzing congestion and capacity impacts from disruptions to critical infrastructures in the rail network. Final report. The National Center for Intermodal Transportation
7. Karoń G (2005) Algorithm of coordination of railway closures in transportation networks. PhD Dissertation, Warsaw University of Technology, Warsaw (in Polish)
8. Karoń G (2003) Application supporting the coordination process of track closures. Scientific Journal of Silesian University of Technology. Publishing House of Silesian University of Technology, Gliwice. Series Transport. No. 1586(47), pp 335–344 (in Polish)
9. Karoń G (2003) Applications supporting the coordination process of track closures. Scientific Journal of Silesian University of Technology. Series Transport. Publishing House of Silesian University of Technology, Gliwice. No. 1604(48), pp 283–290 (in Polish)
10. Karoń G (2002) Coordination of track closures. Scientific Journal of Silesian University of Technology. Series Transport. Publishing House of Silesian University of Technology, Gliwice. No. 1562(44), pp 195–202 (in Polish)
11. Karoń G, Biliński Ł, Pawlicki J (2009) The influence of modern rolling stock and the modernization of the track on time train—computer simulation. In: Woch J, Janecki R (eds) Contemporary transport systems. Selected problems of theory and practice. Publishing House of Silesian University of Technology, Gliwice, pp 53–66 (in Polish)
12. Karoń G, Firlejczyk G (2006) Estimation of expected traffic smoothness on railway section with computer train control system. Scientific Journal of Silesian University of Technology. Series Transport. Publishing House of Silesian University of Technology, Gliwice. No. 1721 (62), pp 237–246 (in Polish)
13. Karoń G, Skrzypek M (2006) Actual method of train timetable construction. Scientific Journal of Silesian University of Technology. Series Transport. Publishing House of Silesian University of Technology, Gliwice. No. 1721(62), pp 247–256 (in Polish)
14. Kozan E, Burdett R (2011) A railway capacity determination model and rail access charging methodologies. Transp Plann Technol 28(1):27–45

15. Lai YC, Barkan C (2011) Comprehensive decision support framework for strategic railway capacity planning. J Transp Eng 137(10):738–749
16. Lai YC, Shih MC, Jong JC (2010) Railway capacity model and decision support process for strategic capacity planning. Transp Res Record J Transp Res Board 2197:19–28
17. Lam WHK, Huang H-J (2002) Modeling and solving the dynamic user equilibrium route and departure time choice problem in network with queues. Transp Res B Methodol 36(3):253–273
18. Liudvinanavicius L, Dailydka S, Sładkowski A (2015) New railway traffic control system possibilities. In: Sładkowski A (ed) Proceedings of VII international scientific conference. IV international symposium of young researchers. Silesian University of Technology. Faculty of Transport. Katowice, pp 341–350
19. Mattsson LG (2007) Railway capacity and train delay relationships. Critical infrastructure. Springer, Berlin, pp 129–150
20. Młyńczak J, Toruń A, Bester L (2016) European rail traffic management system (ERTMS). In: Sładkowski A, Pamuła W (eds) Intelligent transportation systems—problems and perspectives. Series: Studies in systems, decision and control, vol 32. Springer International Publishing Switzerland, pp 217–242
21. Peña-Alcaraz M, Webster M, Ramos A (2011) An approximate dynamic programming approach for designing train timetables. ESD working paper series. ESD-WP-2011-11. Available: http://dspace.mit.edu/bitstream/ handle/1721.1/102831/esd-wp-2011-11.pdf?sequence=1
22. Pouryousef H, Lautala P, White T (2013) Review of capacity measurement methodologies; similarities and differences in the US and European railroads. Available: http://docs.trb.org/prp/13-4502.pdf
23. Pouryousef H, Lautala P (2015) Hybrid simulation approach for improving railway capacity and train schedules. J Rail Transp Plann Manag 5(4):211–224
24. Railways closures guidance (2006) Department of Transport. Transport Scotland
25. Surma S (2010) Role and functions of a local control centre. In: Janecki R, Sierpiński G (eds) Contemporary transportation systems. Selected theoretical and practical problems. The development of transportation systems. Publishing House of Silesian University of Technology, Gliwice, pp 207–212
26. Wendler E (2007) The scheduled waiting time on railway lines. Transp Res Part B Methodol 41(2):148–158
27. Woch J (1998) Shaping the traffic flows in dense transport networks. Szumacher Publishing, Kielce (in Polish)
28. Woch J (2001) Tools of efficiency analysis and optimization of the rail network (South system—a description of the basic software). Publishing House of Silesian University of Technology, Gliwice (in Polish)
29. Woch J (1977) Project MK 122.11 system for assessment of track configuration. The Centre for Research and Technological Development Railway COBiRTK No. 3144/16, Katowice (in Polish)
30. Yaghini M et al (2014) Capacity consumption analysis using heuristic solution method for under construction railway routes. Netw Spat Econ 14(3-4):317–333
31. Żochowska R, Karoń G (2016) ITS services packages as a tool for managing traffic congestion in cities. In: Sładkowski A, Pamuła W (eds) Intelligent transportation systems—problems and perspectives. Series: Studies in systems, decision and control, vol 32. Springer International Publishing Switzerland, pp 81–104
32. Żochowska R (2015) Multicriteria decision support for planning of temporary traffic arrangement in urban network. (Monograph) Publishing House of Warsaw University of Technology, Warsaw (in Polish)

Assessment of Polish Railway Infrastructure and the Use of Artificial Intelligence Methods for Prediction of Its Further Development

Bogna Mrówczyńska, Maria Cieśla and Aleksander Król

Abstract The aim of this chapter is to analyze economic trends in rail freight volume in Poland, based on the analysis of data from the years 2009–2013, for evaluating decisions on planned investments in railway infrastructure envisioned by Poland and the EU at the time the EU was founded. The theoretical analysis presents a trend of functional Polish railways and its impact on investment decisions. In addition, it shows the long-term plans for railway transport in Poland from both the Polish government and the EU perspectives. An analysis of the current investment to support the development of railways in Poland is also elaborated. The research part of the chapter presents an analysis of statistical data on rail freight. Forecasts are precisely presented of selected transport parameters made by the Bayesian network method and Holt-Winters double exponential smoothing using an artificial immune system to determine parameters and initial conditions.

Keywords Statistical forecasting · Railway transport · Artificial immune system · Clonal selection · Bayesian networks · Methods of exponential smoothing · Holt-Winters double exponential smoothing

B. Mrówczyńska (✉) · M. Cieśla
Department of Logistics and Industrial Transport, Faculty of Transport,
Silesian University of Technology, Krasinskiego 8, 40-019 Katowice, Poland
e-mail: bogna.mrowczynska@polsl.pl

M. Cieśla
e-mail: maria.ciesla@polsl.pl

A. Król
Department of Transport Systems and Traffic Engineering,
Faculty of Transport, Silesian University of Technology,
Krasinskiego 8, 40-019 Katowice, Poland
e-mail: aleksander.krol@polsl.pl

A. Sładkowski (ed.), *Rail Transport—Systems Approach*, Studies in Systems, Decision and Control 87, DOI 10.1007/978-3-319-51502-1_9

1 Introduction

A diagnosis of the current state of rail transport in Poland and SWOT [strengths, weaknesses, opportunities, threats] analysis shows that the most significant factor hampering the development of rail transport in Poland is a pronounced deterioration of infrastructure. The most obvious symptom of this degradation is the low maximum speeds on a substantial part of the railway network. These speeds in many cases are smaller than the designed speeds [36] which formerly existed on the individual sections of the line. As a result, speed limits have been introduced more and more often. Reduced maximum speed imposes limitations such that the travel times are being extended considerably in relation to the shorter times achieved in each section in the past. This, in turn, results in lower quality of service regarding transport timeliness.

A systematic decline in the length of railroads, the development of which in the past was not always rationally planned, remains a constant and very unfavorable trend. In 1990, 24.1 thousand km of railway lines were in use, and now only 19.3 thousand km, a decrease of 20%. The last railway line was built in 1987, which, taking into account changes in the technical and technological changes in the global railway industry, puts Poland among the countries with the most neglected and outdated railway infrastructure.

The poor condition of the railway infrastructure is caused by a lack of sufficient financial resources allocated for repair, modernization and maintenance. To counter this trend, an increasing number of projects are being directed to modernization of existing linear and point infrastructure and to infrastructure expansion.

The research goal of this chapter is to verify the rationality and correctness of expenditures on railway infrastructure in relation to future trends in the field of rail freight. The chapter presents an outline of the degradation of the railway infrastructure, as well as recommendations for the appropriate sources and amounts of funding from EU funds for modernization of the railway infrastructure. The rationale for the decisions is verified by the forecast of basic data relating to the quantities of freight carried by rail. Forecasts were made using two methods: exponential smoothing using an artificial immune system for determining the model parameters (modified Holt method), as well as the initial conditions, and by using Bayesian networks method.

2 Prospective Plans for the Railway Transport in Poland

To address unfavorable factors causing the deterioration of the infrastructure, dither Polish State Railways must be transformed or the state-owned enterprise must be delegated to commercial companies. EU directives recommended the separation of railway infrastructure management from transport operations, assuming the coverage of losses as inadmissible passengers profits from freight. The EU Directive pointed to the division of the railway company into three sectors:

- passenger,
- freight,
- infrastructure.

In September 2000 Polish parliament adopted a law on commercialization, re-structuring and privatization of Polish State Railways (Polskie Koleje Państwowe). It established a new company, PKP SA, and four commercial companies:

- PKP Polskie Linie Kolejowe S.A.—infrastructure company,
- PKP Cargo S.A.—freight transport company,
- PKP Intercity Sp. z o.o.—passenger transport company,
- PKP Przewozy Regionalne Sp. z o.o.—regional passenger transport companies.

These statutory solutions, carried out administratively without financial support, did not solve the problem of the railway infrastructure in Poland. Its condition continued to degrade, which caused a further decrease speed throughout the network. The maximum speed has been reduced, and in a number of important connections, journey time was seriously extended, which undermined the competitiveness of rail transport on the market and increased its energy consumption.

2.1 *Polish Plans Related to the Development of Rail Transport*

Modernization of railway lines became possible only when Poland created substantive conditions for the development and improvement of the situation, producing program documents such as:

- National Development Strategy 2007–2015,
- State Transport Policy for the years 2006–2025,
- Master Plan for Rail Transport until 2030.

The National Development Strategy for the years 2007–2015 is the basic strategic document setting the objectives and priorities of socioeconomic development and the Polish conditions ensuring this development. This strategy directly addresses the national railways [32], where it is stated that:

> Increasing the share of rail passenger and freight requires a significant improvement in the quality of rail services, especially in light of the approaching opening of the sector to strong competitive pressure within the Common European Market.

With these assumptions a real support investment was possible, which will result in raising the operational parameters of the main transport routes, including increasing the possible speed of transport and increasing interoperability.

The national strategy envisages support for the construction of a high-speed rail system integrating the Polish metropolises. Investment in railway infrastructure will be primarily aimed at the liquidation of bottlenecks on lines with high traffic, i.e.

between the larger destinations, as well as reconstruction activities and modernization of railway lines.

The necessity of modernizing the existing railway lines, caused by years of under-investment in railways, and the lack of any support from public funds, resulted in the need for financing the state budget funds for the tasks associated with repair and maintenance of railway infrastructure from the European Union funds.

Actions taken in the framework presented by the Ministry of Transport "Master Plan for Rail Transport in Poland until 2030" [30] should lead, among other things, to a fundamental improvement of the infrastructure, and consequently to an improvement in the competitive position of rail transport. Thanks to that, the basis for creating a new quality of transport services in the field of transport of passengers and cargo was established.

At the same time directions of development of the railway infrastructure must be consistent with assumptions set forth in the updated Concept of National Spatial Development [23]. Actions on infrastructure included in this section are related to the following planning periods, coinciding with the planning periods of the European Union:

- 2007–2013,
- 2014–2020,
- 2021–2030.

A key element of the Master Plan for Rail Transport 2030 is to develop a plan and a timetable for the modernization, rehabilitation and expansion of rail infrastructure. Objectives of infrastructure investments provide a comprehensive subset of the specific objectives set for the entire Master Plan, as contained in the introduction to this study, and they must contribute to achieving the main objectives of the Master Plan. They are as follows:

- improving the transport of passengers and freight in the corridors of the trans-European transport network (TEN-T)—the fulfilment of international obligations in the field of standards and bandwidth on modernized lines;
- increasing the efficiency of the rail system, as a result of its reconstruction (including stopping the infrastructure degradation), taking into account the technical standards for interoperability and environmental standards;
- enabling the widest possible use of existing rail infrastructure, especially in two prospective sectors:
  - passenger market: between large urban areas and within large agglomerations,
  - freight market: mass transport of large cargo volumes and intermodal transport;
- facilitating mobility with the use of different modes of transport, in particular for passengers with reduced mobility: rail links to airports, linking with road transport and railway integration with public transport, with particular emphasis on urban agglomerations;

- improving standards of passenger service at railway stations and bus stops, including the adjustment to the needs of people with limited mobility.

The Master Plan framework provides three levels of investment activities, differing in material scope, level of costs and implementation period:

- construction of a new railway infrastructure of a high standard,
- modernization of the existing railway infrastructure, with particular emphasis on lines belonging to the TEN-T,
- investments restoring normal parameters of the railway infrastructure on the lines considered relevant (replacement investments).

A separate group of measures envisaged in the framework of the modernization are investments, including the construction of control systems on lines with small and medium traffic load, with the task of operating the automation and reduction of operating costs of these lines. Investments in infrastructure systems to improve the management of passenger and freight transport are also expected.

The widest range of investment activities will apply to projects relating to the construction of a new railway infrastructure. These investments can be divided into the following groups:

- building connections complementary to significant gaps in the railway network,
- building connections between the centers of large cities and airports supporting these metropolitan areas.

Modernization investments implemented on the Polish railways since the first half of the nineties mainly related to lines in the pan-European transport corridors. These lines are now part of the TEN-T.

However, the new EU perspective for the years 2007–2013 changed the priority of the modernization investments with the support of assistance programs such as the "Infrastructure and Environment" OPI&E and the "Regional Operational Programme" (ROP). The process of modernization of railway lines in these years accelerated, and the scope of modernization diversified and began to depend on the final destination of the line. In the framework of railway line modernization, all railway junctions were reconstructed, especially those that were bottlenecks on the railway network. Modernization included both the expansion of existing track and construction of new control systems and traffic management.

Due to the very poor condition of the rail infrastructure, long overdue for repairs and maintenance, replacement investments were the main measures needed to restore railway network operating parameters to their normal scale, both in terms of the speed of scheduled services and the pressure of the rolling stock axles. Figure 1 shows the values in the range of speeds [31], and Fig. 2 shows the value of the maximum pressure of the rolling stock on the track [36].

The geographical scope of these investments will be far larger of all groups of investment activities. It is understood that the technical scope of replacement investments will concentrate on railway tracks and, in typical cases, will include the repair of main roads (replacement of individual components) or repair the current

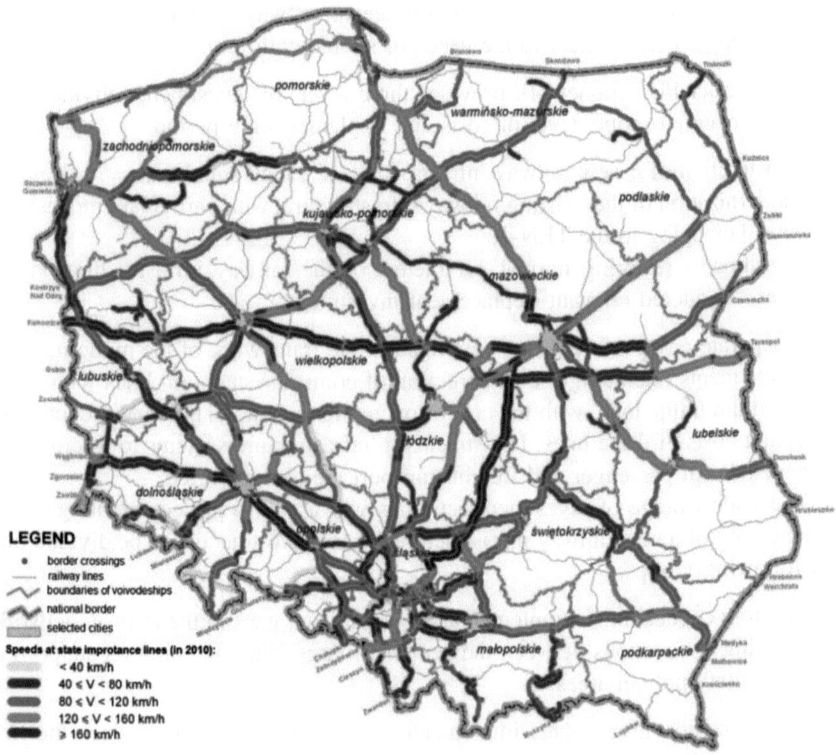

Fig. 1 The maximum scheduled speeds on the tracks of important lines in Poland

extended range of the existing surface. At the same time reconstruction of line should be comprehensive, meaning that reconstruction will include repair of drains and weak spots in the subgrade. It will include reconstruction of damaged or worn-out civil engineering and exploited crossover stations. Reconstruction will also include other work, such as building of automatic vehicular traffic controls at level crossings, which can be introduced as a result of the liquidation of operation speed limits. The aim should be to achieve significant improvement in operational performance without the risk of reducing speed limits.

2.2 *European Plans Related to the Development of Railway Transport in Poland*

The objectives of the development of railway transport in Poland can be achieved only through investment, including the construction of new sections of lines, and the modernization of existing lines of infrastructure that will improve the

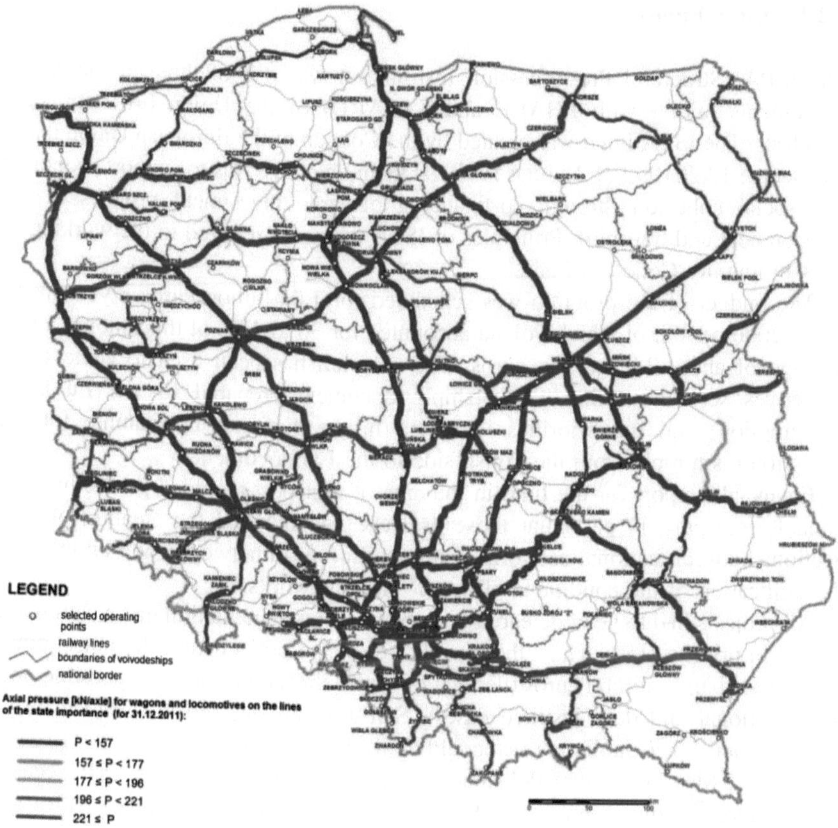

Fig. 2 The maximum axle loads on the track line of importance lines in Poland

operational capacity of the railways and, consequently, help to increase the speed on the tracks. This will also improve traffic safety, increase throughput at hubs, reduction of travel time, and, consequently, improve the competitiveness of rail transport in relation to other modes of transport [37]. Increased railway transport efficiency depends on stable financing of infrastructure and effective management of all its components and systems.

Extensive plans are not feasible without a stable long-term financial plan. Funds for numerous investments come from several sources. First of all, thanks to the Polish accession to the European Union, railways may use funds such as ISPA (until 2004.), the TEN-T Fund, the Cohesion Fund and the European Regional Development Fund. An important role is also played by funds from the European Investment Bank (EIB), which distributes deals with National Economy Bank (NEB) in our country. They allow us to cover insufficiencies in our own matching contribution, which is required to obtain funding from the European Union.

2.2.1 ISPA Funds

The ISPA (Structural Pre-Accession Instrument) was one of three pre-accession instruments (along with PHARE and SAPARD) of the Union's 10 candidate countries. The ISPA Fund was created by the Council of the European Union under the Decree 1267/1999 of 21 June 1999. Its main objective was to support economic and social cohesion through co-financing of large investment projects in the environment and transport. Principles of ISPA measures referred to the working of the Community Cohesion Fund.

The budget of the program was scheduled for 1.04 billion euros a year in the 2000–2006 period, of which Poland accounted for 30 to 37% of this amount, or an average of about 348 million euros. The program has been used to achieve the objectives set out in the "Partnership" (document prepared by the European Commission) and the priorities identified in the National Programme for the Adoption (as a reply from the Polish side to the EU document).

Financial support for the fund in the area of transport included promoting sustainable transport, in particular those projects that included the creation of connections between national networks and trans-European networks, and allowed for the unification of the conditions of use of these networks.

All projects had to be large enough so that their implementation had a significant impact in priority areas. Thus, the overall cost of the project from the outset could not be less than 5 million euros (derogations from this condition were possible only in exceptional and duly justified cases). Self-government, self-government organizations and other public entities could apply for the grant.

Since 01 May 2004, after Polish accession to the European Union, ISPA ceased its operation in our country. In Poland, in accordance with Annex II of the Accession Treaty, all projects that have received the possibility of funding under the ISPA, and which have not been completed, were pursued within the framework of the Cohesion Fund, operating on similar principles.

2.2.2 TEN-T Fund

The aim of the fund was to support projects implemented by Member States, which are called "common interest." These projects have been identified in the Community guidelines for the development of the trans-European transport network. The network is aimed at increasing the efficiency of the functioning of the common market. It was also supposed to fully enable citizens of the Union, economic operators and regional and local communities to benefit from the establishment of an area without internal borders. The European Union aims to extend national transport networks through the development of intermodal transport. The aim is also to ensure access from remote regions or islands to the central regions of the EU and to reduce the high transport costs in these regions.

Trans-European networks include, among others, transport network (TEN-T), for which the EU has allocated a separate pool of funds in the EU budget. For the development of trans-European networks of the 4600 million euros was addressed in the years 2000–2006. For 2007–2013, in order to further develop itself, the TEN-T budget has been established in the amount of 8013 million euro, including targets for all EU Member States.

The beneficiaries of the TEN-T budget include both state actors and private entities operating in the area of public services. The Fund also supports projects conducted in public-private partnership. Projects co-financed from the TEN-T projects were those that lay in the interest of all Member States, i.e. those that:

- contributed to the sustainable development of the transport network across the European Union;
- ensured the coherence of the TEN-T and access to it;
- integrated all modes of transport;
- contributed to the protection of the environment and increased safety standards.

Cross-border projects, especially, were supported. Of key importance were also projects related to environmentally friendly transport (rail, sea, inland waterways) and those carried out by more than one member state. A list of priority investments was adopted in the framework of the fund which placed, among other railway infrastructure projects running through Polish territory:

- Railway axis Gdańsk–Warsaw–Brno/Bratislava–Vienna,
- Railway axis "Rail Baltica" Warsaw–Kaunas–Riga–Tallinn–Helsinki.

At the beginning of 2014 the last competition of the TEN-T was settled. Implementation of the projects will be completed in 2015. After 2014, the TEN-T Fund has been replaced by the CEF (Connecting Europe Facility) instrument. The European Union has allocated a separate pool of funds in its budget for this purpose. These funds will be used in 2014–2020 for investments in the construction and modernization of infrastructure in the fields of transport, energy and telecommunications.

2.2.3 Cohesion Fund

The Cohesion Fund was established in 1993 by the Treaty of Maastricht; the decision to create it had been made a year before at the European Council in Edinburgh. The motivation for the creation of the fund was eliminating disparities in the development of the economically weakest members. At the same time, the newly created Cohesion Fund had to compensate for those countries that could bear potential losses associated with the launch of "single market" economies, and make much more competitive goods available to the rest of the Community.

Unlike the Structural Funds, the Cohesion Funds are allocated to states, not to individual regions. The list of countries eligible for aid shall be determined by the decision of the European Commission, on the basis of gross national product. In order for a country to apply for funding from the Cohesion Fund for the implementation of infrastructure investments in the field of environment or TEN-T transport networks, its gross national income must be less than 90% of the average Gross National Income of the European Union. Another condition is that the beneficiary country Cohesion Fund program must lead to the fulfilment of the conditions of economic convergence. This is called the principle of conditionality. Its failure leads to the suspension of aid, but the state is not obliged to return funds already received.

In the years 2007–2013, Poland received 22.2 billion under the Cohesion Fund, which constitutes 33% of the awarded allocation for this period. As already indicated, the Cohesion Fund finances only infrastructure projects in the field of environment.

2.2.4 European Regional Development Fund

The purpose of the European Regional Development Fund is to increase economic and social cohesion in the European Union, eliminating inequalities between regions. In short, the ERDF finances:

- infrastructure related to research and innovation, telecommunications, environment, energy and transport;
- financial instruments (venture capital funds, local development funds, etc.) to support regional and local development and to foster cooperation between cities and regions.

2.2.5 Railway Fund

The Railway Fund is a targeted element of the government's system-wide approach to the development of railway transport in Poland, the aim of which is to implement transport policy in accordance with sustainable development. The Railway Fund was created by the NEB under the Act of 16 December 2005 on Railway Fund, and started to operate on 9 February 2006.

The task of the Railway Fund is to collect funds for the preparation and implementation of construction and reconstruction of railway lines, repairs and maintenance of railway lines and the elimination of redundant railway lines.

Basic financial sources of the Fund are:

- constant revenues from the fuel tax on motor fuels and gas for motor vehicles; the Railway Fund supplies 20% of the proceeds of this account,
- interest rates available on the Fund's account in NEB and income from bank deposits, free funds and their investment in securities issued or guaranteed or underwritten by the State Treasury,
- income from shares transferred by the State Treasury and the revenue from their sale.

2.2.6 Infrastructure and Environment Operational Programme

The Partnership Agreement developed by the European Commission defines the main directions of support under the Cohesion Policy, which also includes the Operational Programme Infrastructure and Environment (OPI&E). This program (OPI&E), strives for sustainable economic development and competitiveness, which will be possible by supporting the development of technical infrastructure in Poland.

Existing infrastructure upgrades were based primarily on the OPI&E and the ROP. In 2006, the European Commission granted to Poland, from the Structural Funds and the Cohesion Fund, about 35 billion euros for infrastructure improvements to be spent in 2007–2013. These were measures increased funding, though were still insufficient. The current investments in the railway infrastructure are executed in accordance with the "Long-term Railway Investment Programme for the year 2015" adopted November 5, 2013 by the Council of Ministers. Figure 3

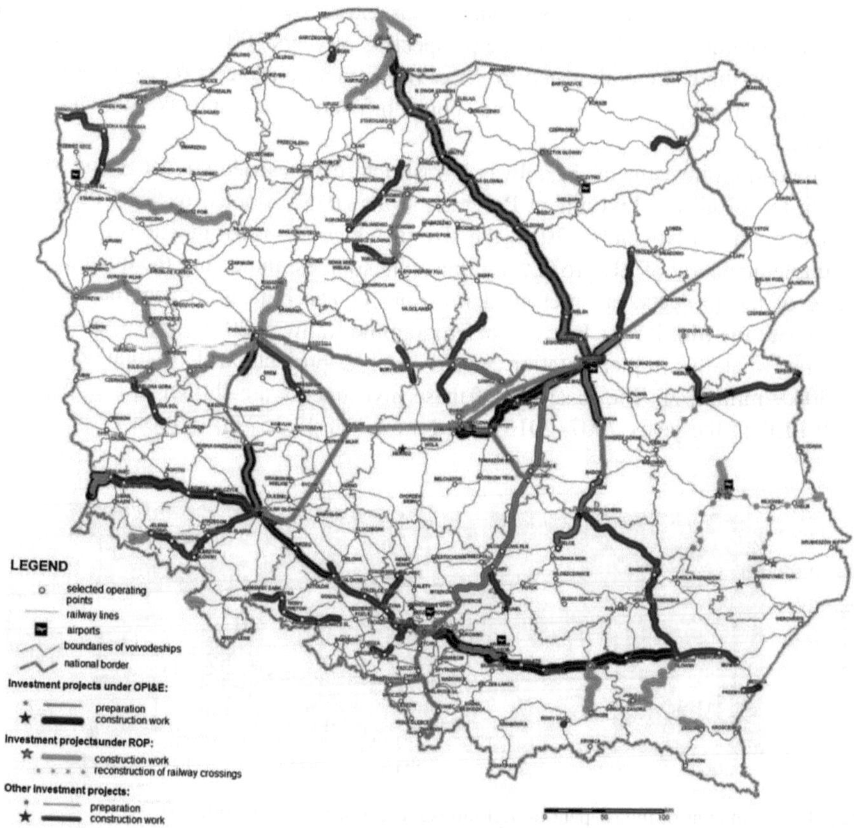

Fig. 3 Investments in Polish railway infrastructure networks in 2011–2013

[31] shows a list of infrastructure investments for the years 2011–2013. The program includes 140 investment tasks under the current financial perspective (2007–2013). Of these, 62 investments are made within the OPI&E and 29 under the ROP.

2.2.7 Regional Operational Programme

The Regional Operational Programme (ROP) is a planning document defining the areas and sometimes specific actions that government bodies of voivodeships take or intend to take to promote the development of the province or region. As the name suggests, this is a document of an operational nature, so it is more detailed and primarily focused on development strategy.

The legal basis for the functioning of the ROP is the Act of 6th December 2006 on principles of development policy.

In total, the 2007–2013 Operational Programme Infrastructure and Environment for railway investments with OPI&E allocated 4.8 billion euros and spent an amount of 23,314 million zloty. The difference between the amount of aid received in euros and the amount of PLN financing was caused by currency fluctuations. The level of spending of EU funds in the perspective of years 2007–2013 are shown in Fig. 4. The funds were allocated to linear infrastructure projects (construction and maintenance of railway lines) and point infrastructure (railway stations and terminals).

The current investments in railway infrastructure are carried out according to the "Long-term Railway Investment Programme for 2015" adopted by the Council of Ministers on 5th November 2013, the size of which is shown in Table 1 [33].

Implementation of EU projects by financing sources, indicating the quantity and value of the projects, are presented in Table 1. The data is in accordance with updates to the Multi-Year Investment Programme Railway 2015 [33], where it was assumed that the implementation would be 142 projects with a total value of 41,304.9 million zł. The locations of these investments are shown in Fig. 5 [33]. PKP PLK in the years 2007–2015 issued a total of 37,157.00 million zł.

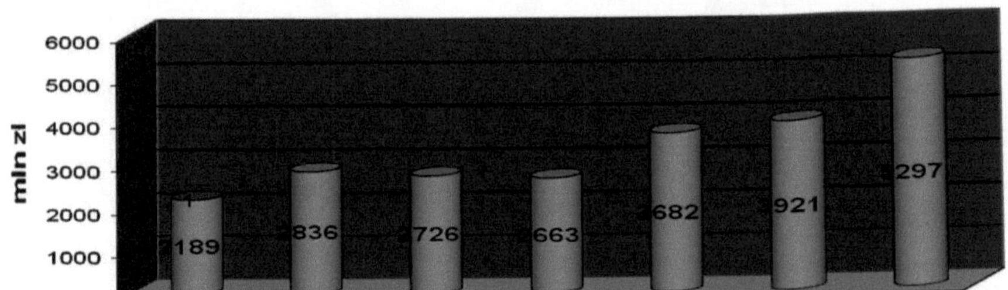

Fig. 4 The amount of funds spent on rail infrastructure in the years 2007–2013

Table 1 Implementation of projects by funding spent on railways in Poland until 2015

Funds source	Number of grants	Value of projects [mln] zl
Operational Programme Infrastructure and Environment	53	26,514.8
Regional Operational Programme	31	9846.7
State budget	37	4133.2
Rail fund	9	143.5
TEN-T fund	4	185.8
Cohesion fund	8	480.9
Total	142	41,304.9

Fig. 5 Infrastructure investments covered by the Multi-Year Programme of Railway Investment for the years 2013—2015

2.3 *The Current State of Modernization of Railways in Poland*

Ongoing and future modernizations of the railway infrastructure in Poland include both national and EU financial funding.

2.3.1 Investments from National Funds

In 2015, PLK announced tenders for the projects of modernization and reconstruction of several important railway lines, to be implemented in 2015–2016. They will be financed from the state budget, the Railway Fund and bonds that support Polish Railway Lines. Their total estimated investment value approximately 1 billion zł. In this group there will be a subtask for endarterectomy detours for those episodes, which will be modernized with EU funds' new financial perspective.

2.3.2 CEF Investments

In 2014–2020, the EU is launching a new program for co-financing major infrastructure investments (not just rail) called CEF. From the sum of 28 billion euros, Poland was awarded 4 billion euros. Poland applied to this program with measures worth at least 3 billion euros. In addition, the railway tasks involved in interfacing with maritime transport and improving access to ports might be able to qualify for an additional 3.5 billion euros. The European Commission will announce recruitment in three stages, the deadline for the first round of projects ended in February 2015. Contenders were announced in September 2015, 2016 and the closure will take place in December 2016. The first call PLK start with six investment projects. These will be assessed by the European Commission at the end of the third and fourth quarters of 2015. The most advanced project is to complete the modernization of the Wroclaw–Poznan section (to complete work on the Rawicz–Czempin section), worth 1.5 billion zł., and a project to modernize the Sochaczew–Swarzędz line for 2.6 billion zł.

There also will be investment in railway lines peripheral to Warsaw (Warsaw section Gołabki/Warsaw West–Warsaw Gdansk)–Warsaw Wlochy–Grodzisk Mazowiecki (line 447).

In total, in the first competition of the project, Polish CEF investors will report to the European Commission projects worth approximately 10 billion zł, by the probable date of announcement of public tenders to 2016, for the implementation of at least two jobs for more than 4.1 billion zł.

2.3.3 The Remaining Pool of CEF

In addition, PLK predicts that CEF funds for projects funded under the "List of offshore projects" will be available. Three tasks to improve rail access to seaports in Szczecin and Swinoujscie and Gdansk and Gdynia will then be reported. All of these projects will involve a total reconstruction of the station ports, adapting them to the specifics of the projected freight traffic and thus enabling the further development of ports.

2.3.4 Funds from OPI&E

As the financing of railway investment projects are based almost exclusively on the state budget and EU funds, the flow of funds for projects of new perspective 2014–2020 will be initiated after the final approval of the new operational programs. Meanwhile, the negotiations with the European Commission on the final shape of the OPI&E are still in progress. Therefore, in contrast to the railways, road builders can already announce a number of tenders, because the roads are financed from other sources, such as the National Road Fund. Polish Railway Lines must wait for the completion of procedures and complete accounting of financing tasks from the perspective of 2007–2013. Not until 2015 was a new program adopted of co-financing national measures of new tasks for 2014–2020. In total, PKP and PLK expect to receive from the new perspective a total of about 7 to 8 billion euros, which is at least 40 percent more than in the perspective of 2007–2013. Undoubtedly, prospects for the "Master Plan" and "Long-Term Investment Plan" look promising. Part of the work included in both programs is in progress, part completed. Implementation of strategic objectives must overcome a variety of difficulties. First of all, implementation of a project is a multidisciplinary task that requires effective coordination and management. This involves obtaining multiple permits, making a lot of arrangements, and most importantly is connected with the necessity of application of the "Public Procurement Law" Act. All this requires the fulfilment of a number of formalities related to the investment. The whole procedure also extends the requirement to obtain administrative decisions. The problem is determination of the sole criterion for choosing the best contractor, because the commonly used criterion of price does not guarantee the investment in the framework of a negotiated amount.

A modern and efficient rail infrastructure is a prerequisite for the development of rail transport and for the country. Rail transport should be regarded as the most ecological and safest mode of transport. It is attractive both for passengers and for businesses. The existing investment plans should, however, be reliably verified in relation to trends in freight volume forecast.

3 Applied Methods of Forecasting

All the phenomena described in the section related to the operation of shipping are non-periodic investigations, close to linear. Some of the waveform charts are decreasing (Figs. 11, 12, 16, 17, and 18), a few are growing or virtually constant (Figs. 18, 22, 24, 25, 26 and 27). In the case of the variation of electric and diesel multiple locomotives (Figs. 15 and 16), decreases in the number of exhaust units are combined with increasing amounts of electrical components. Both graphs are characterized by clear increments in 2006. Generally, the variable course of the collapse in 2009 can be observed in Figs. 20, 21, 24, and 23.

The trend of changes in the near future can be predicted using the exponential smoothing method, which is easy to use and can be accurate enough. For non-periodic phenomena on the course, a Holt-Winters double exponential smoothing is used, which extracts significant changes in observed rail transport phenomena and reduces the influence of random fluctuations. This method was slightly modified in the monograph. The values of model parameters α and β and the initial values F_1 and S_1 are calculated as the optimal values using the clonal selection algorithm. The Bayesian network is second method.

Forecasting, understood as scientific prediction of the development of observed phenomena, plays an important role in planning. Forecasting the volume of transport is of great economic importance for the entire Polish economy.

An interesting approach to the use of forecasting in the supply chain optimization is described in [1]. The article presents the optimization of the supply chain cost using an integer programming method. The data on demand, production and inventory forecasting needed for optimization are obtained by using the exponential smoothing methods.

Mathematical methods are applied to forecasting. A brief history of forecasting is presented in [17]. Early forecasts constituted simple inference from observation; such methods developed especially in the nineteenth century. The situation changed at the beginning of the twentieth century with the proposed treatment of the time series as a realization of a stochastic process in [40]. Brown's [3], Holt's in 1957 [25] and Winters' [38] publications initiated the development of exponential smoothing methods. Viewed state exponential smoothing method for the year 1980 can be found in [21]. Exponential smoothing methods are often used in many fields of science, because they are easy to use and provide forecasts burdened with only small errors.

Smoothing methods continue to develop. The adjustments in the Holt-Winters double exponential smoothing can be read in the article [26]. The choice of model parameters for exponential smoothing by means of empirical performance of two derivative free search methods for solving the problem of minimization is presented in the article [35]. The use of weighted coefficients, moving average and exponential smoothing is written in the [39].

Currently, forecasting methods are often aided by artificial intelligence. For example, to estimate optimal values of coefficients of logarithm support vector regression the immune algorithm was used [29]. Neural networks can be used to reduce the sensitivity to input errors, and the ARIMA method (Autoregressive Integrated Moving Average) was used in [27]. A Bayesian network was used to predict stock price in [20].

3.1 Holt-Winters Double Exponential Smoothing

Looking at the graphs illustrating the course of the various phenomena observed in rail transport, we decided to use the methods of exponential smoothing, and because of the lack of periodic phenomena, we chose the Holt-Winters double exponential smoothings.

The Holt-Winters double exponential smoothing is one of the methods used to smooth the time series of a development trend and random fluctuations [19, 41]. For a time series of length N and the data values $sy_0, y_1, \ldots y_{N-1}$ the following equation are used:

$$F_t = y_t + (1 - \propto)(F_{t-1} + S_{t-1}) \tag{1}$$

$$S_t = \beta(F_t - F_{t-1}) + (1 - \beta)S_{t-1} \tag{2}$$

where

F_t smoothed value of the forecasted variable at time t,
S_t growth trend value at the moment t,

$t = 0, 1, \ldots, N$,
parameters $\alpha, \beta \in [0,1]$.
The equation of prediction for expired periods have the form:

$$y_t^* = F_{t-1} + S_{t-1} \tag{3}$$

for $t = 2, 3, \ldots, N$
and for future periods:

$$y_T^* = F_N + (T - N)S_N \tag{4}$$

where

N the number of periods in the relevant time series
T moment, which forecast, $T > N$

There are many ways of determining the value of the initial F_1 and S_1. Most often it is assumed that $F_1 = y_1$ and $S_I = Y_2 - Y_1$. In turn, parameters α, β are determined by error's minimization of expired forecasts.

In presented calculations some modifications were applied. The parameters α and β and the initial values of F_1 and S_1 are determined by minimizing the error MAPE. A similar solution is also used in [34]. In this chapter we will refer to the method as the "modified Holt method".

MAPE (Mean Absolute Percentage Error) is defined as:

$$\mathrm{MAPE} = \frac{1}{N}\sum_{t=1}^{N}\frac{\left|y_t - y_t^*\right|}{y_t} \cdot 100\% \tag{5}$$

where

N number of observations,
y_t value of the time series for a moment or period of time t,
y_t^* predicted value of y for a moment or a period of time t.

The parameters α and β and the initial values of F_1 and S_1 have to calculate by minimizing the MAPE error. To assess the quality of forecasts Pearson's correlation coefficient was calculated for actual data and expired forecasts.

3.1.1 Artificial Immune System

To determine the forecast using the modified Holt method, an artificial immune system was used. One of the methods of artificial intelligence, the artificial immune system mimic the action of the natural immune system of the human body [4, 28].

The defense system of the human body is made up of physical barriers and chemical properties, such as skin, body temperature, saliva, tears, sweat and mucus. If, in spite of these barriers, microorganisms attacking the body penetrate into the interior, the immune system defends. Its operation is quite complex and the defense involves different cell types. Some of them recognize antigens that attack the body [22].

Recognition is done using antibodies that are produced by the body. The antibodies, which fit to the antigen, are cloned. The clones are mutated. After finding the fitted antibody a proliferation follows. The antibodies are rapidly replicated and released into the blood stream.

After suppressing the number of antibodies is reduced. Some of the remaining antibodies form a memory cells. The next time the immune system has a chance to recognize the enemy faster. The process of searching for better and better antibodies is called clonal selection and its numerical model can be used for optimization.

Figure 6 shows the main stages of the clonal selection numerical algorithm. An antygen is optimal solution, which is searched. Antibodies are the solutions, which are proposed. The inverse of the objective function representing the optimization criterion is the measure of the matching.

The antibody is a sequence of numbers:

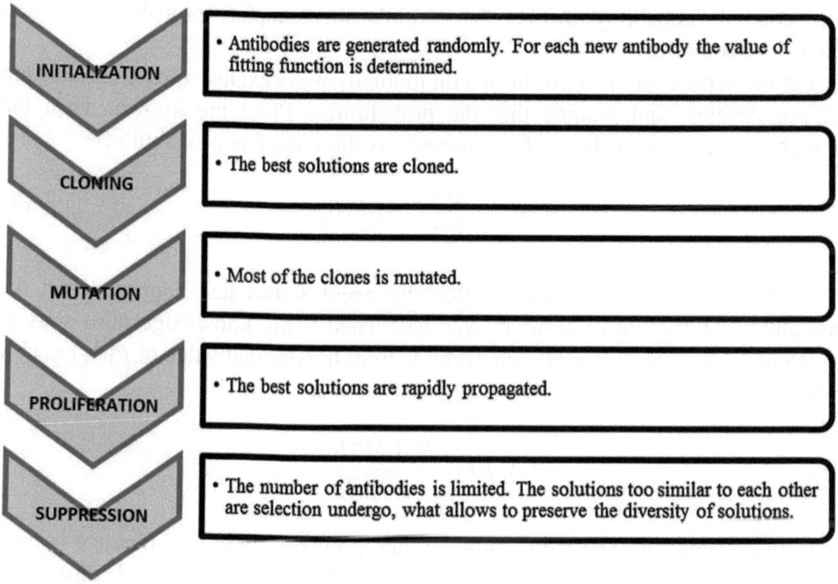

Fig. 6 The main stages of the clonal selection algorithm

$$[F_1, S_1, \alpha, \beta] \tag{6}$$

where F_1, S_1 initial values of Holt model,

α and β are parameters of this model, F_1, $S_1 \in R$, α, $\beta \in [0,1]$.

The inverse of the MAPE error is a measure of matching.

For the calculations, we use our own implementation of the algorithm written in the programming language C++.

3.2 *Forecasting Using Bayesian Networks*

Forecasting the volume of transport and other phenomena in rail transport can also be carried out using a second method, the Bayesian network.

Bayesian network structure corresponds to the relationship of cause and effect in a given set of random variables. As a rule, a Bayesian network topology maps the knowledge of the causes and effects in the area considered. The usefulness of Bayesian networks in practical applications is manifested by the fact that the knowledge of any set of observations (some state variables) allows you to calculate the probability distributions for the remaining unknown variables [2].

The network name is derived from Bayes's theorem, which postulates a revision of the existing beliefs about the probabilities in the light of new facts. Presentation of the rules of operation requires a reminder of two basic and very important theorems of probability theory: the formula for complete probability (7) and Bayes's theorem (8).

Let the some event B occur in several mutually ways exclusive of A_i, exhausting all possibilities, and assume that the probabilities $P(A_i)$ are known. Then the probability of an event B can be expressed as the complete probability:

$$P(B) = \sum_i P(A_i)P(B|A_i) \tag{7}$$

Next, assume that it is known that the event B has just happened and the probability of the above formula was calculated. This knowledge allows us to re-calculate the probability of each event A_k from the original value of $P(A_k)$ to new values of $P(A_k|B)$:

$$P(A_i|B) = \frac{P(A_k)P(B|A_k)}{P(B)} \tag{8}$$

Both formulas, despite their simplicity, are powerful tools for inference in each chain based on probabilities. The inference can be conducted in both directions: from causes to effects and from effects to causes; thus a calculated value can propagate throughout the network [16].

3.2.1 The Structure of a Bayesian Network

A Bayesian network is a directed acyclic graph, where the nodes represent random variables and edges correspond to cause/effect relationships between these variables. With each vertex X is associated a conditional probabilities table, which describes the strength of the relationship. This array contains the conditional probabilities $P(X|P_1, P_2, \ldots)$ of each state of a random variable X with the different states taken by the direct parent P_1, P_2, ... For vertices without parents (so-called root causes) conditional probabilities come down to simple probabilities. Figure 7 shows a simple Bayesian network with three binary nodes (receiving only two values: true and false) [18].

3.2.2 Construction of a Bayesian Network

Knowledge on Bayesian network is included in the topology and in the tables of conditional probabilities associated with the random variables. The first step is to identify those variables, then the relationships between them must be determined. This task is usually performed by an expert, but in rare cases can be performed

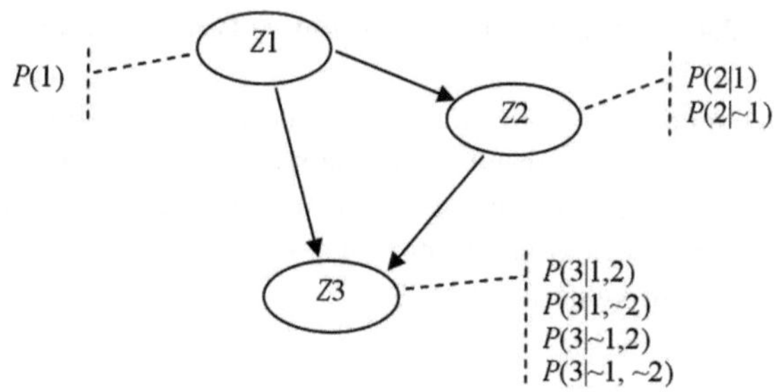

Fig. 7 The structure of a simple binary Bayesian network

Table 2 Modes of inference in Bayesian networks

Inference	Description
Causative	The cause or reasons are known, probability distribution of effects is sought
Diagnostic	The effects are known, potential causes are sought
Mixed	Some effects and some causes are known, possible ways of process occurrence are tested
Intercausative	One cause is known, effect on other potential causes is checked

automatically on the basis of the available data. Values of conditional probabilities can also be introduced by an expert based on his knowledge and experience, but more often the network is "trained" automatically based on historical data [24]. These data do not have to cover all the states of all the variables. A standard algorithm EM (Expectation/Maximization) can be used. EM is an iterative selection of probability and similarities for incomplete data. The algorithm tries to estimate which of the parameters of such a model, relative to the observed data (data provided learning), were the most likely. Each iteration consists of two stages:

- *E* step (expectation)—missing data are adjusted taking into account the known and the current parameters of the model,
- *M* step (maximization)—model parameters are selected so as to maximize the probability on the assumption that the missing data are already known.

3.2.3 Inference in Bayesian Networks

A Bayesian network defining the relationship between random variables allows us to calculate the probabilities of occurrence of events represented by these variables. A practical application of Bayesian networks begins with the information that is currently known. States of some random variables are determined, and then the schedule of probabilities other variables is updated. Possible modes of reasoning are summarized in Table 2.

3.2.4 Application of Bayesian Networks for Forecasting

A mode of causal inference is used in forecasting. It is assumed that the future, unknown state results from the state variables corresponding to past moments that are fixed at the time of prognosis. Thus, the Bayesian network provides the probability distribution of future states (Fig. 8), and the expected value can be treated as a sought value forecast (9).

$$w_{\text{out}} = \sum_i p_i m_i \tag{9}$$

where

w_{out} predicted value,
p_i probability of ith state,
m_i value corresponding to ith state.

An additional state called 'stx' can be seen in Fig. 8. This state is used to indicate improper operation of the network. The variable y_{0+1} is initialized in a way that assigns this state probability 1, and other states are assigned the value of 0. The 'stx' state does not appear anywhere in the training data, thus assigning a non-zero probability in the course of determining the forecast shows an unexpected set of states set for the moments of the past.

One of the simplest Bayesian networks that can be used in a prediction is shown in Fig. 9. The premise is that the value (stocks of random variables) for the present time and the past moment (y_{0-1}, y_{0-2}, y_{0-3}) are known (here, four values), and a

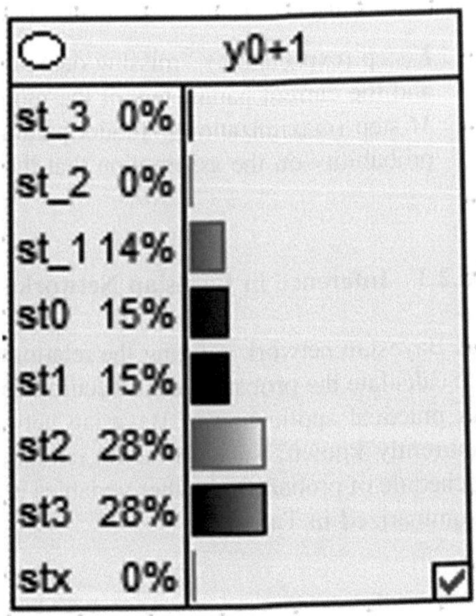

Fig. 8 Forecast as a probability distribution of states

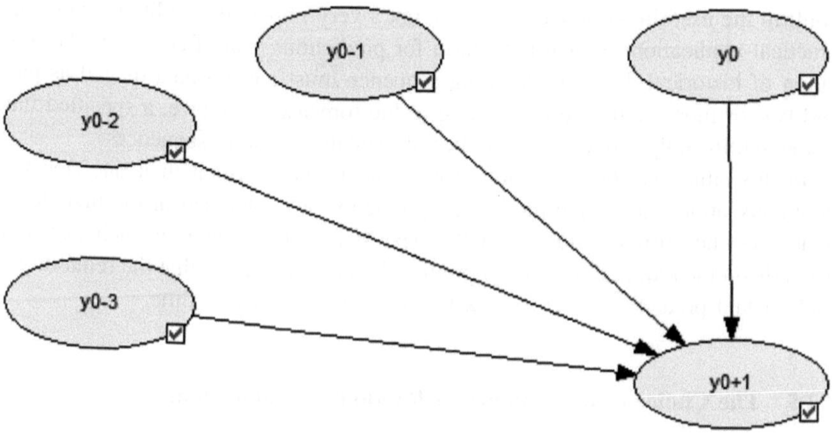

Fig. 9 The simplest Bayesian network used for prediction

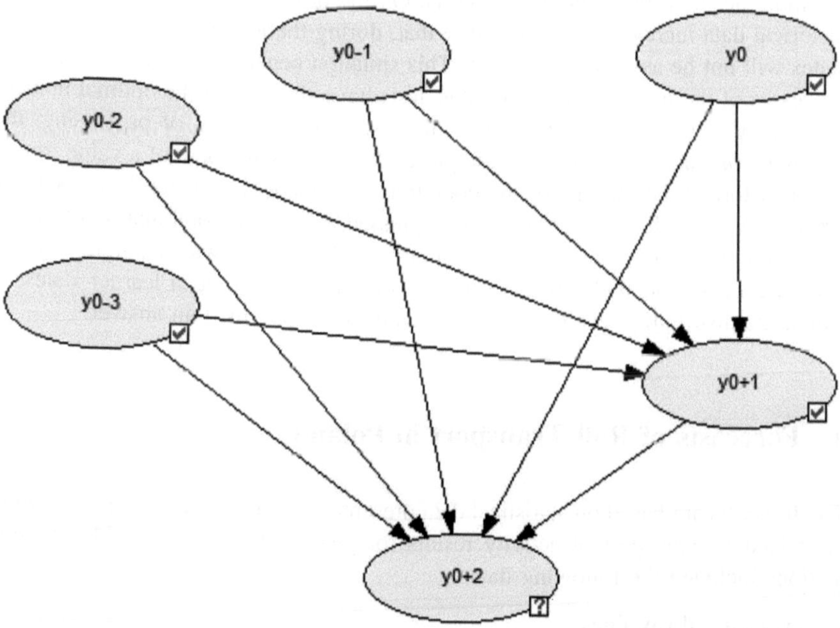

Fig. 10 Bayesian network used for two step prediction

probability distribution for the future moment is sought. There is no relationship between the variables corresponding to the past because they are not needed; all of these variables belongs to the past, and are already known facts. This model allows the calculation of forecasts for the single step, a future moment.

If one needs to perform following forecasts for two or more moments, a Bayesian network can have the structure shown in Fig. 10. The obtained test results confirm the usefulness of such a model, but a very important condition against the practical application of such a solution for predictions year after year is the limitation of historical data. The learning sequence must contain data treated as past, and two or more values corresponding to the forecast. Therefore, a specified time frame significantly reduces the number of available learning sequences.

In this situation, the predictions for moments more distant than one step first iteratively apply the simple model: the prediction y_{0+1} obtained in the first step is treated as the current value y_0 in the next step. The number of such iterations depends on the required number of predicted values. It is clear that the reliability of such distant predictions decreases with increasing distance in time.

3.2.5 The Choice of the Number of Random Variables States

An important issue in the design of a Bayesian network is to determine the appropriate number of states adopted by random variables. The higher the number, the more accurate the obtained forecasts. On the other hand, a limited number of historical data increases the probability that, during the learning, multiple network states will not be assigned any value. This situation negatively affects the learning processes of the network. To avoid this, tests have shown that the optimal number of states is seven. At the same time, to ensure the accuracy of processing, the following procedure was used. In the place of unique assignment of a value from a set of historical data to one of the conditions, we made a variable decomposition process. A single historical value was expressed as a linear combination of values corresponding to the states for each random variable. In this way, a single string of historical data generated a whole set of records containing clean learner states, to which the distribution coefficients of decomposition provided an answer.

4 Forecasts of Rail Transport in Poland

The forecasts are based on statistical data presented by the Central Statistical Office of Poland in Transport of activity results for years 2000–2014 [5–15]. The predictions included the following data:

- operated railway lines,
- standard gauge rolling stock (electric and diesel locomotives, electric railcars, freight cars and coaches),
- cargo rail transportation (in tons),
- cargo rail transportation (in ton-kilometers),
- rail transportation according to types of consignment (export, import, transit),
- total national and international rail transport of containers.

4.1 Operated Railway Lines' Length

Predictions of operated railway line lengths in Poland were based on data presented in Table 3.

As you can see in Fig. 11, the length of railway lines in the coming years decreases slightly. Both forecasts for the next three years provide for a further reduction of railway lines. Using the Bayesian network here is more optimistic. It gives a smaller forecast error MAPE of 0.85%, but the modified Holt's method gives a Pearson correlation coefficient closer to unity, which shows a better correlation of the predicted values with the actual data.

The aforementioned trend of shortening railways applies in particular to railway standard gauge (Fig. 12). Here, again, the first method predicts a smaller reduction of standard gauge lines. Its MAPE forecast error is 0.88%, but the second method gives a higher Pearson correlation coefficient, which is 0.85.

The length of electrified railway lines since 2004 changes slightly and the graph for real data, shown in Fig. 13, differs very little from both forecasts. Both methods show a slight downward trend for the next three years. MAPE error is smaller for the first method at 0.4%. The correlation coefficient is superior to other methods and is equal 0.95, which indicates a strong linear correlation of actual data and forecast results.

4.2 Standard Gauge Rolling Stock Number

Predictions of standard gauge rolling stock in Poland were based on predictions of electric and diesel locomotives and electric railcars, which are presented in Table 4, as well as freight cars and coaches, presented in Table 5.

The number of electric locomotives starts to grow slightly (Fig. 14). Forecasting with Holt method in this respect is more optimistic and predicts greater sales growth than prediction by Bayesian networks. However, the first method has a smaller MAPE error so the prediction should be based on this method. In turn, the second method has a higher correlation of actual data and forecast data for the years 2001–2014. The correlation coefficient is 0.67.

In accordance with predictive analysis shown in Fig. 15, the number of operating diesel locomotives declines. The Bayes prediction expects the number of operating locomotives to stabilize. It can be seen from the chart that this method is more sensitive to the jumping-off data. The Holt's method clearly smooths the data in such a situation. For the first method, correlation of actual data and predicted is almost non-existent. For the Holt method it is 0.24, understood also as weak.

Since 2006, a systematic increase has been observed in the number of electric multiple units, as shown in Fig. 16. The method of forecasting based on the Bayesian network provides stability in the next three years. Holt's method indicates a rapid increase in the number of teams in the coming years. The actual data and the

Table 3 Operated railway line length

Year	Total operated railway lines [km]			Standard gauge operated railway lines [km]			Electrified operated railway lines [km]		
	Real data	Bayesian networks	Modified Holt's method	Real data	Bayesian networks	Modified Holt's method	Real data	Bayesian networks	Modified Holt's method
2000	22,560			21,575			11,905		
2001	21,119			20,134			11,965		
2002	21,073		20,598.97	20,729		21,684.24	12,207		12,606.55
2003	20,665		20,629.07	20,321		20,464.34	12,160		12,296.97
2004	20,250	20,646	20,278.80	19,906	20,313	19,846.82	12,017	12,120	12,113.96
2005	20,253	20,218	1,984,206	19,843	19,878	19,457.02	11,884	12,015	11,877.16
2006	20,176	20,223	20,183.77	19,763	19,828	19,671.73	11,871	11,888	11,748.96
2007	20,107	20,135	20,124.58	19,797	19,742	19,699.80	11,898	11,873	11,851.21
2008	20,196	20,056	20,040.56	20,007	19,763	19,812.44	11,924	11,895	11,924.43
2009	20,360	20,145	20,257.28	20,171	19,973	20,170.10	11,956	11,918	11,950.82
2010	20,228	20,309	20,515.54	20,089	20,137	20,356.91	11,916	11,950	11,987.70
2011	20,228	20,177	20,151.41	20,113	20,055	20,086.94	11,880	11,910	11,880.15
2012	20,094	20,177	20,197.89	19,979	20,079	20,098.69	11,920	11,874	11,842.79
2013	19,328	20,053	19,982.35	19,259	19,952	19,883.64	11,868	11,909	11,955.63
2014	19,240	19,285	18,668.18	19,240	19,230	18,711.49	11,830	11,861	11,822.27
2015		19,200	19,015.25		19,216	18,992.30		11,821	11,790.07
2016		19,170	18,824.23		19,198	18,804.85		11,817	11,750.28
2017		19,123	18,633.22		19,168	18,617.40		11,809	11,710.48
Pearson correlation coefficient		0.7362	0.7787		0.6062	0.8539		0.6737	0.9457
MAPE forecast error		0.0085	0.0133		0.0088	0.0159		0.0040	0.0080

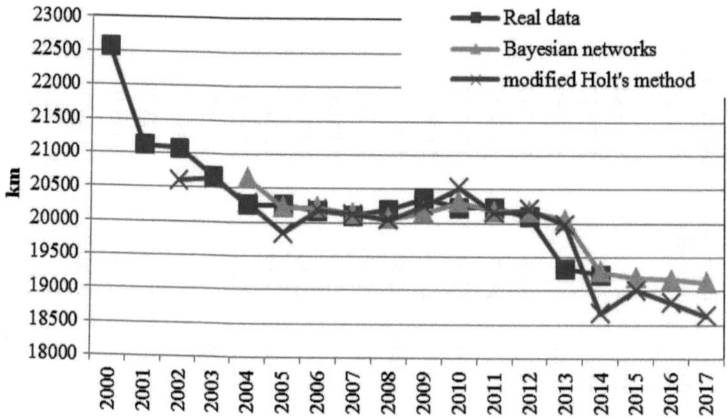

Fig. 11 Predicted total length of operated railway lines

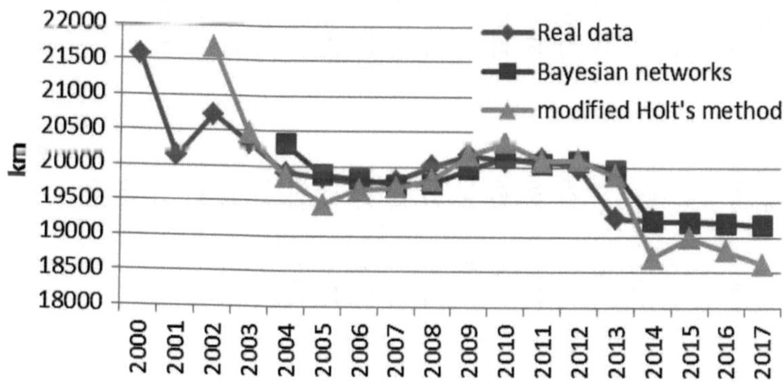

Fig. 12 Predicted length of standard gauge railway lines

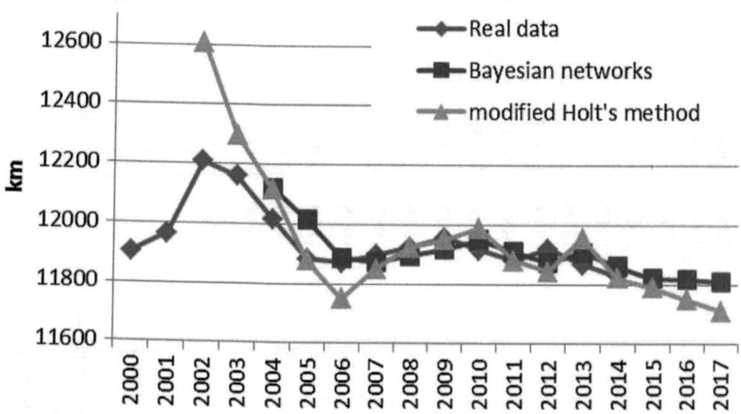

Fig. 13 Predicted length of electrified operated railway lines

Table 4 Standard gauge rolling stock prediction—part 1

Year	Electric locomotives			Diesel locomotives			Electric railcars		
	Real data	Bayesian networks	Modified Holt's method	Real data	Bayesian networks	Modified Holt's method	Real data	Bayesian networks	Modified Holt's method
2000	1774			2120			1187		
2001	1763		2827.07	2112		2247.10	1205		1546.84
2002	1811		2045.20	2314		2318.16	1182		1425.75
2003	1816		1886.29	2405		2422.21	1176		1234.52
2004	1842	1816	1832.17	2554	2367	2520.19	1167	1185	1155.47
2005	1856	1842	1862.83	2520	2466	2636.62	1341	1147	1139.34
2006	1848	1858	1872.23	1969	2569	2695.99	1180	1349	1397.72
2007	1847	1850	1846.44	2580	1927	2483.77	1196	1184	1188.87
2008	1886	1848	1845.00	2602	2551	2498.31	1200	1200	1163.79
2009	1887	1888	1913.72	2531	2552	2531.12	1202	1207	1183.96
2010	1905	1887	1896.80	2358	2545	2537.24	1213	1211	1200.62
2011	1879	1906	1919.80	2301	2330	2469.04	1256	1206	1220.30
2012	1849	1883	1864.53	2264	2269	2377.20	1226	1260	1280.33
2013	1838	1860	1821.85	2194	2231	2281.59	1268	1226	1235.43
2014	1923	1840	1822.01	2220	2158	2178.74	1318	1268	1279.71
2015		1923	1980.74		2209	2115.52		1314	1351.87
2016		1924	2042.01		2198	2041.69		1316	1393.47
2017		1924	2103.28		2187	1967.87		1316	1435.07
Pearson correlation coefficient		0.2481	0.6701		0.0480	0.2885		0.2023	0.2345
MAPE forecast error		0.0133	0.0661		0.0745	0.0597		0.0416	0.0767

Table 5 Standard gauge rolling stock—part 2

Year	Freight cars			Coaches		
	Real data	Bayesian networks	Modified Holt's method	Real data	Bayesian networks	Modified Holt's method
2000	97,811			5781		
2001	98,272		94,388.56	5509		5529.98
2002	119,308		94,847.75	5178		5264.45
2003	111,532		115,951.63	5093		4951.19
2004	107,315	110,795	108,666.60	4895	4987	4784.52
2005	103,234	107,035	104,361.75	4495	4799	4613.88
2006	103,527	102,977	100,254.17	4397	4406	4298.88
2007	104,982	103,141	100,523.31	4270	4291	4118.68
2008	101,528	104,427	102,041.98	4050	4165	3987.34
2009	95,462	100,743	98,677.36	3800	3959	3813.70
2010	89,270	94,743	92,602.37	3795	3702	3595.22
2011	88,928	88,901	86,347.39	3556	3692	3522.18
2012	91,483	88,542	85,937.71	3356	3458	3359.58
2013	87,726	90,791	88,542.11	3083	3262	3175.40
2014	87,538	87,152	84,896.34	2806	2986	2929.56
2015		87,050	84,690.96		2707	2650.75
2016		86,460	81,896.75		2601	2441.75
2017		85,768	79,102.55		2500	2232.75
Pearson correlation coefficient		0.9376	0.7181		0.9863	0.9916
MAPE forecast error		0.0281	0.0419		0.0345	0.0222

forecasts for both methods are correlated poorly. MAPE error for the first is 4.16%, and 7.67% for the second.

Analyzing the data contained in Table 5, it can be said that since 2002 the number of freight cars has been steadily decreasing (Fig. 17). The Bayesian network method forecasts a slight further decline in this number in the coming years. Holt's method predicts a greater decrease in the amount of freight cars. Because the MAPE error for the first method is smaller and is 2.8%, and the correlation of real data and the forecast is as high as 0.94, the forecast for the next few years according to this method seems to be more reliable.

Figure 18 shows the number of passenger coaches, which since 2000 decreases linearly. Both methods of forecasting predict a further linear drop in the number of freight cars. The correlation coefficients of both methods show a very strong linear relationship between the actual data and the forecast. Forecast errors are also small.

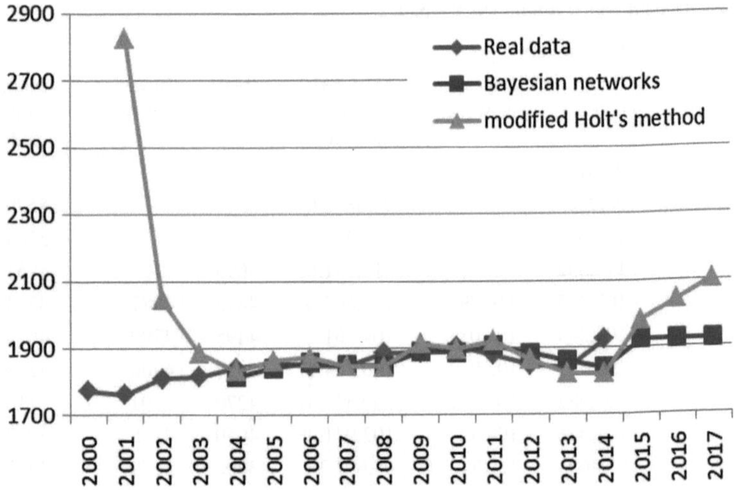

Fig. 14 Predicted number of electric locomotives

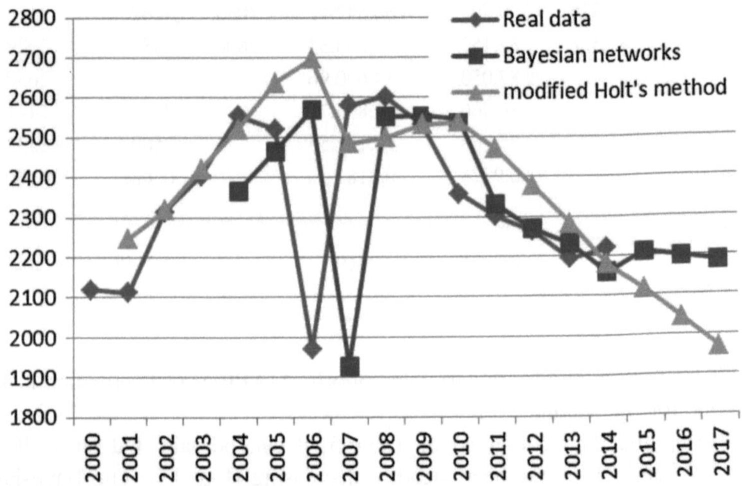

Fig. 15 Predicted number of diesel locomotives

4.3 Transportation of Cargo in Tons

Forecasts for freight transport and goods by rail in tons are based on data contained in Table 6.

There is clearly a growing trend in Poland in transportation of cargo by all means of transport, which is shown in Fig. 19. As for the forecast for the next three years, a modified Holt method predicts a slight decline in traffic, while the Bayesian

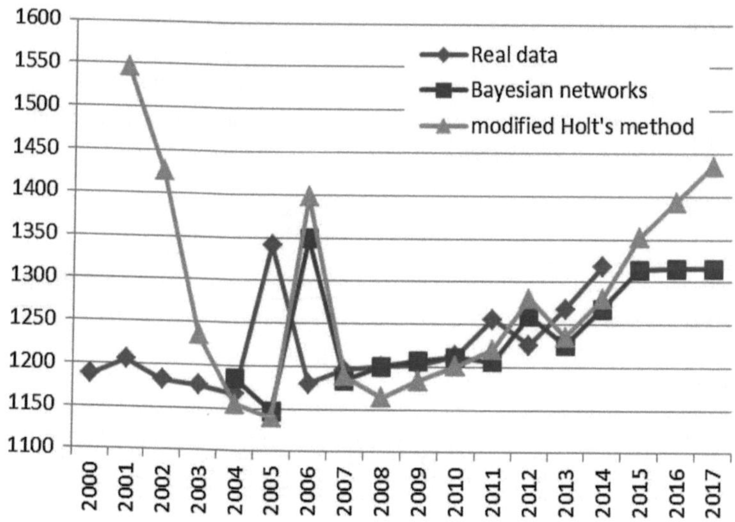

Fig. 16 Predicted number of electric railcars

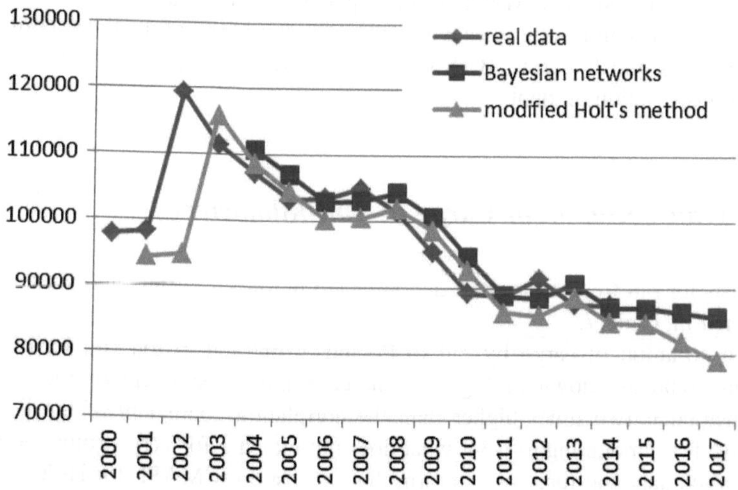

Fig. 17 Predicted number of freight car

network method shows growth. The forecast with the Holt method in this case has a smaller MAPE error, equal to 2.5%. Also, Pearson's correlation coefficient is high, close to unity, indicating a strong correlation of actual data and forecasts. The forecasting method based on Bayesian networks gives much worse results, so it seems safer to assume that there may be drop in freight volume.

According to the diagram in Fig. 20, cargo transportation by rail in the past fifteen years has large fluctuations. Holt's method in this case gives a forecast

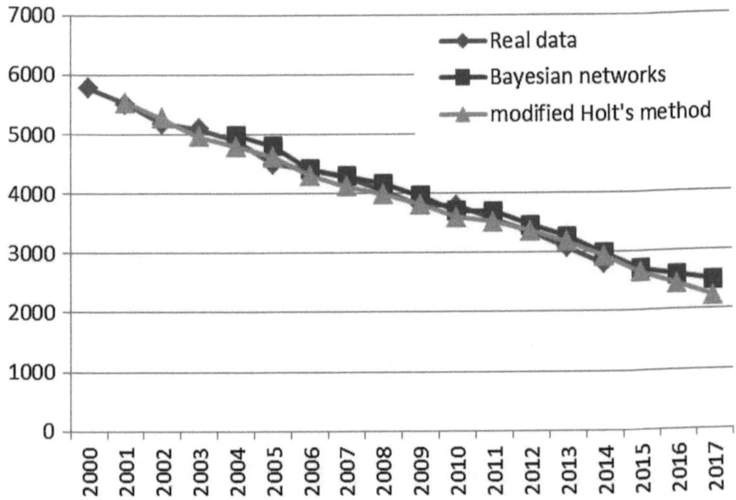

Fig. 18 Predicted number of passenger coaches

burdened with a smaller MAPE error, equal to 5.9%. It also has a correlation coefficient that is almost twice as large, which is about 60% of the linear correlation of actual data and forecasts. Both methods of forecasting predict a fall in freight cargo transportation amount.

4.4 Transportation of Cargo in Ton-Kilometers

Forecasts for rail freight transport denominated in ton-kilometers are based on data contained in Table 7.

Transportation of cargo by rail in Poland, expressed in ton-km, have a clear growing trend as shown in Fig. 21. The correlation coefficient of the modified Holt method is two times higher than the correlation coefficient of the Bayesian network. It is amounting to 0.34, what does not indicate for very strong correlation of observed data and forecasts. In turn, the forecast error MAPE for Holt's method here is lower and amounts to 5.75%. Despite the complicated course of the graph in Fig. 21, Holt's method shows the greatest smoothing of data in 2009.

4.5 Rail Transportation According to Types of Consignment

Forecasts of rail transportation according to types of consignment prediction, by export, import and transit are based on data contained in Table 8. Rail

Table 6 Cargo transportation by mode of transport prediction

Year	Total cargo transportation [thousands of tons]			Rail cargo transportation [thousands of tons]		
	Real data	Bayesian networks	Modified Holt's method	Real data	Bayesian networks	Modified Holt's method
2000	1,347,895			187,247		
2001	1,317,169		1,339,412	166,856		164,934.9
2002	1,304,387		1,296,423	222,908		144,905.2
2003	1,308,802		1,286,852	241,629		215,534.6
2004	1,324,511	1,341,971	1,303,788	282,919	236,534	240,060.5
2005	1,421,959	1,350,063	1,332,082	269,584	285,732	289,660.5
2006	1,480,259	1,443,742	1,480,167	291,420	258,532	273,093.8
2007	1,532,728	1,505,893	1,543,281	245,346	292,554	298,108.3
2008	1,655,965	1,550,995	1,589,938	248,860	237,354	242,409.4
2009	1,691,015	1,673,353	1,749,009	200,820	250,352	246,493.9
2010	1,838,492	1,714,870	1,755,593	216,899	204,383	190,009.1
2011	1,912,178	1,851,026	1,945,689	248,606	214,842	210,557.3
2012	1,844,070	1,931,295	2,005,298	230,878	246,352	249,686.4
2013	1,848,348	1,857,678	1,846,547	232,596	223,707	228,905.6
2014	1,839,961	1,876,631	1,843,273	227,820	231,117	231,086.5
2015		1,870,118	1,833,156		225,321	225,745.3
2016		1,892,679	1,826,175		222,741	223,631.4
2017		1,921,081	1,819,194		220,161	221,517.4
Pearson correlation coefficient		0.9584	0.9730		0.3419	0.6048
MAPE forecast error		0.0319	0.0252		0.1032	0.0589

transportation by different types of consignment in the years 2003–2014 is characterized with great variability in the amount of cargo.

Export of goods by rail hit bottom in 2009. In the whole period 2003–2014, high variability of the freight volume can be observed, see Fig. 22. Data and forecasts are weakly correlated, and the results are burdened with a high MAPE error. The future of export goods by rail is also projected by two different methods. Modified Holt's method predicts the future stabilization of export volume by rail, while the Bayesian network method predicts the decline of transport.

The import of goods by rail is a constant upward trend. In Fig. 23 a strong linear correlation data and forecasts can be seen at the same time. For the next three years a steady growth in freight volume is expected.

Railroad transportation of goods by transit through Poland is characterized with very large fluctuations in the years 2003–2014, as shown in Fig. 24. The modified

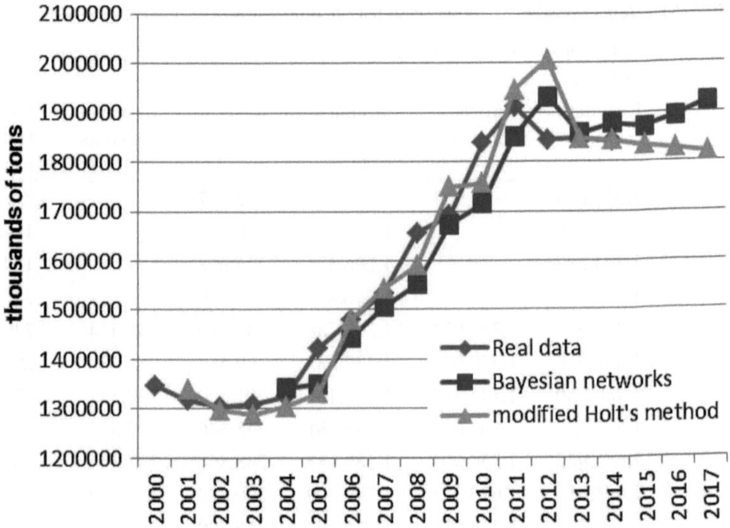

Fig. 19 Predicted total transportation of cargo

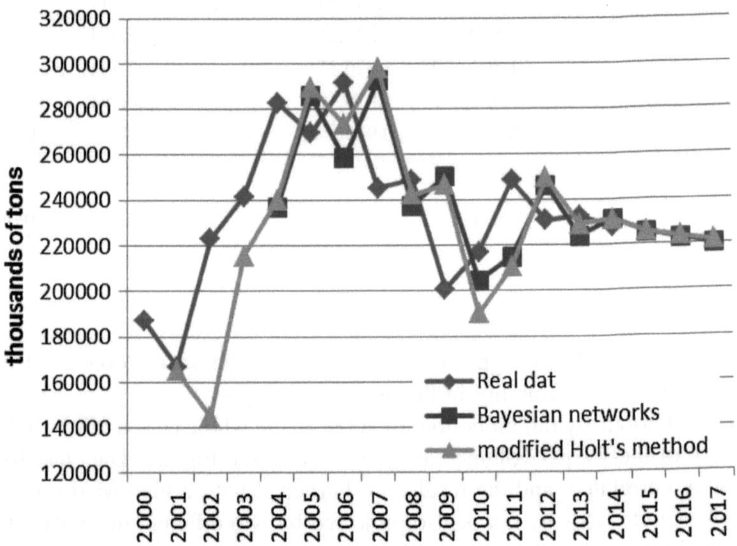

Fig. 20 Predicted amount of freight cargo transportation

Table 7 Cargo transportation by rail

Year	Transportation of cargo by rail [millions of ton-kilometers]		
	Real data	Bayesian networks	Modified Holt's method
2000	54,448		
2001	47,913		54,492.32
2002	47,756		51,150.58
2003	49,549		49,305.28
2004	52,137	49,148	49,258.95
2005	49,779	52,204	50,573.77
2006	53,427	49,894	50,060.65
2007	54,253	53,324	51,671.99
2008	52,043	54,026	52,952.14
2009	43,554	53,062	52,491.93
2010	48,795	44,398	47,879.64
2011	53,746	46,853	48,144.88
2012	48,903	54,469	50,841.70
2013	50,881	49,220	49,779.33
2014	50,073	50,212	50,239.76
2015		49,823	50,071.24
2016		50,187	49,984.93
2017		50,318	49,898.63
Pearson correlation coefficient		0.1598	0.3416
MAPE forecast error		0.0739	0.0575

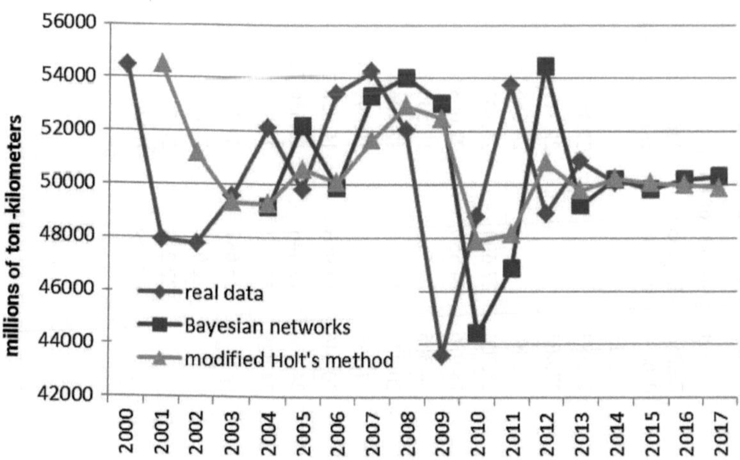

Fig. 21 Predicted amount of total cargo transportation

Table 8 Rail transportation by types of consignment

Year	Export [thousands of tons]			Import [thousands of tons]			Transit [thousands of tons]		
	Real data	Bayesian networks	Modified Holt's method	Real data	Bayesian networks	Modified Holt's method	Real data	Bayesian networks	Modified Holt's method
2003	29,515			2114			3172		
2004	24,583		27,451.76	2426		2435.06	2371		3233.72
2005	27,463		22,711.09	2239		2751.34	2765		3160.89
2006	28,886		24,686.19	5982		2788.71	3363		3129.82
2007	23,704	27,690	26,803.33	8895	6707	4775.46	3702	3251	3182.64
2008	18,936	24,656	22,688.92	11,347	9279	7847.09	3722	3703	3288.49
2009	14,365	19,516	17,535.77	9888	12,188	11,318.60	1797	3873	3399.65
2010	20,872	13,035	12,406.81	12,258	10,612	12,703.47	3203	1751	3208.78
2011	17,235	21,497	17,699.43	13,872	13,541	14,373.67	3075	3432	3209.32
2012	20,446	15,956	15,828.76	19,172	14,394	15,932.75	4591	3188	3189.60
2013	24,778	19,446	18,636.72	18,693	20,101	19,462.57	5022	4311	3404.62
2014	24,429	24,417	23,505.81	21,301	19,378	21,366.64	4206	5150	3703.78
2015		23,687	24,354.23		22,289	23,520.06		4130	3887.20
2016		22,945	24,468.71		22,939	25,697.83		4215	4010.36
2017		22,203	24,583.19		23,670	27,875.60		4228	4133.51
Pearson correlation coefficient		0.2208	0.5497		0.8856	0.9528		0.2698	0.2417
MAPE forecast error		0.2359	0.1753		0.1542	0.1789		0.3154	0.2288

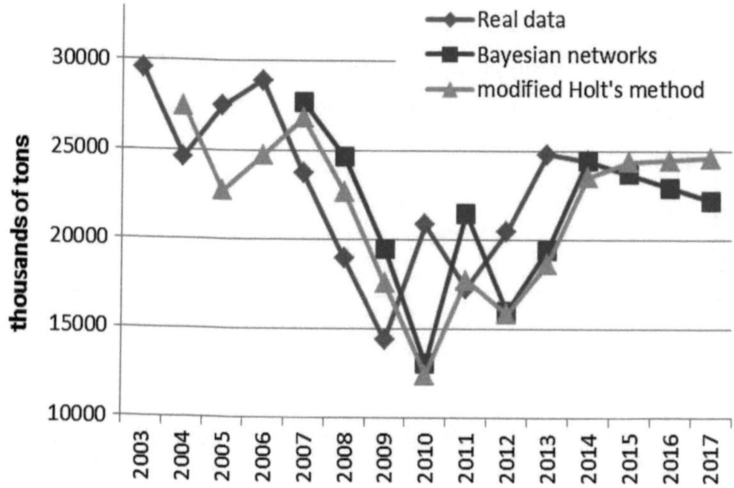

Fig. 22 Predicted export by rail transportation

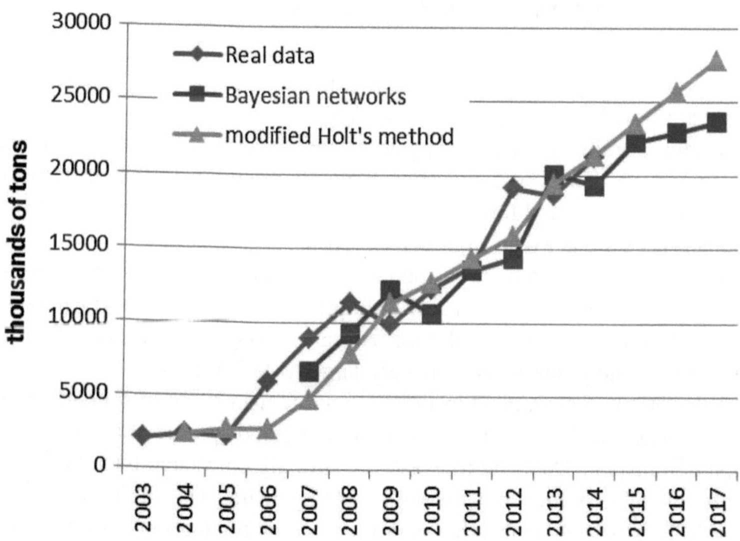

Fig. 23 Predicted imports by rail

Holt method has smoothed the graph. For subsequent years, both methods predict a slight increase in the volume of cargo in transit. The resulting predictions are affected by significant MAPE error. The correlation coefficient indicates a weak link between data and forecasts.

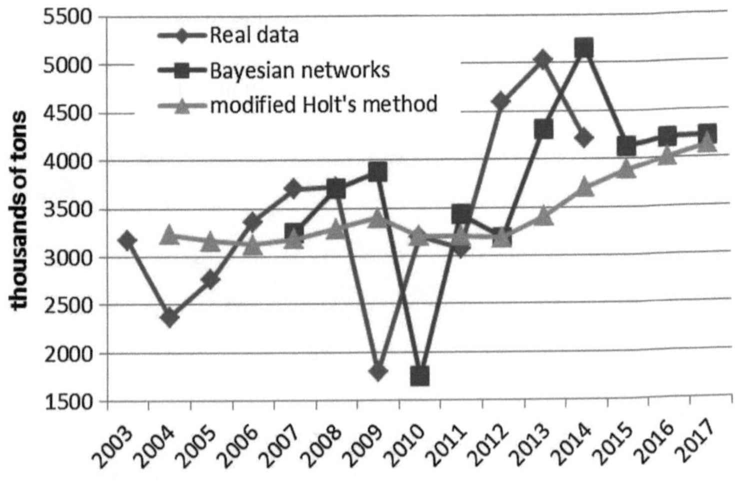

Fig. 24 Predicted transit by rail transportation

4.6 Rail Transportation of Containers

Utilization of railway for the transport of containers has a clear upward trend; the forecasts are based on data contained in Table 9.

Utilization of railway for the transport of containers has a clear growing trend as shown in Fig. 25. The two forecasting methods predict a further increase in this type of transport. Data and forecasts are strongly correlated linearly. The results are, unfortunately, burdened with high MAPE errors.

Transportation of containers in the country is also characterized by an upward trend, as shown in Fig. 26. Both forecasting methods predict a further increase in this type of transport. Data and forecasts are strongly correlated linearly. The results, unfortunately, are subject to very large MAPE errors.

Transportation of containers in international transport are also characterized by an upward trend, see Fig. 27. Both forecasting methods predict a further increase in this type of transport. Data and forecasts are strongly correlated linearly. The results are, unfortunately, burdened with high MAPE errors. The Holt method has the lower MAPE error and, as the chart shows, smooths fluctuations in freight volume.

4.6.1 A Summary of the Calculations

The results and future predictions look believable. The two methods used, modified Holt method and Bayesian networks, give comparable results. The values of MAPE errors and the coefficients of corelations are respectively similar for both methods.

Table 9 Rail transportation of containers

Year	Total containers [thousand tone]			National container transport [thousand tone]			International container transport [thousand tone]		
	Real data	Bayesian networks	Modified Holt's method	Real data	Bayesian networks	Modified Holt's method	Real data	Bayesian networks	Modified Holt's method
2004	2378			201			2177		
2005	2524		2659.83	122		124.63	2402		2621.08
2006	3294		2822.57	249		44.93	3045		3019.76
2007	4084		3506.59	762		224.25	3322		3467.91
2008	4801	4125	4323.93	812	906	879.15	3989	3288	3880.76
2009	3276	5228	5102.61	527	887	922.13	2749	4401	4345.49
2010	4266	3897	3900.53	847	523	534.44	3419	3034	4457.80
2011	5735	5126	4465.84	1390	1093	927.16	4345	3881	4677.53
2012	7874	5938	5855.43	2220	169	1594.89	5654	4270	5038.23
2013	8527	8190	8011.22	2602	2371	2594.12	5925	5866	5593.61
2014	9249	8703	9020.91	2907	2740	2990.02	6342	5880	6093.21
2015		9556	9822.26		3024	3273.86		6699	6577.39
2016		10,028	10,442.97		3067	3639.15		6768	7011.31
2017		10,534	11,063.67		3196	4004.44		7050	7445.23
Pearson correlation coefficient		0.8668	0.9118		0.6667	0.9484		0.7340	0.8868
MAPE forecast error		0.1820	0.1644		0.3522	0.3394		0.1891	0.1336

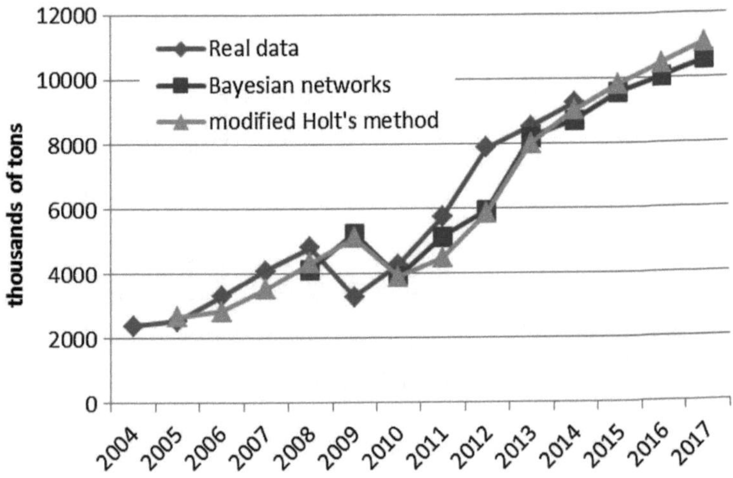

Fig. 25 Predicted total transport of containers by rail

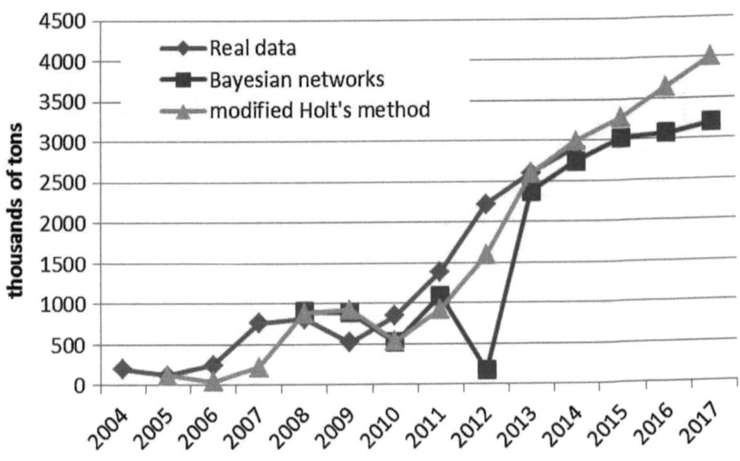

Fig. 26 Predicted national transport of containers by rail

In the case of the Bayesian network, the expired forecasts depict on the graphs of results the slight shift to the right. The maxima and minima of the original time series were preserved. The modified Holt method clearly smooths the random jumps of the original time series.

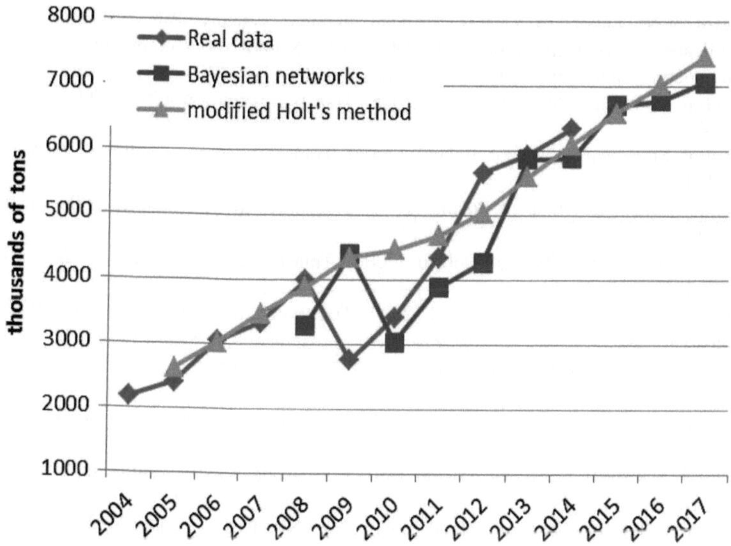

Fig. 27 Predicted international transport of containers by rail

5 Conclusion

On the basis of the calculation results, it is obvious that if nothing changes in the policy of rail transport, it will continue to see reduced length of the railway line, electrified lines and locomotives (although for electrical forecasts predict stagnation or even a slight increase). The number of freight cars and passenger cars is also predicted to be continuously reduced. No wonder that in such a situation, rail services have large fluctuations in the coming years and are predicted to continue to show a slight decrease in traffic, which both methods confirm in extremely similar fashion (Fig. 18). This is alarming because, as shown in Figs. 20, 21, and 22, interest is growing in rail transport. The railway has large share of export, import and transit. An increase in container traffic is observed, which may be the greatest argument in the context of the modernization of infrastructure and investment programs analyzed in the chapter.

References

1. Acar Y, Gardner ES (2012) Forecasting method selection in a global supply chain. Int J Forecast 28(4):842–848
2. Bolstad WM (2013) Introduction to Bayesian statistics. Wiley, London
3. Brown RG (1959) Statistical forecasting for inventory control. McGraw-Hill, New York

4. Castro LN, Zuben FJ (1999) Artificial immune systems, Part I—Basic theory and applications, technical report, TR-DCA 01. School of Computing and Electrical Engineering, State University of Campinas, Brazil
5. Central Statistical Office of Poland (2005) Transport-activity results in 2004. Warsaw (in Polish)
6. Central Statistical Office of Poland (2006) Transport-activity results in 2005. Warsaw (in Polish)
7. Central Statistical Office of Poland (2007) Transport-activity results in 2006. Warsaw (in Polish)
8. Central Statistical Office of Poland (2008) Transport-activity results in 2007. Warsaw (in Polish)
9. Central Statistical Office of Poland (2009) Transport-activity results in 2008. Warsaw (in Polish)
10. Central Statistical Office of Poland (2010) Transport-activity results in 2009. Warsaw (in Polish)
11. Central Statistical Office of Poland (2011) Transport-activity results in 2010. Warsaw (in Polish)
12. Central Statistical Office of Poland (2012) Transport-activity results in 2011. Warsaw (in Polish)
13. Central Statistical Office of Poland (2013) Transport-activity results in 2012. Warsaw (in Polish)
14. Central Statistical Office of Poland (2014) Transport-activity results in 2013. Warsaw (in Polish)
15. Central Statistical Office of Poland (2015) Transport-activity results in 2014. Warsaw (in Polish)
16. Cheng J, Druzdzel MJ (2000) AIS-BN: An adaptive importance sampling algorithm for evidential reasoning in large Bayesian networks. J Artif Intell Res 13:155–188
17. De Gooijer JG, Hyndman RJ (2006) 25 years of time series forecasting. Int J Forecast 22 (3):443–473
18. Dempster AP, Laird NM, Rubin DB (1977) Maximum likelihood from incomplete data via the EM algorithm. J Roy Stat Soc B 39(1):1–38
19. Ditmann P (2008) Forecasting in the enterprise. Methods and application. Cracow (in Polish)
20. Eisuke K, Maasaki H, Takao M (2012) Application of Bayesian network to stock price prediction. Artif Intell Res 1(2):171–184
21. Gardner ES (1985) Exponential smoothing: the state of the art. J Forecast 4(1):1–38
22. Gołąb J, Jakóbisiak M, Lasek W, Stokłosa T (2008) Immunology. Polish Scientific Publishers, Warsaw (in Polish)
23. Government Centre for Strategic Studies (2015) The updated concept of national spatial development. Warsaw (in Polish)
24. Heckerman D, Geiger D, Chickering DM (1995) Learning Bayesian networks: the combination of knowledge and statistical data. Mach Learn 20(3):197–243
25. Holt CC (2004) Forecasting seasonals and trends by exponentially weighted averages. Int J Forecast 20(1):5–10
26. Lawton R (1998) How should additive Holt-Winters estimates be corrected? Int J Forecast 14:393–403
27. Lénárt B (2011) Automatic identification of ARIMA models with neural network. Period Polytech Transp Eng 39(1):39–42
28. Lin KP, Pai PF, Yang SL (2011) Forecasting concentrations of air pollutants by logarithm support vector regression with immune algorithms. Appl Math Comput 217:5318–5327
29. Lin KP, Pai PF, Lu YM, Chang PT (2013) Revenue forecasting using a least-squares support vector regression model in a fuzzy environment. Inf Sci 220:196–209
30. Ministry of Infrastructure (2008) Masterplan for rail transport until 2030, Warsaw (in Polish)
31. Ministry of Infrastructure (2012) Report on the implementation in 2011. Multi-annual investment program of railway until 2013 with the prospect of 2015. Warsaw (in Polish)

32. Ministry of Regional Development (2006) National development strategy 2007–2015. Warsaw (in Polish)
33. Ministry of Transport, Construction and Maritime Economy (2013) Long-term investment rail programme until 2015. Warsaw (in Polish)
34. Mrówczyńska B, Łachacz K, Haniszewski T, Sładkowski A (2012) A comparison of forecasting the results for road transportation needs. Transport 27(1):73–78
35. Pinto R, Gaiardelli P (2013) Setting forecasting model parameters using unconstrained direct search methods: an empirical evaluation. Expert Syst Appl 40:5331–5340
36. PKP Polish Railway Lines SA (2005) Technical requirements of surface maintenance on railway lines. Attachment to the Ordinance No. 14/2005 The Management Board of PKP Polish Railway Lines SA of 18 May 2005. Warsaw (in Polish)
37. Syryjczyk T (2009) The white paper of problems map of the Polish railways. Warsaw. http://www.rynek-kolejowy.pl/pliki/BialaKsiega.pdf. Accessed 12 Sept 2015 (in Polish)
38. Winters PR (1960) Forecasting sales by exponentially weighted moving averages. Manag Sci 6:324–342
39. Yager RR (2013) Exponential smoothing with credibility weighted observations. Inf Sci 252:96–105
40. Yule GU (1927) On the method of investigating periodicities in disturbed series, with special reference to Wolfer's sunspot numbers. Philos Trans R Soc London, pp 267–298
41. Zeliaś A, Pawełek B, Wanat S (2004) Economic forecasting: theory, examples, exercises. Polish Scientific Publishers. Warsaw (in Polish)

Dynamic Optimization of Railcar Traffic Volumes at Railway Nodes

Aleksandr Rakhmangulov, Aleksander Sładkowski, Nikita Osintsev, Pavel Mishkurov and Dmitri Muravev

Abstract A major direction in the development of modern world transport systems involves the concentration of freight and traffic flows within international transport corridors and transport nodes in terminals and hubs. The changing role of rail transport is taking place under these conditions. Increased structural complexity and irregularities in cargo and railcar traffic volumes have been observed, despite the higher levels of transport equipment and technology standardization, the increased container transport volumes and consequent reduction in the cost of intermodal operations, and the interaction between different modes in transport nodes. This is largely due to the increasing need for cargo owners to lower logistics and warehouse costs, which is achieved by reducing the size of freight shipments and ensuring their uniform delivery. Moreover, privatization of the railway industry in certain countries and the sale of rolling stock to operating companies have made the coordinated management of rail fleets more difficult. The demand for improved efficiency of railcar traffic volume management in the case of complex structures is especially relevant for large railway nodes, particularly the transport systems of industrial enterprises. Here, the application of traditional approaches to the management of transportation processes involving individual elements of traffic flow (trains, railcars, locomotives) and transport infrastructure (railway stations, loading areas, rail hauls) leads to additional transport costs as a result of the increased length of time that railcars are located at the railway node. The aim of this study, therefore,

A. Rakhmangulov (✉) · N. Osintsev · P. Mishkurov · D. Muravev
Nosov Magnitogorsk State Technical University, Magnitogorsk, Russia
e-mail: prtrans@gmail.com

N. Osintsev
e-mail: nikita.osintsev@gmail.com

P. Mishkurov
e-mail: mishkuroff@mail.ru

D. Muravev
e-mail: mura4747@gmail.com

A. Sładkowski
Silesian University of Technology, Katowice, Poland
e-mail: aleksander.sladkowski@polsl.pl

A. Sładkowski (ed.), *Rail Transport—Systems Approach*, Studies in Systems, Decision and Control 87, DOI 10.1007/978-3-319-51502-1_10

is to provide improved methods for the management of RTVs at complex railway nodes based on a systematic review of RTVs in conjunction with transport infrastructure and traffic control systems. The authors review the case of a systematic approach to the organization and management of railcar traffic volumes, both for mainline rail transport and at railway nodes and industrial rail transport. This study investigates the impact of irregular railcar traffic volumes on railway node functioning by applying a dynamic simulation model of the transport system of a large metallurgical enterprise. The application of the original methodology for assessing the amount of information in the operational management system for rail transportation allowed us to estimate the effect of structural complexity and railcar traffic volume irregularities on the efficiency of the dispatch control system. The authors propose a new system of parameters and indicators for assessing railcar traffic volumes, taking into account factors of complexity and irregularity, and provide a comprehensive assessment of its effectiveness for railcar traffic volume management. For this purpose, a method of dynamic optimization of these parameters (dynamic programming) is selected. The main hypothesis of the study is that improved accuracy in parameter optimization for irregular railcar traffic volumes is achieved by adjusting the duration of the base periods which constitute the optimization period in the dynamic problem. In this study we have formulated a mathematical model for dynamic parameter optimization of railcar traffic volume on the basis of base periods of variable duration, with an algorithm for model implementation as well. Experimental verification of the effectiveness of the developed model and algorithm is conducted on the constructed process-centric railway node simulation model. Three series of experiments are conducted: without railcar traffic volume optimization, and with dynamic optimization of railcar traffic volume with the application of base periods of constant and variable duration. The experimental results demonstrate increased optimization accuracy with the use of the proposed model, reducing transport costs (time of railcar traffic volume handling) at the railway node by 11% on average. For the realization of the model and algorithm, a method is proposed for their integration in information management and intellectual transport systems.

Keywords Railway node · Railcar traffic volumes · Irregularity · Railcar traffic volumes with complex structure · Information quantity · System of parameters for railcar traffic volumes · System dynamic model · Dynamic optimization · Mathematical model · Process-centric simulation model

1 Main Conditions and Tendencies of Rail Transport Functioning

Transportation science has accumulated much specific knowledge about the various functional components of transport systems since the early 1970s. This has been formed on the basis of documented research, the main method of which has become a systematic approach. A systematic review of the transport process involves the study of the system not as individual features of transport elements, but of these elements as a whole. The set of transportation facilities that constitute the transportation system is the result of diverse relations between elements of the system. The systematic review of technical aspects of transport and technology and of the organization of transportation processes and freight traffic flow, along with economic constraints (costs), has helped in formulating modern theory and developing methods for railcar traffic volume organization.

The basis of modern organizational systems for railcar traffic volume (RTV) on the rail systems of various countries is the train timetable and train formation plan, which are designed to minimize total transportation costs for a set of RTV of known volume and structure. Changes in traffic volumes and RTV structure are considered by periodic adjustment of the formation plan. Formally speaking, current organizational systems comprehensively solve two systemic problems in rail transport. The first is the maintenance of a set of features to preserve the integrity and ability to perform rail transport functions—RTV organization based on train formation plans; while the second is the implementation of a mechanism for maintaining appropriate system properties in terms of the disruptive impact of the external and internal environment—the adjustment mechanism of these plans.

However, the implementation of sophisticated logistics solutions beginning in the early 2000s has led to a dramatic increase in the complexity of operating conditions for rail transport, which must function in these ever-changing operational and economic conditions. Moreover, the greatest influence on operations is from external factors, particularly the needs of freight owners, competition in carrier markets, rolling stock owners, competition from other modes of transport, and price fluctuations.

Existing rail transport systems for traffic organization do not adequately address these external factors, particularly the costs and losses arising from actions taken upon failure to meet freight owner demands for transportation quality (timeliness). In market conditions, such losses not only lead to additional costs in the creation of reserves and reduced operating funds, but also in the loss of competitive advantage.

Two trends have emerged as a result of inadequate orientation of existing RTV systems to the needs of rail transport customers, especially in the Russian Federation. The first is associated with the transition of many small and medium customers' volume shipping on other modes of transport, in particular road transport, which reduces the profitability of the railways. There is a tendency worldwide to increase the share of transportation through the route trains circulating on solid (hard) the graphics thread that enables the most effective use of rail advantages.

In addition, small-lot cargoes and cargoes transported by carload freight shipments are often transported by road. This situation is typical in the USA. Piggyback and roadrailer transportation, combining the advantages of rail transport (low prime cost, regularity, reliability) with the advantages of road transport (flexibility, speed) is actively developing in Europe, where freight train weight and distance of goods delivery are insignificant. This combination has provided the means to satisfy the needs of transport customers in a modern transportation system.

A second emerging trend is driven by freight owners' desire to ensure timely transportation by taking advantage of competition between operators as owners of rolling stock [1]. The lack of coordination between operators on rolling stock utilization often leads to violations of train formation plans, creating the need for more frequent adjustments compared with the existing sequence, or causes operators to suffer losses associated with inefficient utilization of throughput and handling capacity of railway stations.

The realization of logistics solutions, in particular for systems of inventory management, has increased the tendency toward reduced consignment size in order to lower storage costs for transport clients.

The emergence of a large number of low-power jets contributes to the complexity of the RTV structure (Fig. 1). This trend is observed most clearly in Russia, where in accordance with the formation and development of a competitive rail transport environment, the share of small-lot goods is reduced in the total turnover (Fig. 2) [2, 3].

Obviously, the provision of timely goods transportation in small batches on rail transport is related to increased transportation costs. One direction for the conservation of rail transport competitiveness in such conditions is through improved traffic management systems, oriented to meet the needs of transport customers.

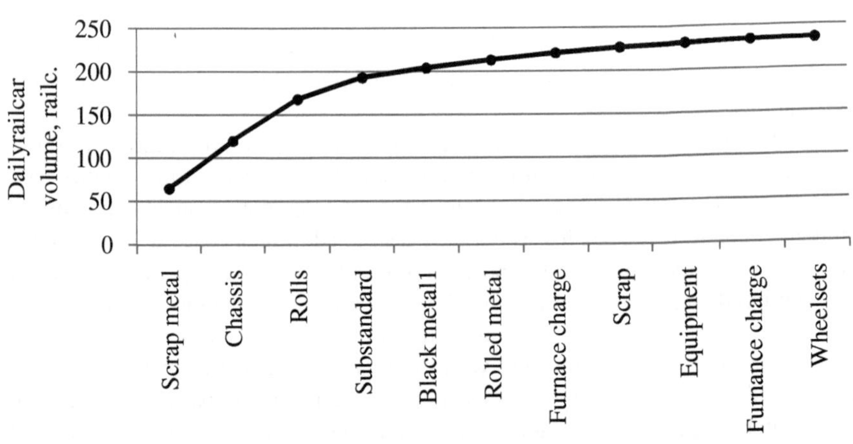

Fig. 1 The example of a Pareto chart for the value of railcar volume on the jets at the industrial railway station

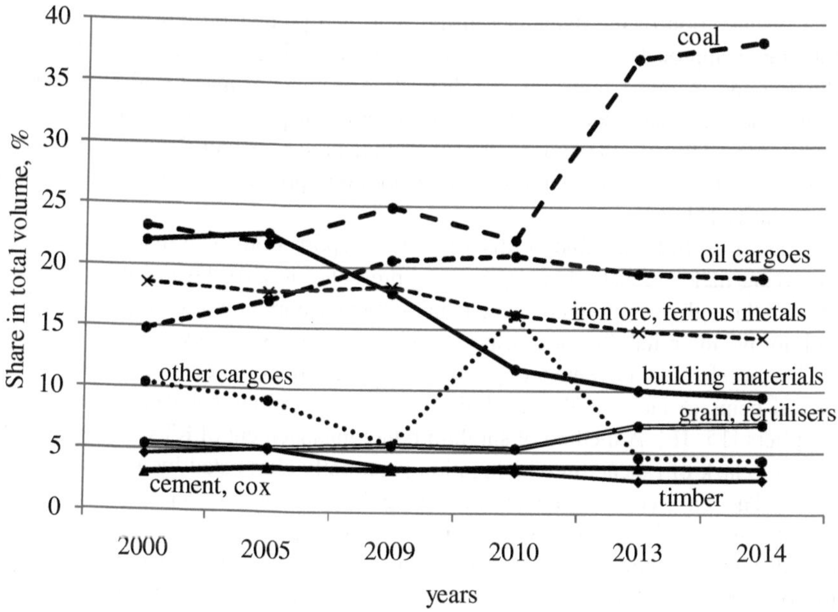

Fig. 2 The structure of volume by main freight type at the mainline rail transport in the Russian Federation, %

2 A Review of Research on System Approaches for the Organization and Management of RTV

Extensive experience in solving problem of efficiency within complex systems dealing with elements of uncertainty can be found in other sciences. Stability, i.e. the ability of a system to maintain defined parameters within certain borders in the presence of disturbance factors, is achieved through a variety of mechanisms.

Management is one such component of complex artificial systems. Management processes have a twofold orientation—strategic and operational. Strategic management addresses the challenge of system development, while operational management is a "working" strategic mechanism. The task is to neutralize the disturbing effect on the system through timely adjustments. Therefore, "hard" organizational forms of transportation processes are ineffective in the context of modern, structurally sophisticated RTV systems, and result in increased irregularity.

Such opinions in relation to the various subsystems of rail transport have been presented in many works, particularly in [4–6].

There are also many studies about the problems of operational management in mainline rail transport, for example [7, 8]. The problem of railcar acceleration is discussed based on methods of operational management of RTV in [9, 10]. The process of industrial transport management, including the subsystem of interaction with the main modes of transport, by its very nature requires a unique approach.

Consideration of the dynamics of transport and production processes, using an integrated approach to management problems affecting the interests of the production and mainline transport, are distinctive features of many studies on operational management issues within industrial rail transport performed in the 1980s and 1990s. The organization of transportation processes and flexible technology have helped to improve reliability in transportation production under conditions of irregularity [11–13].

The most fully flexible forms of rail transport from a systems perspective involving major industrial enterprises are presented in [14]. Dynamic reserves are created through operational management of the transportation process, i.e. the ability to improve the reliability of transport services with the same level of technology.

The accelerated movement of goods through less important consumers in a given period helps to increase the reliability of communications between them and the suppliers [15, 16]. A method of single-product management of RTV is presented in [17]: the lack of railcars in one direction is partially eliminated by a surplus of the other. The balance between production and consumption can thus be quickly adjusted to exploit the adaptive capacity of the transport system [14]. Optimization of industrial transport by considering multi-criteria goals is considered in [18]. The use of technology for calculating optimal interaction within sorting complexes for industrial and mainline transport is presented in [19]. Algorithms for optimal control of shunting work in industrial transport hubs are presented in [20]. These methods expand the potential of industrial RTV systems to adapt to changing conditions.

Nowadays, strategies for improving RTV calculation and adjustment systems for mainline rail transport are geared toward more detailed consideration of the various factors that have a significant impact on the effectiveness of the train formation plan. For example, in [21], consideration of these factors, including conditions of uncertainty, is proposed for implementation on the basis of the aggregation (stream representation) of information in management information systems and the application of algorithms and models of stream optimization. Methods and techniques for adapting to variation in RTV parameter within the development of rail transport are presented in [22].

Gradually, a new trend has emerged in transport science, according to which RTV organization is considered not just a method for volume distribution of sorting operations between railway stations, but as a complex system of interactions between participants of the transportation process. We must consider not only the costs related to the accumulation and processing of railcars, but also many requirements imposed on this system. For example, in [23], proposed criteria for choosing the optimal train formation plan include car-kilometers, delivery time and the required number of locomotives, as well as total energy consumption. This approach requires the development of multi-criteria models and methods of calculation.

The influence of RTV irregularity, including daily irregularity, is considered in [24], focusing on meeting the changing operational needs of rail transport clients with respect to empty railcars.

The need to take into account many factors determining the effectiveness for RTV organization systems and timeliness of rail freight transportation has required the development of new mathematical methods for solving transport problems not previously used for RTV optimization. These methods consider the time factors of maximal dynamic flow [19] and dynamic transport through a network matrix [19].

The dynamic programming method, which is oriented toward the consideration of RTV irregularity, has been shown to be effective [14, 25].

The dynamic coordination of production, transport and consumption is "based on the application of multistage approach, takes into account the dynamics of production volumes and consumption, the irregularity of stocks in final and intermediate points". This method improves the formulation of a linear programming problem to improve the accuracy of the response on the dynamics of transport and production processes. However, determining transport costs only at the intersection of transport and production leads to less accurate optimization; thus it is necessary to take into account losses emerging as a result of random variations in cargo delivery times as well as random deviations from the "planned" rhythm of consumption. A stochastic statement of the dynamic transportation problem with delays and with a random spread of delivery and consumption time [14] compensates for this drawback. The optimization problem in the stochastic statement is formed through analysis of the interaction between shipper and consignee in a random environment. It is premised on the use of a time buffer in organizing the arrival of goods, depending on the magnitude of possible damage, incomplete delivery and storage. The stochastic statement thus increases optimization precision at the crossroads of transport and production. This stochastic formulation has been developed as a model for controlling flow with railcars of different owners. The model is based on an approach for solving the multi-product dynamic transport with delays [26], which is reduced to a static multi-product transport problem on a network with standardized methods for its solution.

Linear programming [14] is a dynamic method used for solving routing problems, while integer programming [27] and the branch and bound method [1, 19] are used for dynamic schedule optimization for vehicle maintenance. The common element among these methods is the presentation of a system for organizing RTV, and transport in general, as a flexible, adaptive system whose operation is optimized at a predetermined time interval.

European researchers have identified mathematical and heuristic approaches for solving optimization problems for RTV movement in rail transport nodes. The discrete mathematical models, algorithms, and implementation and development strategies in planning RTV movement at various speeds for a transportation network were presented in the late 1990s [28–32].

The authors describe mathematical models and algorithms of route adjustment for train movements in rail transport nodes [33–40]. These works propose the adjustment of train routes with the goal of increased responsiveness to changing train schedule parameters. These studies are compiled in [41, 42], which present modern approaches for solving the problems of RTV management in transport nodes. The direction of European RTV management system development is described in studies [43–45], and proves the relevance of this approach from the experience of North American researchers. Heuristic approaches for RTV management optimization in rail transportation hubs are proposed in [46, 47].

Studies [48–50] focus on the practical application of modern heuristic methods based on multiproduct transportation solutions. These methods were developed in [33, 43, 51, 52], which describe transport optimization through minimization of costs for separate railcars in transport systems, in rail transport nodes and in industrial transport systems as well.

The implementation of an adaptive RTV structure from a technological perspective was presented in [14]. The structural technology is the scope of the technological methods enabling RTV distribution between elements of the transport system (infrastructure elements) in the operational mode so as to ensure the required intensity of RTV movement (timeliness) and a rational workload of the transport elements. In other words, changing the parameters and traffic intensity at different stages of the transportation operations ensures that railcar downtime due to transportation element delays is minimized, thus reducing unproductive downtime of infrastructure elements and vice versa. Modern structural technology strategies are analyzed in greater detail with respect to industrial rail transport, and generalization to conditions of mainline railways are examined in [11, 14].

The connection between the structural elements in online interaction "…of the various devices of the system is expressed by the technology of the transportation process" [14]. The value alignment of available throughput of stations happens as a result of technology adjustment, for example, a consistent increase in train groups formed on an "overloaded" sorting station and their subsequent reformation to less loaded stations. The implementation of structural technology requires sustainable information links in the traffic control system. These links enable the adaptation of information transport systems to changing RTV parameters, and can be considered as a structural connection through which to "transfer" throughput and capacity between elements of the transport infrastructure [1, 14].

However, the application of mathematical models has known limitations regarding the adequacy of the results obtained in real situations, because they do not take into account many disturbance factors that have a significant impact on the transport operations. Consideration of these factors requires complex mathematical tools for calculation of the parameters of an RTV organizational system and for optimal development of transport infrastructure. Each of the methods for analysis of transport facilities has advantages and disadvantages and is oriented for solving a certain range of problems. The best method for an assigned problem is one which can be applied systematically, and should combine the advantages of the individual models. The idea is to create a "synthesis model" representing complex simulation models [6, 15] and one or more optimization models [53].

The results obtained with "abstract optimization" models can rarely be applied in their "pure" form in practice. Therefore, some authors have advised using "a simplified model (fast algorithm) [to] conduct the rejection of possible alternatives and [form] the variety of options, which is presented for simulation experiment on the complete model" [19]. This approach was tested on a transport system in [14].

However, the construction of analytical and simulation models to address problems of RTV parameter determination will require additional research and refinement.

Railway and industrial transport systems occupy a special place in flexible, adaptive RTV management systems. A particular challenge of such systems at the interface between the mainline rail transport and the consignee or manufacturing system is related to the need to adapt in operating mode to changes in RTV parameters of mainline rail transport and to the fluctuations in consumer demand for cargo transportation.

As mentioned before, mainline rail transport seeks to reduce costs by merging streams of RTV and maximizing utilization of throughput and carrying capacity based on the routing. This objective is fundamentally contrary to freight owners' need for timely transportation. Rail nodes and industrial transport have traditionally played the role of a "buffer" in coordinating the work of the mainline rail transport and consumer warehouses (technological units), thereby reducing total transport and storage or logistics costs.

The problem of RTV organization and management at railway nodes is extremely complex and multifaceted, requiring a systematic approach and a wide range of system analysis methods.

On the other hand, the implementation of an adaptive approach for RTV management and adaptive solutions in response to changing RTV structures and movement in the node has also been discussed. Such an approach is based on the application of software to create information management systems or intelligent transportation systems of railway nodes, using a variety of mathematical and simulation models for operational management of RTV parameters.

A mathematical programming apparatus for optimizing railcar movement in industrial transport systems based on the identification of regularities for each concrete enterprise, determining the rhythm of supply, unloading, loading and product shipment to the consumer, is presented in [54–57]. Methods for the formation and direction of computerized RTV control systems in railway nodes are described in [51, 58–60], and general construction principles of an automated traffic control system in [61]. In [62–64], the authors describe a two-level optimization system as a basis for implementing modern information technology at railway nodes [65, 66].

The dynamic programming method is a mathematical approach that can reflect the complex dynamics of RTV parameters in the railway node, thus providing justification for operational decisions on the management structure and traffic routes of RTV in the nodes. However, this method offers only a general direction for solving optimization problems, and cannot provide the accuracy needed for operational management of RTV. Optimization models using classical dynamic programming are not able to respond to the system uncertainty of RTV associated with the changing parameters and duration of technological cycles.

The dynamic approval method eliminates these optimization accuracy disadvantages [14, 19, 25].

The dynamic coordination of production, transport and consumption addresses the problem of optimization by taking into account the dynamics of production and

Table 1 The results of a comparative analysis of methods for dynamic parameter optimization of RTV conducted by Russian researchers

Indicators		Method of dynamic alignment	Stochastic statement of the dynamic transportation problem with delays	Multi-transport problem
RTV parameters	Traffic route	–	–	–
	Travel time on the route	+	+	+
RTV parameters	Mass	+	+	+
	Complexity of structure	–	–	+
Limitations of objective function	Throughput reserve	+	+	+
	Handling capacity reserve	+	+	+
Components of the objective function	Mass of RTV	+	+	+
	Costs for RTV movement	+	+	+
	Costs for RTV delay	+	+	+
Response to changes of actual RTV parameters	Irregularity of RTV	+	+	–
	Infrequency of RTV	–	–	–
Planning mode	Current	+	–	+
	Operative	–	+	–

consumption volumes and inventory imbalance at final and intermediate points [14]. A dynamic coordination approach has been applied to the stochastic formulation of a dynamic transportation problem with delays [14] and in solving multi-product transportation problems [19]. The results of a comparative analysis of methods for dynamic parameter optimization of RTV are presented in Table 1.

The main disadvantage of dynamic programming methods is insufficient orientation to irregular RTV. RTV irregularity arises as a result of separate streams received in the railway node at random times. If the interval between these times is less than the duration of the base period of the dynamic problem, the accuracy of the dynamic optimization results is reduced. The shorter base period duration leads to exponentially increased complexity of dynamic programming, and makes it impossible to implement in modern intelligent transportation systems. The fixed borders (duration) of base periods have limited the application of dynamic optimization in the management of RTVs under conditions of high irregularity at railway nodes, as presented in [27]. RTV optimization methods for modern computerized control systems are needed for improving transportation efficiency, as evidenced by works of foreign authors [45, 48, 51, 52] and Russian researchers [62, 64, 67, 68].

A compilation of studies on the composition and structure of information management systems at railway nodes is presented in [68], where a simulation method for dynamic alignment (I-MDA) and the results of its application in nodal automated transport systems is described. We suggest that nodal automated systems with an I-MDA optimization module could form the basis for the realization of modern information technology at railway nodes. A simulation method for dynamic coordination was developed in study [1], where the author described the principles of automated calculation for transport service technology in tight train scheduling. Recommendations are proposed on the basis of a more accurate reaction to RTV dynamics with the I-MDA application for developing technological processes at railway nodes and the industrial transport system's "final beats".

The analysis of scientific works on the theory of transport system development, process management and transport logistics showed that in transportation process management, control of RTVs is necessary from the time they appear in the node until the moment of their departure on the mainline rail network, taking into account the RTV parameter dynamics in the application of RTV management tools in intelligent transportation systems.

The effectiveness of the integration of "block optimization" into existing information management systems has been proven in studies [17, 68]. However, in the case of the appearance of irregular RTVs at railway nodes, the application of the well-known method of dynamic optimization leads to delays in RTV movement, as this method does not accurately account for their parameters in operational mode.

Thus, improving the operational management of RTVs in the railway node using dynamic optimization of RTV parameters will entail improving the effectiveness of RTV management and reducing railcar dead time at railway nodes.

3 The Impact Study of RTV Irregularity on the Functioning of Railway Nodes

Analysis of the impact of RTV irregularity on the functioning of railway nodes showed that dead time increased RTV by an average of 20% (Fig. 3). The main reasons for this increase were RTV structural complexity, an increase in the number of low-power streams among total RTV, and a significant increase of RTV irregularity in terms of power and time. In general, incoming RTVs at industrial enterprises today should be characterized as irregular. The system of RTV operational control currently adopted in Russian industrial transport does not provide effective handling of irregular RTVs because it considers separate railway stations, trains and railcars, and not RTVs as a general element. As a result, inefficient route planning for irregular RTVs leads to increased railcar dead time.

RTV irregularity is a major cause of increased railcar dead time. For example, the coefficient of daily irregularity of the total input flow can reach as high as 1.85 (Fig. 4).

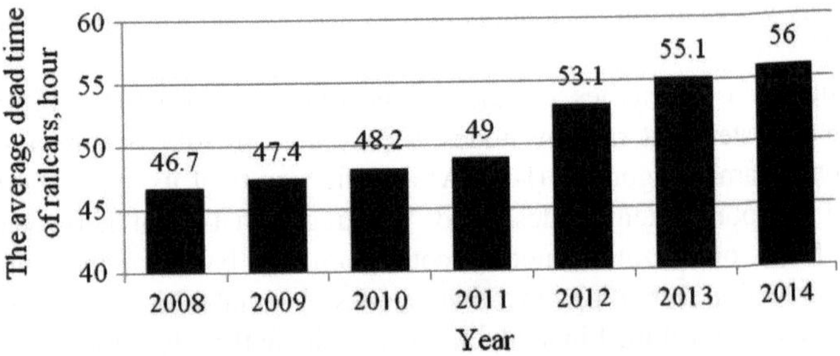

Fig. 3 The dynamics of average dead time of railcars at an industrial railway station of a metallurgical enterprise

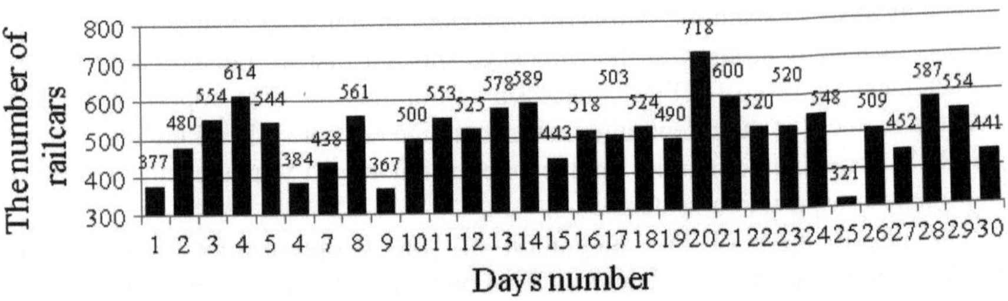

Fig. 4 The monthly dynamics of numbers of railcar arriving at the railway node

A system dynamics simulation model of railway tracks of a major industrial enterprises was built to study the impact of RTV irregularity on railway node functioning. The model was built in AnyLogic software [69], with input data consisting of actual intensity and irregularity RTV data and handling intensity at railway stations [70]. RTV in the model is described by the flow of the system dynamics, characterized by intensity and irregularity. Flow irregularity is established by functions generating random values according to certain laws of distribution. RTVs of different cargo types are described by separate flows (jets). Dynamic change in flow intensity of each jet is established with the aid of cyclic "events"—instruments of action planning (changing the values of variables). The stations and rail hauls are described as a set of two types of stock (input and output), connected to each other by internal flow (Fig. 5).

The input stock of the station is used to mimic railcar delay while waiting for handling at the station, and the output stock is used to mimic railcar dead time, waiting for rail haul availability. The use of two types of stock for reserve dynamics allows us to evaluate the use of railway station capacity. If the stock reserve (number of railcars) exceeds the maximum capacity of the station, then the input flow is delayed, which results in an increase in stock at the previous station. A general view of the system dynamics model of industrial rail transport at a metallurgical enterprise (Fig. 6) is presented in Fig. 7.

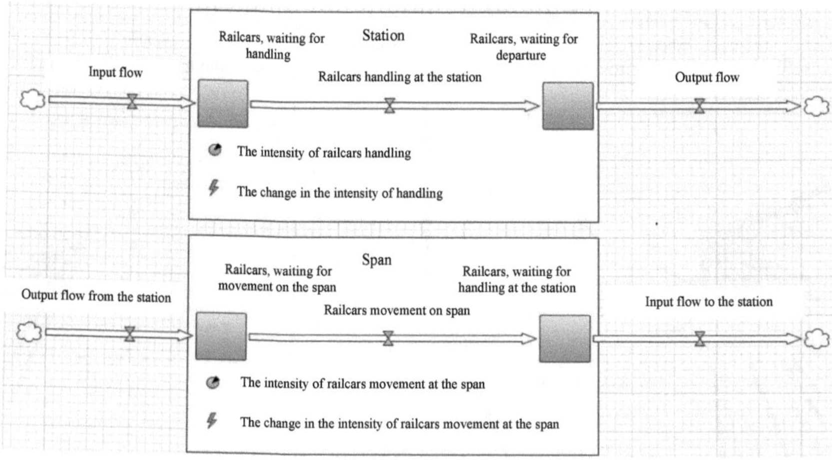

Fig. 5 Schematic presentation of railway stations and spans in system dynamics simulation model

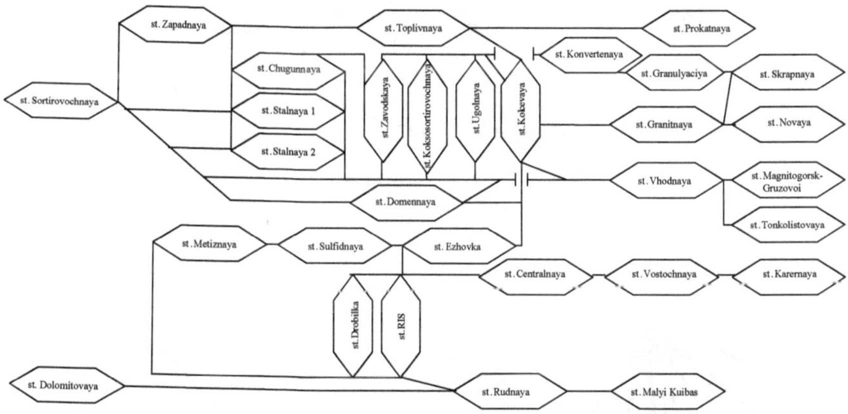

Fig. 6 The scheme of railway stations and spans of the studied industrial transport system at a metallurgical enterprise

The workload dynamic graphics of throughput, handling capacity of station and hauls, and capacity utilization are presented in Figs. 8, 9, 10, 11, 12, 13 and 14 [70]. Analysis of these graphics allowed us to identify typical and recurring changes in stock status, characterizing the loading of railway stations. The experiments revealed four characteristic types of changes to the intensity of RTV handling at the stations (internal flow intensity) over 24 h: decrease in intensity (Fig. 8), increase in intensity (Fig. 9), brief increase followed by a decrease (Fig. 10) and intensity fluctuation (Fig. 11).

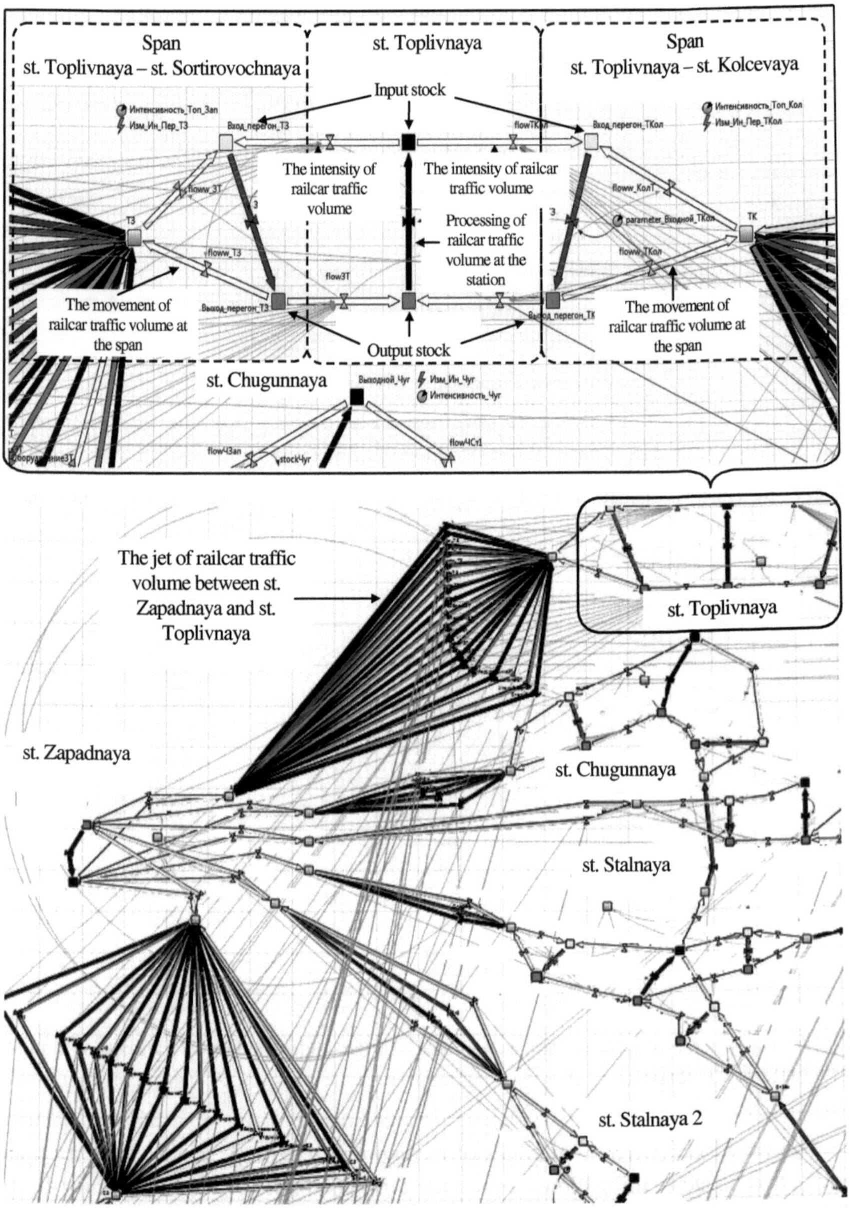

Fig. 7 The fragment of a system dynamics model in AnyLogic software of an industrial transport system functioning at a metallurgical enterprise

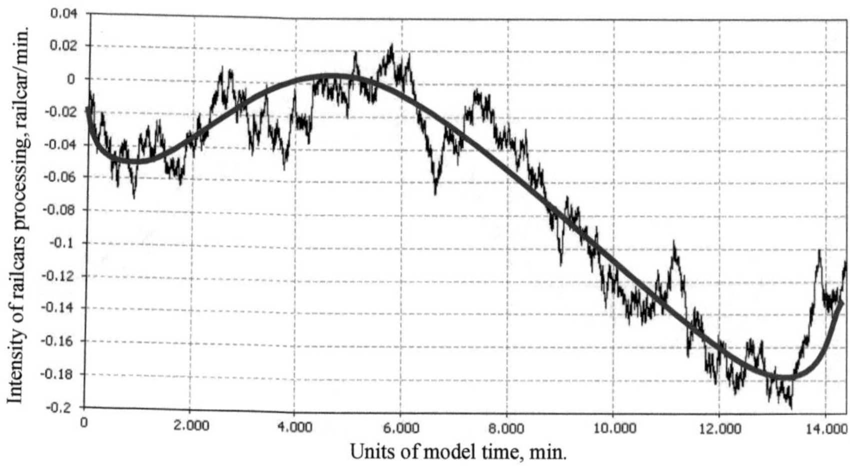

Fig. 8 The dynamics of railcar processing intensity at a railway station serving electric arc furnace production with a handling capacity equal to 120 railcars per day

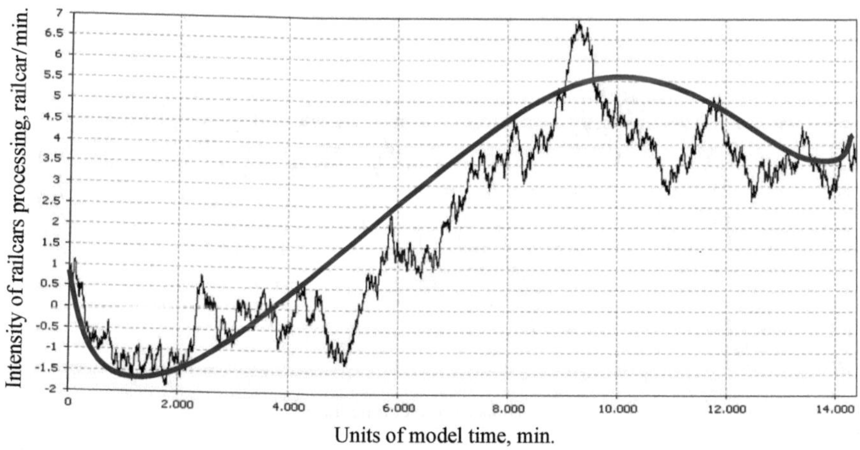

Fig. 9 The dynamics of railcar processing intensity at a railway station serving blast furnace facilities with a handling capacity equal to 220 railcars per day

These characteristic changes connected with the main parameters of industrial railway stations are defined by the intensity and structure of RTV handled and the set of technological operations performed at the station. Typification of industrial railway stations is proposed on the basis of identifiable patterns providing different dynamics of railcar handling. Typification of industrial railway stations is implemented based on four parameters:

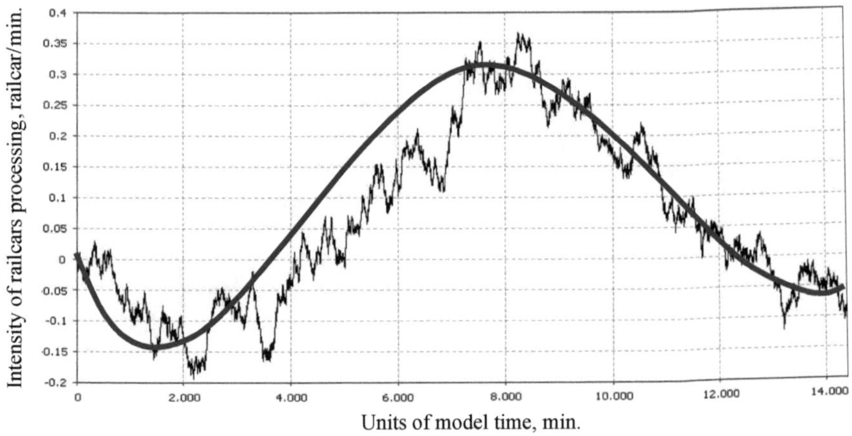

Fig. 10 The dynamics of railcar processing intensity at a railway station serving oxygen converter facilities with a handling capacity equal to 540 railcars per day

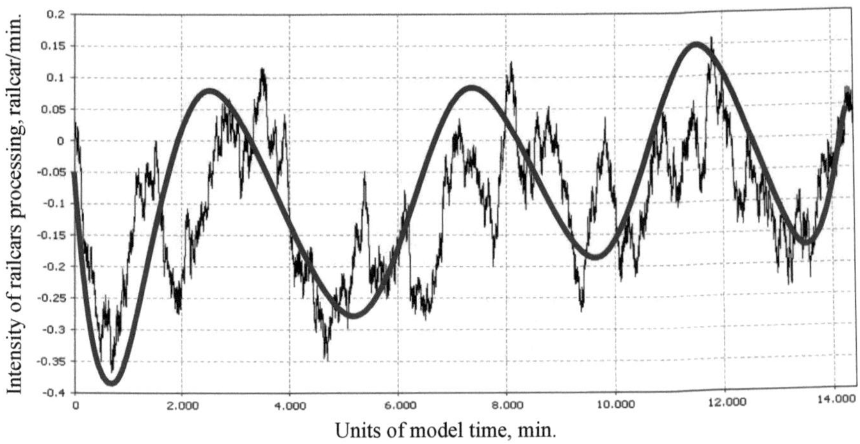

Fig. 11 The dynamics of railcar processing intensity at a railway station serving rolling shop facilities with a handling capacity equal to 440 railcars per day

- railcar share from general RTV which has freight operations (the coefficient of freight work);
- share of transit RTV (the coefficient of transition);
- structural complexity of RTV (the coefficient of RTV structural complexity determined in accordance with the empirical Pareto principle, as the ratio of the number of jets whose combined intensity is 20% of total flow intensity, to the number of jets whose combined intensity is 80% of total flow intensity);
- RTV irregularity (the coefficient of RTV irregularity).

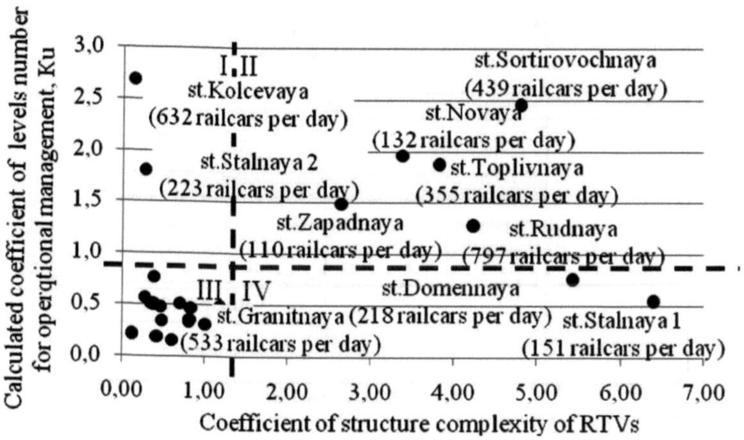

Fig. 12 Clustering of industrial railway stations for a metallurgical enterprise by RTV management efficiency with different degrees of complexity. *I*—stations with low-efficiency operational management systems; *II*—stations with low-efficiency operational management systems and complex irregular flows; *III*—stations with effective operational management systems; *IV*—stations with effective operational management systems and complex regular flows

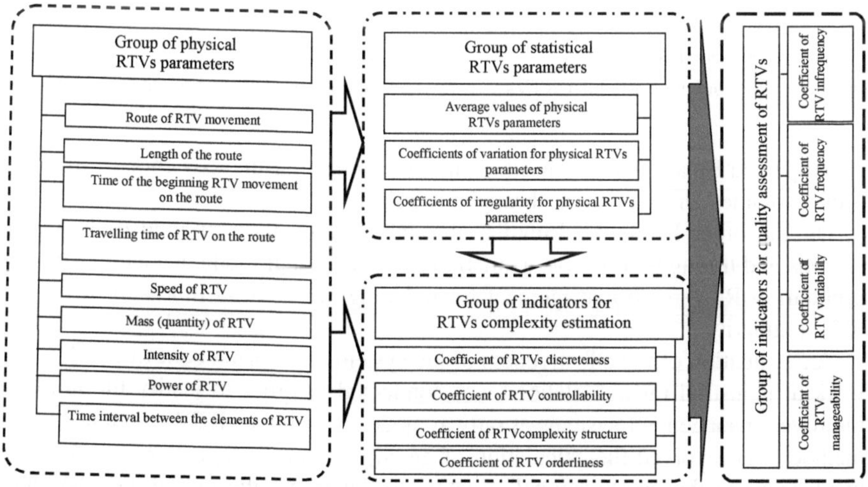

Fig. 13 The scheme of the system for RTV parameters and indicators

The effects of typification of industrial railway stations on the selected parameters are presented in Table 2. Typification enables the prediction of industrial railway station utilization and planning of operational work on RTV management with regard to utilization dynamics. For example, the work level varies from 76 to 143% at the stations with mostly irregular RTVs (fourth type), causing a bottleneck of movement. Furthermore, reduced RTV handling intensity at one station is

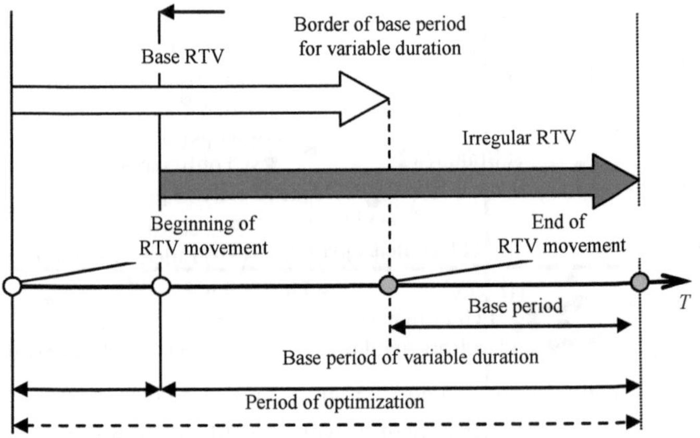

Fig. 14 The scheme of base periods of variable duration

Table 2 Typification of industrial railway stations by the parameters of processed RTVs

Group	Station type	Parameter	Value
1	Freight stations	Coefficient of freight work	0.7–1
2	Transit stations	Coefficient of transition	0.7–1
3	Stations with complex structure of processed RTV	Coefficient of RTV structural complexity	>8
4	Stations with irregular RTVs	Coefficient of RTV irregularity	>1.3

accompanied by an increase at other stations, indicating the existence of irregular traffic volumes and inefficient throughput utilization.

On the other hand, the irregular throughput utilization at industrial railway stations and throughput reserve availability in operational mode makes relevant the problem of RTV route management for reducing railcar dead time and increasing RTV regularity.

An evaluation of the effectiveness of the system of dispatch control for shunting work at industrial railway stations was conducted in order to justify the need for improved management systems at railway nodes in terms of RTV irregularity. The method [71] involved the calculation of an adjustment coefficient Ki for management levels in the organizational structure to determine the system effectiveness. This coefficient shows the difference between the actual and required (regulatory) number of management levels, which occurs as a result of overloading (underloading) of the management system for information flow processing. We assume that overloading of the dispatch control system of a railway station occurs as a result of processing irregular RTV with complex structure.

The normative number of management levels for each station is calculated from the amount of information flows processed at each level. This value is determined according to a methodology based on evaluating the regularity of the arrival of each jet of RTVs at the station [71].

The value of the adjustment coefficient *Ki* is calculated by the formula:

$$Ki = e^{(N_n - N_a)}, \tag{1}$$

where N_n is the normative (estimated) number of management levels and N_a is the actual number of management levels.

The normative number of management levels N_n is calculated by the formula:

$$N_\mathrm{n} = \frac{\lg(S_\mathrm{inf}/s_a)}{\lg K_\mathrm{com}} \tag{2}$$

where S_inf is the total number of information flows handled by the operational management system for the railway station, s_a is the average number of information flows received on each management level of the organizational structure, and K_com is the coefficient of information compression.

The coefficient of information compression is calculated by the formula:

$$K_\mathrm{com} = (s_2 + s_4)/(s_1 + s_3), \tag{3}$$

where s_2, s_4 is the amount of information transmitted on the lower and higher management levels, and s_1, s_3 is the quantity of information transmitted on the specific management level from lower and higher levels.

It is necessary to define the structure of information flows generated in the technological operations process in order to calculate the quantity of information transmitted between levels of management, for example, train reception and disbandment, formation and departure of the train, service of loading areas, and railcar loading and unloading operations.

The number of messages transferred between levels of management (s_2, s_4 and s_1, s_3) is calculated for each informational connection (information flow) in carrying out one technological process.

The number of messages transferred between levels of management during the adjusted base period—for example, one day—depends on the number of technological processes performed during this period, i.e. the number of trains or supplies handled, and is calculated by the formula:

$$s_\omega = \sum_{\varphi=1}^{\beta} s_\varphi N_\varphi \tag{4}$$

where s_φ is the number of transmitted or incoming messages on each level of management station, N_φ is the average daily number of technological processes performed at the station, ω is the index showing the direction of information flow (2, 4, 1, 3), β is the number of technological operations performed at the station, and φ is the technological process index.

Since the likelihood of implementing each technological process performed at the station is equal to 1, the amount of information s_2, s_4 and s_1, s_3 can be conditionally calculated as equal to the number of relevant messages.

The total number of information flows S_{inf} (formula 2) handled at the railway station during the day is calculated by the formula:

$$S_{\text{inf}} = F_{\max} b_{\text{rc}}, \tag{5}$$

where $F_{\max}$ is the average maximum daily railcar volume, in number of railcars, and b_{rc} is the amount of information in the message about the arrival or departure of one railcar.

The amount of information in the message about the arrival or departure of one railcar can be calculated with the use of a statistical (probabilistic) method combined with the thesaurus method, by the formula:

$$b_{\text{rc}} = T_{\text{th}} K_{\text{cmp}} \sum_{k=1}^{r} b_k, \tag{6}$$

where T_{th} is the coefficient of thesaurus (qualification) recipient information (changes from 0 to 1), which can be calculated for each concrete subject of the process of information processing with testing, expert evaluations or fuzzy sets theory; r is the jet number of traffic railcar volumes handled at the station; K_{cmp} is the coefficient of railcar traffic volume complexity; b_k is the amount of information about the arrival or departure of traffic railcar volume, belonging to the kth jet of traffic railcar volume, in bits;

$$b_k = -A_k \log_2(A_k), \tag{7}$$

where A_k is the probability of railcar arrival (departure) during the day, belonging to the kth jet of traffic railcar volume.

The average number of incoming information flows at each level of management is determined by the formula:

$$s_a = \frac{b_{\text{msg}}(s_2 + s_4 + s_1 + s_3)}{4}, \tag{8}$$

where b_{msg} is the amount of information contained in a single message, in bits.

$$b_{\text{msg}} = -\frac{F}{N^{\text{per}}} \sum_{\alpha=1}^{s} A_\alpha \log_2(A_\alpha) \tag{9}$$

where P_z is the probability of a single message transmission between the levels of dispatch control for work at the station during the runtime of separate technological operations, N^{per} is the average daily number of operations of the technological process at the station; s is the quantity of messages transmitted between levels of dispatch management for the station during the execution of individual technological operations, and α is the number of messages transmitted between levels of dispatch management during the execution of individual technological operations.

It is also necessary to take into account the value of the transit railcar volume passing through the station during the base period, using the coefficient of transit in order to compare the adjustment coefficient for the number of management levels and the value of daily RTV at the station with the coefficient of RTV complexity (calculated with Pareto charts):

$$K_{\text{tr}} = \frac{F_{\text{tr}}}{F_{\text{vl}}} \tag{10}$$

where F_{tr} is the transit railcar volume passing through the station, and F_{vl} is the total railcar volume at the station.

The coefficient of RTV complexity with transit railcar volume is calculated by the formula:

$$K_{\text{cmp}}^{\text{tr}} = K_{\text{cmp}}(1 - K_{\text{tr}}) \tag{11}$$

The evaluation of the effectiveness of dispatch control using the method developed herein revealed that the highest adjustment factors for the number of levels in the operational management system organizational structure were observed at stations with RTVs characterized by a high coefficient of complexity. However, the number of operational levels deviated from regulatory control in some cases, conditioned by high values of coefficients of data compression.

The results of calculations of adjustment coefficients for levels of dispatch control at industrial railway stations are presented in Fig. 12. The clustering of stations based on values of adjustment coefficients and coefficient of RTV structural complexity allows us to define the stations requiring priority changes to RTV organizational and management systems. These include the cluster II stations (Fig. 12). For stations in cluster I, improvements to management systems should be implemented using traditional methods (motivation, training, stimulation, etc.). The operational management systems for stations of cluster III do not require reorganization, nor do stations of cluster IV. Indeed, despite the RTV structural complexity for stations in cluster IV, these RTVs are characterized by high regularity; their movement is organized in accordance with tight scheduling, with trains circulating in agreed-upon schedules within the enterprise.

The investigation of the influence of structural complexity and RTV irregularity on the effectiveness of industrial transport system functioning showed the need to improve the system of RTV organization and management, because the current system cannot provide the necessary railcar dead time on industrial track or timeliness of transport services production.

4 Systematization of RTV Parameters and Indicators

In light of the complexity of RTVs and the current economic conditions in Russia, an improved system is needed for evaluating RTV parameters and indicators, particularly in accounting for RTV irregularity and management system effectiveness. The implementation of an evaluation system for complex RTVs is proposed herein based on a system of parameters and indicators developed for logistical material flows, presented in studies [72, 73].

The parameters of RTVs are grouped on the physical and statistical parameters; indicators are grouped on the indicators of complexity and quality of RTVs.

1. **The group of physical parameters of RTVs** includes parameters characterizing their properties in space and time [1, 72, 73].

   - *The route of RTV movement* (path, *M*) is the vertex sequence l of the transport network with coordinates x, z, which passed material flow during the movement from the starting point to the final point of the chain:

     $$M\{(x_1, z_1), \ldots, (x_g, z_g), \ldots, (x_q, z_q)\} \tag{12}$$

     where g is the number of the vertices of the transport network.
   - *The length of the route* (path, *L*) is the total distance (the sum of the vector lengths $g - 1$, component of the route), passed by flow element at movement on the route:

     $$L = \sum_{g=2}^{u} \sqrt{(x_g - x_{(g-1)})^2 + (z_g - z_{(g-1)})^2} \tag{13}$$

     If the lengths of arcs connecting the vertices of the transport network are settled in general view (not the coordinates of vertices), the length of the route is calculated by the formula $L = \sum_{g=1}^{u} l_{\lambda_g g}$, where λ_g is the number of the vertex predating gth at the RTV movement on the route M, $l_{\lambda_g g}$ is the length of the arc connecting the vertices λ_g and g, and u is the number of arcs of the transport network included in the route of RTV movement. The value $l_{\lambda_g g}$ is called the arc's evaluation of the transport network. The arc's evaluation of the network is not only the distance; it is also costs arising within the movement of RTV from vertex λ_g to vertex g. Each vertex g th of the transport network is then characterized by evaluation l_g, equal to the route length from the starting vertex to the gth, i.e. $l_g = L_g$, such that the length of the total route will be equal to the evaluation of the last vertex in the route of RTV movement.
   - *The travel time of RTV on the route* (t_r) is time consumed for RTV on the route M. The RTV travel time will be equal to the evaluation of the last vertex of the route or the sum of the arc lengths of the transport network if

the arc evaluation satisfied the travel time of RTV between the arcs of the transport network.

- *The speed of RTV* (*V*) is the ratio of the route length:

$$V = L/t_r \tag{14}$$

- *The mass* (*quantity*) *of RTV* (*f*) is the total number of railcars in the RTV located in the movement on the route (moving on arcs of the transport network):

$$f = \sum_{i=1}^{u} f_{\lambda_g g} \tag{15}$$

where $f_{\lambda_g g}$ is the number of railcars in the RTV moving in an arc $(\lambda_g g)$ of the transport network.

- *The intensity of RTV* (*I*) is the number of railcars in the RTV passing per unit time through the section of the channel (communication) presented by the arc of the transport network. The intensity of the RTV is calculated as the ratio of RTV mass to the travel time of the RTV on the route:

$$I = f/t_r \tag{16}$$

- *The power of RTV* (*W*) is the product of the intensity of the RTV and its speed:

$$W = I \cdot V \tag{17}$$

- *The time interval between the elements of RTV* (Δt_{RTV}) is calculated as the ratio of the travel time of RTV on the route to the mass of RTV:

$$\Delta t_{\text{RTV}} = t_r/f \tag{18}$$

2. **The group of RTV statistical parameters** includes the parameters characterizing the pattern of change in the physical parameters with the passage of time [72, 73]:

   - *The average values of the RTV physical parameters* are calculated by the accumulated values of the RTV physical parameters during a certain period. The period of observation is defined as the time required for the passage of RTVs on the route. The physical parameters of RTVs are calculated for the route as a whole. If the observation period is equal to the value of RTV travel time on the route, the physical values of the RTV parameters are averaged on the separate arcs of the transport network.

- *The coefficient of variation of the physical parameters of RTVs* (v) is the ratio of the standard deviation of the physical RTV parameter from its average value during the observation period.
- *The coefficients of irregularity of the physical parameters of RTVs* (k_{irr}) are defined as the deviations of the RTV physical parameters from their average values. The coefficients of irregularity are calculated as the sum of units and coefficient of variation v of RTV parameters.

3. **The group of indicators of RTV complexity estimation**:

- *The coefficient of RTV discreteness* (k_{dis}) is the ratio of time interval between elements of RTV to the minimum interval level different from zero when RTV is considered to be continuous [72, 73]:

$$k_{\mathrm{dis}} = \Delta t_{\mathrm{RTV}} / \Delta t_{\mathrm{RTV,min}},\ \Delta t_{\mathrm{RTV,min}} > 0 \tag{19}$$

- *The coefficient of complexity of the RTV structure* (k_{scm}) characterizes the separation of the RTV among the jets following the route, depending on the properties of its elements (e.g. type of cargo, type of rolling stock). We assume that the greater the number of jets with relatively small mass (quantity) in general RTV, the more complex the RTV structure will be. The structural complexity of the RTV determines how difficult it is to manage. The coefficient of complexity is calculated in accordance with the empirical rule of Pareto, as the ratio of the number of jets of the RTV in which the total mass is not more than 20% of the total RTV mass, to the number of jets of the RTV having at least 80% of the total RTV mass. The calculation of the coefficient of complexity of the structure is carried out by the following algorithm: the total sum of all jets f forming the RTV is calculated; the mass of the separated jets of RTV f_r is calculated; the jets of RTV are sorted in ascending order of their mass values; the number of jets $r^{20\%}$ is calculated such that the total mass does not exceed 20% of the f value (calculated in order of increasing jet mass); the value of the coefficient of complexity of the structure of the RTV is calculated as [72, 73]:

$$k_{\mathrm{scm}} = r^{20\%} / (r_s - r^{20\%}) \tag{20}$$

where r_s is the total number of jets in the RTV.

- *The coefficient of differentiability of the RTV* (k_{dif}) is the ratio of the number of jets of the RTV at the final point of a route (r_β) to the number of jets at the starting point (r_1):

$$k_{\mathrm{dif}} = r_\beta / r_1 \tag{21}$$

- *The coefficient of RTV orderliness* (k_{ord}) is the ratio of the weighted average number of jets of the RTV with the same speed to the total number of jets. The coefficient of RTV orderliness is calculated by the following algorithm: the value of the travel speed of the separated jets of the RTV (or intervals of speed values) N_v is defined; the number of jets having the same speeds r_μ is calculated; and the coefficient of RTV orderliness is calculated by the formula [72, 73]:

$$k_{\text{ord}} = \frac{\sum_{\mu=1}^{\mu=N_v} r_\mu V_\mu / \sum_{\mu=1}^{\mu=N_v} V_\mu}{r_s} \quad (22)$$

where μ is the index of RTV jet.

- *The coefficient of RTV controllability* (k_{con}) is the ratio of information quantity elements, which are messages about quality (in accordance with established normative values), *managerial* commands (s_c), to the quality of information and control messages. The complex of methods based on indicator rationing of RTV and determination of degree of achievement of these indicators can be applied for the quality assessment to perform managerial commands [72, 73].

4. **Group of indicators for RTV quality assessment**:

- *The coefficient of RTV infrequency* (k_{inf}) is the ratio of the number of time intervals different *from* the average value between elements of RTV for the base period, i.e. the probability of deviation of the time interval between elements of RTV from the mean value of the interval:

$$\begin{aligned} k_{\text{inf}} &= \sum_{\tau-1}^{\ell-1} \delta_\tau/(\ell-1), \\ \delta_\tau &= 1, \; npu \; \Delta t_{\text{RTV}} \neq \overline{\Delta t}; \\ \delta_\tau &= 0, \; npu \; \Delta t_{\text{RTV}} = \overline{\Delta t}, \end{aligned} \quad (23)$$

where τ is the number of intervals between elements of RTV, ℓ is the number of elements of the RTV passed on the route during the period of observation, and $\overline{\Delta t}$ is the average value of the time interval between the elements of the RTV.

- *The coefficient of RTV frequency* (k_{fr}) characterizes the patterns of change in the physical parameters of RTV, generally the mass (quantity) of RTV. The study of periodicity and calculation of the corresponding coefficients is conducted with methods of time series analysis [72, 73]. The coefficient of frequency for the multiplicative time series model is calculated as the ratio of the product of the number of levels for regular components of the time series (trend Y_T, seasonal Y_S and cyclical components Y_C) to the number of levels belonging to the random component Y_E, i.e. $k_{\text{fr}} = Y_T Y_S Y_C / Y_E$.

The coefficient of frequency for the additive time series model is calculated as the difference between the sum of the number of levels for regular components and number of levels belonging to the random component $k_{fr} = (Y_T + Y_S + Y_C) - Y_E$.

- *The coefficient of RTV variability* (k_{var}) is the reverse value to the maximal during the observation, in which the duration of the continuous period within the value of the variability coefficient of each physical parameter for RTV does not exceed a set value. The correlation of the parameters and indicators of RTVs in the system is presented in Fig. 13.

The application of the proposed system for RTV parameters and indicators makes it possible to fully evaluate them, normalize the indicators, identify the causes of deviations from specified parameters and indicators, and choose ways to eliminate these deviations. A mathematical model and algorithm of dynamic optimization of RTVs developed through dynamic programming are proposed in this work. The model is relevant for complex RTVs, and quantitative assessment of deficiencies in management becomes possible with the application of this system.

5 The Method of Dynamic Optimization of RTV Based on Base Periods of Variable Duration

The mathematical model and implementation algorithm based on the dynamic optimization method are developed to improve the system for managing irregular RTVs in transport nodes. The model features the application of base periods of variable duration, which allow for more clearly optimized parameters of complex irregular RTVs—unlike the well-known dynamic programming methods with estimated periods of constant duration—and ultimately leads to a reduction in total costs of their movement in the railway nodes.

It is necessary to introduce a number of concepts for identification of the base period concept:

- Dynamic programming is a method involving the separation of dynamic processes on the base periods, each of which has parameters for optimization, regardless of the other parameters in the base periods.
- The problem of dynamic optimization for RTV parameters is in minimizing the total costs for movement, delay and change in RTV structure based on parameters calculated using dynamic optimization.
- The period of parameter optimization for RTVs is the period of time during which the problem of dynamic optimization for RTVs is solved (Fig. 14).
- The base period in the problem of dynamic optimization of RTV parameters is the constant part of the period (on duration) for optimization of RTV parameters.

We designate the moment of time for the start and end of the base period as the border of the base period. If actual RTVs different from the calculated mass and time of the start of their movement are observed within the borders of the base period, then these actual RTVs are irregular. The implementation of management of RTVs at railway nodes functioning in conditions of high irregularity is proposed on the basis of change in the borders (duration) of base periods. The decision is made on the basis of cost comparisons of irregular RTV delay and adjusted parameters for total RTVs. Thus, the base period varies in duration based on the period of parameter optimization for the RTVs, with its duration changing in accordance with optimal cost ratios of RTV delay and adjustment of their parameters. The application of base periods of variable duration enables greater accuracy in calculations for problems of dynamic optimization of RTV parameters at railway nodes functioning in highly irregular of RTVs.

The problem of dynamic optimization for RTV parameters based on base periods of variable duration is formulated in the following way. Let $Gr = \{P, E\}$ be a connected, undirected graph with no loops, where $P = \{p_i | i = 1, \ldots, \eta\}$ is vertex variety (η is vertex quantity), $E = \{e_{ij} | i, j\}$ is arc variety (e_{ij} is the arc of the transport network limited by vertices p_i and p_j), $[t, t_0]$ is the base period of variable duration and $[0, T]$ is the optimization period.

As opposed to an ordered sequence of arcs $\lambda_g g$ in a route used in the system of RTV parameters (Formula 15), in describing the developed model we will use unordered sets of vertices P and arcs E.

The procedure *route_d*() is employed for their ordering in the future (Figs. 19, 21 and 22) to determine optimal routes of RTV movement on the transport network.

The representations $d_{ij}(t)$ and $d_i(t)$ are referred to as the reserve of throughput for each arc e_{ij} and handling capacity of each vertex p_i on the moment of time t for base period, $d_{ij}(t), d_i(t) \in R^+$, i.e. defined on the variety of real nonnegative numbers. Then $Gr = \{P, E\}$ is the transport network (Fig. 15).

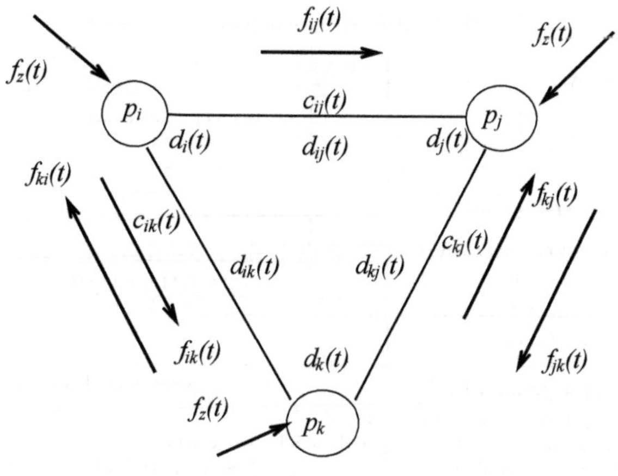

Fig. 15 The scheme of the transport network fragment. $f_z(t)$—external RTV, incoming into the transport network

The representation $f_{ij}(t)$ called RTV in arc i from starting vertex p_i to the final vertex p_j, $F_{ij} = \{f_{ij}(t)|i = [1,\eta], j = [1,\eta]\}$ is the RTV variety in arcs of network G, where $f_{ij}(t) \in R^+$—is the value of RTV mass, defined on the variety of real non-negative numbers at some point of time t, and $c_{ij}(t)$ is the costs for the movement of RTV mass unit $f_{ij}(t)$.

Let the value of RTV mass $f_{ij}(t)$ have random variation; then at each time t the RTV mass will be equal to $f_{ij}(t) \pm y_i(t)$, where $y_i(t)$ is a random value. If the RTV mass is equal to $f_{ij}(t) + y_i(t)$ (Fig. 16), losses of revenue will arise as a result of irrational utilization of transport infrastructure.

Therefore, it is advisable to adjust the RTV parameters, taking into account the condition $f_{ij}(t) \le d_{\min}$, where $d_{\min}$ is the border of the tolerance field, i.e. minimum reserve of throughput $d_{ij}(t)$ and handling capacity $d_i(t)$ for the ability of arcs and vertices of the transport network on each route of RTV.

Thus, if the RTV mass is $f_{ij}(t) - y_i(t)$, then the following can be observed:

1. A lack of throughput reserve on the RTV route, resulting in additional costs arising for the adjustment of parameters (Fig. 16).
2. Loss of dead time for irregular RTV with mass $y_i(t)$ as a result of dividing the actual RTV (Fig. 17).
3. Costs for irregular RTV movement with mass $y_i(t)$ at the transport network (Fig. 18).

In the problem of dynamic optimization for RTV parameters, the minimum sum of costs for movement, dead time and change in RTV structure must be found, taking into account the additional costs for adjusting RTV parameters. The target function of this problem will be:

$$J = J_1 + J_2 + J_3 + J_4 \to \min \tag{24}$$

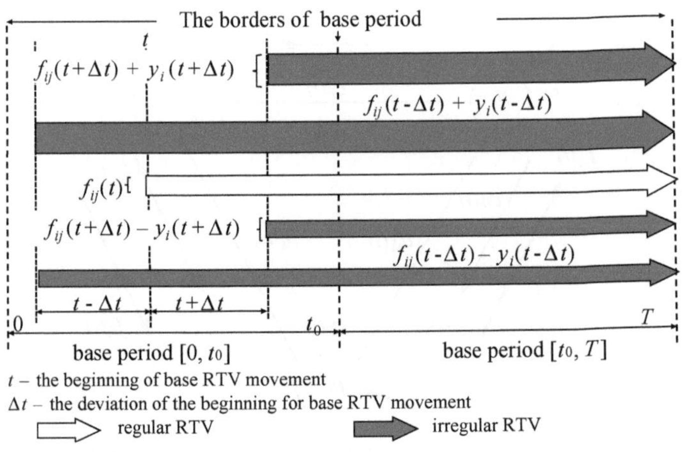

Fig. 16 The scheme of RTV variants with different mass at the beginning of the movement

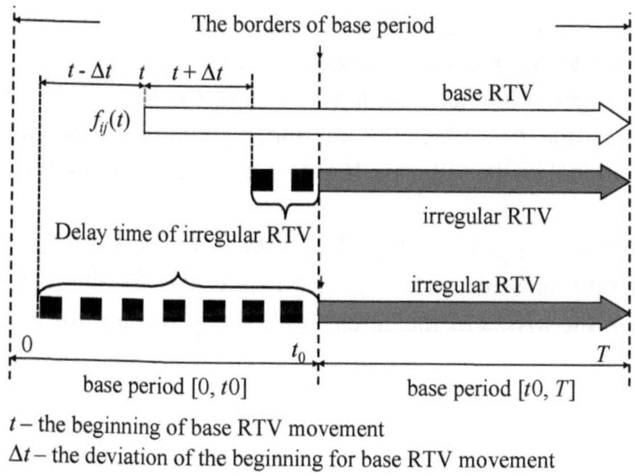

Fig. 17 The scheme of irregular RTV delay

Fig. 18 The scheme of border change for the base period

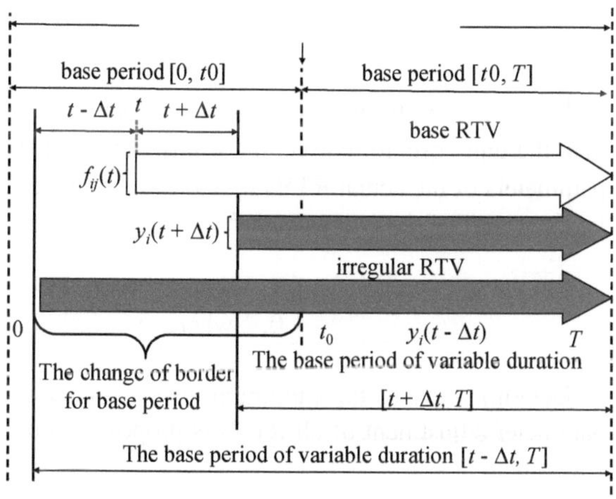

where J_1 is the transport cost calculated by the formula

$$J_1 = \sum_{t=0}^{T} \sum_{p_i}^{p_j} c_{ij}(t) f_{ij}(t) \tag{25}$$

where J_2 is the cost for actual RTV delay in the starting vertex of its route:

$$J_2 = \sum_{t=0}^{T} \sum_{p_i}^{p_j} c_i(t) f_{ij}(t) \tag{26}$$

For cost evaluation J_3 and J_4 we will introduce the following concepts. We designate the border of the base period by moment t_0. If the moment of RTV occurrence t will occur sooner than t_0 (i.e. $t < t_0$), the losses from RTV delay will arise during the time $t_0 - t$ waiting for the beginning of the next base period. Otherwise, additional costs will arise for the adjustment of parameters for the actual RTV.

If we assume that $c_{1,(i)}$ is a single loss from actual RTV delay waiting for the next period of optimization, and $c_{2,(ij)}$ is a single loss for the adjustment of RTV parameters, then the losses of the actual RTV delay waiting for base period J_3 are defined as

$$J_3 = \sum_{0}^{t_0} f_{ij}(t) c_{1,(i)}(t_0 - t), t = [0; t_0] \tag{27}$$

and the cost for parameter adjustment of actual RTV J_4 will be

$$J_4 = \sum_{t_0}^{T} f_{ij}(t) c_{2,(ij)}(t - t_0), t = [t_0; T] \tag{28}$$

If a single loss from the delay of actual RTV waiting for base period $c_{1,(i)}$ is more than the unit costs for correction of this parameter $c_{2,(ij)}$ it is advisable to adjust the parameters of the actual RTVs by the change in duration for the base periods. Let e *be the result of the cost comparison*:

$$e = \begin{cases} 1, & \text{where } c_{1,(i)}(t_0) > c_{2,(ij)}(t_0), \\ 0, & \text{where } c_{1,(i)}(t_0) \leq c_{2,(ij)}(t_0), \end{cases} \tag{29}$$

Determination of the minimum cost caused by the delay of irregular RTV or parameter adjustment of all RTVs is proposed to be implement on the basis of the equation:

$$C(t_0) = e(t_0)\, c_{2,(ij)} + |e(t_0) - 1| c_{1,(i)}(t_0) \tag{30}$$

Target function (24) has the following limitations: the equation of dynamic changes for RTV mass $f_{ij}(t+1) = f_{ij}(t) + y_i(t+1) - y_i(t)$; the initial and final conditions of the task: $f_{ij}(0) = 0$; $y_i(0) = 0$; non-negativity constraints on variables: $f_{ij}(t) \geq 0$; $i \neq j$; $y_i(t) = 0$; limitations of throughput reserves: $0 \leq f_{ij}(t) \leq d_{ij}(t)$, $0 \leq f_{ij}(t) \leq d_i(t)$, $0 \leq f_{ij}(t) \leq d_{\min}(t)$.

The algorithm for calculating the RTV parameters and algorithm for parameter adjustment are developed for the practical utilization of the developed mathematical model within the framework of an information management system (Fig. 19) [75]. Algorithms of calculation and parameter adjustment of RTVs are described with an object-oriented approach for programming [74] in Figs. 20 and 21. These algorithms have the following designations.

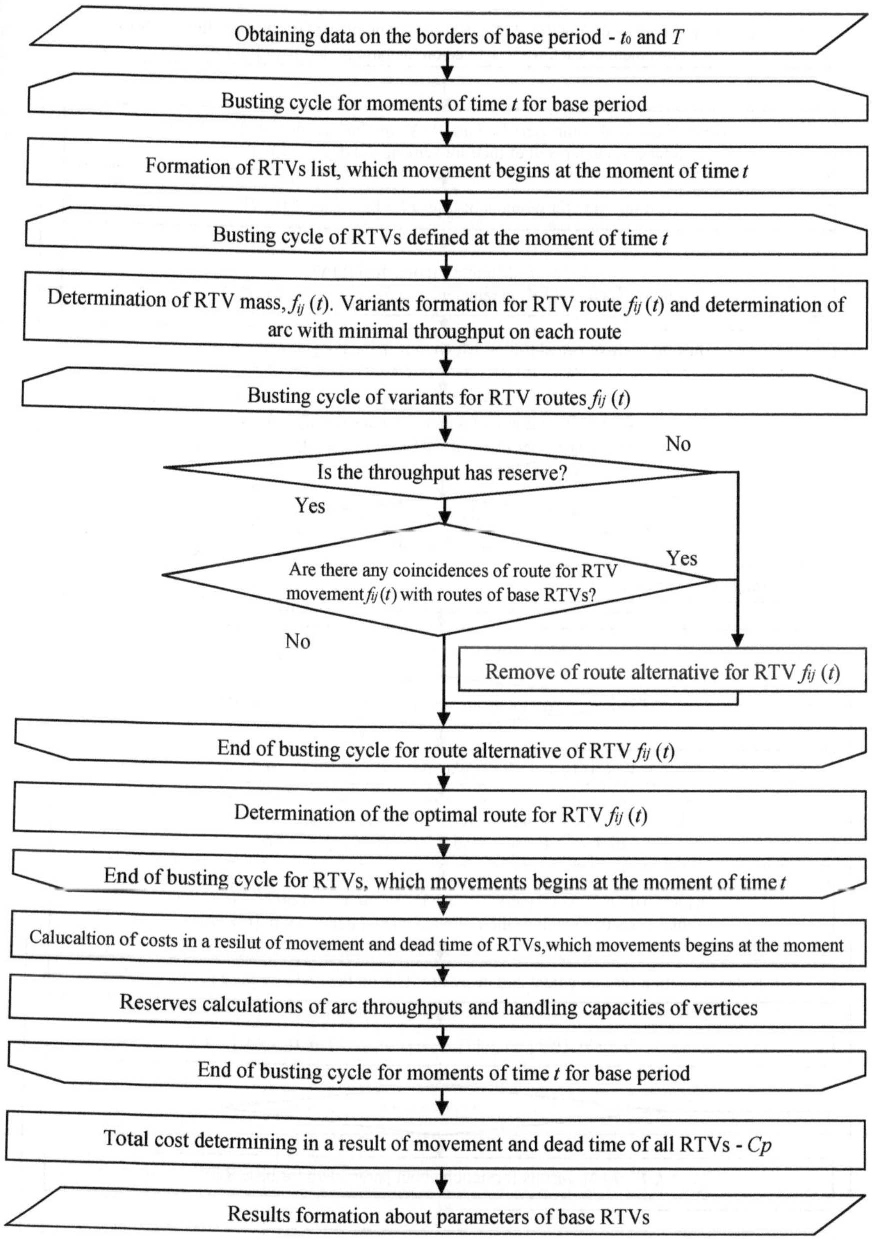

Fig. 19 The algorithm for RTV parameter calculation

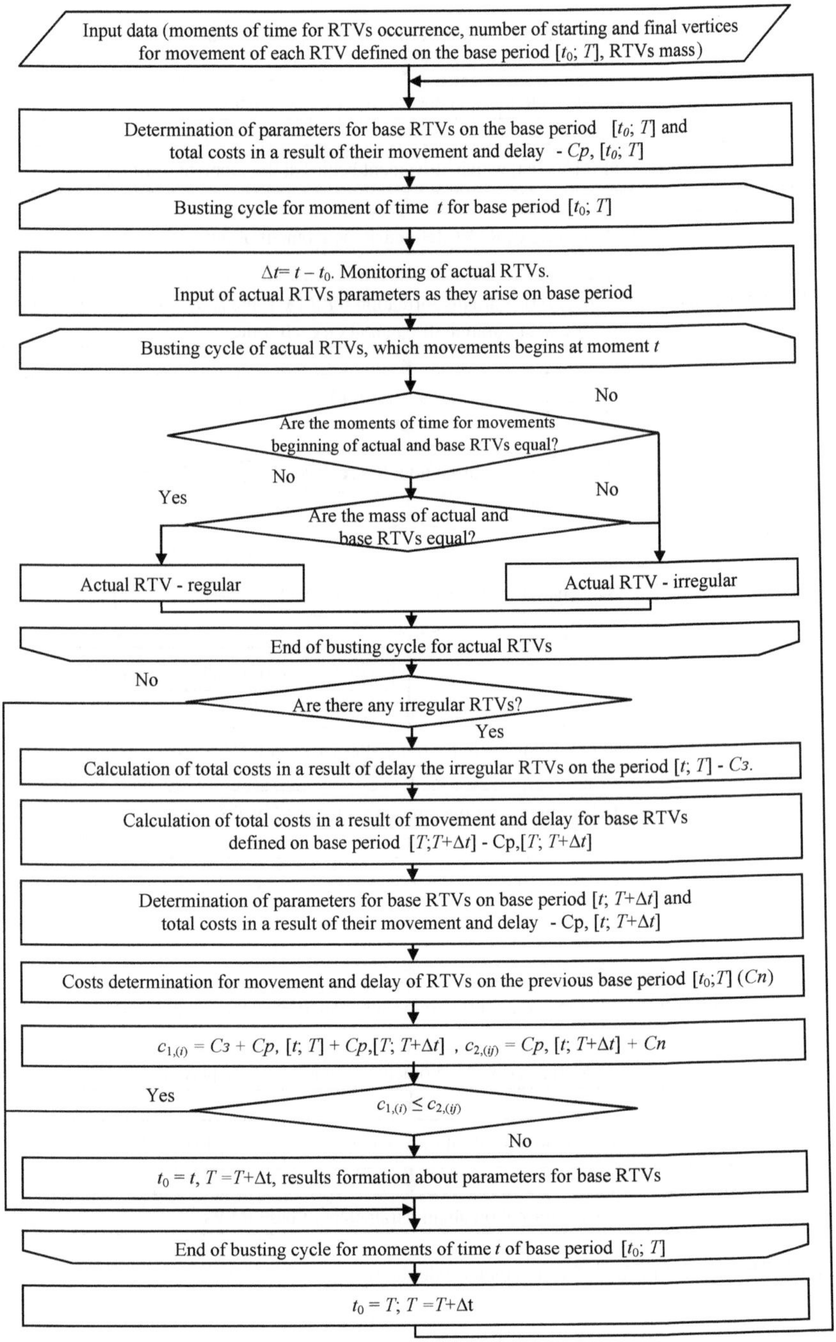

Fig. 20 The algorithm of RTV parameter adjustment

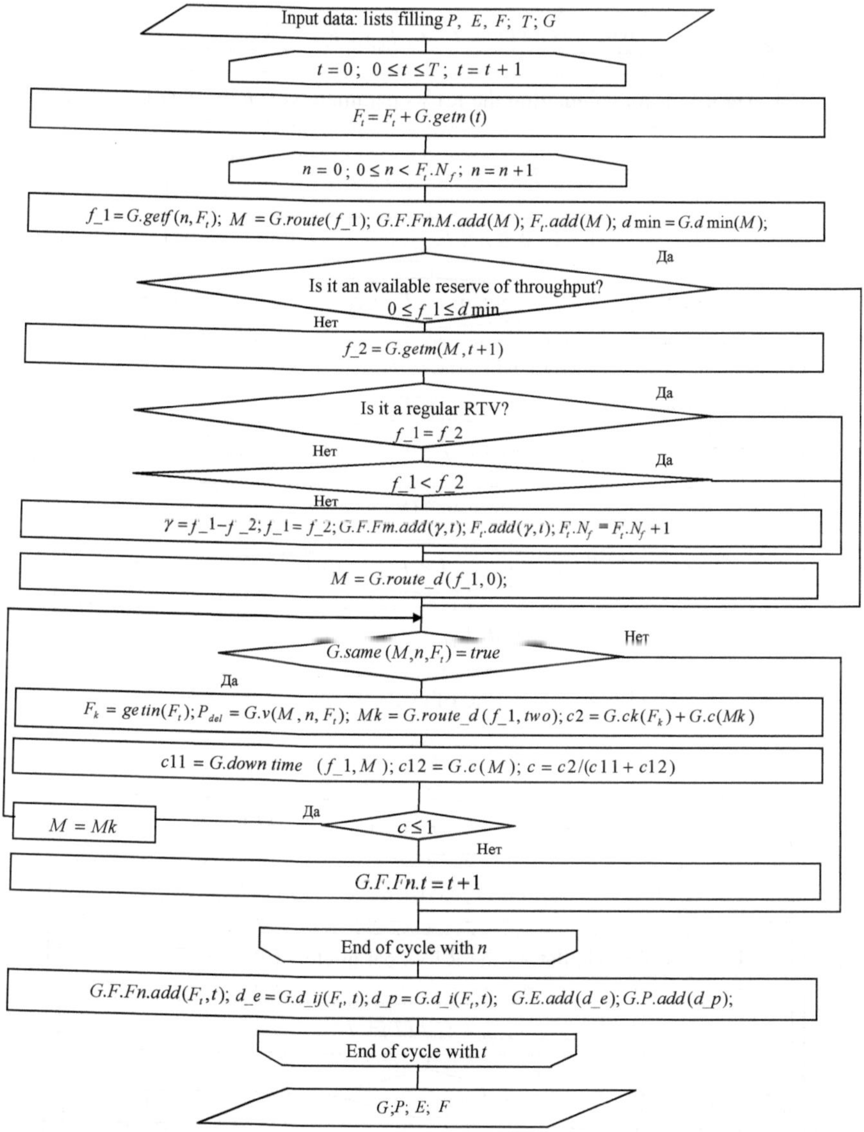

Fig. 21 The algorithm of RTV parameter calculations based on the an object-oriented approach [74]

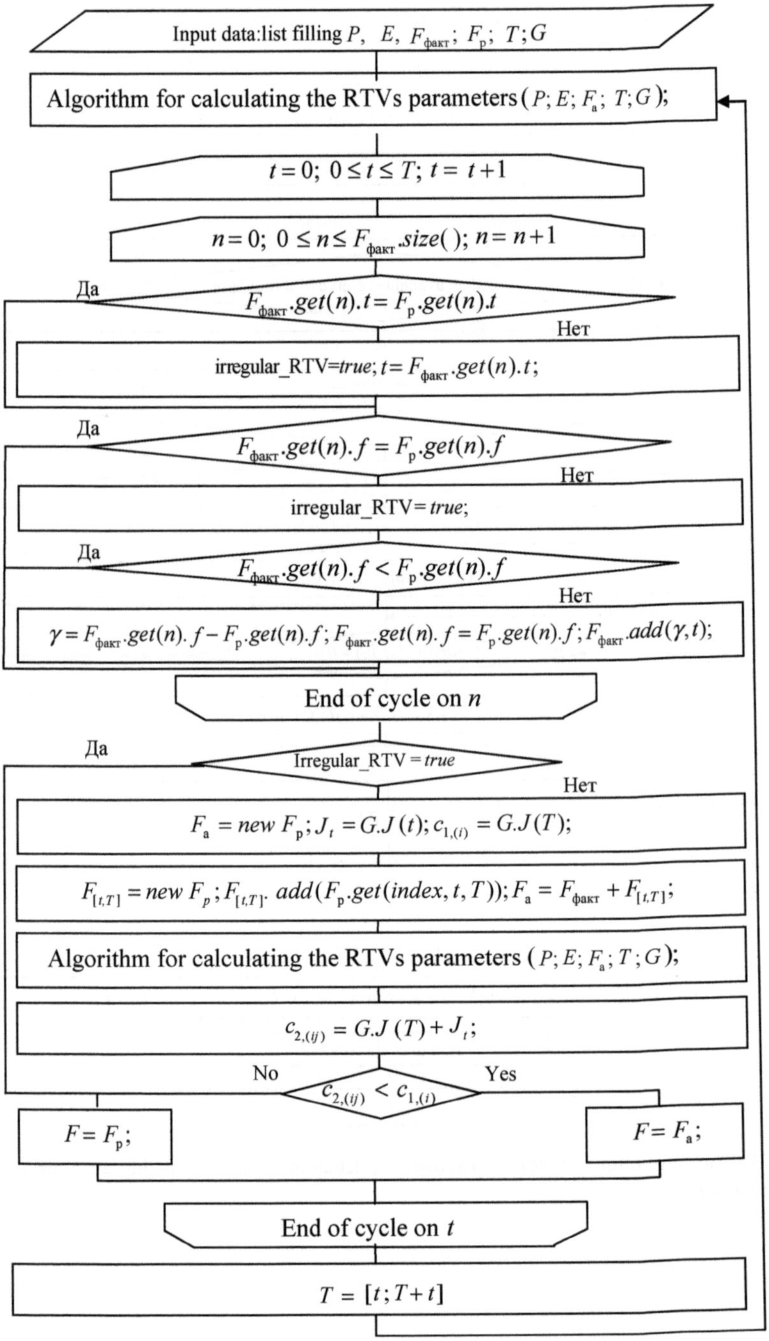

Fig. 22 The algorithm of RTV parameter adjustment based on an object-oriented approach

Transport network presented as a class (object G) described by the combination of data and methods

The list of vertex $P(p)$, where p is the vertex.

The data of vertex p:

- i is the vertex number;
- list $D(d_i, t)$, where d_i is the throughput reserve of vertex at the time t;
- list $C(c_i, t)$, where c_i is the cost of one railcar delay in vertex p_i at the time t;
- arc list $E(e)$, where e is the arc.

The data of arc e:

- i is the number of the starting vertex of the arc, j is the number of the last vertex of the arc;
- list $D(d_{ij}, t)$, where d_{ij} is the reserve of the arc throughput at the time t;
- list $C(c_{ij}, t)$, where c_{ij} is the cost for RTV mass unit movement on the arc e_{ij} at the moment of time t;
- t_{ij} is the time of RTV movement on the arc;
- RTV list $F(Fn)$, where Fn is railcar traffic volume.

The data of railcar traffic volume Fn:

- n is the RTV number;
- f is the RTV mass;
- t is the moment of time;
- $M(P_f, c)$ is the RTV route, where P_f is the list of arc numbers, where RTV is moving, and c is the cost for movement of mass unit for RTV on the route.

The functions of object G:

- $getn(t)$ is the RTV list receiving, whose movement begins at moment of time t;
- $getf(n, F_t(N_f))$ is the calculation of mass value of the nth RTV, where $F_t(N_f)$ is the list of RTVs having structure similar to the list $F(Fn)$, and N_f is the quantity of RTVs;
- $getm(M, t)$ is the calculation of mass value for RTV, whose movement begins at the moment of time t on the route M;
- $route(f)$ is the search for the optimal route for RTV with mass f according to the criterion of minimum cost «c»;
- $route\_d(f, P_{\text{del}})$ is the search for the optimal route for RTV with mass f according to the criterion of minimum cost «c» with limitations for throughput d, where P_{del} is the vertex lists which are not included in the route;
- $d\min(M)$ is the determination of minimum reserve of arc and vertex throughput of route M;
- $same(M, n, F_t)$ is the match search of vertex numbers of route M with nth RTV with vertices of RTV routes from the list F_t;
- $v(M, n, F_t)$ is the list receiving of the identified matches for vertex numbers of route M with nth RTV with vertices of RTV routes from the list F_t;
- $F_k = getln(F_t)$ is getting of RTV lists with coincident vertex numbers;

- $downtime(n, M)$ is the cost calculation as a result of the nth RTV delay in the starting vertex of route M;
- $c(M)$ is the determination of costs for RTV movement on the route M;
- $ck(F_k)$ is the calculation of costs for adjusting RTV movement F_k;
- $J(t)$ is the calculation of costs for movement, dead time and change in RTV structure, taking into account the additional costs of RTV parameters adjustments at moment of time t;
- $d\_p = d\_i(F_t, t)$ is the determination of current throughput reserves for vertices of RTV route F_t in the moment of time t, where p is the list of vertices having a structure similar to list $P(p)$;
- $d\_e = d\_ij(F_t, t)$ is the determination of current throughput reserves for arcs of RTV route F_t at the moment of time t, where e is the arc list having a structure similar to list $E(e)$.

The correctness of the developed model and algorithm was assessed on the base example of a transport network (Fig. 23). Let the known values of the RTV mass $f_{12}(0) = 4,\ f_{12}(10) = 2,\ f_{12}(20) = 2,\ f_{32}(0) = 2,\ f_{23}(13) = 2,\ f_{31}(21) = 1$. The period of optimization will be $[0; T] = 30$.

The solution to the problem starts with the last (third) base period $[t_{0,2}; T]$. RTVs at this period will be $f_{12}(20) = 2$; $f_{31}(21) = 1$. We assume that RTVs $f_{12}(20)$ and $f_{31}(21)$ are irregular.

Checking for limitations of the availability of reserves for throughput of elements of the transport network on the RTV route:

$$d_{\min,12} = \min(3) = 3;\ 0 < f_{12}(20) < 3;$$
$$d_{\min,31} = \min(1) = 1;\ 0 < f_{31}(21) = 1.$$

Since the mass of RTVs $f_{12}(20)$ and $f_{31}(21)$ does not exceed the throughput, it is not necessary to search for alternate routes. As a result of movement of RTVs $f_{12}(20)$ and $f_{31}(21)$ transport costs for the third period of optimization will be

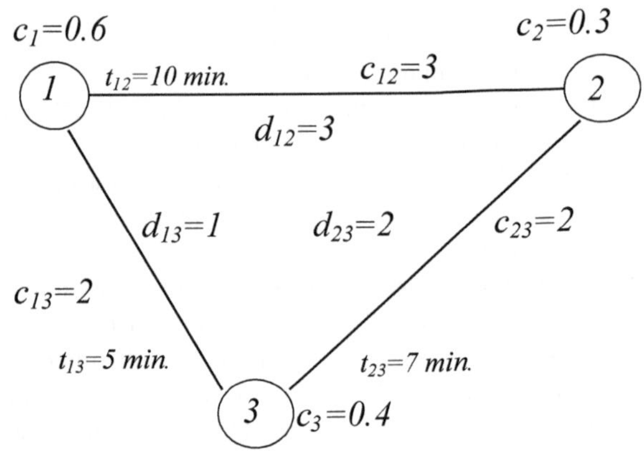

Fig. 23 The calculation scheme of the transport network

$J_1 = 6+2 = 8$, the costs of delay for the regular RTV in the starting vertex of its route $J_2 = 0$, dead time losses of the actual RTV waiting for the next period of optimization $J_3 = 0$, and the costs for adjustment of parameters for the actual RTV $J_4 = 0$. The total costs for RTV movement on the third period of optimization will be $J = 8+0+0+0 = 8$.

The RTV parameters are similarly calculated for the second calculation period, $[t_{0,1}; t_{0,2}]$. The total costs, taking into account the result at the third calculation period, will be 18.

At the first base period $[0; t_{0,1}]$ RTV $f_{12}(0) = 4$ is irregular. We will check the limitation for availability of throughput reserves $f_{32}(0) = 2$; $d_{\min,32} = \min(2) = 2$; $0 < f_{32}(0) = 2$. If $f_{12}(0) = 4 \neq f_{12}(10) = 3$, RTV $f_{12}(0)$ is irregular. We propose dividing the RTV into $f_{12}(0) = 3$ and $y_{12}(0) = 4 - 3 = 1$. In this case, RTV $f_{12}(0)$ becomes planned. As a result, the route $f_{12}(0)$ matches the route of similar RTVs at the previous base periods, where $d_{\min,12} = \min(3) = 3$; $0 < f_{12}(0) - y_{12}(0) = 3$.

Let us consider the possible variants for the movement of irregular RTV $y_{12}(0)$.

Variant 1: Movement on the standard route, scheduled on the first base period on the arc (ij). RTV route $f_{12}(0)$ is planned on the first period; therefore, for each RTV $y_{12}(0)$ $d_{\min,12} = 0$. RTV movement $y_{12}(0)$ is possible only in the second period, considering the limitation for throughput reserve. Consequently, as a result of the delay of RTV $y_{12}(0)$ waiting for the next base period, total costs will be:

$$J_1 = 3 \cdot 3 + 2 \cdot 2 = 13;\; J_2 = 0;\; J_3 = (0.6 \cdot 10) \cdot 1 = 0.6;$$
$$J_4 = 0 \Rightarrow J = 13 + 6 = 19.$$

Variant 2: Movement on the alternative route. We will check the limitation for availability of throughput reserves $d_{\min}, 132 = \min(1; 2) = 1$, $0 < y_{12}(0) = 1$.

The total time of RTV movement $y_{12}(0)$ is $t_r = t_{13} + t_{23} + t_3 = 5 + 7 + 2 = 14$ min. RTV time movement $y_{12}(0)$ is more than the established value of the base period of 10 min; as a result, there are two variants for solving the problem.

Variant 2.1: The additional delay of irregular RTV $y_{12}(0)$ on vertex 3 waiting for RTV movement $f_{23}(13)$ on the arc (23). If we select this variant, the border of the base period is unchanged, and RTV $y_{12}(0)$ is delayed in vertex 3 waiting for the next period.

As a result of the delay of irregular RTV $y_{12}(0)$ total costs will be calculated as:

$$J_1 = 3 \cdot 3 + 2 \cdot 2 = 13; J_2 = 0; J_3 = 0;$$
$$J_4 = 2 \cdot 1 + (0.4 \cdot (2 + 13)) \cdot 1 + 2 \cdot 1 = 10 \Rightarrow J = 13 + 10 = 23.$$

Variant 2.2: RTV delay $f_{23}(13)$ in vertex 2 waiting for the movement of irregular RTV $y_{12}(0)$ on the arc (23). The border of the base period will change in this variant; as a result, the accuracy of problem solving will increase. Thus, the total costs are calculated as follows:

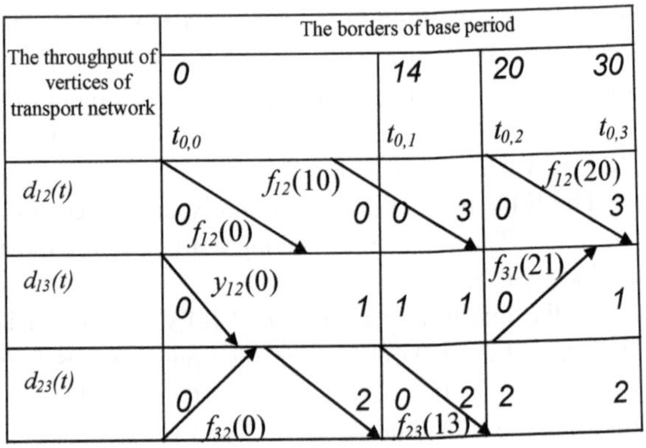

Fig. 24 The results of the control example for dynamic optimization problem of RTV parameters

$$J_1 = 3 \cdot 3 + 2 \cdot 2 = 13;\ J_2 = (0.3 \cdot 1) \cdot 2 = 0.6;\ J_3 = 0,$$
$$J_4 = 2 \cdot 1 + (0.4 \cdot 2) \cdot 1 + 2 \cdot 1 = 4.8 \Rightarrow J = 13 + 4.8 + 0.6 = 18.4.$$

The total costs of RTV movement on the transport network will be 37 when selecting variant 2.1 as the optimal solution to the problem, where RTV $y_{12}(0)$ is delayed in vertex 3 waiting for the next base period. However, when the problem is solved using base periods of variable duration (variant 2.2), the accuracy of its solution will increase. The increased accuracy is the result of increasing the borders of base period $t_{0,1}$; in this case the RTV $y_{12}(0)$ delay is reduced in vertex 3. When variant 2.2 is selected, the total costs for RTV movement on the transport network will be 36.4.

Thus, we will adopt the following as the optimal solution to the problem: separation of the actual RTV $y_{12}(0)$; adjustment of route for irregular RTV $y_{12}(0)$; change of the border of the base period $t_{0,1}$. The results of problem solving for dynamic optimization are presented in Fig. 24, where total costs for RTV movement on the transport network are 36.4.

The utilization of the developed algorithms for calculation and adjustment of RTV parameters is the basis for the implementation of the developed mathematical model and its integration into information management systems of industrial rail transport or intelligent transport systems.

6 A Combined Simulation Model of Railway Node Functioning

The combined simulation model of railway node functioning in operative mode with utilization of the simulation modeling instrument AnyLogic was developed to assess the proposed mathematical model. The combined simulation model was constructed on the basis of the combined use of discrete and agent approaches in one model and algorithms developed for calculating and correcting the RTV parameters as well. There are currently many software tools for building simulation models and conducting experiments. The most commonly known include AnyLogic, Arena, Plant Simulation, Business Studio, Aimsun, GPSS, Extend and Witness. Software packages such as AnyLogic, Arena, Extend, GPSS and Witness are universal, while others are focused on simulation in specific subject areas [69, 76]. The most universal and modern simulation systems with the potential for constructing a model of railway node functioning are the AnyLogic (formerly XJ Technologies) and Arena (Rockwell Automation) packages.

A comparative analysis of these packages was conducted in [1, 77] to determine the more suitable software tool for model construction of railway node functioning. The AnyLogic system was chosen for this study because of the ability to use it in different approaches and paradigms of simulation modeling within one model. The combination of discrete event and agent-based approaches for modeling the handling of complex, irregular RTVs at railway nodes satisfied the need for the model to display technological processes of industrial railway stations using the discrete event approach [78] and handling of separate jets of RTV (railcars and groups of railcars) based on the agent approach [53].

In accordance with the agent-based approach, industrial stations, hauls and moving RTV (Fig. 25) are presented by agents (Table 3).

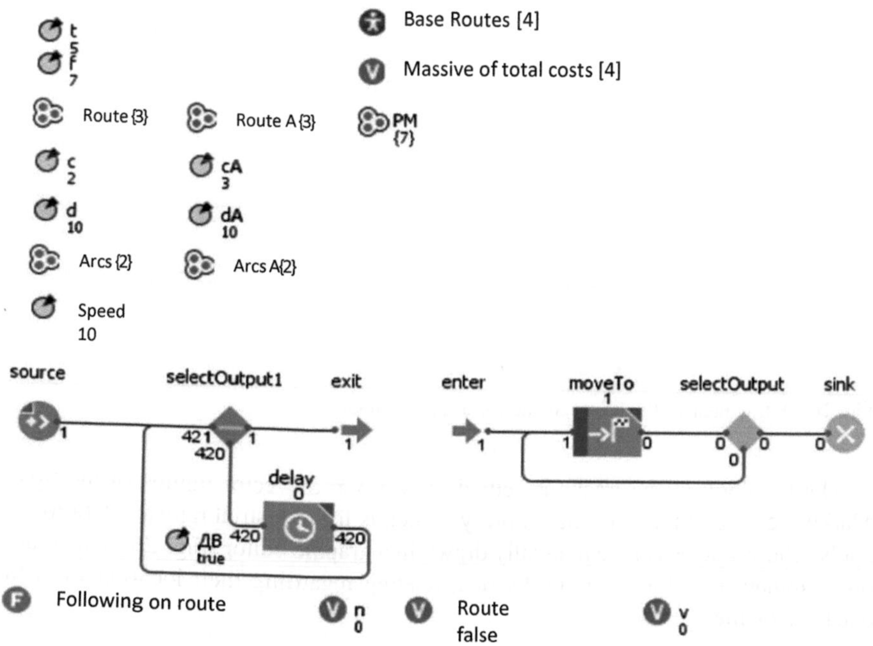

Fig. 25 The structure of agent «RTV» in a combined simulation model of rail node functioning

Table 3 The composition and parameters of the agents for a combined simulation model of the private railway track functioning

Name of agent	Parameters of the agent
Station	Number of stations
	Reserve handling capacity
	The value of losses from the delay of a unit mass of RTV
	The costs for handling an RTV unit mass
Haul	Number of starting stations
	Number of final stations
	Reserve of throughput
	Costs for movement of an RTV unit mass
Railcar traffic volume	Route
	Moment of time for movement on the route
	Time for movement on the route
	Speed
	Mass of RTV
	Costs for movement of an RTV unit mass
	Minimum reserve of throughput

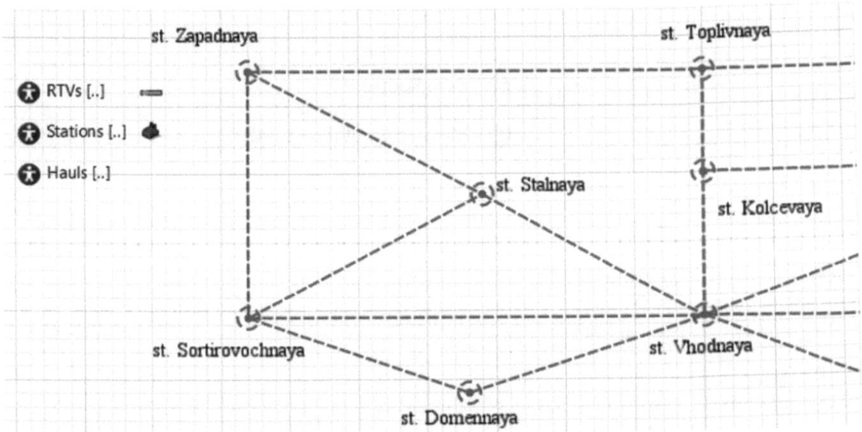

Fig. 26 A fragment of the base scheme for a railway node

The topology of the tracks is settled by a group of vector figures of the Space Markup feature of the AnyLogic library, which is for industrial railway stations and hauls. These shapes can be manually drawn in a graphic editor (Fig. 26) and created programmatically, for example, by data reading regarding their location from the database or file.

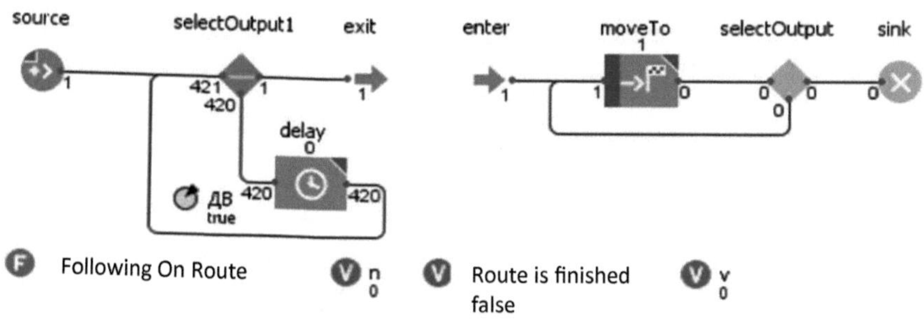

Fig. 27 The general view of the flow diagram in AnyLogic software, simulating train movement

The RTV movement in the model is mimicked by a flow diagram (Fig. 27), which has objects from the AnyLogic Process Modeling Library.

The flow diagram of train movement on the route consists of two objects, «moveTo» and «selectOutput», function «FollowingOnRoute» and variables «n» and «Routeisfinished». The object «moveTo» simulates the agent movement «RTV» up to the destination. The destination is «point node» (railway station), which is determined from the list «Route» by the function call *get*(*n*), where *n* is the position of the element in the list. In the field «Exit action» of object «moveTo» the function «FollowingonRoute()» function is called, which determines the final route, if *n* is the last position in list «Route» and returns the value of the variable «Routeisfinished» as «true».

In turn, object «selectOutput» defines further movement of the train. If the value of variable «Routeisfinished» is equal to «false», then the agent enters the object «moveTo» and moves to the next railway station according with the list «Route»; otherwise, the agent has reached the destination. The list «Route» of consecutive numbers of railway stations is a result of problem solving to determine the optimal train route. The Action Charts of AnyLogic represent a realization of the developed algorithms of calculation and parameter adjustment for RTVs (Figs. 28 and 29).

The experiments with the constructed combined simulation model were conducted to evaluate the effectiveness of the mathematical model and algorithm of dynamic optimization for RTV parameters. Three series of experiments were conducted: without optimization of the RTV parameters (model verification), and dynamic optimization of RTVs with base periods of variable and constant duration. The results of the experiments (Table 4) showed that the utilization of known methods for dynamic optimization within operative management of RTV parameters in terms of their irregularity reduced average railcar dead time at the railway node by 8%, while the application of the proposed model reduced dead time by 11% (Fig. 30).

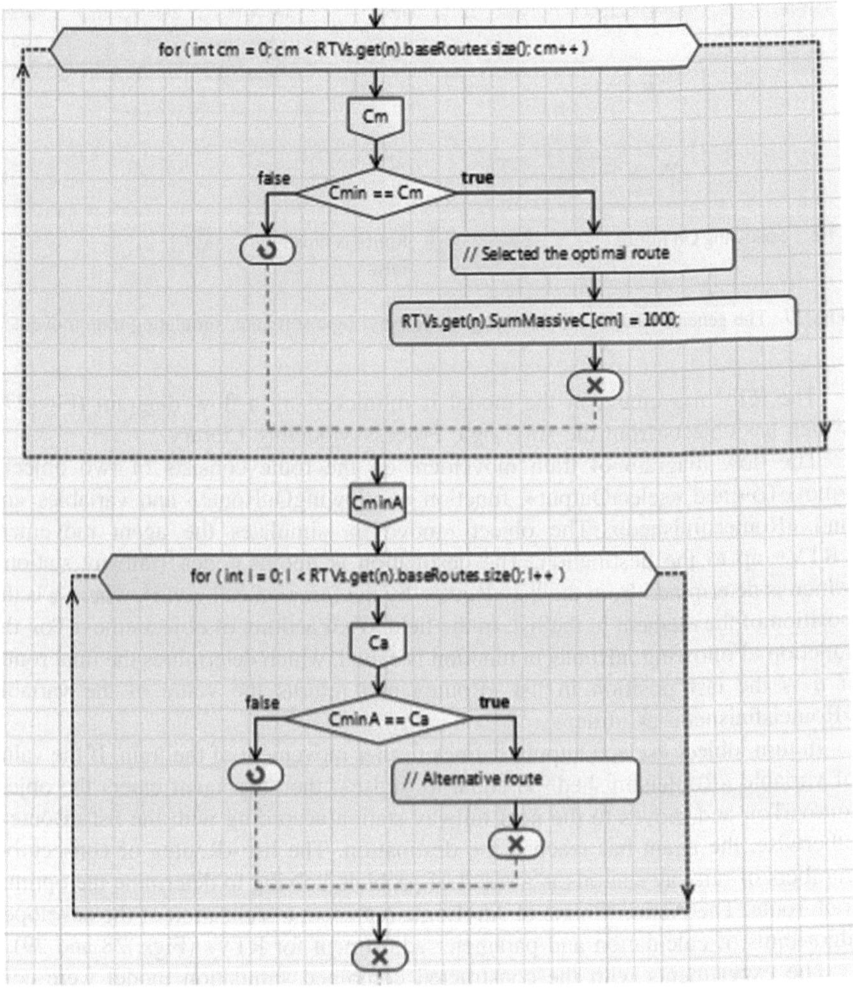

Fig. 28 The action charts in a combined simulation model

7 The Method for Integrating the Model for Dynamic Optimization of RTV Parameters in an Intelligent Transport System of a Railway Node

We recommend the integration of a mathematical model of dynamic optimization of RTV parameters in a functional structure of information management systems [79] using the proposed method in order to improve the responsiveness of the existing systems to the high irregularity of RTVs in industrial rail transport.

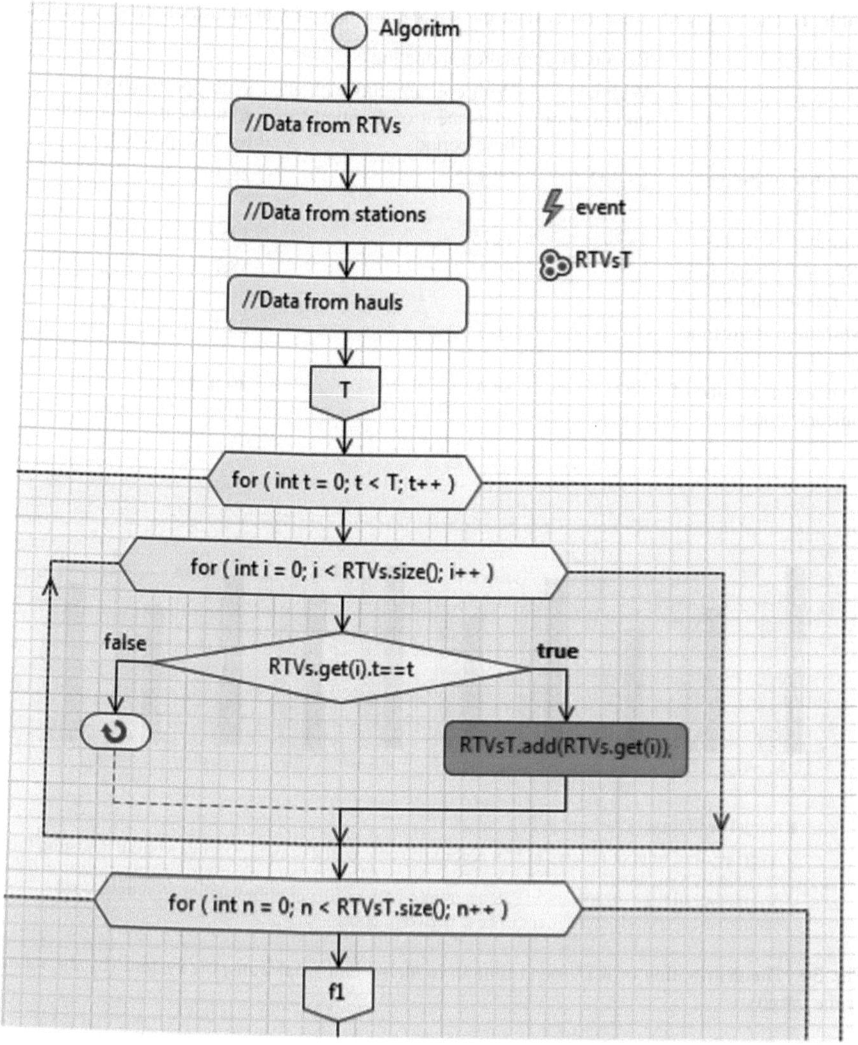

Fig. 29 The fragment of action charts for combined simulation model

The method for integrating the dynamic optimization model includes the following main stages:

1. Research of railway nodes regarding the availability of reserves for the handling capacity of railway stations with high loads at neighboring stations. This is implemented by constructing a system dynamics simulation model and conducting experiments. The study determines the availability of handling capacity reserves of industrial railway stations and the possibility of using these reserves to manage irregular RTVs.

Table 4 The results of experiments with the simulation model of an industrial transport system

Parameters	Results of simulation modeling		
	Without optimization	With optimization without adjustment of duration for base periods	With optimization and adjustment of duration for base periods
Number of railcars handled	28,392	28,392	28,392
Number of routes	1432	1275	1317
Maximum dead time of railcars, hours	66	60.1	58.7
Minimum dead time of railcars, hours	27	26.1	26.1
Average dead time of railcars, hours	56	51.5	49.8

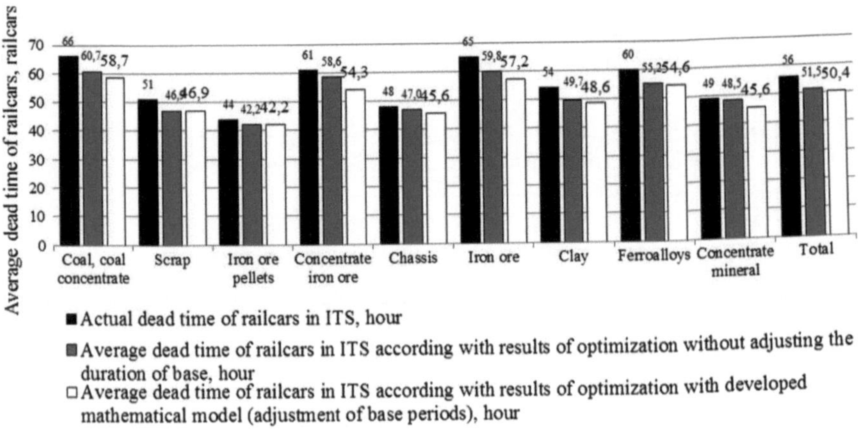

Fig. 30 The distribution of RTV dead time in modeled industrial transport system (by the nature of the cargo)

2. The construction of a combined simulation model of rail node functioning based on the joint application of discrete event and agent-based approaches in the same model. The combined simulation model is needed for predicting RTV parameters and industrial station workload in operational mode. The data generated by the simulation model are used, along with the actual data of existing information systems, as a source for solving dynamic optimization of RTVs.
3. The collection of input data in information management systems to solve the problem of RTV dynamic optimization in accordance with the proposed system of their parameters. Input data are in the form of information management system databases and registry parameter values for elements of the transport network and parameter values reflecting the physical properties of the actual RTVs in space and time (Fig. 31) at the beginning of the optimization period:

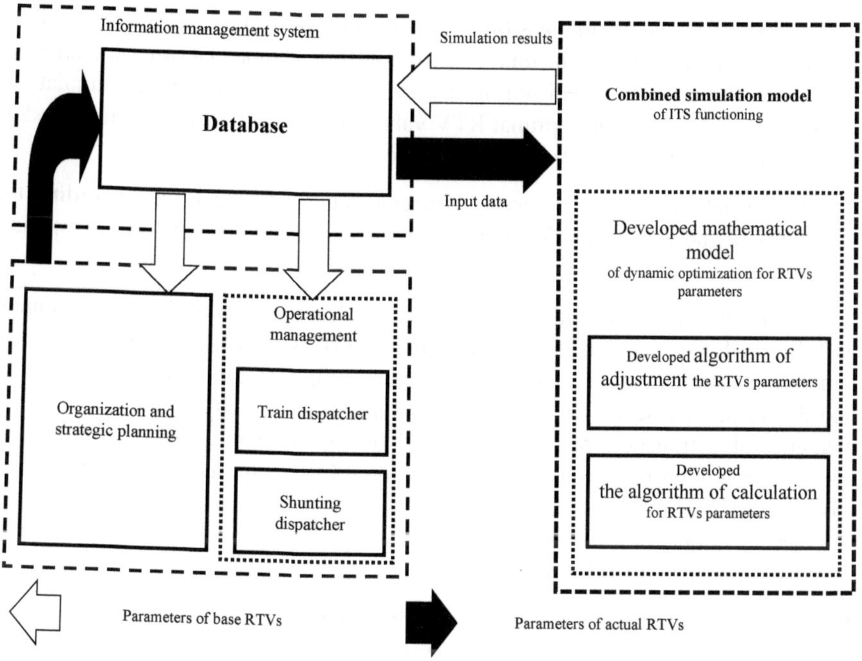

Fig. 31 The scheme of integration of the developed mathematical model in the information management system of indusial rail transport

- for station: number of station; handling capacity reserve; value of losses in the case of a delay of one unit mass of RTV;
- for haul: number of starting station; number of last station; reserve of throughput;
- for railcar traffic volume: moment of time for movement on the route; RTV mass; number of starting vertex for route; number of last vertex for route.

The generation of initial data for solving the problem of RTV dynamic optimization is carried out at specified intervals depending on the refresh rate of values for parameters of the actual RTVs in the information management system database.

4. Implementation of software in the information and control system of the developed algorithm to adjust the RTV parameters.

The adjustment of parameters for actual RTVs with the developed algorithm is implemented at the end of the run of the combined simulation model (Figs. 21 and 22). The algorithm of RTV parameter adjustment is constructed using an object-oriented approach to computer programming, and as a result can be implemented in industrial rail transport information management systems.

Analysis of the economic feasibility of parameter adjustment of the actual RTVs is performed by comparing the total cost for railcar dead time in a simulated railway node with the costs obtained during a previous run of the combined simulation model. The data array with optimal RTV values in the combined simulation model is formed according to the results of the analysis.

5. The formation of data sets with parameter values of base RTVs according to results of the run of the developed combined simulation model in operational mode.

The accumulation of results of the combined simulation model is implemented as follows. Each element of the transport network is presented in the combined simulation model, with comparable parameters, reflecting the number of elements and the value of reserve for throughput and handling capacity. RTVs, passing on elements of a transport network presented by agents with a set of parameters, averter the unique properties and behavior of RTVs in the combined simulation model.

Over the course of the experiment, the developed combined simulation model is filled with an array of RTV values:

- for stations: number of station, reserve handling capacity;
- for hauls: number of starting station, number of final station, throughput reserve;
- for railcars: route, moment of time for the beginning of movement on route, travel time on route, speed, RTV mass, costs for RTV movement on route, minimum throughput reserve.

6. The organization of operational managers informs the transportation process regarding parameters of base RTVs and the creation of a monitoring system for compliance with these parameters.

The process of informing the operational managers of the transportation process regarding parameters of the base RTVs is based on the scheme of information flow movement in operational mode presented in Fig. 32.

The array data formed from the results of a combined simulation model is transferred to an industrial rail transport information management system database [80].

The monitoring of deviations of parameters for actual RTV from the estimated RTVs is implemented on the basis of comparison of registry values of specified parameters. In the case of deviations, detection starts in combined simulation model of railway node functioning, with input data of parameters for the actual RTVs.

The application of the developed method will allow adjustment of actual parameters for RTVs on the basis of the combined simulation model of railway node functioning, with subsequent transfer of the design parameters for operational RTV managers of the transportation process.

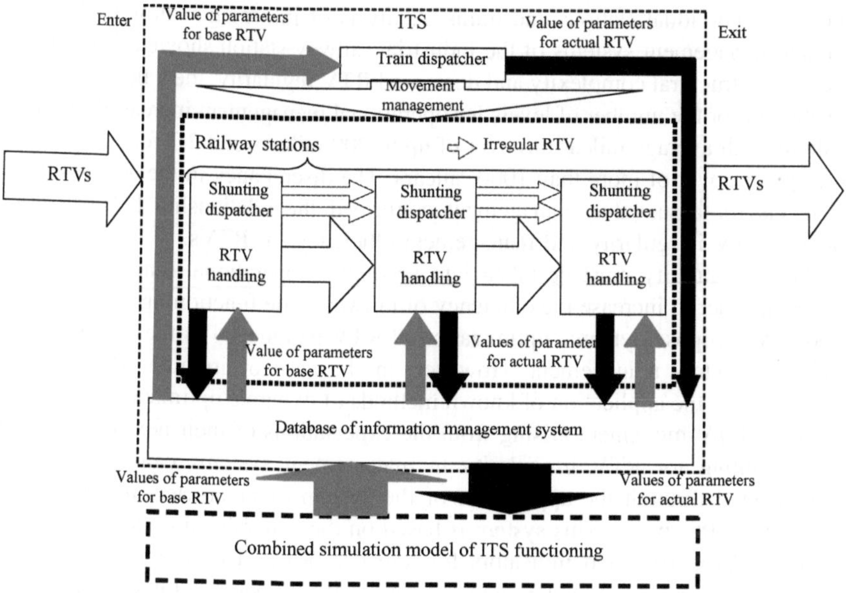

Fig. 32 The scheme of information flow movement in operational mode for RTV management

8 Conclusion

The analysis of railway node functioning and industrial transport systems identified a number of factors with a significant impact on RTV irregularity in these systems (the value of the daily coefficient of irregularity reaches as high as 2.5) and the time of their movement in a node (exceeding average values by 3.95 times), which can account for 75% of RTV jet irregularity at the nodes. The external factors identified include irregularity of commodity flows, dynamics of production, climatic factors, and the development of a competitive rail transport environment (in particular, the emergence in Russia of companies/operators with their own railcars). Internal factors include an increasing number of RTV jets with low power and lack of consistency in decisions regarding the operational management of RTVs.

We have noted that the increased RTV irregularity leads to uneven station workloads and hauls in the node. This irregularity has a cyclical character because of inconsistent rhythms of production processes. The results of this study of a constructed system dynamic simulation model of an industrial transport system showed the presence of workload oscillations at railway stations ranging from 76 to 100%. Uneven utilization of throughput and handling capacity of stations and hauls increased RTV dead time an average of 20%.

The main reason for the increased dead time is inadequate accounting for RTVs in the existing RTV management system focused on management in operational mode by individual stations and trains. Analysis of information flow in the operational management systems of the industrial railway station showed that with the increased structural complexity and decreased RTV regularity, the value of the base coefficient for the number of levels for operational management increased to 1.1 for stations with average railcar turnover of up to 200 railcars and 2.7 for stations with average turnover of more than 1000 railcars. The discrepancy in the actual number of levels for base value management confirmed the relationship between complexity, RTV irregularity and management efficiency for RTVs.

RTV irregularity must be taken into account in operational management processes in order to increase the efficiency of railway node functioning in operational mode. We propose dynamic optimization of RTV parameters as the instrument for RTV parameters management. However, in the case of irregular RTVs at the railway node, the application of known methods of dynamic optimization increased delays in RTV movement arising from the expectations of their next base period.

Accounting for RTV irregularity in management systems can be efficiently implemented through the application of the system of parameters and indicators discussed in this paper. This system is based on the generalization of classification criteria of properties and indicators for logistics flows. The system includes the following groups of parameters and indicators: the physical parameters reflecting the properties of RTV in space and time, statistical parameters characterizing the regularity of changes in physical parameters of RTVs over time, indicators of RTV complexity and indicators of RTV quality. Application of the proposed system will take into account RTV irregularity, thereby improving upon known dynamic optimization methods.

In this work, we have developed a mathematical model of dynamic optimization of RTV parameters and an algorithm for adjusting these parameters, taking into account RTV irregularity in operational management mode. The main difference between the proposed method and known methods of dynamic optimization is the application of base periods of variable duration, depending on the optimal ratio of the costs of RTV delay and the adjustment of their parameters.

An effectiveness evaluation of the developed method of dynamic optimization for RTVs was conducted through the construction and application of a combined simulation model of an industrial transport system functioning with daily volume of up to 2500 railcars. The results of comparative model experiments applying known methods of dynamic optimization and mathematical models established that accounting for calculation periods of variable duration led to an average reduction in railcar dead time of 6.2 h (11%).

We described a method for integrating the combined model of dynamic optimization of RTV parameters into the intellectual transport system of an industrial enterprise. The application of this method will take into account RTV irregularity on the basis of utilization of the system for RTV parameters and adjustment of the actual RTV parameters by calculating the results of the combined simulation model of railway node functioning.

In light of the general tendency toward structural complexity and increased RTV irregularity, the realization of the proposed set of methods and models for managing RTV parameters at railway nodes will increase management efficiency and parameter optimization accuracy, reduce the time that railcars are located at railway nodes, and improve the timeliness of freight transportation and quality of transport service for cargo owners.

References

1. Ostrovskiy AM, Druzhinina MG, Kuzmin AA (2011) Interaction of operator companies with industry and railroad. Rail Transp (2):61–63 (in Russian)
2. Rakhmangulov AN, Mishkurov PN, Kopylova OA (2014) Railway transport technological systems: organization of functioning: monograph. Magnitogorsk, Nosov Magnitogorsk State Technical University (in Russian)
3. Okulov NE (2014) Methods and ways for improvement the interaction of production and transport. PhD thesis, Ural State University of Railway Transport (in Russian)
4. Marfin MA, Kozlov PA, Bugaev AV (1986) Simulation model increased the throughput. Ind Transp (12):8–9 (in Russian)
5. Levin DY (1988) Optimization of train flows. Transport, Moscow (in Russian)
6. Rezer SM (1982) Integrated management of transportation process at transport nodes. Transport, Moscow (in Russian)
7. Persianov VA (1983) Stations and nodes in modern transport system. Rail Trans (3) (in Russian)
8. Sotnikov IB (1967) Theoretical foundations of interaction in the work of train departure parks of stations and surrounding areas. Moscow State University of Railway Engineering (in Russian)
9. Clausen U, Rotmann M (2014) Measurement and optimization of delivery performance in industrial railway systems. In: Efficiency and innovation in logistics. Lecture Notes in Logistics, pp 109–120
10. Clausen U, Voll R (2013) A comparison of North American and European railway systems. Euro Transp Res Rev 5(3):129–133
11. Rakhmangulov AN, Trofimov SV, Kornilov SN (2004) Methods for development the systems of industrial rail transport in a changing environment activities of the enterprises, Magnitogorsk. Nosov Magnitogorsk State Technical University (in Russian)
12. Trofimov SV (2004) Scientific-methodical bases of functioning and development of industrial transport systems. Doctoral dissertation, Moscow State University of Railway Engineering (in Russian)
13. Osintsev NA, Rakhmangulov AN (2013) Railcar traffic volumes management in industrial transport systems. Vestnik of Nosov Magnitogorsk State Techn Univ 1(41):16–20 (in Russian)
14. Kozlov PA (1988) Theoretical basis, organizational forms, methods of optimization for flexible transport services of black metallurgy factories. Doctoral dissertation, Moscow (in Russian)
15. Rakhmangulov AN, Osintsev NA, Mishkurov PN, Kopylova OA (2014) Intellectualization of transport service of the metallurgical enterprises. Steel (4):115–118 (in Russian)
16. Popov AT (1984) Optimization of interaction process for railway transport and production (on example of steel plant). Candidate dissertation, Moscow (in Russian)
17. Komarov AV (ed) (1983) Problems of transport development in USSR. Comprehensive operation. Transport, Moscow (in Russian)

18. Geraets F, Kroon L, Schoebel A, Wagner ZC (eds) (2004) Algorithmic methods for railway optimization, 320 p (in Russian)
19. Reggiani A, Schintler LA (2005) Methods and models in transport and telecommunications cross Atlantic perspectives. Springer, Berlin 364 p
20. Cascetta E (2009) Transportations systems analysis: models and applications. Springer, Berlin, 681 p (in Russian)
21. Aleksandrov AE (1995) Flexible control technology of inroad loop routes. Candidate dissertation, Moscow (in Russian)
22. Baturin AP, Borodin AF, Panin VV (2010) Railcar traffic volumes organization into same group of trains. World Transp 5(33):72–77 (in Russian)
23. Osminin AT (2000) Rational organization of railcar traffic volumes based on the methods of multi-criteria optimization. Doctoral dissertation, Samara (in Russian)
24. Bodyul VI (2006) Improving the rhythm and efficiency of transport production through reduction of daily irregularity of freight traffic on the railways. Doctoral dissertation, Moscow (in Russian)
25. Aleksandrov AE, Yakushev NV (2006) Stochastic formulation of the dynamic transport problem with delays and the random spread of delivery time and time consumption. Manag Big Syst (12–13):5–14 (in Russian)
26. Kozlov PA, Vladimirskaya IP (2009) Systems construction of automatic control for flows of railcars of different owners. Vestnik Railway Res Inst (6):8–11 (in Russian)
27. Rakhmangulov AN, Mishkurov PN (2012) Problems of method application of dynamic programming for operational management for railcar traffic volume. Mod Prob Russ Transp Complex (2):279–285 (in Russian)
28. Carey M, Lockwood D (1995) Model, algorithms and strategy for train pathing. J Oper Res Soc 46(8):988–1005
29. Carey M (1994) A model and strategy for train pathing with choice of lines, platforms, and routes. Transp Res Part B 28(5):333–353
30. Carey M (1994) Extending a train pathing model from one-way to two-way track. Transp Res Part B 28(5):395–400
31. Dorfman MJ, Medanic J (2004) Scheduling trains on a railway network using a discrete event model of railway traffic. Transp Res Part B: Methodol 38(1):81–98
32. Rakhmangulov A, Kolga A, Osintsev N, Stolpovskikh I, Sładkowski A (2014) Mathematical model of optimal empty rail car distribution at railway transport nodes. Transp Prob 9(3):125–132
33. Caprara A, Kroon LG, Monaci M, Peeters M, Toth P (2007) Passenger railway optimization. In: Barnhart C, Laporte G (eds) Handbooks in operations research and management science, vol 14, pp 129–187
34. Caimi G, Chudak F, Fuchsberger M, Laumanns M, Zenklusen R (2011) A new resource-constrained multicommodity flow model for conflict-free train routing and scheduling. Transp Sci 45(2):212–227
35. Blum J, Eskandarian A (2002) Enhancing intelligent agent collaboration for flow optimization of railroad traffic. Transp Res Part A: Policy Pract 36(10):919–930
36. Teornquist J (2007) Railway traffic disturbance management an experimental analysis of disturbance complexity, management objectives and borderations in planning horizon. Transp Res A: Policy Pract 41(3):249–266
37. Carey M, Crawford I (2007) Scheduling trains on a network of busy complex stations. Transp Res B: Methodol 41(2):159–178
38. Erlebach T, Gantenbein M, Heurlimann D, et al. (2001) On the complexity of train assignment problems. In: Algorithms and computation, 12th international symposium, ISAAC 2001 Christchurch, New Zealand, Proceedings, vol 2223 of Lecture Notes in Computer Science, pp 390–402
39. Meng X, Jia L, Chen C, Xu J, Wang L, Xie J (2010) Paths generating in emergency on China new railway network. J Beijing Inst Technol 19(2):84–88

40. Lee Y, Chen C (2009) A heuristic for the train pathing and timetabling problem. Transp Res B 43(8–9):837–851
41. Pellegrini P, Marli`ere G, Rodriguez J, Marliere G (2014) Optimal train routing and scheduling for managing traffic perturbations in complex junctions. Transp Res Part B 59:58–80
42. Lusby RM, Larsen J, Ehrgott M, Ryan DM (2013) A setpacking inspired method for real-time junction train routing. Comput Oper Res 40(3):713–724
43. D'Ariano A (2008) Improving real-time train dispatching: models, algorithms and applications. PhD thesis, Delft University of Technology, Delft, The Netherlands
44. Goverde RMP, Hansen IA (2000) TNV-prepare: analysis of Dutch railway operations based on train detection data. In: Allan J, Brebbia CA, Hill RJ, Sciutto G, Sone S (eds) Computers in railways VII. WIT Press, Southampton, UK, pp 779–788
45. Goverde RMP (2005) Punctuality of railway operations and timetable stability analysis. PhD thesis, Delft University of Technology, Delft, The Netherlands
46. Fugenschuh A, Homfeld H & Schulldorf H (2009) Single car routing in rail freight transport. In: Barnhart C, Clausen U, Lauther U, Mohring R (eds) Dagstuhl seminar proceedings 09261, Schloss Dagstuhl - Leibniz-Zentrum fr Informatik, Deutschland
47. Fugenschuh A, Homfeld H, Huck A, Martin A, Yuan Z (2008) Scheduling locomotives and car transfers in freight transport. Transp Sci 42(4):1–14
48. Barnhart C, Jin H, Vance PH (2000) Railroad blocking: a network design application. Oper Res 48(4):603–614
49. Jha KC, Ahuja RK, Sahin G (2007) New approaches for solving the block-to-train assignment problem. Networks 51(1):48–62
50. Ahuja RK, Jha KC, Liu J (2007) Solving real-life railroad blocking problems. Interfaces 37 (5):404–419
51. Hailes S (2006) Modern telecommunications systems for train control. In: The 11th institution of engineering and technology professional development course on railway signalling and control systems, Manchester, UK, pp 185–192
52. Kauppi A, Wikström J, Sandblad B, Andersson AW. Future train traffic control: control by re-planning. Cogn Technol Work 8(1):50–56. Available at: https://it.uu.se/research/project/ftts/reports/ControlByReplanning.pdf
53. Muravev DS, Rakhmangulov AN, Mishkurov PN (2013) Application of simulation modeling to evaluate handling capacity of sea ports and justification of the need for dry port construction. Mod Probl Russian Transp Complex 4(4):66–72 (in Russian)
54. Turanov HT, Chuev NP (2012) Construction of differential models of the rolling stock movement on non-public places. Transp Sci Technol Manag 7:13–18 (in Russian)
55. Turanov HT, Chuev NP, Portnova OU (2013) Mathematical modeling of freight railcars movement on driveway tracks of the enterprise. Sci Technol Transp (1):26–42 (in Russian)
56. Turanov HT, Chuev NP, Portnova OU (2013) Numerical simulation of the freight ralcars movement on driveway tracks of industrial enterprises in maple. Transp Sci Technol Manag (12):7–14 (in Russian)
57. Rakhmangulov AN, Kolga AD, Osintsev NA, Stolpovskikh IN, Sładkowski AV (2014) Mathematical model of optimal empty rail car distribution at railway transport nodes. Transp Prob 9(3):125–132
58. Hernando A, Roanes-Lozano E, Maestre-Martínez R, Tejedor J (2010) A logic-algebraic approach to decision taking in a railway interlocking system. Ann Math Artif Intell 65 (4):317–328
59. Corman D, D'Ariano A, Pacciarelli D, Pranzo D (2010) Railway dynamic traffic management in complex and densely used networks intelligent infrastructures. Intell Syst Control Autom: Sci Eng 42:377–404 (in Russian)
60. White TA (2007) The development and use of dynamic traffic management simulations in North America. In: Hansen IA, Radtke A, Pachl J, Wendler E (eds) Proceedings of the 2nd international seminar on railway operations modelling and analysis, Hannover, Germany

61. Sotnikov EA, Poplavskiy AA (2007) General principles of system construction for operational management of the transportation process. Transp Sci Technol Manag (2):80–87 (in Russian)
62. Kozlov PA, Vladimirskaya IP (2009) A method for optimizing the structure of transport system. World Transp 26(2):84–87 (in Russian)
63. Kozlov PA (2007) Information technologies in transport. The modern stage. Transp Russ Fed (10):38–41 (in Russian)
64. Kozlov PA, Vladimirskaya IP (2009) Optimization method of interaction in the production and transport systems. Modern Prob Sci Educ (2–6):17–18 (in Russian)
65. Aleksandrov AE (2008) The application of models within the calculation and optimization of systems for rail transport. Sci Technol Transp (2):54–56 (in Russian)
66. Kutirkin AV (2004) Development of models and algorithms for solving functional tasks of the management for transport systems and production. Doctoral dissertation, Moscow (in Russian)
67. Potzsche C, Heuberger C, Kaltenbacher B, Rendl F (eds) System modeling and optimization (2014) 26th IFIP TC 7 conference, CSMO 2013, Klagenfurt, Austria. Springer, Berlin, 359 p (in Russian)
68. Kozlov PA, Osokin OV, Permikin VU (2013) Automated control of the processes in the transport node. Vestnik Rostov State Univ Railway Eng 2(50):118–122 (in Russian)
69. Borshchev A (2013) The big book of simulation modeling: multimethod modeling with AnyLogic 6. AnyLogic, North America
70. Mishkurov PN (2012) Typification of industrial railway stations. Mod Prob Russ Transp Complex. (2):143–151 (in Russian)
71. Gromov NN, Persianov VA, Uskov NS (2003) Management on transport. Academy, Moscow (in Russian)
72. Gavrishev SE, Dudkin EP, Kornilov SN, Rakhmangulov AN, Trofimov SV (2003) Transport logistics. St. Petersburg (in Russian)
73. Kaigorodcev AA, Rakhmangulov AN (2013) Factors of efficiency for logistics distribution centers. Vestnik transporta Povolzhya 2:11–19 (in Russian)
74. Booch G, Maksimchuk R, Michael W, Bobbi Y, Jim C, Houston K (2008) Object-oriented analysis and design with applications. E-Book, 720 p
75. Pepevnik A, Belsak M (2011) Information system in the function of railway traffic management. Transp Prob 6(1):37–42
76. Treiber M, Kesting A (2013) traffic flow dynamics. Data, models and simulation. Springer, Berlin, p 506
77. Ran B, Boyce D (1996) Modeling dynamic transportation networks. An intelligent transportation system oriented approach. Springer, Berlin 356 p
78. Rakhmangulov AN, Mishkurov PN (2012) Features of a simulation model building for technology of the railway station in the AnyLogic system. Collect Sci Works SWorld 2(4): 7–13 (in Russian)
79. Sładkowski A, Pamuła T (eds) (2016) Intelligent transportation systems—problems and perspectives, Studies in Systems Decision and Control, vol 32. Springer, Switzerland, 303 p
80. Muravev D, Aksoy S, Rakhmangulov A, Aydogdu V (2016) Comparing model development in discrete event simulation on Ro-Ro terminal example. Int J Logistics Syst Manag 3 (24):283–297

MIX
Papier aus verantwortungsvollen Quellen
Paper from responsible sources
FSC® C105338

If you have any concerns about our products,
you can contact us on
ProductSafety@springernature.com

In case Publisher is established outside the EU,
the EU authorized representative is:
Springer Nature Customer Service Center GmbH
Europaplatz 3, 69115 Heidelberg, Germany

Printed by Libri Plureos GmbH
in Hamburg, Germany